AF894610

Temperate Fruit Crops in Warm Climates

Temperate Fruit Crops in Warm Climates

Edited by

Amnon Erez

Institute of Horticulture,
The Volcani Center,
Bet Dagen, Israel

KLUWER ACADEMIC PUBLISHERS
DORDRECHT / BOSTON / LONDON

Library of Congress Cataloging-in-Publication Data

Temperate fruit crops in warm climates / edited by Amnon Erez.
p. cm.
ISBN 0-412-63290-X (alk. paper)
1. Tropical fruit. 2. Fruit-culture. I. Erez, Amnon.

SB359 .T39 2000
634'.0913--dc21

00-047808

ISBN 0-412-63290-X

Published by Kluwer Academic Publishers,
P.O. Box 17, 3300 AA Dordrecht, The Netherlands.

Sold and distributed in North, Central and South America
by Kluwer Academic Publishers,
101 Philip Drive, Norwell, MA 02061, U.S.A.

In all other countries, sold and distributed
by Kluwer Academic Publishers,
P.O. Box 322, 3300 AH Dordrecht, The Netherlands.

Printed on acid-free paper

All Rights Reserved
© 2000 Kluwer Academic Publishers
No part of the material protected by this copyright notice may be reproduced or utilized in any form or by any means, electronic or mechanical, including photocopying, recording or by any information storage and retrieval system, without written permission from the copyright owner.

Printed in the Netherlands.

TABLE OF CONTENTS

PREFACE

As a member of the working group (WG) on "Temperate Zone Fruit Trees in the Tropics and Subtropics" of the International Society for Horticulture, I was aware of the lack of readily available information needed in many warm-climate locations where temperate fruit crops are grown. The founder of this WG, Frank Dennis, Jr., was motivated to encourage knowledge transfer by sharing knowledge with many developing countries. We shared his drive and in presenting this book we believe we are doing a service to all persons interested in temperate fruits, but especially to those in tropical and subtropical countries, many of which are developing countries interested in growing these crops and lacking the knowledge needed. In this book, we have collected information covering a variety of different aspects of growing temperate fruit crops in warm climates.

As this is the first time such an evaluation of these species has been done, interesting and novel aspects of tree development and fruiting are presented, with stress on elements like dormancy and irrigation that are not of such basic concern in the natural habitat of the temperate zones. We are living in a transition age; horticultural studies are changing and expertise such as can be found in the array of participants in this book is probably not going to be easily found in the future.

I hope that this book will broaden our understanding of the fruiting Temperate Zone tree in general and of its adaptation to warm climates, in particular.

I would like to thank the participants who agreed to write the chapters that represent their activities. Actually, many of the contributors summarized a major part of their life work in their respective chapters and I would like to thank them for their meticulous work and effort put in the book.

I would like to thank Kluwer Academic Publishers for their editorial work, particularly Gloria Verhey, and thanks also to Patrick Dumont for helping in bringing this book to its present stage.

Special thanks are due to Frank Dennis, Jr., who helped in editing several chapters in the book.

Bet-Dagan 17.4.2000 A. Erez

INTRODUCTION

The motivation to devote a book to temperate fruit crops in warm climates arose from the continuing and increasing interest in these fruit crops in various locations around the world. The spread of commercial growing of fruit crops of cold-climate origin in subtropical and tropical countries is impressive. From the FAO reports it is clear that growing temperate fruit trees in warm climates has developed rapidly in the last 10 years. Especially noted are the increases in areas planted with apples in Brazil and Egypt, both of which are marginal climatically. These two countries represent two entirely different climates: southern Brazil, a humid country with acidic soils contrasts with Egypt, with its dry desert climate and high soil pH. Nevertheless, the trend towards increasing development points to the interest in increasing production all around the globe. This rapid increase has raised many problems, and our aim in this book is to deal with them.

Considerations that had not caused any commercial problems became major concerns when temperate zone species were transferred to warm climates leading to creation of new cultivars in most species, that are better adapted to such climates. This enabled us to evaluate the enormous flexibility among the temperate-zone fruit species in their climatic adaptation. Side by side with adaptation of cultivars, adaptation of rootstocks became a major concern as new soil-related constraints appeared in new locations. In addition, new agrotechnical means, appropriate to the new conditions were developed, especially with regards to overcoming dormancy problems and excessive top and root temperatures. The success of these means has led to the establishment of commercial temperate fruit tree orchards in many subtropical and tropical countries.

A clear indication of the rising interest is the wide participation in the working group of the International Horticultural Society for "Temperate Zone Fruit Trees in The Tropics and Subtropics" that attract people from all corners of the globe.

The warm climates, especially in the tropics, opened new hitherto unknown potential for growing these crops in a continuous system, with almost no growth arresting during a dormant period, thus enabling the commercial production of the same cultivar throughout the year. This "curiosity" is actually a basis for commercial production in all tropical countries where temperate fruit crops are grown.

In this book we aimed at a collective endeavor that brought together experts in all the major temperate fruit crop species, to address the problems of growing them under warm climates. It cannot be ignored that the fingerprints of the climatic conditions of the various authors' countries affected his/her way of thinking, and the resulting emphasis in their respective chapters.

The term "warm climates" encompasses a large range of types of climates, from the humid and hot tropics to the arid subtropics. Although not similar, all these types of climate pose difficulties in temperate fruit production. We address here problems posed by warm climates in the various phases of fruit tree development. Mostly tropical and subtropical climates are considered, but excessively high temperatures during the growing period in temperate - continental climates are also relevant.

In the past and also in the present, temperate fruit crops have attracted great efforts as they are the species found in and adapted to countries where horticultural research was and is active and strong. The diversion of part of such activity towards tropical and subtropical countries has led to many of the achievements described in this book. This effort represents one of the most important forms of support provided to many developing nations looking for plant material and know-how. On the other hand, the world-wide spread of species has revealed traits in many species that were not known to breeders in colder climates. Thus the interaction became fruitful to all concerned parties.

In addition, the possibility that climate may change in the future because of global warming resulting from human activity, may turn theoretical considerations into practical ones for temperate fruit production in cool climates. This book may hold the answer to such unfortunate developments.

The book is divided into two parts. The first deals with disciplines in fruit production under warm climates that are specific or which carry a special weight and differ from the demands of cooler climates. The presentation there is wide enough to cover related conditions in different climates. The second part deals specifically with commodities. We tried to cover all the important species and to discuss both the genetic pool and available cultivars, as well as dealing with specific cultural practices needed for successful production.

The reader may find some overlap among chapters, especially between the discipline and the commodity sections. Subjects like dormancy, nutrition and irrigation are touched by all commodity reports. For a general view the reader is referred to the discipline section. The choice of authors from different locations and growing conditions balanced the subjective approach of each, so that an overall balanced result seems to have been obtained. A large part of the information presented in this book is new, either intrinsically or in the context within which it is written.

The book is aimed at all readers who have an interest in the subject, especially those that are involved in growing temperate fruit crops in warm locations but also for all those who want to broaden their knowledge of the potential for adaptation of known species to widely different types of climates. It is our hope that it will serve as a textbook as well as a reference book for growing temperate fruit trees in warm climates.

1 LIGHT REGIMES IN TEMPERATE FRUIT-TREE ORCHARDS GROWN AT LOW LATITUDES

JOHN E. JACKSON

Horticultural Research Centre, P.O. Box 810, Marondera, Zimbabwe

1 Introduction

Solar radiation is the energy source for photosynthesis and hence all fruit tree growth and cropping. Potential gross productivity is thus determined by total light interception which, in turn, depends on the length of the growing season, the light energy available at different times through it and the amount and distribution of the foliage available to capture this light. A proportion of the carbohydrate produced by photosynthesis is, however, used up in respiration, which is temperature dependent so total growth depends on the balance of photosynthetic gain against respiratory loss. Light climate thus cannot be adequately considered in isolation from its effects on temperature, which may be positive with respect to many aspects of growth as well as negative through its effects on respiration (Lakso 1994). It must also be noted that high levels of incident radiation may actually reduce fruit growth at certain developmental stages apparently through effects on fruit water relations (Loreti *et al.*, 1993).

Light levels also influence a number of processes which may be only partially dependent on photosynthesis or even effectively independent of it. The main example of this is in anthocyanin formation in fruit skin (Jackson 1980, Lakso 1994) but there is also evidence that low light may have a disproportionately adverse effect on fruitfulness in comparison with growth *per se* (Jackson 1980) and that high exposure to solar radiation can cause sunscald on fruits.

These different factors result in the opportunity to maximise yields of good quality fruits by modifying light interception and distribution through choice of cultivar, planting density, tree size and arrangement and pruning. The factors controlling the influence of latitude in these respects are discussed below.

2 Available solar energy

2.1 Annual pattern

The amount of short-wave radiation, which includes photosynthetically active radiation (P.A.R.) arriving at a particular place on the earth's surface depends on latitude, season of the year, time of day and degree of cloudiness or dust in the atmosphere.

The effects of the first three of these factors depend on the position of the earth in relation to the sun. The earth revolves around the sun once a year and rotates daily on

Amnon Erez (ed.), Temperate Fruit Crops in Warm Climates, 1-15.
© 2000 *Kluwer Academic Publishers. Printed in the Netherlands.*

its axis. This axis is inclined at $66^1/_2^\circ$ to the orbital plane and always points into space in the same direction. At midsummer (northern hemisphere) the north pole is tilted towards the sun and six months later away from it. The effects of this can be visualised by thinking of a single 'beam' of light from the sun. If this hits the earth's surface at right angles the area which receives the radiation is minimal and the intensity maximal. If it hits at a lower angle the same amount of radiation is distributed over more surface at lower intensity. The effect of this on the amount of solar radiation (R_A) which is received at the outer limit of the atmosphere (*i.e.* would be received at the earth's surface if the atmosphere was totally transparent and had no cloud) in relation to latitude is given in Table 1.

Table 1. Yearly variation of the horizontal surface daily insolation at the outer limit of the atmosphere MJ m^{-2} day^{-1} assuming I_{sc} = 1367Wm^{-2}

	Lat. in degrees	*JAN*	*FEB*	*MAR*	*APR*	*MAY*	*JUNE*	*JUL*	*AUG*	*SEP*	*OCT*	*NOV*	*DEC*
	60	3.5	8.6	17.4	27.8	36.7	41.0	38.9	31.3	21.3	11.7	5.0	2.3
	50	9.2	14.7	22.9	31.7	38.5	41.6	40.3	34.4	26.3	17.6	10.8	7.7
	40	15.3	20.6	27.8	34.8	39.7	41.7	40.7	36.6	30.4	23.1	16.9	13.8
North	30	21.4	26.0	31.9	36.9	40.0	41.1	40.4	38.0	33.7	28.0	22.7	20.0
	20	27.1	30.8	34.9	38.0	39.3	39.5	39.2	38.2	35.9	32.1	28.0	25.8
	10	32.1	34.7	37.0	37.9	37.5	37.0	37.1	37.4	37.0	35.2	32.7	31.2
	0	36.3	37.5	37.9	36.8	34.8	33.5	33.9	35.6	37.1	37.3	36.5	35.7
	10	39.6	39.3	37.7	34.5	31.1	29.2	29.9	32.7	36.0	38.3	39.3	39.4
	20	41.9	40.0	36.3	31.2	26.6	24.2	25.1	28.9	33.8	38.2	41.0	42.1
South	30	43.0	39.6	33.9	27.1	21.4	18.7	19.8	24.3	30.6	37.0	41.7	43.8
	40	43.2	38.0	30.4	22.1	15.8	12.9	14.1	19.1	26.5	34.6	41.2	44.5
	50	42.4	35.5	26.0	16.6	10.0	7.2	8.3	13.4	21.6	31.3	39.9	44.2
	60	41.1	32.1	20.8	10.8	4.5	2.1	3.0	7.7	16.1	27.2	37.8	43.6

Abridged from Iqbal 1983 Tables 4.2.2 and 4.2.3.

In fact the atmosphere is not totally transparent and the amount of radiation reaching the earth's surface will depend on the length of the beam path through the atmosphere and the amount of radiation-intercepting material in the atmosphere. Again, the length of the beam path through the atmosphere will be minimal if the beam is at right angles to the earth's surface and correspondingly greater as the angle is lowered. The effect of latitude on radiation receipt at the surface of the earth, other factors being equal, is shown in Table 2. In general, the effect of latitude on radiation receipt on the earth's surface is greater than on that at the upper limits of the atmosphere because of the higher solar altitudes therefore shorter beam paths through the atmosphere in the tropics.

Table 2 gives values for direct radiation only. The diffuse sky radiation reaching the ground under cloudless conditions at any latitude at any date can be calculated as

0.5 (0.91 I_0 - I) where I_0 is the radiation at any latitude and date above the earth's atmosphere and I the direct solar radiation reaching the ground (List 1958). The total radiation reaching the ground at any latitude and date therefore will fall between the figures in Table 1 and those in Table 2.

Table 2. Daily direct solar radiation reaching the ground with different atmospheric transmission coefficients (0.6 upper table, 0.8 lower table) at different latitudes MJm^{-2} day^{-1}

	Latitude	*March 21*	*May 6*	*June 22*	*Aug 8*	*Sept 23*	*Nov 8*	*Dec 22*	*Feb 4*
	60	5.0	13.0	17.0	12.9	4.9	04		0.4
	50	8.5	15.8	18.9	15.6	8.3	2.4	0.7	2.4
	40	11.8	17.8	20.0	17.6	11.6	5.4	3.1	5.5
North	30	14.7	19.0	20.1	18.8	14.5	8.9	6.4	9.0
	20	16.9	19.2	19.5	19.0	16.7	12.3	9.9	12.4
	10	18.3	18.6	17.9	18.4	18.0	15.3	13.5	15.5
	0	18.7	17.1	15.6	16.9	18.4	17.7	16.6	17.9
	10	18.3	14.8	12.7	14.6	18.0	19.3	19.1	19.5
	20	16.9	11.8	9.3	11.7	16.7	19.9	20.8	20.1
South	30	14.7	8.6	6.0	8.5	14.5	19.7	21.5	19.9
	40	11.8	5.2	2.9	5.2	11.6	18.5	21.3	18.6
	50	8.5	2.3	0.7	2.3	8.3	16.4	20.2	16.6
	60	5.0	0.4		0.4	4.9	13.5	18.2	13.7
	60	10.1	21.8	27.2	21.6	10.1	1.8	0.2	1.8
	50	14.9	24.8	28.7	24.5	14.7	5.7	2.7	5.7
	40	19.1	26.9	29.7	26.6	18.8	10.3	6.9	10.4
	30	22.6	28.0	29.6	27.7	22.3	15.1	11.6	15.2
North	20	25.2	28.0	28.4	27.8	24.8	19.4	16.5	19.6
	10	26.8	27.2	26.4	26.9	26.4	23.3	21.1	23.5
	0	27.3	25.2	23.5	25.0	26.9	26.2	25.1	26.4
	10	26.8	22.4	19.7	22.2	26.4	28.2	28.2	28.5
	20	25.2	18.7	15.5	18.5	24.8	29.1	30.4	29.3
South	30	22.6	14.5	10.9	14.4	22.3	29.1	31.6	29.3
	40	19.1	10.0	6.4	9.9	18.8	27.9	31.7	28.1
	50	14.9	5.5	2.5	5.5	14.7	25.6	30.7	25.9
	60	10.1	1.8		1.8	10.1	22.6	29.1	22.8

Calculated from List 1958, Table 135.

Two other factors need to be considered. In most of the tropics, temperate zone fruits are grown at higher altitudes than they are in the temperate zone for reasons of temperature in general and winter chilling in particular. Where this is so, the transmission through the atmosphere to the ground can be enhanced because the path of the sun's rays through the atmosphere decreases and atmospheric transmission increases. Becker and Boyd (1957) cited by Reifsnyder and Howard (1965), calculated that at a latitude of 40 the percentage increase in solar radiation intensity at 610, 914, 1219, 1524, 1829, 2134 and 2438 metres (originally calculated for 2000 to 8000 feet) relative to that at sea level on 21 December was 3, 7, 9, 11, 13, 14 and 15% and on 21 June was 7, 12, 15, 18, 21, 23 and 24% respectively. The June figure will be more representative of altitude effects at low latitudes. The second special factor is cloudiness. Whereas on a clear day 75% or so of solar radiation can reach the earth's surface only about 20% does so on days of complete cloud cover (Wilson 1993). List (1958) quotes data from Haurwitz (1948) showing the percentage of clear radiation is 80-85% under Cirrus cloud, 29-35% under Strato cumulus, 15-25% under Nimbo stratus and 17-19% under fog. In general there is least cloud above desert regions, which tend to be found in the subtropical regions rather than tropical or temperate ones (if one excludes the cold deserts which are usually unsuited to fruit production). As Wagenmakers (1994) has pointed out, the major effect of latitude between 35° and 55° on light at measured sites is a result of the differences in cloud cover and is much more pronounced than that in radiation levels above the atmosphere. There is also the contrast between times of cloudiness in winter and summer rainfall regions which can result in the incident radiation during a summer growing season being much less in the summer-rainfall tropics and subtropics in relation to adjacent, higher-latitude, Mediterranean regions than would be expected from Table 2. Additionally there are localised effects of distance from the sea, prevailing winds etc.

Measured values at a number of sites are given in Table 3.

2.2 Diurnal pattern

Any given daily or monthly level of summer solar radiation will obviously involve higher light intensities at low latitudes than high latitudes because of the shorter days (Table 4). Some diurnal patterns at different latitudes are given in Table 5.

2.3 Consequences of changes in solar energy through the season

By far the most important of these is the length of season over which there is adequate solar radiation for good fruit crop production. This is not to say that availability of photosynthetically active radiation, which makes up approximately 45% of extraterrestrial solar radiation, is the main determinant of the length of growing season (in fact this is most frequently determined by temperature factors, including frost, or by water availability) but that if the amount of P.A.R. received during a normal four- to five-month temperate-zone growing season is regarded as being adequate for deciduous fruit production then the P.A.R. levels in subtropical and tropical climates are adequate for cropping over most or all of the year. This means that subject to other environmental factors and appropriate cultivars and management practices:

Table 3. Yearly Solar radiation MJ m^{-2} day^{-1} at a number of fruit growing sites at different latitudes

Site	*Latitude N S*	*Jan Jul*	*Feb Aug*	*Mar Sept*	*Apr Oct*	*May Nov*	*Jun Dec*	*Jul Jan*	*Aug Feb*	*Sept Mar*	*Oct Apr*	*Nov May*	*Dec Jun*
East Malling U.K.[1]	51° 16'N	2.1	4.4	11.1	12.4	16.5	17.9	16.8	14.5	10.8	6.1	3.1	1.6
Nelson New Zealand[1]	41° 17'S	6.4	8.9	13.6	18.9	23.0	25.0	25.3	22.1	16.1	11.0	7.5	5.7
Elgin S. Africa[2]	34° 08'S	9.2	13.3	17.2	22.0	28.0	23.9	23.5	19.1	14.1	10.6	9.8	6.1
Apple-Thorpe Australia[3]	28° 13'S	11.8	15.2	19.6	22.4	22.9	22.3	24.7	19.6	15.8	15.7	14.0	9.5
Marondera Zimbabwe	18° 31'S	18.2	21.4	24.6	22.3	25.2	20.9	21.4	15.7	20.7	22.0	17.3	16.3

Data supplied by Dr J.W. Palmer (1), Mr R.C. Saunders (2) and Dr S.G. Middleton (3). The results are not necessarily from long runs of data but are typical.

Table 4. Mean daily duration of maximum possible hours of sunshine for different months and latitudes

Latitude North South	*Jan Jul*	*Feb Aug*	*Mar Sept*	*Apr Oct*	*May Nov*	*Jun Dec*	*Jul Jan*	*Aug Feb*	*Sept Mar*	*Oct Apr*	*Nov May*	*Dec Jun*
50°	8.5	10.1	11.8	13.8	15.4	16.3	15.9	14.5	12.7	10.8	9.1	8.1
40°	9.6	10.7	11.9	13.3	14.4	15.0	14.7	13.7	12.5	11.2	10.0	9.3
30°	10.4	11.1	12.0	12.9	13.6	14.0	13.9	13.2	12.4	11.5	10.6	10.2
20°	11.0	11.5	12.0	12.6	13.1	13.3	13.2	12.8	12.3	11.7	11.2	10.9
10°	11.6	11.8	12.0	12.3	12.6	12.7	12.6	12.4	12.1	11.8	11.6	11.5
0°	12.1	12.1	12.1	12.1	12.1	12.1	12.1	12.1	12.1	12.1	12.1	12.1

Abridged from Water Requirements Irrigation and Drainage Paper 24 (United Nations FAO Rome 1975)

Table 5. Diurnal pattern of solar radiation at mid summer at different latitudes

	Hourly totals of radiation MJ/m^{-2}		
Latitude	*42° 53 N*	*34° 08 S*	*17° 48 S*
Site Hours	*Geneva, N.Y*	*Elgin, RSA*	*Harare*
	July 1995	*4 Jan 1994*	*23 Dec 1988*
0500	0.03	0.0	0.00
0600	0.34	0.3	0.05
0700	0.79	1.4	0.68
0800	1.43	2.3	1.57
0900	1.98	3.0	2.42
1000	2.74	3.6	3.15
1100	2.93	3.9	3.66
1200	3.21	4.0	3.90
1300	3.22	4.0	3.89
1400	3.12	3.6	3.37
1500	2.68	3.1	2.68
1600	2.22	2.4	1.89
1700	1.63	1.4	1.28
1800	0.97	0.8	0.75
1900	0.36	0.2	0.08
2000	0.02	0.0	0.00
Total	27.67	34.0	29.37

Data supplied by Dr A.N. Lakso, Mr R.C. Saunders and Met Office, Harare. All data were from bright sunny days approaching maximum radiation.

a. Tropical and subtropical areas are suitable for cultivars with a long growing season requirement *i.e.* 'late' cultivars.

b. Tropical and subtropical climates can permit the production of several crops consecutively over the season. Examples of this are at 13° N in India (Javaraya 1943), 8° S in Indonesia (Saure 1973, Janick 1974), and 18° S in Zimbabwe (Jackson 1990) in each of which cases two crops a year were obtained, and in Peru at latitudes between 9° and 18° South where two or even three crops a year can be grown (Bederski 1989). It should be noted in this respect that the necessary time from blossom to harvest is in itself temperature dependent so that a 'late' or long-season cultivar in a cool climate may become an 'early' or short-season cultivar in a warm climate. For example, Denne (1963) showed that Cox's Orange Pippin apple requires about 5 months to mature in England but only 3.5 months in the warmer climate of New Zealand while Luton and Hamer (1983) found that the annual Cox harvest in England is about 3.5 days earlier for each increase of 1°C in mean temperature during June, July and August. As a result, although in warm, sunny, low-latitude climates the available light energy per day may be greater than that at higher latitudes, the available energy during a season of fruit growth and development may not be correspondingly greater for any specific cultivar in a single growing season because the fruit matures more quickly so the effective growing season is shorter. To take full advantage of the tropical/sub-tropical climate, it is necessary to use cultivars requiring a very long growing season or to obtain several crops within a year.

The second point is that given the same level of cloudiness and altitude, solar radiation levels in the tropics as a whole are not higher than those in the months of highest radiation in the temperate zone up to latitude 40° (List, 1958). When average intensity is calculated by adjusting for daylength the peak intensities are, again, very little different between the tropics and the subtropics with the highest values around the actual Tropic of Cancer and Tropic of Capricorn, not the equator, and peak values being but little lower at 40°. This leads to the important conclusion that the major differences in light regimes in orchards in tropical and warm-temperate climates are a consequence of differences in canopy development rather than above-orchard radiation levels unless there are major altitudinal and cloudiness effects (see below).

3 Light interception and distribution

3.1 Interception of available solar energy at different latitudes

Primarily for reasons of management and fruit quality, orchards are grown as discontinuous canopies, with individual trees or rows of trees physically separate from each other. Fruit tree growth and yield is, however, linearly dependent on the amount of solar radiation intercepted (Jackson 1978, Barritt 1989, Palmer 1993, Wagenmakers 1994, Lakso 1994). The amount of solar radiation transmitted to the orchard floor (T) consists of two components, that which would reach the orchard floor even if the trees were totally opaque (T_f) and that which reaches the ground only after passing through the orchard canopy (T_c). The latter depends on L' which is the Leaf area index divided by the mean daily shadow area and a light extinction coefficient (K). The relevant equations are:

$$T = T_f + T_c \tag{1.1}$$

$$T_c = (1 - T_f)\, e^{-KL'} \tag{1.2}$$

The latter equation can be re-written in terms of fractional proportion of the available light intercepted (F) as follows.

$$F = F_{max} - F_{max}\, e^{-KL'} \tag{1.3}$$

where F_{max} is the proportion of light which would be intercepted if the trees were solid (Jackson 1980).

F_{max} is the cast shadow area as a proportion of the total ground surface integrated over a chosen period of time and at any one time depends, for a hedgerow orchard, on height of hedge in relation to width of alley, hedge orientation and geometry and solar altitude and azimuth. Because of this, it is influenced by latitude and Jackson and Palmer (1972) calculating light interception by 'solid-model' hedgerows found that at the lowest latitude studied (34°) light interception by East-West hedgerows was much less than that of hedgerows of similar dimensions and orientation at 45° or 51.3° or than of north-south hedgerows at all three latitudes. Palmer (1989) using a much more detailed computer model and assuming constant LAI values showed E-W rows to intercept significantly less light than N-S at 30° N of latitude as compared with 51.3° N and Lakso (personal communication) using the Jackson and Palmer model showed that over the latitudes 10°-25° E-W rows intercepted significantly less light than N-S ones overall. There is therefore greater benefit from having N-S rows than E-W ones at low latitudes than high latitudes in terms of total light interception.

The higher sun at low latitudes also, in general, results in less effective light interception by 'vertical' hedgerow systems with wide alleyways than at high latitudes. There is thus likely to be a greater potential advantage from canopy systems which approximate to a continuous canopy at low latitudes, *e.g.* with open-centre trees, with multirow planting systems and with V-trellis systems in which the canopy can almost close over the top of the alleyway.

Latitude may also have a significant effect on light interception through its effect on leaf area index hence $L^{/}$ in the above equations. Many cultivars which are not properly adapted to warm winters, with inadequate winter-chilling for bud break, show the disorders called delayed foliation when grown in the tropics. This results not only in delayed bud break but also in many of the vegetative as well as fruit buds remaining dormant through the season. Table 6 shows the relative leaf areas on 25/10/1995 on branches of similar diameter (6 branches per cv/treatment) on trees in a trial at HRC Marondera (latitude 18.11, altitude 1631 m) in which the relative responses to dormancy-breaking sprays (hydrogen cyanamide) of different cultivars as an index of adaptation to tropical conditions have recently been determined (Jackson and Bepete 1995). Anna is well adapted and does not require a dormancy breaking spray, NJ46 is relatively poorly adapted, Ohinemuri very poorly adapted.

Table 6. Foliage area per apple branch as affected by inadequate winter chilling for different *cvs* at the Horticultural Research Centre, Zimbabwe

Cultivar	*Dormancy breaking sprays 1988-1995*	*Mean branch diameter (mm)*	*Leaf area per branch dm²*
Anna	None	32.7 ± 2.57	185
NJ46	None	39.7 ± 5.78	24
NJ46	Cyanamide	31.7 ± 3.52	160
Ohinemuri	None	33.5 ± 2.53	5
Ohinemuri	Cyanamide	33.3 ± 3.71	227

The leaf area per branch of the Anna and of those trees of NJ46 and Ohinemuri which had been treated with cyanamide to break bud dormancy was similar to that found by Verheij (1972) for Golden Delicious and other workers in temperate zones (Jackson 1980) although in the lower part of the range. The very low leaf areas recorded on untreated trees of NJ46 and Ohinemuri are well below expectation for branches of the same size in the temperate zone.

Reduced development of foliage under tropical conditions can be extremely disadvantageous but in a moderate form may even be beneficial. For example, very vigorous, heavy cropping trees of Sunlite nectarines at Marondera (18° 11'S) do not show the excessive shoot growth and large leaves of the same cultivar in the Cape Province of South Africa (latitudes 34° 08'S) which necessitate heavy and frequent summer pruning to avoid excessive within-tree shading in the latter area (Oosthuizen, personal communication).

3.2 Light penetration and distribution in tree canopies

The general equation governing light intensity at any given depth (expressed in leaf area index terms) in a continuous canopy like that of a wheat field is:

$$I_L / I_o = e^{-KL} \qquad (1.4)$$

where I_o is light intensity above the canopy, I_L is light intensity below a leaf area index (LAI which is m^2 leaf per m^2 ground) of L and K is a measured extinction coefficient calculated from measurements of I_L, I_o and L under relevant conditions.

In a discontinuous canopy like an orchard, where light can come from the side as well as above, this equation cannot be used to estimate light at any point in the canopy from I_o and vertically summed LAI but the underlying relationship still applies and the amount of canopy, L_I, above a certain irradiance level I (expressed as a fraction of radiation above the canopy) can be calculated (Jackson 1980) as:

$$L_I = F_{max} (\ln I / -K) \qquad (1.5)$$

This can, alternatively, be written

$$L_I = F_{max} \frac{(\ln I_0 - \ln I_L)}{K} \qquad (1.6)$$

Where I_0 and I_L are irradiance levels above and below the canopy respectively. The canopy volume in which light intensity is more than a given fraction of that above the canopy will therefore be affected by latitude in the same way the latter affects F_{max}, *i.e.* effects will be small excepting for that at low latitudes hedgerow orientation will be more important than it is at high latitudes.

This, however, refers to light intensities as proportions of light intensity above the canopy. Latitude also influences the canopy volume and leaf area which receives light above different threshold levels in absolute terms (*e.g.* M. Joules/m^2) through its effect on levels of solar radiation above the canopy of the type discussed earlier.

The effect of above-canopy radiation level can be illustrated using the radiation figures given by Wagenmakers (1994). Davis, California (38° N) has a high light climate with

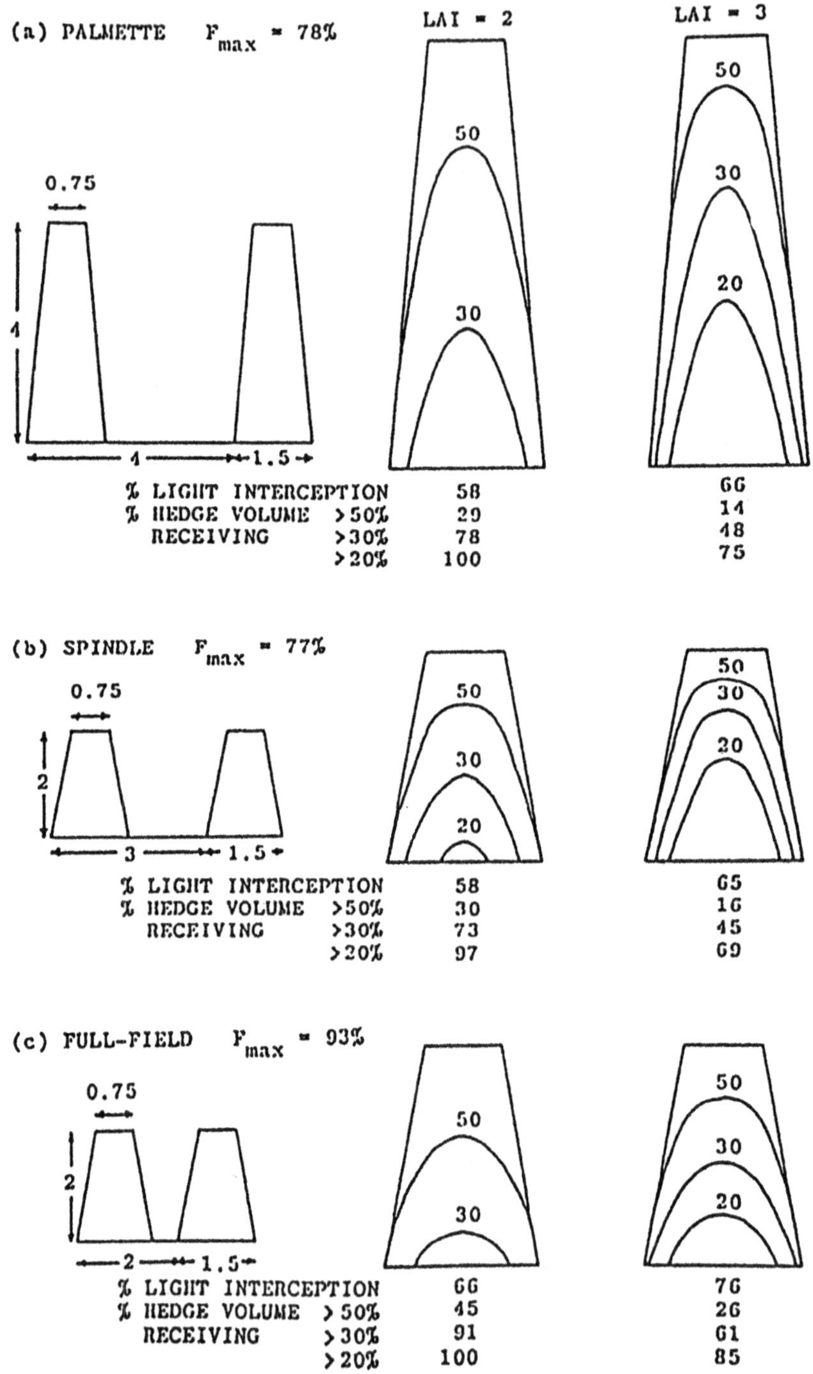

Figure 1. Calculated light profiles as % of light above the canopy within individual sections of hedge of given dimensions (m) and two LAIs for three orchard systems. From Palmer (1980)

4.13 GJm^{-2} over a 5-month growing season whereas Wilhelminadorp, Netherlands has an average of 2.5 GJm^{-2} over a similar period. If 30% of above canopy radiation at Wilhelminadorp is taken as the critical level for 'good' fruit bud formation and fruit size development the same light intensity within the canopy at Davis will be attained with only 18% of the above canopy radiation. Assuming F_{max} to be 70% and K to be 0.6 then the leaf area index, LAI, which receives more than this 'critical' level will be only 1.4 at Wilhelminadorp (with $I_L/I_o = 0.30$) whereas it will be 2.0 at Davis where the same light intensity will be found at an I_L/I_o value of only 0.18. This type of calculation of LAI above 'radiation contours' in a fruit tree can be carried out very simply and has been shown to give very similar estimates at a range of latitudes, especially for north-south oriented hedgerows, to those obtained using a detailed computer model (Palmer 1989). Data showing detailed radiation contours within trees published by Palmer (1980, 1989) can also be used to determine the effect of above-canopy radiation level on well-illuminated canopy volume. For example in Fig. 1, taken from Palmer (1980) the within-hedgerow radiation contour for 30% full sun at Wilhelminadorp would in energy terms, be approximately the same as that for 20% full sun at Davis and the 50% full sun contour at Wilhelminadorp corresponds to the 30% full sun contour at Davis. With an LAI of 3 with tall palmettes only the 14% of the canopy at the very top of the trees would receive more than 50% full sun at Wilhelminadorp, which is clearly undesirable as this would largely be made up of purely vegetative current years shoots, whereas under Davis conditions 48% of an identical canopy would receive a similar level of solar radiation in absolute terms. The greater volume of canopy (and LAI) which can be adequately illuminated in a high-energy than a low-energy environment presumably explains the general use of taller tree systems in the former area than the latter. This does not imply that physically larger trees are necessary for full production under such conditions although this is obviously one way in which to make use of the potential higher LAI under high radiation conditions. However, the limits even under these conditions must be recognised. Although discontinuous canopies such as orchards permit light to reach low down on the sides of trees without passing through foliage (cf Fig. 1) the general equation governing light transmission through canopies, $I_L/I_o = e^{-KL}$, with a transmission coefficient of 0.6 for apple, indicates the following correspondences for vertically summed LAI and for radiation expressed as a proportion of that above the canopy: LAI = 1, $I_L/I_o = 0.55$; LAI = 2, $I_L/I_o = 0.3$, LAI = 3, $I_L/I_o = 0.16$; LAI = 4, $I_L/I_o = 0.09$; LAI = 5, $I_L/I_o = 0.05$. The LAI above any radiation level will be less for discontinuous canopies (it should be multiplied by the percentage cast shadow area expressed as a decimal fraction). Systems with too dense canopies will therefore fail to produce good quality fruits under even high light conditions at intermediate or low latitudes as has been confirmed by recent experience in New Zealand (Tustin *et al.*, 1989, 1993).

4 Special features of light climate for fruit production at low latitudes

The foregoing has been concentrated on the effects of low latitudes on radiant energy availability, interception and penetration into fruit tree canopies, including the indirect effect of latitude on the radiation-intercepting canopy through the general association of low latitudes with inadequate winter chilling hence poor foliage development. There are several other latitude-specific features that influence light climate in orchards or the responses to it.

4.1 Temperature effects associated with high solar radiation

These are as important in warm-temperate and subtropical regions as in the tropics *per se* because of the high radiation levels which can be encountered. High direct radiation may lead to considerable heating of fruits. Thorpe (1974) showed that at a constant wind speed (3 m sec^{-1}) and with a radiation flux density up to 800 W m^{-2} the equilibrium surface temperature was a linear function of irradiance for both a green Golden Delicious apple and a red Jonathan apple, the latter showing a 7% greater warming effect. Increasing windspeed at a constant irradiance reduced the warming effect. Under field conditions, in England on a cloudless day with an air temperature of 27°C and windspeed less than 1m sec^{-1}, apples in full sunlight (630 ± 30 W m^{-2}) were 13-14°C above air temperature and the cool sides of these fruits were 3°C above air temperature. Thorpe suggested that these temperature effects had a positive effect on fruit growth but under Israeli conditions Lavee and Rauchberger (1993) reported that radiant heating of grape fruits exposed to sunlight, which raised temperatures up to 140% of ambient, had a deleterious effect on their growth. That this was an effect of temperature not irradiation *per se* was shown by the better development of grapes subjected to high light but cooling winds. It should be noted in this context that the growth of apple fruits and shoots is positively related to temperatures below about 25-30°C but may be limited by other factors at higher temperatures (Lakso 1994). Another adverse effect of radiant heating is the 'sunscald' which is observed in apples under conditions of high exposure.

Although under temperate zone conditions high temperatures in summer and autumn have favourable effects on the stage of bud development reached before winter dormancy and hence on the ability of the flowers to produce fruits in the following year (Luckwill 1974), excessive temperatures can have adverse effects on flower bud differentiation (Dennis, this book).

It should be noted that excessive temperatures associated with high solar radiation levels can be controlled either by reducing the available radiation by artificial shade, or by pruning and training the trees so as to create some shade where needed, or by direct cooling by water-sprinkling. These technologies are likely to become increasingly important at low latitudes, being made more viable in economic terms by the potential use of netting support structures for protection against hail and of over-tree sprinklers for evaporative cooling to prevent de-chilling of fruit buds in winter.

4.2 Effects associated with high altitudes

Although a common perception of the tropics is of a climatic zone with high temperatures maintained through the year, this is not true for the major areas of deciduous fruit production at low latitudes. Although this situation may change with the increasing availability of cultivars with little requirement for winter chilling, at present much of the tropical production is at high altitude to meet, or partially meet, the chilling requirements of mainstream temperate zone cultivars. This has a marked effect on temperatures. Thus in a major new apple-growing area in Brazil in Santa Catarina at 26° 46'S and an altitude of 960 m the mean annual temperature is 16.6°C and that over the months of the actual fruiting season ranges from around 15 to about 21°C (Bernardi 1988). At Applethorpe, Queensland, latitude 28°C 37'S and an altitude of 870 m the monthly mean temperatures from October to April inclusive are 15.0, 17.5, 19.7, 20.5, 19.9, 18.6 and 15.2°C respectively (Middleton, personal communication) and at Marondera, Zimbabwe, latitude 18° 11'S, altitude 1631 m they are 19.3, 19.8, 19.5, 19.8, 19.5, 19.0, and 17.3 respectively (Agritex 1989).

In Guatemala 1400-2000 m is considered a low elevation for apple growing, with other areas going up to 3000 m (Williams and Menegazzo 1988). In Equador Niegel (1988) cites Tungurahua as the most important deciduous fruit production area with a representative weather station at 2.770 m and latitude 01° 10' north, and mean monthly temperatures throughout the year ranging from 11.6 to 13.5°C.

These temperature regimes are significant in relation to the effectiveness of utilisation of solar radiation for actual production. Wagenmakers (1994) has emphasised that much of the benefit of higher solar radiation at low latitudes is lost because of the effect of the higher temperatures there on respiration. Where temperatures are moderated by altitude, and insolation is probably increased by this, the balance of photosynthesis to respiration is likely to be much more positive. This argument applies both to the tropical highlands which dominate southern Africa and the Andean countries and Mexico and even to major areas of production in the warm-temperate zone such as the Alto-Adige (Süd Tirol) in Italy and the Pyrenean foothills in Spain.

4.3 Cloudiness

Presumably for reasons of chilling and other low temperature requirements of many deciduous fruits, appreciable production of these in the tropics is carried out in environments with a high incidence of cloud or fog. In Peru, between 9° and 19° S, one substantial apple growing area is at low altitude below a dense layer of fog at 400 m altitude for 4 to 5 months of the year. Production is therefore under conditions of low light intensity (Bederski 1988). In Colombia, where production takes place at latitudes of 1800 to 2900 m, temperatures are relatively low with maxima seldom exceeding 20°C and solar radiation limited, with less than 4 hours a day during the March-September dormant season and no more than around 6 hours a day in the production season (Silva 1988). This could result in the same effect as high latitude leading to reduced respiration, due to relatively low tree temperature.

5 Conclusions

Light climates at low latitudes, as compared with those in traditional, temperate-zone, areas of deciduous fruit production, offer special opportunities in terms of the length of the potential cropping season. Higher solar elevations and differences in the solar track during the day have some implications for orchard design, making, especially, E-W hedgerows relatively less effective in light interception and adding to the relative advantages of high-ground-cover systems such as open-centre trees, multirows and V-trellises as compared with widely-spaced vertical hedgerows. Light intensities above the canopy are not, in general, higher than those at many centres of fruit production in the temperate zone so concepts and practices dependent on this should be transferable from appropriate sites. Major differences in light penetration into canopies may result from limited bud-break and reduced canopy density if cultivars with high requirements for winter chilling are grown at low latitudes. The frequent production of deciduous fruits at high altitudes in the tropics, to avoid conditions of sub-optimal winter chilling, results in exposure to increased light intensities under clear sky conditions or reduced ones because of prevalent cloudiness. Production at high altitudes/low latitudes also tends to shift the balance between light-controlled photosynthesis and temperature-controlled respiration in favour of the former.

Acknowledgements

Thanks are due to A.N. Lakso, S.G. Middleton, J.W. Palmer and R.C. Saunders for providing data and to Mrs M. Kurwakumire for assistance in preparing the manuscript.

6 References

Agritex (1989). *Agro-climatological summaries and analysis* Volume 2. Department of Agricultural and Technical Services, Department of Meteorological Services Harare, Zimbabwe.

Barritt, B.H. (1989). Influence of orchard system on canopy development, light interception and production of third-year Granny Smith apple trees. *Acta Horticulturae* 243, 121-130.

Bederski, K.A. (1989). Apple growing in the coastal areas of Peru. *Acta Horticulturae* 233, 51-55.

Bernardi, J. (1988). Behaviour of some apple cultivars in the subtropical region of Santa Catarina, Brazil. Acta *Horticulturae* 232, 46-50.

Denne, P.M. (1963). A comparison between fruits of Cox's Orange Pippin from Kent, England and Auckland, New Zealand. *New Zealand Journal of Botany*, 1, 295-300.

Haurwitz, B. (1948). Insolation in relation to cloud type. *Journ. Meteorol.* 5, 110-113.

Iqbal, M. (1983). *An introduction to solar radiation.* Academic Press Toronto.

Jackson, J.E. (1978). Utilization of light resources by high density planting systems. *Acta Horticulturae* 65, 61-70.

Jackson, J.E. (1980). Light interception and utilization by orchard systems. *Hortic Rev.* 2, 208-267.

Jackson, J.E. (1990). Cropping and multiple cropping of apples in medium altitude tropical Zimbabwe. 23rd International Horticultural Congress. *Intern. Soc. Hortic. Sci. Firenze* 1990 Abstr. No.4325.

Jackson, .J.E. and Bepete, M. (1995). The effect of hydrogen cyanamide (Dormex) on flowering and cropping of different apple cultivars under tropical conditions of sub-optimal winter chilling. *Scientia Horticulturae* 60, 293-304.

Jackson, J.E. and Palmer, J.W. (1972). Interception of light by model hedgerow orchards in relation to latitude time of year and hedgerow configuration and orientation. *J. appl. Ecol.* 9, 341-357.

Janick, J. (1974). The apple in Java. *HortScience* 9, 13-15.

Javaraya, H.C. (1943). Biannual cropping of apple in Bangalore. *Indian J. Hort.* 1 (1), 31-34.

Lakso, A.N. (1994). Apple in: *Handbook of environmental physiology of fruit crops*. Edited by Bruce Schaffer and Peter C. Andersen. Vol. 1. Temperate Crops, Chapter 2, 3-42.

Lavee, S. and Rauchberger, S. (1993). The effect of bunch location in relation to light and temperature on fruit development and maturation of cv. Perlette (Vitis vinifera) vines in two training systems. *Acta Horticulturae* 349, 133-137.

List, R.J. (1958). Smithsonian Tables. *Smithsonian Institute Publication* 4014. 527 pp.

Loreti, F., Morini, S., Muleo, R., Vitagliano, V. and Masetti, C. (1993). Effect of solar radiation on some growth parameters of peach fruits. *Acta Horticulturae* 349, 117-121.

Luckwill, L.C. (1974). A new look at the process of fruit bud formation in apple. *Proceedings of the XIX International Horticultural Congress* 3, 237-245.

Luton, M.T. and Hamer, P.J.C. (1983). Predicting the optimum harvest dates for apples using temperatures and full bloom records. *Journal of Horticultural Science* 58 (1), 37-44.

Niegel, W. (1988). History, actual advance and future of the Equadorian apple growing. *Acta Horticulturae* 232, 67-73.

Palmer, J.W. (1980). Computed effects of spacing on light interception and distribution within hedgerow trees in relation to productivity. *Acta Horticulturae* 114, 80-88.

Palmer, J.W. (1989). The effects of row orientation, tree height, time of year and latitude on light interception and distribution in model apple hedgerow canopies. *Journal of Horticultural Science* 64 (2), 137-145.

Palmer, J.W. (1993). Recent developments on light and fruit tree canopies. *Acta Horticulturae* 349, 99-109.

Reifsnyder, W.F. and Howard, W.L. (1965). Radiant energy in relation to forests. U.S.D.A. *Forest Service Technical Bulletin* No 1344.

Saure, M. (1973). Successful apple growing in Indonesia. *Fruit Var. Jour.* 27, 44-45.

Silva, B.E. (1988). Algunas consideraciones del clima en Colombia relacionadas com el rompimiento de la dormencia. *Acta Horticulturae* 232, 90-94.

Thorpe, M.R. (1974) Radiant heating of apples. *Journal of Applied Ecology* 11, 755-760.

Tustin, D.S., Hurst, P.M., Warrington, I.J. and Stanley, C.J. (1989). Light distribution and fruit quality through multilayered trellis apple canopies. *Acta Horticulturae* 243, 209-216.

Tustin, S., Hirst, P.M., Cashmore, W.M., Warrington, I.J. and Stanley, C.J. (1993). Spacing and rootstock studies with central leader apple canopies in a high vigour environment. *Acta Horticulturae* 349, 169-177.

Verheij, E.W.M. (1972). Competition in apple as influenced by alar sprays, fruiting, pruning and tree spacing. Meded. *Landbouw hogeschool Wageningen* No. 72-4, 54 pp.

Wagenmakers, P.S. (1994). Light relations in orchard systems. *Thesis Wageningen.*

Williams, W.T. and Menogazzo, G. (1988). Apple culture in the highlands of Guatemala. *Acta Horticulturae* 232, 57-66.

Wilson, E.M. (1993). *Engineering Hydrology*. Fourth Edition. Macmillan.

2 BUD DORMANCY; PHENOMENON, PROBLEMS AND SOLUTIONS IN THE TROPICS AND SUBTROPICS

AMNON EREZ

A.R.O., The Volcani Center, Institute of Horticulture
P.O. Box 6, Bet-Dagan, Israel

1 Introduction

Marked increases in production of temperate fruit crops have occurred in many subtropical and tropical countries (Table 1). One of the major limitations of production in warm countries is overcoming the dormancy period typical of all temperate zone fruit trees.

Table 1. Production (1000 MT) of selected temperate fruits in tropical and subtropical countries in 1990 and 1996*

Country	*Apple*		*Pear*		*Peach (+ nectarine)*		*Grapes*	
	1990	*1996*	*1990*	*1996*	*1990*	*1996*	*1990*	*1996*
Egypt	57	455	55	93	41	60	577	740
Libya	98	1	1	11	9	36	30	
Madagascar	66	1	2	7	7	10	10	
Morocco	255	360	36	35	29	41	249	270
Algeria	54	74	42	37	35	45	262	132
Tunisia	42	70	31	53	35	48	85	107
Zimbabwe	76			1	1	2	3	
South Africa	485	500	194	220	158	138	1528	1670
Guatemala	10	28					5	6
Mexico	497	645	26	28	146	140	487	535
Bolivia	89	4	4	31	35	19	21	
Brazil	509	653	17	19	96	135	19	21
Ecuador	27	24	7	10	5	14		
Venezuela					9	9	12	14
Paraguay	11			2	2	22	23	
Peru	122	177	7	7	38	35	59	94
India	1118	1200	131	130	71	85	398	600
Pakistan	257	600	32	36	24	40	34	72
Iraq	72	80	3	3	27	25	428	300
Yemen	12			4	3	139	98	
Jordan	645	1	2	4	9	36	84	
Israel	118	134	18	30	40	59	87	86
World	53672		13093		10409		57410	

* FAO Production Yearbook No. 50 1996

Amnon Erez (ed.), Temperate Fruit Crops in Warm Climates, 17-48.
© 2000 *Kluwer Academic Publishers. Printed in the Netherlands.*

In this chapter I will use the definition of dormancy proposed by Lang *et al.* (1987). Paradormancy, or ectodormancy for the initial stages of dormancy development when the buds fail to develop because of an outside influence, primarily correlative inhibition by the apex or leaves; endodormancy (ED) or rest, for the period during which bud break can not be induced by practices such as pruning and defoliation; and ecodormancy, when the bud can not develop because of unfavorable environmental conditions normally low temperature. According to Faust *et al.* (1997) endodormancy is further divided to d-endodormancy and to s-endodormancy. The s-endodormancy is the stage during which dormancy breaking chemicals can substitute for chilling as against d-endodormancy when no substitution for chilling is possible. The term "dormancy development" is defined as the series of events that takes place in the bud from dormancy induction to dormancy breaking. "Depth of dormancy" is an expression related to the chilling requirement needed. The deeper the dormancy, the higher are the number of chilling hours required to break it.

Without doubt, overcoming dormancy is the most critical element in growing temperate fruit crops in climates lacking the chilling required to overcome dormancy naturally.

I will discuss dormancy of all bud types. Great variations exist between lateral and terminal buds and vegetative and reproductive buds with regard to their responses to environmental conditions. The integrated response of all buds reflects the behavior of the tree.

2 The phenomenon

Dormancy of buds of temperate-zone fruit trees, is a basic phase in their annual developmental cycle allowing survival under extreme cold winter climate. Dormancy was natures' development to enable survival of the perennial plant during a climatically hostile period. Dormancy allows adaptation of the plant to its environmental conditions by coordinating growth and development to the appropriate annual periods. By preparing the plant to the unfavorable conditions, it enables the tree to accumulate reserves, mostly carbohydrates, to drop sensitive organs, *e.g.* leaves, to develop organs to protect the meristems and to resist harsh conditions by develop cold hardiness. Once the bud is in the dormant state, a time measuring mechanism sets in operation that dictates when growth will be resumed. The basic limit in these plant strategies is the irreversible direction of events, once a decision is taken the sequence of events must proceed forward. This strategy allows for no mistakes.

The basic decisions for the plant regarding dormancy are: dormancy induction and dormancy release. In the first case, the critical element is not to respond too late, as this may result in damage to unprepared tissues, while in the second case the critical element is not to respond too early, for early bud break may result in freeze injury. On the other hand, if dormancy induction will occur too early or if bud break will occur too late, the maximal developmental potential of the tree will be reduced and with that its ecological competition ability.

Keeping this in mind will help in understanding the natural choice of the elements of control: short days and low temperature for induction, and chilling accumulation for time measurement during dormancy.

A typical aspect of bud dormancy is the development of isolated entities that lose their normal interconnections as xylem and phloem movement are extremely reduced and even plasmodesmata are disrupted between the meristem and surrounding tissues (van der Schoot 1996). This development leaves the bud to respond independently to the environment. This results in variation in response of the buds due to: 1. variation in depth of dormancy among the buds and 2. variation in exposure to environmental conditions affecting dormancy.

2.1 Dormancy induction

Although dormancy can be induced under optimal growth conditions of long days and warm temperatures (Samish 1954), short days (Nitsch 1957) and low temperatures (Lavarenne *et al.*, 1975, Crabbe 1994, Faust *et al.*, 1997) will hasten ED induction. Thus, an endogenous mechanism induces dormancy while environmental factors seem to modulate it. Therefore the pace of induction is more rapid under temperate climates where environmental changes in autumn are much quicker than in the subtropics and tropics.

Dormancy development in axillary buds is gradual. It starts with paradormancy, mostly due to apical dominance that prevents laterals from breaking. The level of apical dominance is species-related, with stronger polarity or acrotony in pome fruits and sweet cherries than in peaches plums and apricots. However, within each species, polarity is cultivar-dependent. In the case of apples, four groups were defined by Lespinasse and Delort (1986) as to their pattern of growth from upright growth of spur types to wider open pattern of cultivars like ‘Granny Smith’. This characteristic is becoming of special importance when management practices are used in warm climates (see section 2.6.2.d). As to lateral buds: as they are formed successively on the growing branch, they enter ED consecutively. The modulating effects of cold temperatures and short days make this response more uniform but at lower latitudes the individual behavior of the bud is maintained. This is especially conspicuous with lateral buds on long growing shoots. The first to form are the basal buds and their entrance into ED is therefore, earliest. Late forming buds may remain paradormant side by side with endodormant basal buds.

The terminal bud reacts differently, it may continue to grow much later in the season and is thus more affected by environmental conditions that retard its entering into dormancy. Typically, under warm climates with otherwise non-limiting conditions, longer terminal growth occurs which may accentuate paradormancy of the lateral buds and affect their chilling requirement.

Flower buds of low chill cultivars will not enter ED or will enter it very slowly when autumn temperatures remain high. Vegetative buds, on the other hand, will respond to short days in spite of high temperatures (Lerner 1990). This may lead to autumn bloom, sometime very abundant, under warm conditions when bud break is stimulated by loss of leaves, irrigation after a long period of drought or dormancy-breaking chemical sprays (Erez 1987b). In many subtropical countries this response of the tree is very unfavorable, as the fruits that may set normally, would not be able to reach commercial maturity because of their small size and low quality, which result from a shortage of carbohydrates. On the other hand, flowers that open in autumn do so at the expense of the spring bloom which consequently may be sparse.

It is a puzzling problem that low temperature, the same factor that will be responsible for breaking dormancy later in the season, plays the opposite role of first promoting its establishment.

3 The climatic requirements of dormant buds

3.1 Temperature and light

The study of climatic effects on buds, is hampered by the difficulty in moving trees into controlled conditions. This has led researchers to analyse the responses of trees in the field, and to evaluate their response to the natural variation of climate among years. With the development of techniques of producing viable, container-grown, small trees (Couvillon and Erez 1980, Erez 1983), investigation under controlled conditions became possible.

Two climatic elements affect the dormant bud and hence its response: light and temperature. Light is perceived by dormant peach vegetative buds. Dormancy induction is enhanced by short days while flower buds are not affected (Erez *et al.,* 1966). Long exposure to darkness prevent vegetative buds from breaking even after exposure to optimal temperatures that break dormancy (Erez *et al.,* 1966). This effect was demonstrated in a few other species although species appear to differ in this respect (Erez *et al.,* 1998). A few woody species respond positively to light and long days are the critical factor for breaking bud dormancy and causing continuous growth as in Weigela florida (Nitsch 1957). In peach trees long days enhance bud break in spring (Erez *et al.,* 1966). During winter it was shown that limitation of light or even total darkness have a positive effect on spring bud break providing light is applied during bud break (Erez *et al.,* 1968). Thus the natural change of daylength during winter and spring seems to be perceived by the vegetative bud and to contribute to its bud break response (Fig. 1).

Although under certain experimental conditions light has a considerable effect, temperature is the major element in the climatic requirement of the dormant bud. Even though chilling may first enhance dormancy induction as mentioned above, it soon becomes the major controlling factor in the development of the dormant bud (Coville 1920, Chandler and Tufts 1934, Samish 1954).

Once ED has been induced, exposure to low temperature is needed to break dormancy and to allow growth resumption. Based upon many studies, the low temperature effect is optimal around 6-8°C (Erez and Lavee, 1971, Richardson *et al.,* 1974), lower and higher temperatures are less effective. An upper threshold has been demonstrated at 14°C (Erez and Couvillon 1987) (Fig. 2). While an accurate lower threshold was never reported experimentally it is clear that temperatures lower than 0°C may have little effect. However, the effect of temperature is more complex, and an interaction of temperatures was found when low and high temperatures were cycled (see below). For a long time it was known that high temperatures have a negative effect on

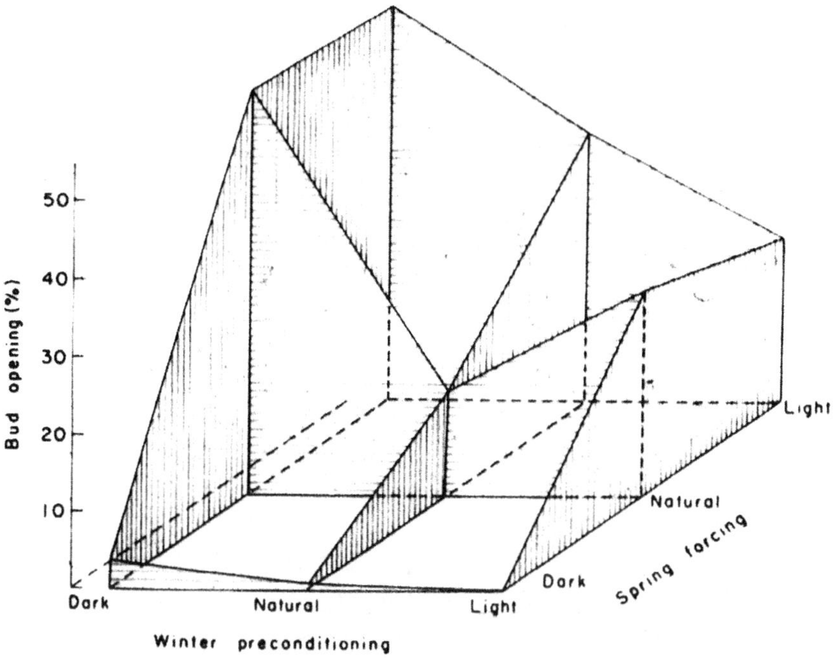

Figure 1. Effect of daylength on vegetative bud break of potted 'Redhaven' peach trees grown in Bet-Dagan, Israel. Trees were exposed to total darkness, natural daylength (10.5 h in December, 12 h in March) or to long days by extending the natural day with 6 h of incandescent light from dusk. First exposure period (winter preconditioning): December-February; Second exposure period (spring forcing): March-April. Nine trees per treatment underwent the first exposure; later divided to three per treatment in the second exposure.

dormancy development (Bennett 1950, Overcash and Campbell 1955) similarly to other temperature-related phenomena such as devernalization by high temperatures (Chouard 1960). In fact, this is one of the reasons why cloudy and foggy weather has been found very advantageous for dormancy breaking, as the absence of direct sunshine keeps buds cool thus avoiding the negation of chilling by high temperatures. Erez *et al.*, (1979a) and Couvillon and Erez (1985b), studying the effect of temperature on rooted peach cuttings under controlled conditions, showed that alternating high and low temperatures in a daily cycle negated chilling, when the high temperature exceeded 19°C. Chilling negation by high temperatures depended on the temperature of exposure and its duration in the cycle (Fig. 3). The longer the duration and the higher the high temperature in a daily cycle, the stronger the negating effect (Couvillon and Erez 1985b). The length of the cycle was also found to be critical: short cycles, of 24 hours, were the most negating, while with longer ones the negating effect disappeared (Erez *et al.*, 1979b) (Fig. 3). This was interpreted as evidence for a fixation effect of chilling after a certain period of time, calculated to be about 30 hours at 6°C. Applying overhead sprinkling during warm clear winter days will reduce bud temperature by evaporative cooling (Gilreath and Buchanan 1981a, Erez and Couvillon 1983). Depending on the relative humidity, this method can reduce bud temperature by up to 13°C allowing the accumulation of night chilling that otherwise would be negated by excessive day temperature, and thereby improving bud break. This method has been used commercially in Israel (Erez *et al.*, 1993, Erez 1995), and South Africa (P. Allan, Personal Communication). Interestingly, a promotive effect of high temperature may be evident after exposure to partial chilling. This was first shown for peaches by Erez and Lavee (1971) and verified recently (Linsley Noakes *et al.*, 1994).

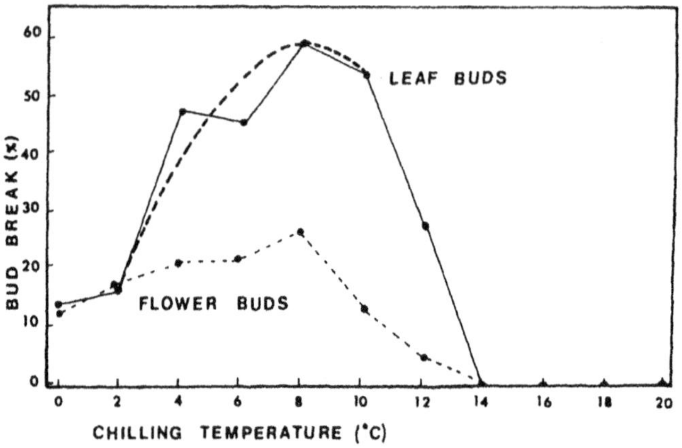

Figure 2. Chilling efficiency of temperature. Rooted cuttings of 'Redhaven' peach were held for 1200 h at temperatures of 0 to 20°C and then at continuous 22°C for 30 days. (After Erez and Couvillon 1987)

A B

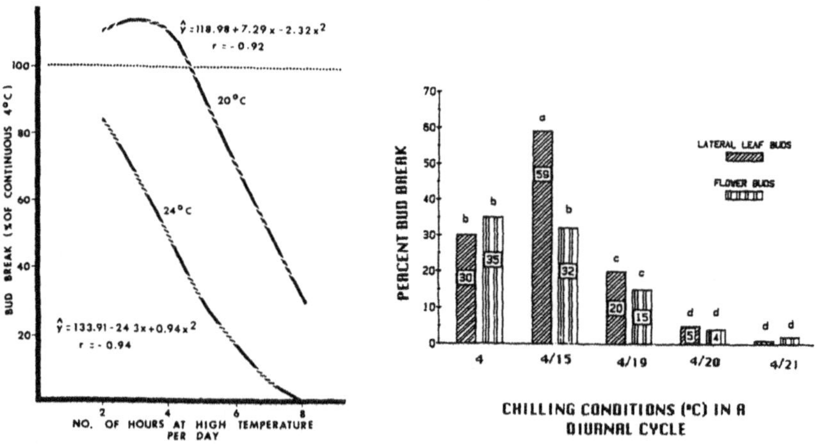

Figure 3. Chilling negation by high temperatures. A. 'Redhaven' peach plants were exposed to a total of 1200 chilling hours at 6°C with different diurnal durations of intermittent heating at 20 or 24°C. B. 'Redhaven' peach plants were exposed to 1200 hours at 4°C continuously or alternating with temperatures of 15,19, 20 and 21°C. In the alternating temperature regimes, the high temperature was for 8 h per day. C. 'Redhaven' peach plants were exposed to continuous chilling at 4°C (Control) or to temperature cycles of 1, 3, 6 or 9 days durations, alternating between 6 and 24°C. In all cyclical treatments plants were held at 6°C for two-thirds of the period and at 24°C for one third. In all cases Bud break was tested after 30 days at 22°C. (A, B after Couvillon and Erez 1985b; C after Erez *et al.* 1979b).

C

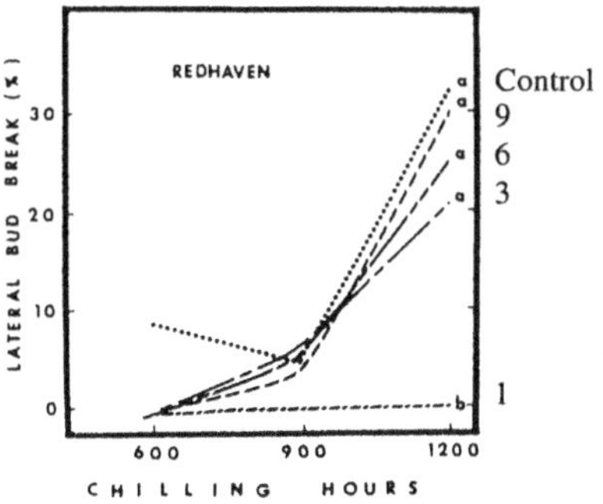

Figure 3. Continued.

An additional effect of temperature was found for moderate temperatures. Synergistic effects were obtained in peach when "neutral" moderate temperatures (15°C) were cycled with chilling (Erez and Couvillon1987), (Fig. 3B).

3.2 Climatic requirements of different buds

Response to chilling varies with type of bud. Generally the terminal bud has a lower chilling requirement, which may be interpreted as having a shallower dormancy. This is especially evident under conditions where chilling is marginal. Under such conditions, the terminal bud breaks first; with little competition from other buds, growth may be vigorous. Floral or mixed buds generally have lower chilling requirement allowing bud break under marginal conditions (Fig. 4). In Filberts a differential chilling require-

Figure 4. Poor bud break in apple (left) and apricot (right) exposed to insufficient chilling. Note mostly terminal bud break in apple and developing fruits with no leaves in apricot.

ment is observed between male (lateral) and female (terminal) flowers. Catkins flower much earlier than female flowers testifying to their lack of dormancy or to having a very shallow dormancy.

In a few cases competition between developing buds is evident. When all vegetative buds start together, growth is limited, leading to spur production in pome fruit species. With uniform heavy bloom in peaches, almonds and cherries flowers may set heavily. On the other hand, less uniform bud break may lead to a lower rate of fruit set as the late flowers must compete with stronger sinks already active.

3.3 Heat effects

The question of the interaction of chilling and heat in overcoming dormancy is very puzzling. Clearly there is a dynamic change regarding temperature effect with dormancy development. Erez and Lavee (1971) working with potted 'Redhaven' peach trees, reported that long warm periods after partial chilling not only did not negate the chilling effect, but actually improved the level of bud break. Linsley Noakes *et al.* (1994) evaluated the effect of winter climate on three nectarine cultivars, in five locations in South Africa, over three years, and noted that there was a positive component of the effect of heat in addition to the chilling effect. In addition, Sparks (1993) claims that for pecan trees, chilling interacts with heat and both may affect bud break.

Rageau *et al.* (1998), working in Clermont-Ferrand at 45° N latitude, observed that 'Redhaven' peach trees protected from chilling in a greenhouse at temperatures of 15-30°C, exhibited a high and rather uniform level of vegetative bud break by mid July 1995. They further verified that long days and high light intensity were needed to obtain this effect. Dormancy breaking was in this case specifically an effect of high temperature. "High" here refers to temperatures of 25-30°C not to extreme temperatures above 45°C that are well known to break bud dormancy (Chandler 1960). Thus, the application of long period of heating induces a slow effect that may lead to bud break. Note that: a. simple floral buds will not survive these conditions; b. secondary dormancy will soon appear in the growing vegetative buds that will result in rosetting. It is not clear how mixed buds would react to such condition, or whether normal fruit would develop. This is also a matter of concern with nut crops that have a different requirements for the terminal female inflorescence and the lateral male catkins.

All stone fruit species are highly sensitive to high winter temperatures that may not only delay bud break, but also cause abnormal development, even if the temperatures drop later. Typical responses are the appearance of double or triple pistils and development of abnormal mummy fruits with aborted embryos. This effect on flower development needs more explanation. While the simple vegetative bud is anatomically complete in the fall, the floral bud does not complete its structural development until late in winter or even until the start of bud swell, when micro and macro sporogenesis occur (see chapter 5). This renders the floral bud sensitive to temperature during its development. Warm conditions during winter may cause abnormal flower development, mostly abnormal ovary development or severe drop of floral buds. This is not the case with pome fruits in which bud drop is rare.

3.4 Climatic models

The need for quantifying chilling is important for two independent reasons: to define the chilling requirements of a variety, and to define the available active chilling in a specific location.

The phenomenon of dormancy has attracted continuous interest because of its special characteristics. Evaluation of chilling has led to many misconceptions that still have an impact in the horticultural community. The reason for this is the difficulty in determining the actual chilling requirements of buds on trees. The first evaluations of chilling requirements were done in the field by comparing climatic data over long periods with bud break levels; trials with potted trees under strict controlled conditions were carried out only much later. Weinberger (1950) explored the temperature effect on the breaking of peach bud dormancy by examining climatic data over a period of 50 years in Fort Valley, Georgia, USA and comparing them with phenological observations. He concluded, on the basis of the best correlation obtained, that the chilling effect can be expressed as the sum of hours at and below a temperature threshold of 45°F (7.2°C). This led to the definition of a "chill hour" as one hour at or below this threshold. This definition, entered the horticultural literature and practice. Weinberger established two important principles:

1. that chilling requirement is quantitative and can be measured as exposure to a certain total number of chilling hours; and
2. that a critical temperature threshold exists below which chilling is effective. Based on his suggestion, site and climatic evaluation, as well as cultivar requirements for chilling hours, were determined. Breeders still use the term "chilling hours" below 7.2°C as a measure of chilling requirements of new and old cultivars.

Challenge to the validity of Weinberger's model came from observers in warm locations where temperate fruit trees are grown. They noted that good bud break could be obtained even when winter temperatures were always above 7.2°C. Erez and Lavee (1971), working with potted trees under controlled temperatures, demonstrated peach bud dormancy breaking at 10°C although it was only half as effective as that at 6°C. They suggested the term "weighted chilling hours" to evaluate chilling accumulation at different chilling temperatures. They also have shown quantitatively the negation effect of high temperature in a daily cycle but not when applied in long durations of a few days.

The Utah model developed by Richardson *et al.* (1974) based on Erez' and Lavee's (1971) data, adopted a wide optimum between 3°C and 9°C with reduced effect at higher and lower temperatures. The Utah model provided a measure of response at intervals of 0.1°C but had no experimentation to support it. The basic elements of the model are:

1. A set of values from negative to positive is given for a range of temperatures. Positive and negative hourly values are accumulated and net values are summed to obtain a specific chill unit requirement of a given cultivar.
2. Accumulation of chill units begins from the time the maximal level of negative values is recorded. The contributions of the Utah model were the introduction of the following two basic principles:

1. there is an optimum curve to chilling effect and therefore a weighted value to every temperature.
2. Chilling can be negated by high temperatures.

This model performed well under temperate conditions but failed to predict the end of dormancy under subtropical conditions. Variations of the model were suggested for different conditions and cultivars. (Gilreath and Buchanan, 1981b, Seeley 1996).

In the Utah model negative values were assigned to temperatures higher than 16°C. No explanation is given for choosing this specific temperature. In view of the sharpness of the changes in the effect of temperature on dormancy breaking, as the temperature changes from 12°C (still breaks dormancy) to 15°C (showing synergy with chilling) to 18°C (neutral temperature) to 21°C (strong negative effect) the choice of 16°C was not appropriate. At the lower end, the Utah model assumes, again without experimental support, that 0°C has no promotive effect (Richardson *et al.,* 1974). In work under controlled conditions, Erez and Couvillon (1987) found that 0°C had low but significant effect. By extrapolation no effect was obtained at -2°C.

Although the model appears to be valid in cooler regions of the temperate zone, it is not applicable in sites where high winter temperatures are experienced, *i.e.*, in subtropical and tropical locations (see Partridge and Allan 1980). Thus, the weaknesses of the Utah model are related mainly to the part dealing with chilling negation rather than with chilling accumulation.

Another aspect in the Utah model that deserves attention is its lack of sensitivity to the different effects of high temperatures if experienced in a short (daily) *vs.* a long cycle. The different responses observed under different cycle lengths deserves differentiation in the climatic model. A variation of the Utah model was suggested in South Africa by Linsley Noakes *et al.* (1994) and Linsley Noakes and Allan (1994). They suggest that chilling negation by high temperatures be calculated on a daily basis only and refer to "daily positive chill units". This led to a much improved model under their conditions.

A model which integrates all the known effects of temperature on bud dormancy, based on experimentation with potted peach plants under controlled conditions, was developed by Fishman *et al.* (1987a,b) and was termed the Dynamic model. This model comprises the following elements:

1. Chilling has a maximum effect on breaking bud dormancy at 6°C with reduced effects reaching a value of zero at about -2°C and at 14°C.
2. A two-step system operates in chilling accumulation. The first step, which is reversible, leads to accumulation of an intermediate while high temperature that negates chilling reduces the level of the intermediate. The second step produces a stable fixed product, is induced automatically at temperatures >4°C, once the level of the intermediate reached a critical level. Below 4°C, this transfer is temperature-dependent.
3. At every temperature a specific steady-state level of the intermediate is obtained, it represents the equilibrium between its production and destruction. The lower the temperature, the higher is the steady-state level of the intermediate. Temperature level, its duration in the daily cycle, and the length of the cycle affect the accumulation of the intermediate. Accumulation of chilling results in successive quanta of the intermediate being transferred to the final stable product. A quantum is obtained after exposure to *ca.* 30 hours of continuous chilling at 6°C and is termed a chilling portion. (Fig. 5)

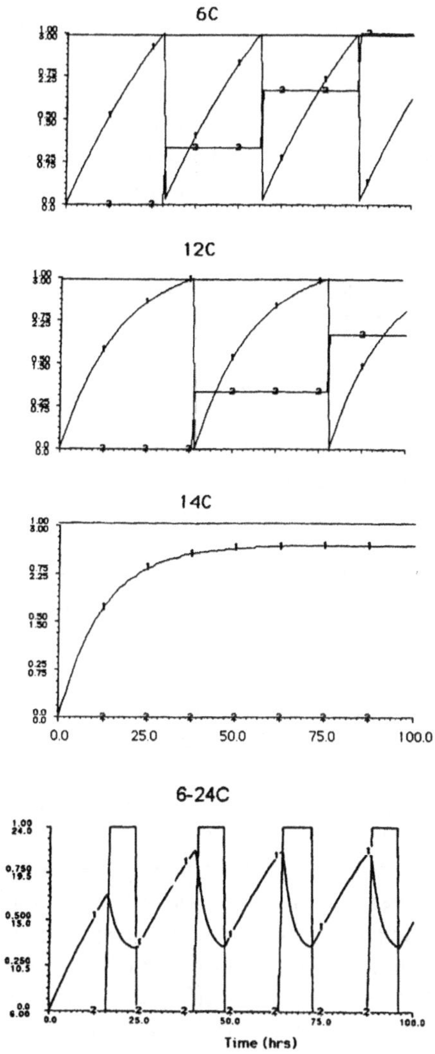

Figure 5. Illustration of the effect of efficient (6°C), less efficient (12°C) and inefficient (14°C or 6-24°C) temperatures during a 100-h period, on breaking bud dormancy, on the march of the intermediate accumulation (1) and the portions obtained (2) according to the dynamic model. The upper value of 1 designates the critical level the intermediate that has to be reached in order to accumulate one portion. Note the change with temperature in the intermediate level in the 6-24°C temperature. In 6-24 (16/8 h at 6/24°C) and at 14°C no accumulation of chilling portions is evident. Simulation of model run on Stella program by S. Fisherman.

4. Negation of chilling by high temperatures affects only the portion of the intermediate that has not been fixed. (Fig. 5)
5. A correction for temperature effect was introduced at 4°C to simulate the effect of cooler temperatures on transition between the intermediate and the product.

6. The range of chilling requirements runs from 12 chilling portions for peach and nectarine cultivars with very low chilling-requirements to 70 portions for apples and sweet cherries with high chilling-requirements.

The dynamic model accounts for the negation of chilling by high temperatures in term of diminution in the steady-state level of the intermediate as the temperature rises (Fig. 5). Temperatures of 13°C and lower will cause bud break because their steady-state level of the intermediate is higher than the critical level, which leads to the accumulation of chilling portions (Fig. 5).

The dynamic model also accounts for the synergism between moderate temperatures and chilling. Moderate temperatures have their intermediate steady-state set very close to the critical level (Fig. 5) so that no chilling accumulation will occur without additional exposure to a cooler temperature, but the amount of added chilling needed to reach the critical level is very low. Exposure to a cooler temperature will raise the intermediate to the critical level, and a chilling portion will accumulate.

The contribution of the dynamic model were:

1. Introducing the fixation element of chilling and thus the dynamic element of temperature effects depending on former conditions;
2. introducing a synergistic effect to moderate temperatures with chilling;
3. proposing a complete theory of temperature effects on dormancy based on experimentation.

Model parameters may change with exposure to chilling. Real Laborde working with apple seeds, suggested that high temperatures may have a more positive effect later in dormancy, an element introduced into the modified Utah model (cited by Seeley 1996). Erez and Couvillon (1987) demonstrated this effect with rooted peach cuttings, for which the last third of the chilling period was more responsive to higher temperatures. Hanninen (1995) suggested that the effect was caused by the change of responsiveness of the dormant bud to higher temperatures as it approaches bud break.

With the increased interest in the mechanisms that control dormancy, it is believed that understanding of the pattern of temperature responses may be helpful in clarifying the mechanisms controlling the responses of the dormant bud to the environment.

An important question is whether the difference in response to temperature between high and low chill cultivars is quantitative or qualitative. Do low chilling requiring cultivars (low chill) respond more positively to higher temperatures than do high chill ones? Gilreath and Buchanan (1981b) reported a slightly different response of low chill cultivars compared to high chill using a modified Utah model. However, even with high chill cultivars similar effects of temperatures were obtained (Erez and Couvillon 1987). Thus, the major difference among cultivars of the same species seems to be quantitative, while the specific effects of temperature remains the same. Does this hold for other species? With most of the Rosaceae tree fruit species, the response seems similar, the Utah model actually works equally well under cool climate with a variety of species (*e.g.* for apple Shaltout and Unrath 1983) still it should be verified experimentally. A special case may be pecan which seems to respond entirely different (Sparks, 1993). A comparison among the chilling requirements of a few deciduous fruit trees, according to the Utah and the dynamic models is presented in Table 2.

Table 2. Comparison of chilling requirements of a few selected cultivars of deciduous fruit tree species as measured by chill units (Utah model) and chilling portions (dynamic model)

Species	*Cultivar*	*Chill units*	*Chilling portions*
Peach	Flordaprince	150	8
	Earligrande	200	12
	Maravilha	200	12+18 *
	Babcock	350	-
	Redhaven	750	75
Nectarine	Aprilglo	150	12
	Mayglo	250	18
	Flavortop	550	35
	Fantasia	600	42
Apricot	Canino	450	35
Plum	Santa Rosa	420	-
Apple	Anna	150	-
	Golden Delicious	1200	50
Cherry	17GE580		18
	Lapins		35
	Rainier		45
	Burlat		53
	Sam		70

* floral buds + vegetative buds

4 The problem of an incomplete dormancy release

The impact of an incomplete dormancy release on a modern fruit production system is very heavy. Incomplete dormancy release affects tree behavior in three main ways: a late bud break; a low level of bud break; and lack of uniformity of leafing and bloom.

The three parameters of time, level and uniformity of bud break have special significance for particular crops and various locations.

Time of bud break

Complete dormancy release results in relatively early bud break. This is especially important for precocious cultivars whose economic return depends very much on their time of ripening. Early bloom will lead to early maturation that has a special premium value. In addition, bloom time may be important for cross-pollination purposes. Quite often, two cultivars that cross pollinate each other have different chilling requirements leading, when chilling is insufficient, to delay in bloom of the higher chiller and thus to lack of bloom coincidence. Early bloom may be important also to avoid a specific climatic condition that might occur later.

High level of bud break

For obtaining high yield and good foliage cover, a high level of bud break is needed. This will enable an adequate photosynthesis to support the tree and its fruits. When fruit set conditions are not optimal, increasing the number of flowers will increase yield. In young trees this is one way to achieve early bearing. By obtaining a high level of vegetative bud break, many growing apices will result, that will grow moderately and produce short vegetative growth or spurs that will differentiate easier to floral buds than on more vigorous growth that may result from few growing apices.

Uniform budbreak

This parameter of good dormancy breaking is very important in the modern orchard that depends on the application of various treatments, such as chemical thinning, at a specific stage of fruit development. If the fruit is not developing uniformly, the efficiency of such treatments will be reduced. Furthermore, where there is a non-uniform bloom, the late bloom wave always has a low chance of setting because of competition with stronger established sinks leading to a reduced yield. On the other hand, sometimes spread of bloom and hence ripening over an extended period is desirable especially if a single cultivar has been grown.

In most cases all three aspects of bud break interact. A low level of bud break coupled with late bud break exposes the leafless tree to long periods of direct insolation, and under warm conditions sun scald and drying out of main branches may lead to a gradual deterioration of the tree. The poor leafing may lead to limited leaf area and hence to limited photosynthesis. If this situation repeats year after year the tree may be lost. The lack of uniformity may stretch the bloom period over a period of months. Most of the late blooms normally do not set even in self fruitful cultivars and species. In self-unfruitful cultivars and species, such a long duration of bloom markedly reduces the potential for fruit set. Stone fruit species, are sensitive to extreme high temperatures in winter as discussed above (3.3) responding in lack of fruit set or producing abnormal fruit. Thus assuring good bud break in warm climates is a major concern of growers of temperate fruit crops.

4.1 Functionality of developing buds

The effect of dormancy release has further effects subsequent to the actual bud breaking. The fact that a bud is breaking and either a flower or a vegetative growth is emerging does not secure a normal development. Typically, warm winter climate may lead to abnormal floral development especially in stonefruits resulting in small flowers with smaller than usual petals, and abnormal ovule development leading to flower drop, mummies formation or excess June drop resulting from aborted embryos.

With vegetative buds, a typical rosette formation testifies to a non complete dormancy release. Fuchigami and Nee (1987) and Nee and Fuchigami (1992) suggested that the breaking of rest involves 2 distinct processes: bud release and stem elongation, and that the action of some growth regulants may affect only one of the two processes involved in dormancy release. This phenomenon indicates two stages that can be easily separated *i.e.*: The actual bud break and the second stage of elongation of the axis and further development of new leaves and internodes.

5 Other related phenomena

5.1 Do roots have chilling requirements?

Trials by Young and Werner (1985) indicated with apple that the root system is affected by temperature. Exposure of the roots to warm conditions, even when the top was chilled, resulted in poor development of the top. However, this was not verified with peaches creating a puzzling situation. In addition, scientific data as well as practical experience indicate that the rootstock influences the dormancy of the top. There are rootstocks that start to grow before others, and rootstocks that stimulate early bud break to the grafted scion. The question is to what extent the rootstock behavior represents response to dormancy. It should be stressed that only meristems can enter dormancy while other tissues are affected by their dependence on the meristems. There could be differences among roots meristems in the effect of temperature on their beginning of active growth and this might have an indirect effect on the behavior of the top. As roots are important sources of growth regulators especially cytokinins, which are known for their dormancy breaking effect (Wang *et al.*, 1986), early activity in the roots might promote an earlier dormancy break in the top. Under mild winter conditions, roots may grow throughout the winter while the top remains totally dormant. Furthermore, the technique of rooting of hardwood cuttings applied to many deciduous fruit trees, is based on the principle that root initiation and development are possible during the period that the top is dormant. In conclusion: the general dogma does not recognize the roots as a primary site for dormancy control.

5.2 Growth rate and productivity

When dormancy is not completely broken, poor vigor may ensue in the developing buds leading to quick setting of a terminal bud due to secondary dormancy. Total productivity is normally low. The low level of bloom and photosynthesizing area leading to poor set of late bloom or to poor coincidence of bloom when cross pollination is needed resulting in low yield. It is quite common to find micro element deficiency. It seems that under poor bud break conditions, poor uptake of micro nutrients will occur leading to micro element deficiencies. This seems to result from poor root growth. Another common feature is scaffold damage by sun scald. Lack of foliage cover leads to exposure of tree frame to heat. Low foliage cover affects also the transpiration stream that cools normally the tree frame as well. Therefore, one of the risks of poor foliation is trunk and scaffold heating leading to sun scald damage, enabling attack by insects and eventually to tree decline.

The following sections discuss separately the specific problems and their solution in the tropics and subtropics.

6 The subtropics

6.1 The poor bud break syndrome

When insufficient chilling occurs under subtropical conditions, poor bud break will occur typically characterized by an earlier break of the terminal buds, scattered non-uniform bloom and lateral leafing. The few buds that break will grow vigorously, leading to low production of spurs in species like apple pear plum and cherry. As vigorous growth always carry buds having a high chilling requirement, the poor bud-break cycle will continue. Trees bearing on one-year-old wood such as peach will be able to produce enough buds for cropping and their response will depend on flower bud break.

The increased availability and new releases of germplasm of temperate-zone fruit tree species with lower chilling requirements will much improve future production (See relevant chapters in this book). Low chilling cultivars are found in all species and are being exploited by current active breeding work. In the meantime we have a number of different chemical and physical means to improve performance of the existing germplasm.

6.2 Dormancy breaking means

a. Chemical means

The dormancy-breaking activity of chemicals was first noted through the chance observation of the response of buds to insecticides. Thus, it was found that oils containing dinitro chemicals would break dormancy (Chandler 1957, Samish 1945). This observation has led to the development and adoption of various chemicals as regular commercial means of dormancy breaking under marginal winter conditions. Many chemicals showed activity in breaking bud dormancy but only a few gained commercial acceptance. The main characteristics of such chemicals are: strong effect, low cost, and minimal toxicity to plants and humans. Artificial means may compensate fully for unfulfilled chilling requirement, prior to entering ED, under certain climatic conditions, as described below (7.1). Once the bud becomes D-endo-dormant, no single treatment or combination of available treatments can fully replace all the required chilling; there is compensation for only part of the actual chilling requirement, and it can be obtained only after the buds had been exposed to partial chilling. However, these chemicals have great potential and could have great economic importance (Fig. 6).

Flower buds are more sensitive than other tree organs to most of the dormancy breaking chemicals, and this sensitivity is manifested in flower bud phytotoxicity and loss of flowers. This may frequently lead to loss of yield after treatment, even, in may cases, when there is extremely good vegetative bud break. On the other hand, good vegetative bud break will lead to better spur production in the subsequent year that may affect the yield favorably for several years.

Oils. This group of chemicals is the longest established in commercial use. The powerful combination with dinitrophenols group acting as phosphorilation

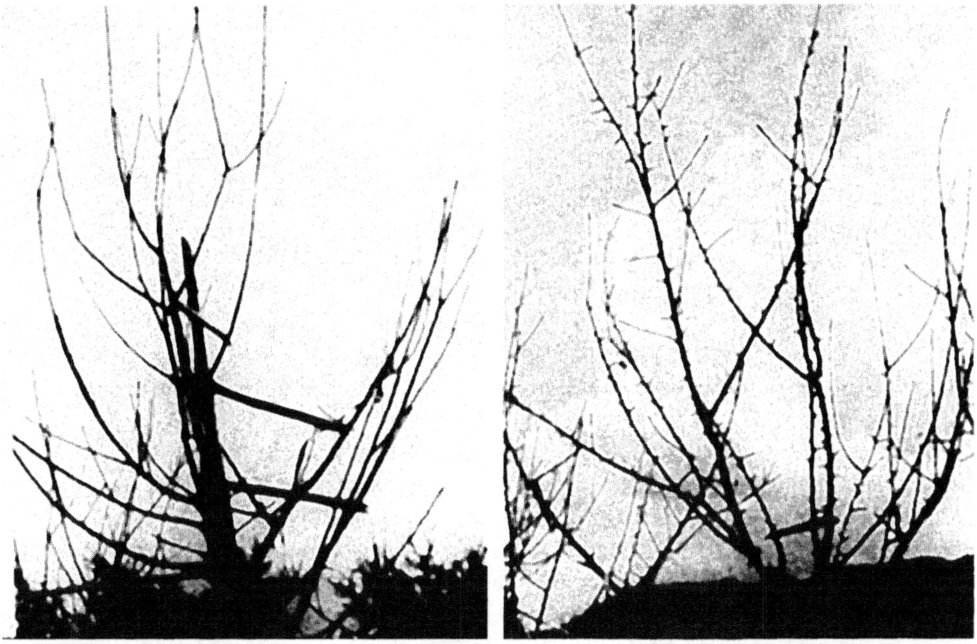

Figure 6. The effect of a single dormancy-breaking treatment applied in the early spring of the previous year, on lateral budbreak and spur production in apple. Left untreated control, right treated. Santa Catarina, Brazil.

uncouplers gained acceptance in the mid-thirties (Samish 1945, Erez and Zur 1981), and is still used in several countries world wide. Apart from their dormancy-breaking effect, oils are also used as insecticides and miticides.

In the last decade, the increased awareness of its human toxicity (Nehez *et al.*, 1981) has stimulated a search for alternatives to dinitro ortho cresol as an adjuvant to the oil, either by finding a substitute for it, or by replacing the oil altogether. Certain dinitro compounds are less toxic and may be accepted as replacements; an example is Carathane, known as Waicap in South Africa (Honeyborne and Rabe 1993).

Oil-DNOC treatment stimulates respiration, therefore, its effect is dependent on the prevailing temperatures, during spraying and in the following seven to ten days until the oily layer deteriorates. It has been shown that a high day-time temperature is essential for a good effect (Erez 1979). The cause for the effect is a temporary anaerobic condition in the buds, resulting from oxygen deprivation by the oil cover that leads to ethanol production which actually is responsible for dormancy breaking (Erez *et al.*, 1980). This explains the sensitivity of treated trees to excessive temperatures, that enhance respiration markedly, and to waterlogged soil. In both cases, excessive anaerobiosis may result leading to fermentation and tissue death. Under normal conditions, the phytotoxic risk is low, so that spraying can be applied even at the stage of bud swell. It has been shown that at this stage swelling terminal buds are inhibited (by DNOC) and this may increase the tendency of the laterals to break.

Cyanamide. There is no doubt that during the last decade, cyanamide has become the leading dormancy-breaking chemical. Calcium cyanamide has long been known to act as a rest-breaking agent; it was first used in Japan (Kuori *et al.,* 1963), at rather high concentrations, but its low water solubility prevented its wide use. The discovery that hydrogen cyanamide, which can easily be sprayed in a solution, acts as a rest-breaking agent (Shulman *et al.,* 1983) opened the way to a wide series of trials on many deciduous species.

This chemical is extremely effective on grape vines (Shulman *et al.,* 1983, George *et al.,* 1988) and kiwi fruit (Hampton and Parker 1992, Henzel and Allison 1993), and it has also been found effective on apple, plum, apricot, high-chilling-requiring peach (Erez, 1987a, De Benito 1990), pear (Lin and Lin 1992), raspberry (Snir 1983) and fig (Weizman *et al.,* 1985). From the very beginning it was evident that vegetative buds respond readily to cyanamides. However, damage to flower buds has been reported for various species (Nee and Fuchigami 1992, George and Nissen 1993); the more protected ones (grape, kiwi) are not damaged, while the less protected ones (such as the simple buds of stone fruits) are more sensitive. Interestingly, sweet cherries are far less sensitive to cyanamides, probably because their bud initials have better protection than those of other stone fruit species, and they respond very favorably to the chemical (Snir and Erez 1988, Nicolas and Bonnet 1993).

In the case of pome fruits, a better effect on the appearance of the trees than on the crop was noted in initial trials with this chemical. Flowers within the mixed bud were damaged, and the bud opened as a vegetative bud, which led to yield reduction.

Contradictory results were obtained for different years and different locations, from the use of the chemical on a given species or even cultivar. The main cause for the variable results seems to reside in the level of ED of the buds. Resistance to the phytotoxic effect of the chemical declines rapidly, following ED release. Thus, in order to avoid damage to flower buds, it is essential to avoid late application, unless it is clear that dormancy was not released. This is one of the main reasons why it is so important to monitor the development of dormancy in the buds prior to treatment.

Too early applications have their own drawbacks. As with other chemicals, cyanamides will not compensate for more than an estimated 30% of the chilling requirement. On the one hand, late application may allow more chilling to accumulate but on the other hand, it may damage buds that had already been released from ED. Late application of oil-DNOC does not generally cause any damage, and it is possible to rely on visual symptoms, such as terminal bud swelling in apple, as a good timing indicator and to apply the chemicals when the weather is warm. In the case of cyanamide, however, we have to apply the chemical before any visual changes are obvious, and the timing of its application must be based on information regarding dormancy development. At present, the best means is to monitor chilling accumulation and compare that with the chilling requirement of the cultivar using climatic models.

Another important question is whether we really need the maximal level of bud break or bloom. Under conditions which promote high fruit set, and under which thinning is usually practised (peach, apple), a reduced bloom level may be beneficial. On the other hand, with small fruits or nuts, such as sweet cherries and almonds, which normally need cross pollination, a maximal bloom level is desirable.

The excess production of fruitlets in peaches and in kiwi led to consideration of the possibility of obtaining a small, controlled level of damage, in order to improve the return in the orchard. Thus with kiwi, it was found that at 3%, Dormex (SKW, Germany, 49% a.i. hydrogen cyanamide) induced improved bud break, but drop of the side flowers of the triplets led to a thinning effect and larger fruits were obtained (Henzel and Allison, 1993, Hampton and Parker 1992). In peaches, mild damage to flower buds could reduce laborious hand thinning and improve orchard economics. Until now, no commercial recommendation can be made regarding these responses.

Another important use of cyanamide is to advance fruit maturation. In this area, the need for early application in order to obtain earlier bud break goes hand in hand with the precautions needed to prevent phytotoxicity. Reports from southern France (Nicolas and Bonnet 1993) indicate an advance of the maturation of the precocious 'Burlat' sweet cherry by 7-10 days when Dormex was applied at 2.5%, 50 days prior to normal bloom. In Israel we have recorded a 12-day advance of bloom in the sweet cherry 'Rainier' (Snir and Erez 1988), while George and Nissen (1993) in Australia achieved a 19-day harvest advance by applying hydrogen cyanamide to peach. Concentrations of Dormex sprays used mostly are 3-5% in grapes kiwi and sweet cherry and 1.5-2% for other species.

Competition between vegetative and reproductive development may develop, especially under conditions that would advance leafing. Excessive vegetative bud break may have negative effects on fruit set, because of sink competition (Erez *et al.*, 2000). A marked advance of leafing compared with bloom is a typical effect of high cyanamide concentration on stonefruit species.

Dormancy breaking agents can also be used to improve bloom coincidence when cross pollinating cultivars do not bloom together, by advancing the late blooming cultivar. Application of hydrogen cyanamide not closer than 30 days prior to expected bud break is recommended. This pattern of application is good in relatively warm climates and with cultivars which require relatively high chilling, but with other cultivars, especially stone fruits, the physiological stage of the buds is more important, and this is not defined at all accurately by timing according to the average expected time of bud break.

To summarize, the toxicity of hydrogen cyanamide is dormancy-stage dependent. With more resistant species, this chemical is excellent and with optimal timing it may have a major impact on production and on the crop economy. With more susceptible species it may be too risky to use it, and a reduced concentration in combination with other chemicals is recommended.

Effects of oil-cyanamides. Petri *et al.* (1990) reported on the efficiency of the combination of oil and hydrogen cyanamide in breaking apple bud dormancy. More recently (not yet published) they have studied the relative efficiency of various oil-hydrogen cyanamide combinations applied to apple during a 3-year period; they obtained the best results with a spray containing 2% oil and 0.25% hydrogen cyanamide. North (1992), examined various combinations of oil and hydrogen cyanamide in South Africa and found several combinations with low cyanamide concentrations to be as good as or better than oil-DNOC. In Israel, the combination of 4% oil and 0.25% hydrogen cyanamide was also as good as or better than oil-DNOC in breaking apple bud dormancy during a 3-year trial; tank mix application of these two components was as effective as separate applications (Ringwald, 1994). Petri and Pola (1992) found that the oil-cyanamide treatment was better than oil-DNOC treatments in retaining its effectiveness in breaking apple bud dormancy under low temperatures. A comparison of the two chemicals is presented in Table 3.

Table 3. Comparison of the effects of oil-DNOC and Cyanamide on fruiting in 'Summerset' peach

Treatment	*Fruits/tree*	*Increase over control (%)*
Control	122 b	
KNO_3+oil 5.6% +DNOC 0.11%*	144 b	15.5
Dormex (H-Cyanamide) 2%	242 a	98.3

* Trees were sprayed with KNO_3 followed by oil +DNOC

Other chemicals

Thiourea. This chemical has been found to be very effective in breaking vegetative bud dormancy and it was found very suitable for both pome and stone-fruit species in combinations with KNO_3 and oil- DNOC (Erez *et al.,* 1971, Erez, 1987). Unfortunately, it was banned from use because of its toxicity to humans and therefore no further information is available.

KNO_3. This chemical has a rather mild effect even at a concentration of 10%. It is effective in increasing bud break especially of flower buds (Erez *et al.,* 1971, Erez, 1987a, George and Nissen, 1993) and can be used in combination with other chemicals to enhance their effects. Based upon recent work, limited penetration could be one of the reasons for its mild effects (see below).

Growth regulators. Gibberellic acid (GA) and cytokinins break bud dormancy (Erez 1987a, Wang *et al.,* 1986, Lloyd and Firth 1993) but the concentration needed are 100 to 400 mg/l of GA and benzyladenine (Erez 1987a), rendering these rather expensive chemicals non economical. A strong effect was reported for Thidiazuron, a synthetic cytokinin at lower concentrations (Wang *et al.*). Recent developments may change the situation in the future as described below.

Growth retardants. The introduction of paclobutrazol as a growth retardant in fruit tree orchards revealed that, apart from its vegetative growth retardation effect, it also has a specific effect in advancing bud break. This effect was manifested in trees treated in the previous growing season (Erez 1985, George *et al.*, 1992, George and Nissen 1992, 1993). Its mechanism of action appears to be the inhibition of the biosynthesis of GA, which stimulated growth and thereby increases depth of dormancy.

Armobreak. A unique group of fatty amines was introduced recently by AKZO of the Netherlands. These chemicals enhance cuticle penetration carrying with them other chemicals. The agent, named "Armobreak," may change the situation in several ways: it could reduce the cost of treatment by lowering the concentration of expensive rest breaking agents like hydrogen cyanamide and GA; and it could boost the effect of a mild agent like KNO_3 and allow the use of relatively safe and powerful combinations.

The first reports on a combined effect of KNO_3 with Armobreak on apples came from South Africa (North 1992) indicating a strong enhanced effect of KNO_3. These data were confirmed by experiments carried out in Israel since 1994. Because of the mild winter conditions of Israel, many of the deciduous fruit trees grown in the low lands experience poor bud break. The enhanced effects of chemicals to which Armobreak was added, was evident for gibberellic acid on vegetative buds and for hydrogen cyanamide on both floral and vegetative buds of 'Flavortop' nectarine (Erez 1995). The enhancing effect on KNO_3 was evident only in a combination with oil-DNOC applied to vegetative buds but not in combination with oil alone. A tendency to reduce the bloom level may indicate that toxic levels were reached. With 'Canino' apricots, a marked enhancement of leafing and yield was noted when Armobreak was combined with cyanamide, compared with cyanamide alone, applied with a spreader (Erez 1995). Recent trials with Armobreak and thidiazuron also proved a marked dormancy breaking effect on peaches (Erez, unpublished data).

When does chilling accumulation end, while a chemical treatment is being applied? This is a relevant question when chemicals such as hydrogen cyanamide are applied a long time (up to 6 weeks) prior to bud break. Does chilling continue to accumulate and contribute to dormancy breaking, after the treatment? Field observations indicate that the answer is positive, and under marginal conditions this has to be taken into account by using previous climatic data to forecast the expected chilling accumulation.

b. Physical means

The effect of extreme high temperature in breaking dormancy has been known for a long time, from observations that buds break much better following exposure to intense heat from fires in or near the orchard. This was first described by Chandler (1960) who showed that there was a strong positive effect of temperatures up to 45°C even for exposures of only a few hours. Practical use of such treatment in the orchard is very difficult and therefore it did not gain any wide acceptance.

Whenever night temperature falls below 13°C, chilling accumulation may occur (see section 2.3.1). In many marginally subtropical locations, night temperatures are cool enough to break dormancy but day temperature exceeds, sometime by far 19°C, leading to complete negation of chilling. The only means to lower bud temperature under field conditions is by evaporative cooling using over-tree sprinkling (Gilreath and Buchanan 1981a, Erez and Couvillon 1983, Erez *et al.*, 1993, Nir *et al* 1988).

The emitter has to have uniform droplets and a rather dense wetting pattern. Formation of very small droplets of mist is not desirable as they will evaporate before reaching the tree limbs and therefor will not have a cooling effect. In a typical operation on peaches (Erez *et al.,* 1993), over the tree mini sprinklers, one unit per tree, emitting 55 l/h were used. Intermittent operation was controlled by a small computer, with a thermostat set at 19-20°C. The timing of on and off cycles was determined experimentally in the field so as to avoid total drying of the buds. It was shown experimentally that the bud temperature could be reduced from 24°C to 16°C, thereby not only preventing the negation of night chilling accumulation, but also benefitting from the enhancing effect of moderate daytime temperatures (see section 2.3). The use of water to cool the plant canopy is limited, firstly by water availability, secondly by its quality and thirdly by the need for a low relative humidity to allow evaporative cooling. Damage caused by salt deposits on tree branches and buds has been noted (Erez 1995), and high levels of Cl may be detrimental to young branches; the sensitivity seems to depend on the age of the branch, young non-lignified shoots being the most sensitive. Slight damage to shoot tips or even more severe injury to the distal part of the shoot may not reduce yield in peach, which normally has excess flower bud setting, but with other species it may lead to yield loss.

c. Combined effects of chemical and physical means

As was mentioned before, chemical treatment can not compensate for severe lack of the chilling required to break dormancy. On the other hand, evaporative cooling, if climatic conditions are appropriate, may actually expose the buds to further chilling, enabling them to respond much better to chemical agents. This practice has been used, with good results, in Israel in order to advance bud break of peaches and nectarines (see also chapter 9).

d. Other means to improve bud break

In addition to the chilling requirement of the cultivar, certain elements influenced by orchard management have a strong modulating effect that is not stressed enough in the literature. Four such factors are:

Control of tree vigor. Methods that reduce chilling requirement have been found to improve bud break, since the more vigorous the growth, the deeper the ED (Saure 1985). As a result, young trees generally express a higher chilling requirement than older ones. This may explain the positive effects of growth retardants, that check vegetative growth, in facilitating bud break (Erez 1985b, George *et al.,* 1992) and the effect of summer water stress in advancing bloom (George and Nissen 1992). Thus, dwarfing, vigor-controlling rootstocks have been found advantageous under marginally warm growing conditions, over and above their advantages they have under temperate climates. The warm subtropical and tropical conditions with their very gradual change in daylength during the growing period favor a long period of vegetative growth, which leads to a more vigorous plant, under conditions where water and nutrients are not limiting. It is thus common to see better bud break under marginal conditions in poorly kept orchards low in vigor, than in well irrigated and fertilized ones. Thus, in the tropics and subtropics, special attention should be focused on restricting vegetative growth, as this element is a major factor in enabling control of bud dormancy.

Branch orientation. Bud break occurs much more readily in horizontally oriented branches than in upright ones. This effect definitely stems from the change in balance of hormones in the buds and from the reduced vigor if bending is done during the growing period. It also reflects a reduction of apical dominance that tends to prevent bud break on most of the laterals on a vertical branch, leading to the poor bud break syndrome discussed above. Apical dominance is a major concern in warm-winter climates. With upright branches, earlier break of the terminals, which have lower chilling requirements, will induce a strong correlative inhibition thereby further reducing the level of lateral bud break. Thus branch bending techniques are used in many warm countries.

Many training techniques for deciduous fruit trees are based on a horizontal or near-horizontal canopy; the open center systems, the cordon systems of the "Lincoln canopy", the Ebro system and the Solen are all based on a rather horizontally bent skeleton. Although their aim was for a more productive system, since bending induces earlier flower bud differentiation and reduces vegetative competition, their negative effect on vegetative vigor should be especially beneficial under warm-climate conditions.

Time and type of pruning. Heading back under marginal winter conditions, especially if done early (mid-winter), will stimulate growth of the uppermost one or two buds as a consequence of the local dormancy-breaking effect of the wound. This will lead to a resumption of apical dominance which will prevent the opening of other laterals that are lagging. The later the pruning the less its negative effect on bud break of the laterals. Summer pruning is more desirable in the subtropics, since it has a much smaller invigorating effect; it may also help to reduce the size of the tree.

Time of leaf drop. A few reports in the literature (Walser *et al.*, 1981) indicate an effect of time of leaf drop on depth of dormancy. Early defoliation leads to reduced chilling requirement and vice versa. A possible cause for this lies in the movement from the leaves into the buds, of chemicals that control the depth of dormancy. Buds formed on vegetative growth late in the season will require more chilling to support their growth than those formed on early spring growth. This phenomenon was especially conspicuous in meadow peach orchards that started their new growth as late as May or June and were always later in bloom than conventionally trained trees that began growing in February (Erez 1982). Chemical defoliation in early autumn may thus be beneficial if it is not applied so early as to cause dormancy avoidance unintentionally (see next section).

7 The tropics

7.1 Dormancy avoidance: the backbone of temperate fruit crop production

Is rest an obligatory requirement of the dormant bud? Under warm-climate conditions dormancy is imposed gradually. It seems clear that dormancy may be avoided so that normal development takes place, if a strong stimulus is applied prior to induction of ED (Edwards 1987, Erez and Lavi 1984, Erez 1987b, Erez 1990b).

The entry into ED, albeit autonomous, is markedly affected by short days and low temperatures. The effect of these two climatic factors cannot be appreciated in temperate zones but can be evaluated in subtropical and, especially, tropical climates.

Under subtropical conditions, bloom often occurs in autumn because of rest-avoidance prior to the onset of winter. This bloom is mostly of normal flowers capable of setting fruit and occurs because of premature loss of leaves due to desiccation or pest damage followed by irrigation or rain. The flower buds that are not in ED will continue their development to completion without exposure to chilling while those that are endodormant will open only after exposure to chilling. As a result, two crops per year can be obtained, with partial overlapping. In the subtropics, no change in the timing of the production season is possible. The advantage of obtaining a crop by dormancy avoidance is the off-season fruit production. In Israel (31-33° N latitude) peach flower buds are weakly affected by photoperiod and do not enter ED as late as December if temperatures remain high (minimum temperature >10°C). A very high level of bloom may be obtained in the fall by combining the three stimulating factors: defoliation, desiccation followed by irrigation, and dormancy-breaking agents (Erez, 1987b, Erez and Lerner 1990). Fruit can set, develop normally and reach maturity, providing their photosynthates demand is satisfied. Vegetative buds, on the other hand, are influenced mostly by photoperiod, and they enter into ED around late October. Earlier stimuli, too, may induce leafing (Lerner 1990). The short days and cool weather in fall and winter induce a secondary *ED* in the apices of the vegetative growth, leading to a small leaf area available for photosynthesis, and to a rapid leaf ageing. However, a new flush of leaves, that may support the developing fruit, and of flowers, from buds that did not break in the fall, is obtained in spring (Lerner 1990). The new flowers will open in spring in the normal time or a little later. This will lead to two crops with a certain overlapping period. Thus, the peach cultivars 'Maravilha' and 'Earligrande' produced their first crop in March and the second in May (Lerner 1990). The relatively harsh climatic conditions in winter necessitate at least a net protection against hail.

In the tropics, which excel in their regular year-round climate, it has been found in practice that defoliation will induce bud break if applied at the right time in the growing cycle, prior to ED. This was first found and reported for grapes. In parallel, probably by trial and error, it was found that similar forms of treatment may work well with other species, which led to the establishment of rather large production centers of temperate-zone fruits in the tropics (Saure 1973, Janick 1974, Edwards 1987, 1990, Edwards *et al.,* 1990). In Peru it was found that a long period of desiccation followed by abundant watering will break efficiently bud dormancy even with no exposure to cold (Bederski 1987). Today temperate fruit production areas may be found scattered all over the tropics, and is especially important in Indonesia for apples, in Venezuela, Colombia, Ecuador and Peru for apples and peaches, and Kenya, Thailand, Lower Taiwan and northern Brazil for grapes. The system used is dormancy-avoidance, based on the finding that prior to the onset of deep dormancy (ED), buds, if fully developed, may respond to a strong stimulus by concluding their development and by breaking without passing through a deep dormant phase.

In order to achieve this, the following pre-requisites have to be met:

1. To grow the trees under rather cool climatic conditions which, in the tropics, may be identical with higher elevations. This is mostly needed for quality fruit development and for normal flower bud differentiation. To avoid extremely high locations because, although they may have the low temperatures needed for breaking bud dormancy, under tropical conditions of no seasonality, they will have similar weather during the growing season, which will thus be too cold for quality fruit production.
2. To initiate a growing cycle artificially, prior to the onset of ED in the buds. In most cases this is between 6 and 8 months after the beginning of the previous growing cycle. This means a production cycle shorter than one year, leading to more than one crop per year. The means for initiating a new growing cycle are some or all of the following treatments:
   - complete tree defoliation;
   - desiccation followed by irrigation;
   - application of dormancy-breaking chemicals.

Grapes are less sensitive to high temperatures and can be grown in warmer locations. In addition, severe pruning practiced in grapes is a major stimulus for bud breaking. As a result of the above considerations novel options become available for deciduous fruit production in the tropics, which are not available in other locations. Year-round production of the same cultivar is possible. It is achieved by means of planned production in plots separated in their bearing by 1 month, in a two crops per year system (Fig. 7). This limits the need for post-harvest fruit storage.

The most suitable cultivars will be those having a low chilling requirement. However, it is certain that a breeding program aimed at growing temperate-zone fruit trees in the tropics will have a high potential for improving tree performance there. For details see the relevant chapter in this book.

7.2 Climatic limitations for production of temperate fruit crops

Production of temperate fruit crops under tropical conditions is severely influenced by available climatic conditions that may limit or eliminate the potential for growing these crops. One basic requirement is the availability of the appropriate temperature regime: on the one hand, year-round high temperatures would disable production because of problems with flower bud differentiation and difficulty in achieving the optimal response to dormancy avoidance treatments; on the other hand, extremely cool conditions found at high elevations must be free from frost during the whole year.

Figure 7. A series of pictures of the 'Anna' apple in an orchard in Ecuador 2400 m elevation, 0.5° N latitude, taken on the same day. From left to right: Manual defoliation; Full bloom; small fruit stage; growing fruit; ripe fruit.

The best conditions are those found in the tropical highlands, with year-round moderate temperatures. These temperatures will not supply chilling but will allow normal development of tree and fruit (Edwards 1987, Edwards *et al.*, 1990).

Continuous year-round production, with very short non-productive periods leads to a considerable increase in tree productivity, since photosynthesis continues for 11 months per year. Furthermore, by canopy renewal every 6-8 months new leaf area is being produced at a time when photosynthesis is normally on the decline.

The extra load imposed on the tree by carrying more than one crop per year necessitates optimal conditions for photosynthesis. Keeping the tree active for almost the whole year, compared with 6-7 and 8-9 months a year in the temperate and subtropical zones respectively (Erez 1986) enables the tree to recover, providing optimal conditions for growth exist. In many tropical sites this is not the case, even in the best adapted places, because of rain seasonality. There is no doubt that dry growing conditions are preferred if water is available for irrigation. Severe disease problems may develop, especially under the heavy rains that are frequently experienced in the tropics. The dependence on continuous cropping excludes the possibility of skipping over a season of unsuitable weather, and sometimes leads to fruit production under inappropriate conditions. This can be corrected only marginally, by controlling the initiation of the developmental cycles.

There is no doubt that the potential for growing temperate-zone fruits under subtropical and, especially, tropical conditions, and its economic exploitation merits further research.

8 Conclusions

Dormancy is induced slower in lower latitudes than in temperate-zones but once endodormancy has been established, chilling is indispensable for its release. High temperatures, especially in a daily cycle, antagonize chilling and are a major factor in poor chilling accumulation in warm climates. A few climatic models were developed to quantify the temperatures' effect on dormancy release. Chilling requirement differs among buds with vegetative laterals having the highest requirement. Terminal buds and, in many cases, floral buds require less chilling. Evaporative cooling is a method that may improve chilling accumulation mostly by lowering chilling-negating day temperature. Various chemicals are in use to replace chilling. Only part of the missing chilling can be compensated by chemicals. A few important modulating factors may change the chilling requirement of the buds, the most important are diminished vegetative vigor and horizontal growth orientation. While in the subtropics, means to compensate for lack of required chilling are the predominant ways to overcome incomplete dormancy release, in the tropics, avoiding dormancy is the preferred system used. By timing the initiation of the new vegetation and bloom cycle, prior to entering into endodormancy, dormancy can be avoided and new developmental cycle initiated. The length of the growing cycle in the tropics is always shorter than a year and typically of 6-8 months duration.

9 References

Allan, P., Linsley Noakes, G.C., Matthee, G. and Rufus, G. (1996). Winter chilling models in a mild subtropical area and the effect of constant 6°C chilling on peach bud break. *Acta Horticulturae* 409, 9-17.

Anderson, J.L., Richardson, E.A. and Kesner, C.D. (1986). Validation of chill unit and flower bud phenology models for "Montmorency" sour cherry. *Acta Horticulturae* 184, 71-78.

Bauer, M., Chaplin, C.E., Schneider, G.W., Barfield, B.J. and White, G.M. (1976). Effect of evaporative cooling during dormancy on 'Redhaven' peach wood and fruit bud hardiness. *Journal of the American Society for Horticultural Science* 101, 452-454.

Bederski, K. (1987). Selection and dormancy management of temperate zone deciduous fruit tree cultivars in coastal valleys of Peru. *Acta Horticulturae* 199, 33-38.

Bennett, J.P. (1950). Temperature and bud rest period. *California Agriculture* 4, 11.

Chandler, W.H. (1957). *Deciduous orchards.* Lea and Febiger, Philadelphia.

Chandler, W.H. (1960). Some studies of rest in apple trees. *Proceedings of the American Society for Horticultural Science* 76, 1-10.

Chandler, W.H. and Tufts, W.P. (1934). Influence of the rest period on opening of buds of fruit trees in spring and on development of flower buds of peach trees. *Proceedings of the American Society for Horticultural Science* 30, 180-186.

Chouard, P. (1960). Vernalization and its relation to dormancy. *Annual Review of Plant Physiology* 11, 191-238.

Couvillon, G.A. and Erez, A. (1980). Rooting, survival and development of several peach cultivars propagated from semihardwood cuttings. *HortScience* 15, 41-43.

Couvillon, G.A. and Erez, A. (1985a). The influence of prolonged exposure to chilling temperatures on bud break and heat requirement for bloom of several fruit species. *Journal of the American Society for Horticultural Science* 110, 47-50.

Couvillon, G.A. and Erez, A. (1985b). Effect of level and duration of high temperatures on rest in the peach. *Journal of the American Society for Horticultural Science* 110, 579-581.

Coville, F.V. (1920) The influence of cold in stimulating the growth of plants. *Journal of Agricultural Research* 20, 151-160.

Crabbe, J. (1981). The interference of bud dormancy in the morphogenesis of trees and shrubs *Acta Horticulturae* 120, 167-172.

Crabbe, J. (1994). Dormancy. *Encyclopedia of Agricultural Science* 1, 597- 611, Academic Press.

Crossa-Raynaud, P. (1955). Effets de hiver doux sur le comportement des arbres fruitiers a feuilles caduques. *Annalles de Agriculture Tunisie* 29, 1-22.

Darrow, G.M. (1942). Rest period requirements for blueberries. *Proceedings of the American Society for Horticultural Science* 41, 189-194.

De Benito, J. (1990). Dormex, nuevos horizontes para la fruticultura. *Fruticultura Profesional* 30, 119-121.

De Villiers, G.D.B. (1947). Winter temperature and fruit yield. *Farming in South Africa* 22, 638-644.

Del Real Laborde, J.I. (1986). Estimating chill units at low latitudes. *HortScience* 22, 1227-1231.

Edwards, G.R. (1987). Production of temperate-zone fruit at low latitudes, avoiding rest and the chilling requirement. *HortScience* 22, 1236-1240.

Edwards, G.R. (1990) Ten year's experience with temperate fruits in the tropics. *Acta Horticulturae* 279, 47-51.

Edwards, G.R., Sinclair, E.R. and Chapman K.R. (1990). Computer assisted selection of locations in South-East Asia for the continuous cropping of apples and peaches. *Acta Horticulturae* 279, 61-66.

Erez, A. (1979). The effect of temperature on the activity of oil + dinitro-cresol spray to break the rest of apple buds. *HortScience* 14, 141-142.

Erez, A. (1982). Peach meadow orchard: Two feasible systems. *HortScience* 17, 138-142.

Erez, A. (1983). Container-grown peach orchard. *HortScience* 18, 582.

Erez, A. (1985a). Defoliation of deciduous fruit trees with magnesium chlorate and cyanamide. *HortScience* 20, 452-453.

Erez, A. (1985b). Growth control of peaches by paclobutrazol. *Acta Horticulturae* 160, 217-224.

Erez, A. (1986). The significance of the leafless stage in deciduous fruit trees on fruit production. Proceedings of Symposium on *"Regulation of photosynthetic efficiency in fruit trees."* p 23. N.Y. State Agricultural. Experimental Station, Geneva N.Y.

Erez, A. (1987a). Chemical control of bud break. *HortScience* 22, 1240-1243.

Erez, A. (1987b). Use of the rest avoidance technique in peaches in Israel. *Acta Horticulturae* 199, 137-144.

Erez, A. (1990a). Container grown trees: a new option for an intensive culture of temperate-zone fruit trees in the tropics. *Acta Horticulturae* 279, 159-166.

Erez, A. (1990b). Off-season production of deciduous fruits by manipulation of the rest period, pp. 1-9, in *Off Season Production of Horticultural Crops* (ed. J. Bay-Peterson). FFTC Book Series 41 Taiwan.

Erez, A. (1995). Means to compensate for insufficient chilling to improve bloom and leafing. *Acta Horticulturae* 395, 81-95.

Erez, A. and Couvillon, G.A. (1983). Evaporative cooling to improve rest breaking of nectarine buds by counteracting high daytime temperatures. *HortScience* 18, 480-481.

Erez, A. and Couvillon, G.A. (1987). Characterization of the influence of moderate temperatures on rest completion in peach. *Journal of the American Society for Horticultural Science* 112, 677-680.

Erez, A., Couvillon, G.A. and Hendershott, C.H. (1979a). Quantitative chilling enhancement and negation in peach buds by high temperatures in a daily cycle. *Journal of the American Society for Horticultural Science* 104, 536- 540.

Erez, A., Couvillon, G.A. and Hendershott, C.H. (1979b). The effect of cycle length on chilling negation by high temperatures in dormant peach leaf buds. *Journal of the American Society for Horticultural Science* 104, 573-576.

Erez, A., Couvillon, G.A. and Kays S.J. (1980). The effect of oxygen concentration on the release of peach leaf buds from rest. *HortScience* 15, 39-41.

Erez, A., Fishman, S., Gat, Z. and Couvillon, G. A. (1988). Evaluation of winter climate for breaking bud rest using the dynamic model. *Acta Horticulturae* 232, 76-89.

Erez, A., Fishman, S., Linsley-Noakes, G.C. and Allan, P. (1990). The dynamic model for rest completion in peach buds. *Acta Horticulturae* 276, 165-174.

Erez, A., Fishman, S., Gat, Z. and Couvillon, G.A. (1989). The use of the dynamic model for evaluation of winter climate with respect to bud break in deciduous fruit trees. (in Hebrew) *Hassadeh* 70, 250-253.

Erez, A. and Lavee, S. (1971). The effect of climatic conditions on dormancy development of peach buds: I. Temperature. *Journal of the American Society for Horticultural Science* 96, 711-714.

Erez, A., Lavee, S. and Samish, R.M. (1968). The effect of limitation in light during the rest period on leaf bud break of the peach (*Prunus persica*). *Physiologia Plantarum* 21, 759-764.

Erez, A., Lavee, S. and Samish, R.M. (1971). Improved methods to control rest in the peach and other deciduous fruit species. *Journal of the American Society for Horticultural Science* 96, 519-522.

Erez, A. and Lavi, B. (1984). Breaking bud rest of several deciduous fruit tree species in the Kenyan highlands. *Acta Horticulturae* 158, 239-248.

Erez, A. and Lerner, H.(1990). Means to improve leafing using rest-avoidance technique in peaches in Israel. *Acta Horticulturae* 279, 239-246.

Erez, A., Samish, R.M. and Lavee, S. (1966). The role of light in leaf and flower bud break of the peach (Prunus persica). *Physiologia Plantarum* 19, 650-659.

Erez, A., Yablowitz, Z. and Korcinski, R. (2000). Temperature and chemical effects on competing sinks in peach bud break. *Acta Horticulturae* 514, 51-58.

Erez, A. and Yablowitz, Z. (1981). Rooting of peach hardwood cuttings for the meadow orchard. *Scientia Horticulturae* 15, 137-144.

Erez, A., Yablowitz, Z. and Korcinski, R. (1993). High density plantings for protected cultivation of fruit crops; Plastic cover for peach and nectarine. *Acta Horticulturae* 349, 95-98.

Erez, A., Yablowitz, Z. and Nir, G. (1989). Container grown peach orchard. *Acta Horticulturae* 254, 231-236.

Erez, A. and Zur, A. (1981). Breaking the rest of apple buds by narrow-distillation-range oil and dinitro-o-cresol. *Scientia Horticulturae* 14, 47-54.

F.A.O. (1996). Production yearbook, No. 50

Faust, M., Erez, A., Rowland, L.J., Wang, S.Y. and Norman, H.A. (1997). Bud dormancy in perennial fruit trees: Physiological basis for dormancy induction, maintenance and release. *HortScience* 32, 623-629.

Fishman, S., Erez, A. and Couvillon, G.A. (1987a). The temperature dependence of dormancy breaking in plants: Simulation of processes studied under controlled temperatures *Journal of Theoretical Biology* 126, 309-322.

Fishman, S., Erez, A. and Couvillon, G.A. (1987b). The temperature dependence of dormancy breaking in plants: Mathematical analysis of a two-step model involving a cooperative transition. *Journal of Theoretical Biology* 124, 473-483.

Fuchigami, L.H. and Nee, C. (1987). Degree growing stage model and rest breaking mechanisms in temperate woody perennials. *HortScience* 22, 836-844.

George, A.P. and Nissen, R.J. (1992). Effects of water stress, nitrogen and paclobutrazol on flowering, yield and fruit quality of the low chill peach cultivar, 'Flordaprince'. *Scientia Horticulturae* 49, 197-209.

George, A.P. and Nissen, R.J. (1993). Effects of growth regulants on defoliation, flowering and fruit maturity of the low chill peach cultivar Flordaprince in subtropical Australia. *Australian Journal of Experimental Agriculture* 33, 787-795.

George, A.P., Lloyd, J. and Nissen, R.J. (1992). Effect of hydrogen cyanamide, paclobutrazol and pruning date on dormancy release of the low chill cultivar Flordaprince in subtropical Australia. *Australian Journal of Experimental Agriculture* 32, 89-95.

George, A.P., Nissen, R.J. and Baker, J.A. (1988). Effects of hydrogen cyanamide in manipulating budburst and advancing fruit maturity of table grapes in south-eastern Queensland. *Australian Journal of Experimental Agriculture* 28, 533-538.

Gilreath, P.R. and Buchanan, D.W. (1981a). Floral and vegetative bud development of 'Sungold' and 'Sunlite' nectarine as influenced by evaporative cooling by overhead sprinkling during rest. *Journal of the American Society for Horticultural Science* 106, 321-324.

Gilreath, P.R. and Buchanan, D.W. (1981b). Rest prediction model for low-chilling "Sungold" nectarine. *Journal of the American Society for Horticultural Science* 106, 426-429.

Hampton, E. and Parker, B. (1992). Cyanamide essential for kiwifruit economics. *The Orchardist* Aug. 1992, 49-54.

Hanninen, H. (1995). Effect of climatic change on trees from cool and temperate regions: an ecophysiological approach to modelling of bud burst phenology. *Canadian Journal of Botany* 73, 183-199.

Henzel, R. and Allison, P. (1993). Reducing side flowers for increasing export production. *New Zealand Kiwifruit* Aug/Sep. 1993 16-17.

Honeyborne, G.E. and Rabe, E. (1993). Evaluation of two mineral oil based artificial rest breaking compounds on Golden Delicious apples. *Deciduous Fruit Grower* 43, 206-210.

Howell, G.S. and Dennis, F.G. Jr. (1981). Cultural management of perennial plants to maximize resistance to cold stress, pp.176-204. In *Analysis and Improvement of Plant Cold Hardiness*.(eds C.R. Olien and M.N. Smith), CRC Press Inc. Boca Raton, Florida.

Janick, J. (1974). The apple in Java. *HortScience* 9, 13-15.

Kobayashi, K.D., Fuchigami, L.H. and English, M.J. (1982). Modeling temperature requirements for rest development in *Cornus cerecea L. Annals of Botany* 52, 205-215.

Kuori, I., Shiraishi, Y. and Imano, S. (1963). Studies on breaking the dormancy of grapevine: I. Effect of lime nitrogen treatments for shortening the rest period of glass house grown grapevine. *Journal of the Japanese Society for Horticultural Science* 32, 175-180.

Lang, G.A., Early, J.D., Martin, G.C. and Darnell, R.L. (1987). Endo-, para- and ecodormancy: Physiological terminology and classification for dormancy research. *HortScience* 22, 371- 377.

Lavarenne, S., Champagnat, S.P. and Barnola, P. (1975). Influence d'une meme gamme de temperature sur l'entree et sortie de dormance des bourgeons du *Frene (Fraxinus excelsior L.) Physiologie Vegetal* 13, 215-224.

Lerner, H. (1990). Out of season peach production. *M.Sc. thesis*, the Hebrew University, Jerusalem. (In Hebrew with English summary).

Lespinasse, J.M. and Delort, F. (1986). Apple tree management in vertical axis: appraisal after ten years of experiments. *Acta Horticulturae* 160, 139-155.

Lin, H.S. and Lin, C.H. (1992). Enhancement of budbreak of container-grown 'Shinseiki' pear in Taiwan's lowlands by split application of cyanamide. *Gartenbauwissenschaft* 57, 235-237.

Linsley-Noakes, G.C., Matthee, G. and Allan, P. (1994). Modification of rest completion prediction models for improved accuracy in South African stone fruit orchards. *Journal of the South African Society for Horticultural Science* 4, 13-15.

Linsley-Noakes, G.C. and Allan, P. (1994). Comparison of two models for the prediction of rest completion in peach buds. *Scientia Horticulturae* 59, 107-113.

Lloyd, J. and Firth, D.J. (1993). Effect of hydrogen cyanamide and promalin on flowering, fruit set and harvest time of 'Flordaprince' peach (*Prunus persica* L. Batsch) in subtropical Australia. *Journal of Horticultural Science* 68, 177-183.

Nee, C. and Fuchigami, L.H. (1992). Overcoming rest at different stages with hydrogen cyanamide. *Scientia Horticulturae* 50, 107-113.

Nehez, M., Paldy, A., Selypes, A., Scheuflert, H., Berencsi, G. and Freye, A. (1981). Teratogenic and mutagenic effects of dinitro-o-cresol containing herbicides on the laboratory mouse. *Ecotoxicology and Environment Safety* 5, 38-44.

Nicolas, J. and Bonnet, E. (1993). Cerises, avance de maturitè avec cyanamide hydrogene. Trois Annees dessais. *L'Arboriculture Fruitiere* 458, 17-20.

Nir, G., Klein, I., Lavee, S., Spieler, G. and Barak, U. (1988). Improving grapevine budbreak and yields by evaporative cooling. *Journal of the American Society for Horticultural Science.* 113, 512-517.

Nitsch, J.P. (1957). Photoperiodism in woody plants. *Proceedings of the American Society for Horticultural Science* 70, 526-544.

North, M.S. (1992). Alternative rest-breaking agents to DNOC/oil for apples. *South African Plant and Soil* 9, 39-40.

Norvell, D.J. and Moore, J.N. (1982). An evaluation of models for estimating rest requirements of highbush blueberries (*Vaccinium corimbosum* L.). *Journal of the American Society for Horticultural Science* 107, 54-56.

Overcash, J.P. and Campbell, J.A. (1955). The effect of intermittent warm and cold periods on breaking the rest period of peach leaf buds. *Proceedings of the American Society for Horticultural Science.* 66, 87-92.

Partridge, C.J. and Allan, P. (1980). Further studies on the rest period of low chill peach cultivars. *Crop Production.* 9, 191-195.

Petri, J.L. and Pola, A.C. (1992). Influencia de temperaturas baixas e altas na eficiencia do oleo mineral mais cianamida hidrogenada na quebra de dormencia da maciera. *Revista Brasiliera Fruticultura* 14, 133-136.

Petri, J.L., Pola, A. C. and Stuker, H. (1990). Effect of mineral oil and hydrogen cyanamide on dormancy breaking of apples. *Proc. 23rd International Horticultural Congress* Vol. 2, 4326.

Rageau, R., Bonhomme, M., Richard, J.P. and Erez, A. (1998). The climatic determinism of vegetative bud break on peach trees with no exposure to chilling: some experimental results. *Acta Horticulturae* 465, 511-520

Richardson, E.A., Seeley S.D. and Walker, D.R. (1974). A model for estimating the completion of rest for "Redhaven" and "Elberta" peach trees. *HortScience* 9, 331-332.

Ringwald, S. (1994). Alternatives for oil-DNOC which is used for bud breaking spray on apple trees. (In Hebrew with an English summary). *M.Sc. thesis*, Hebrew University Jerusalem. 65pp

Samish, R.M. (1945). The use of dinitro cresol mineral oil sprays for the control of prolonged rest in apple orchards. *Journal of Pomology and Horticultural Science* 21, 164-179.

Samish, R.M. (1954). Dormancy in woody plants. *Annual Review of Plant Physiology* 5, 183-203.

Saure, M.C. (1973). Successful apple growing in Indonesia. *Fruit Varieties Journal* 27, 44-45.

Saure, M.C. (1985). Dormancy release in deciduous fruit trees. *Horticultural Reviews* 7, 239-300.

Seeley, S.D. (1996). Modelling climatic regulation of bud dormancy. pp. 361-376. In: *Plant Dormancy; physiology, biochemistry and molecular biology.* (G.A. Lang ed.) Cab International Wallingford, Oxon UK.

Shaltout, A.D. and Unrath, C.R. (1983). Rest completion prediction model for "Starkrimson Delicious" apples. *Journal of the American Society for Horticultural Science* 108, 957-61.

Shulman, Y., Nir, G., Fanberstein, L. and Lavee, S. (1983). The effect of cyanamide on the release from dormancy of grapevine buds. *Scientia Horticulturae* 19, 97-104.

Snir, I. Chemical dormancy breaking of red raspberry. *HortScience* 18, 710-713.

Snir, I. and Erez, A. (1988). Bloom advancement in sweet cherry by hydrogen cyanamide. *Fruit Varieties Journal* 42, 120-121.

Sparks, D. (1993). Chilling and heating model for pecan budbreak. *Journal of the American Society for Horticultural* Science 118, 29-35.

Van der Schoot C. (1996). Dormancy and symplastic networking at the shoot apical meristem. In: *Plant Dormancy; physiology, biochemistry and molecular biology.* (G.A. Lang ed) Cab International Wellingford Oxon UK.

Walser, R.H., Walker, D. R. and Seeley, S.D. (1981). Effect of temperature, fall defoliation and gibberellic acid on the rest period of peach buds. *Journal of the American Society for Horticultural Science,* 106, 91-94.

Wang, S.Y., Steffens, G.L. and Faust, M. (1986). Breaking bud dormancy in apple with a plant bioregulator, thidiazuron. *Phytochemistry* 25, 311-317.

Weinberger, J.H. (1950). Chilling requirements of peach varieties. *Proceedings of the American Society for Horticultural Science* 56, 122-128.

Weinberger, J.H. (1954). Effect of high temperatures during the breaking of rest on 'Sullivan Elberta' peach buds. *Proceedings of the American Society for Horticultural Science* 63, 157-162.

Weizman, Z., Erez, A., Gur, A. and Shulman, Y. (1985). Breba figs- a commercial orchard: 1. Enhancement of flower bud break by cyanamide. (In Hebrew with English summary). *Hassadeh* 66, 503-505.

Young, E. and Werner, D.J. (1985). Effect of shoot, root and shank chilling during rest in apple and peach on growth resumption and carbohydrates. *Journal of the American Society for Horticultural Science* 110, 769-774.

3 IRRIGATION OF TEMPERATE FRUIT TREES IN DRY AND WARM CONDITIONS

BEN-AMI BRAVDO

Faculty of Agriculture
The Hebrew University of Jerusalem
P.O.B 12 Rehovot, Israel

1 Introduction

Water consumption is an essential process for plants, arising from their need to lose water to the atmosphere through the stomata. The rate of flow which is dictated by the evaporative demands of the atmosphere may expose plants to dehydration in hot climates on the one hand while granting them some benefits such as mineral transport and evaporative cooling of the leaves, on the other. Water is lost from the leaves to the atmosphere through the stomata, the major function of which is to regulate gas exchange between the leaves and the atmosphere. The exchange of CO_2 and H_2O enables the maintenance of the three fundamental processes of plant metabolism: photosynthesis, transpiration and respiration. The productivity of agricultural crops greatly depends on these processes which govern vegetative and reproductive growth, whereas crop quality is mainly dependent on the interaction between these processes and the formation of primary (various forms of sugars) and secondary metabolites (terpenes, phenols, alcohols, etc.) products of the photosynthetic process. These products serve for energy utilization and quality - compound formation, respectively. The movement of water through the soil - plant- atmosphere continuum (SPAC) is mainly passive, driven by water potential gradients and regulated by resistances along this continuum (Passioura, 1982, Kramer and Boyer, 1995). The gradient between the water potentials at both ends of the continuum, *i.e.* the soil and the atmosphere, determine the water status of the plant. Extremely low water potentials at either of these two end segments induce water stress in the plant tissue. Since the atmospheric water potential variations are up to three magnitude higher than those of the soil and in most cases are uncontrollable, the major agricultural means of affecting plant water status is irrigation.

Irrigation provides means for controlling soil water availability and thereby plant water status at various stages of growth and development. Due to recent developments in plastics and computer technology the increasing demand for high fruit production and quality, the use of different forms of irrigation has spread to include more varied climatic zones, including temperate zones such as the east coast of the US and many regions in Europe where summer rains normally prevail. The introduction of micro - irrigation methods, which consist of partial wetting of the root zone with a combination of fertilizer injection into the irrigation water, has broadened our understanding of plant responses to soil water and mineral soil content. It also provides a new means of controlling the root environment, particularly with regard to mineral concentration and

Amnon Erez (ed.), Temperate Fruit Crops in Warm Climates, 49-76.
© 2000 *Kluwer Academic Publishers. Printed in the Netherlands.*

water potentials, thereby affecting the whole plant responses and increasing productivity and fruit size. This chapter will attempt to elucidate and sum up modern approaches of temperate orchard irrigation rather than review general topics of orchard irrigation, and will focus on modern approaches to orchard irrigation under dry and hot climate.

1.1 The soil plant atmosphere continuum (SPAC) system – pathways and conductances

An Ohms law analog was proposed to describe and analyze the path of water flow from the soil through plants into the atmosphere (Huber, 1924, Gradmann 1928, van den Honert, 1948). This water flow pathway running through a series of gradients and resistances (resistance =1/conductance) is termed soil plant water atmosphere continuum (SPAC). The analogy equates water flux to an electrical current, water potential gradients in the liquid phase and vapor pressure gradients in the gaseous phase to the electromotive force, and resistance to either liquid or gaseous diffusion water flux to electrical resistance analog. The system can be presented as a series of gradients divided by resistances (Equation 1) or multiplied by their reciprocals-conductances (not shown).

Equation 1

$$E = \frac{\Delta\Psi_{\text{soil- root surface}}}{R_{\text{soil}}} = \frac{\Delta\Psi_{\text{root surface -xylem}}}{R_{\text{root}}} = \frac{\Delta\Psi_{\text{xylem}}}{R_{\text{xylem}}} = \frac{\Delta\Psi_{\text{leaf - atmosphere}}}{R_{\text{leaf}}}$$

E- Water flux, Ψ- Water potential, R- Resistance

The SPAC pathway involves four major phases;

1. water movement in the soil towards the roots.
2. water movement into the roots and through the conducting tissues to the stems
3. water movement through the stems to the leaves.
4. water movement in the leaves to the evaporative sites in the intercellular spaces and through the stomata to the atmosphere.

The last segment involves a liquid to gaseous phase transition followed by a steep change in water potential gradient up to three orders of magnitude greater in the gaseous vs liquid phase (Fig. 1). Since the diurnal changes in the tree's water content are normally smaller than the amount of water flowing through the tree, the influx and outflux of water are almost equal. Consequently, the water potential gradient:resistance ratios of each segment of the catenary SPAC system (Equation 1) are equal. The highest potential gradient and therefore also resistance is located in the fourth (leaf to air) segment (Fig. 1). The application of the Ohm's law analogy to the SPAC system is an oversimplification because it assumes steady - state isothermal flow and constant resistance - conditions which seldom prevail. It is also important to note that the flux throughout the gaseous phase is linearly related to the vapor pressure gradient from the evaporating surfaces within the intercellular spaces in the leaves to the external atmosphere, rather than to the potential difference (Kramer and Boyer, 1995).

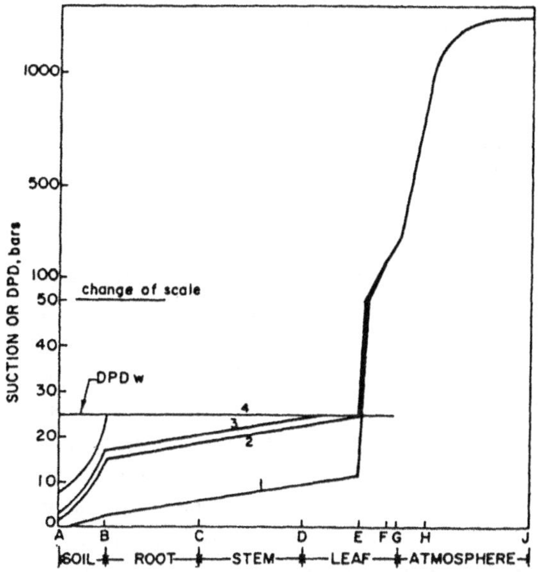

Figure 1. Biophysical model of the transpiration path depicting the energy status at various points along the path (After Philip 1957). A, soil (at boundary of region of influence of the root); B, surface of the root; CD, stem; D, leaf veins; E, mesophyl cells; EF, intercellular space and substomatal cavity; FG, stomatal pore; GH, stationary air layer adhering to leaf; HF, turbulent boundary layer and free atmosphere. Curve 1 = High soil water content; curves 2, 3 and 4 = represent conditions with decreasing soil water content.

Because of the non - linearity of the water potential:vapor pressure relationship, the magnitude of the resistance in this phase is usually not as great as it may appear (Passioura, 1982). In any case, quantitative calculations of the liquid and gaseous phase resistance or conductance must consist of hydraulic and diffusive units in the liquid and gaseous phases respectively.

Tree fruits rootstock x scion combinations differ in their ability to withstand varying water potential gradients between the two end units of the continuum, namely the soil to root segment and the leaf to atmosphere segment. In other words, they vary between their roots ability to efficiently extract water from various soil layers, canopy water loss and gas exchange from leaves. This is one of the main reasons for the discrepancies found among the many published reports on water requirements and water use efficiency among tree fruits. For example, differences in stomatal and non - stomatal conductances among and within a few stone fruit species were well demonstrated by Dejong (1982, 1983). When dealing with the irrigation of fruit trees, it is important to analyze the effects of various physiological processes on economic success in terms of productivity and quality. The economic orchard's profit margin is a product of many factors such as flower bud differentiation, fruit set, and rate of leaf area growth which greatly affects fruit size, coloration, sugar content, flavor - compound synthesis, etc. Variations in turgor potential and in the modulus of elasticity of the cell walls at various stages of growth and development may act as key factors governing the mechanisms of these processes (Salisbury and Ross, 1984, Kramer and Boyer, 1995). However, quantitative information regarding these mechanisms in tree fruits is still lacking. The factors governing the rate of water consumption can be divided into two major categories: - environmental and plant factors.

2 Environmental factors

2.1 General

Evaporative demand is affected by climatic factors, which are the outcome of the macro- meso and microclimates (Geiger 1961, Jackson and Lombard, 1993). The macroclimate is determined by parameters such as latitude, altitude and the distance from the ocean, whereas the mesoclimate by the topography, namely steepness and slope orientation, width, length and shape of a valley, etc. The microclimate is related to the environment in the immediate vicinity of the fruits and buds and is a function of the canopy size and structure, training system, plant spacing, etc. for any given macro, and mesoclimate. Several methods are employed to evaluate the evaporative demands of an orchard. These methods are based on either soil water or climatic measurements. Values obtained from soil water measurements represent the actual evapotranspiration (ET) whereas climatic methods require correction factors related to specific properties, such as stomatal and cuticular transpiration, plant spacing, training system, canopy architecture, crop load etc. for any given crop species.

2.2 Soil water measurements

Soil water measurements provide reliable information on ET for orchards under conventional irrigation where water is applied to the entire soil surface. However, under microirrigation regimes, namely drip or microsprinklers, where only part of the soil in the orchard is irrigated, soil water measurements provide limited quantitative information due to the partial and uneven distribution of the water in the soil. In other words it is very hard to quantitatively determine the amount of water present in the soil or extracted by the plant at any given time. Nevertheless, continuous or periodic monitoring of soil water at fixed positions (distances and soil depths relative to the emitters) can be used to adjust the irrigation regimes. Maintenance of soil water potentials at fixed positions in the irrigated soil volume indicates that irrigation is applied at the rate of consumption. Soil water content data can be collected by either gravimetric measurements or periodic or continuous soil water monitoring using devices such as neutron meters or TDR (time domain reflectometry), gamma - ray attenuation, tensiometers, gypsum blocks, water marks, etc. (Campbell and Mulla, 1990, Kramer and Boyer, 1995). These kind of data can be used for either manual or automated irrigation operations. Additional details concerning the use of soil water monitoring for operating automated irrigation systems will be further detailed later in this chapter. Soil water sensors can be used for irrigation scheduling in two major ways:

1. as a monitoring tool for irrigation systems based on ET evaluation and irrigation at constant intervals,
2. as a main tool for irrigation scheduling based on maintanance of constant water potential thresholds.

In the first case, ET coefficients are adjusted to provide constant water potential readings whereas in the second, intervals are determined by the rate of soil water depletion and ET coefficients are adjusted according to the soil sensors readings. In both cases, it is important to periodically or continuously monitor soil water potential at the 90-120 cm depth in order to ensure the absence of water percolation to layers below the root zone.

When soil water measurements are used as a major tool that determines both the intervals and the amount of water per irrigation. Since microirrigation systems are characterized by relatively rapid depletion of the limited irrigated soil volume and require frequent irrigation, constant intervals and adjustments of the amounts of water applied per irrigation according to weekly soil moisture measurements is normally preferred by growers. Microsprinkler irrigation are normally applied at longer intervals because this method uses the soil as a water reservoir to a larger extent than does drip irrigation.

Irrigation at constant intervals is common when using conventional irrigation systems, whereas irrigation according to soil water potential thresholds is more common in automated - computerized drip or microsprinkler irrigation systems based on continuous soil water potential monitoring.

The term evapotranspiration refers to the total evaporation rate from a field or orchard and is affected by both the evaporative demands and the specific properties of the plants concerned. The reference term for the ET of any particular field, Potential ET (Etp) was defined by Penman (1956) as "The amount of water transpired in a unit time by a short, green crop completely shading the ground, of a uniform height, and never short of water". A common reference crop for ET in California is alfalfa and its water consumption is monitored by either lysimeter or soil moisture measurements (Fereres and Goldhamer, 1990, Hatfield, 1990). This reference is in general use for fruit trees as well whereas the actual ET = ETa is the rate of water loss for a given orchard under current meteorological conditions.

2.3 Class A pan evaporation

The U.S. weather Bureau "class A pan evaporation" is a standard size pan covered by a 5-cm mesh wire screen. The evaporation rate from the pan is measured and used for the estimation of ETp. The necessary adjustments for surrounding area, wind speed and air humidity are detailed and discussed elsewhere (Pruitt 1966, Jensen 1973, Bosman, 1978, Hatfield. 1990). Daily and periodic manual readings or continuous computerized monitoring are in use as well (Phene and Campbell, 1975, Assaf *et al.* 1988, Bravdo *et al.* 1992a, Bravdo, 1993). Measurements obtained by class A pan evaporation are arbitrary and provide a rough estimate of the evaporation rates and their changes due to climatic factors. Practical use for orchard irrigation requires the use of a crop factor for various species, and orchard properties and design. The calculation of crop factors is described later in this chapter. Though the class A pan use has many shortcomings, its major advantages are simplicity and standardization which enable comparisons between various climatic regions around the world. Daily information of ET rates or pan evaporation measured by various methods is provided via the media, phone, fax, internet etc. in many irrigated regions worldwide.

2.4 Weather stations

Continuous and simultaneous measurements of air humidity, wind velocity, solar radiation and air temperature are used to calculate the rate of evaporation. Computerized weather stations are widely used for this purpose providing various formulas based on models such as those of Penman (Penman, 1948, 1956) - Penman Monteith (Monteith, 1965), Jensen - Haise (Jensen and Haise, 1963), Van Bavel (Van Bavel 1966), Szeicz (Szeicz *et al.* 1973) and Monteith,(Monteith, 1985) formulas. The values obtained are multiplied by the crop factor and the percent canopy cover, similar to the use made by class A pan evaporation rates.

2.5 Crop factors

The ETp obtained by either the standard cover crop, class A evaporation pan is multiplied by a factor Kc - a coefficient which takes into account the transpiration characteristics of the relevant plants and is presented as percentage of ET from the same area of a standard cover crop field when grown under similar climatic and soil water conditions. The Kc factor is also used for correcting evaporation data obtained by weather stations and class A pan evaporation. Further corrections take into account the ground surface covered by the canopy or the leaf area index (Howell, 1990). However no quantitative data are available for the crop load effects of various fruit trees. There are experimental evidences that fruiting trees have higher stomatal conductance and photosynthetic rates than deblossomed trees (Hansen, 1971, Tunssuwan and Buneman, 1973, Dejong, 1986, 1986a, Flore, 1989, Lenz, 1986). Naor *et al.* (1997) showed that transpiration rates and stomatal conductance of apple trees increase with the number of fruits per tree, the later having been altered by manual fruit thinning. However, increasing irrigation rates in high crop loaded trees, increased fruit size up to a certain irrigation level above which factors other than water supply were limiting. Assaf *et al.* (1982) reported similar interactions between crop load and water relations effects on apple fruit size. Quantitative data related to the effect of various crop loads on water consumption of fruit trees are still not available and are normally not included in recommended ET crop factors. Nevertheless, it is common for growers to increase the crop factor in proportion to the crop load.

3 Automated irrigation systems

3.1 Class A pan evaporation

Pan evaporation can be used for controlling an irrigation system (Phene and Campbell, 1975). An electronic device is used to continuously measure the level of the water in the pan and a control unit can be set to turn on the irrigation system whenever a predetermined amount of water evaporates and to automatically replenish the water loss using an appropriate factor. Simultaneous or periodic soil water content measurements may be used in order to adjust the coefficients to the rate of consumption.

3.2 Soil water sensors

The use of soil water sensors for automatic irrigation control is aimed at keeping the soil water content or potential within a range of a predetermined threshold and saturation. Commercially available computerized irrigation systems are capable of continuously monitoring the soil water content or potential, averaging them and turning the irrigation system on and off at predetermined threshold levels. The sensors are placed at sites where most of the root activity is likely to occur - the optimal site varies according to the method of irrigation. In drip irrigation systems for example, a 40cm vertical and horizontal distance from the drippers is appropriate for most common soils. Maintenance of high water availability and nutrients in this zone, a most favorable environment for root growth, stimulates root development and ensures continuous activity. In fact, there is an interaction between the application of water and minerals and root development on the one hand, and efficient depletion of available water and minerals on the other (Bravdo *et al.* 1992, Bravdo and Proebsting, 1993). Thus, a cycle of root development and water and mineral uptake is continuously maintained. Wider spacing of the sensor relative to the dripper may cause over and under shootings and consequently wide fluctuations due to the relatively slow movement of the water from the emitter to the site of the sensor. Deeper sensor insertion may result in percolation of water and fertilizers to layers below the root zone. With microsprinkler, microjet, or spray systems any site within the irrigation zone range is suitable though mounting the sensor in the most active root zone is most suitable in terms of increasing the sensitivity and accuracy of the system. A graphic example in which different levels of soil water potential were automatically maintained in a Cabernet Sauvignon vineyard is presented in Figures 2 and 3. The values are averages of the continuous monitoring of three sensors (electrotensiometers) with the irrigation system being turned on at 12, and 16 KPa for each of the two treatments, respectively, and off at 9 KPa - a value representing soil water potential at saturation. The irrigation intervals increase with a reduction in the predetermined soil water potential threshold values whereas the amount of water applied per irrigation is similar (Figs 2 and 3). The frequency of irrigation also varies with the atmospheric evaporative demands, canopy size, leaf area index and crop load (Bravdo *et al.* 1992, Bravdo and Proebsting, 1993). Irrigating according to a predetermined soil water potential not only supplies water at the rate of consumptive use but also determines that rate, since the consumptive use increases with the soil water threshold (Fig. 4). Although no quantitative data is yet available regarding the precise relationship between crop load and water requirements, a computerized irrigation system of this kind provides means for studying such issues.

Irrigation regimes consisting of computerized systems based on soil water potential sensors and drip systems (Figs 2 and 3) maintain a relatively constant volume of irrigated soil with minor swelling and shrinkage as long as the predetermined threshold value is kept constant. Minor changes proportional to the range of soil water potential fluctuations at the sensor sites might occur. It should be noted that when soil water potential at a given point within the irrigated volume of soil is proportionally related and stays almost constant due to the sensors control. Any other point along the water potential gradients in the bulb - shaped volume of soil stays constant as well. The smaller the fluctuations in irrigated

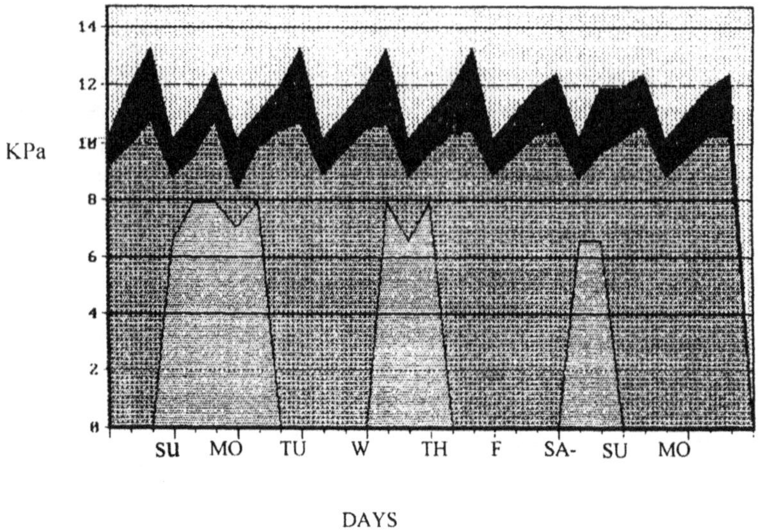

Figure 2. Control of soil matric potential by an automated computerized system. Irrigation was turned on and off whenever soil matric potential reached 12 and 9 Kpa respectively. Values are averages of three electrotensiometers located at 40 cm horizontal and vertical distance from the dripper in a Cabernet Sauvignon vineyard. (Bravdo, unpublished).

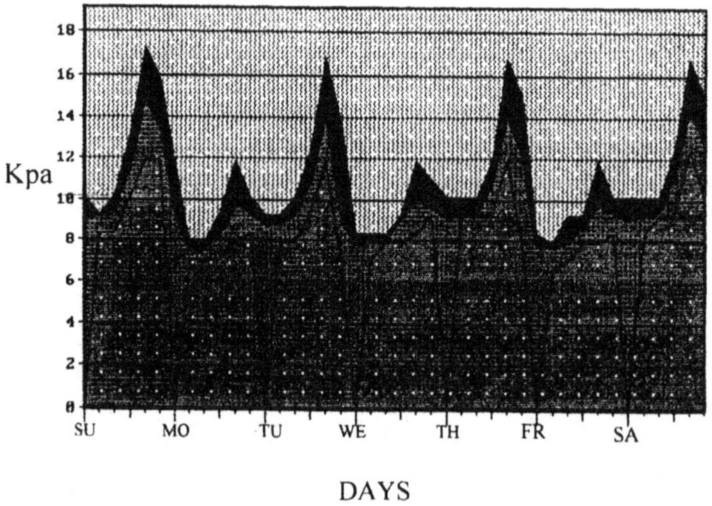

Figure 3. Control of soil matric potential by an automated computerized system. Irrigation was turned on and off whenever soil matric potential reached 16 and 9 Kpa respectively. Values are averages of three electrotensiometers located at 40 cm horizontal and vertical distance from the dripper in a Cabernet Sauvignon vineyard. (Bravdo, unpublished).

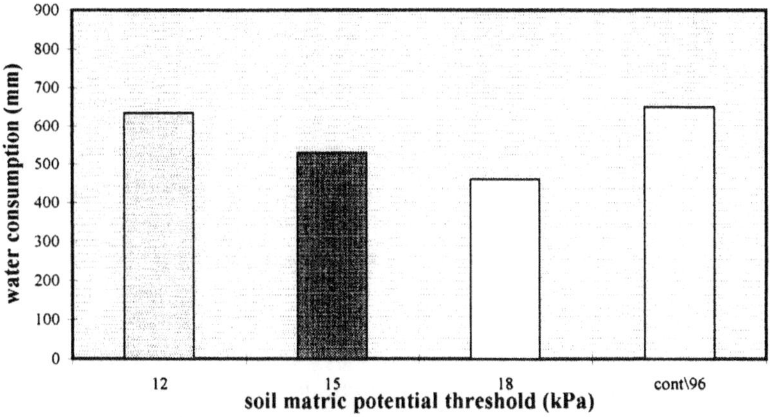

Figure 4. Water consumption at three threshold soil matric potential levels by automated computerized irrigation system as compared to the commercial recommended rate. Details as in Fig. 2. (Bravdo unpublished).

soil volume, the closer is the accomplishment of the concept of "Irrigation according to the rate of consumptive use". Due to these qualities, an automated computerized irrigation system provides means of controlling irrigation depth and thereby eliminates percolation of water and fertilizers to soil layers bellow the main root zone. The concept of irrigation according to the rate of consumptive use does not necessarily mean that water is supplied according to the plant's needs. Frequent applications of water in amounts smaller than the potential ET, force the plants to readjust to lower water availability by reducing vegetative growth and leaf area (Richards, 1986, Bravdo *et al.* 1992, 1992a, Williamson *et al.* 1992). Thus, irrigation rates according to consumptive use may be determined to a certain extent by the grower. It is important to note that in drip irrigation systems, water availability and the size of the irrigated soil volume are closely linked. The technique of drip irrigation opens up possibilities for controlling the rate of consumption by plants at various stages of growth and development and adjusting their water consumption by the growers need for any given growth stage.

Combined treatments of irrigation and fertigation may be used by growers to regulate vegetative and reproductive growth at various developmental stages. For example, optimal water availability is normally beneficial at the beginning of the season in order to establish the desired leaf area before the commencement of fruit ripening processes (Bravdo and Hepner, 1987). In the case of vigorous trees or vines with low crop load, reduced irrigation can be used to restrict excessive vegetative growth at the beginning of the growing season, thereby contributing to a proper vegetative - to -reproductive balance. Intensive irrigation during fruit ripening may induce excessive vegetative growth and hence sink competition for assimilates between the developing fruits and the shoot tips; the consequent expected reduction in fruit quality under such circumstances is reflected in low sugar content, poor coloration, fruit firmness etc. (Assaf *et al.* 1974, 1978, 1985, Reich-Gelfat *et al.* 1974, Safran *et al.*1975, Bravdo and Hepner, 1987, Naor *et al.* 1993). Cutting back or withholding irrigation before harvest increases fruit and wine quality under conditions of excessive growth. Under conditions of appropriate crop load, irrigation of winegrapes before harvest does not impair fruit quality and the best results were obtained in winegrapes at maximum irrigation which enables optimal physiological function without promoting vegetative growth before harvest (Naor *et al.* 1993, 1994,1998, Bravdo and Naor, 1996). Regulated deficit irrigation (RDI) was suggested for stone fruits which exhibit a double sigmoid growth pattern as well as for pome fruits (Chalmers 1989, Behboudian and Mills, 1997). the concept of RDI is still under debate (Fereres and Goldhamer, 1990) and succeeds however mainly under conditions of excessive growth or late in the season when new growth may not become harden before the winter. In hot climates, under high evaporative demands and heavy crop load, deficit irrigation may reduce yields and fruit size even though the number of fruits per shoot may increase in many fruit tree species due to the positive effects of drought on flower bud differentiation (Bravdo, unpublished).

Operating an automated irrigation system at various threshold values as shown in Figures 2 and 3 results in various irrigated soil volumes and consequently, various rates of water consumption and irrigation frequencies. Due to the frequent irrigations commonly practiced in most automated, computerized drip irrigation systems, critical levels of soil water potential in the immediate vicinity of the roots, which may cause incipient wilting under conditions of high evaporative demand are eliminated. Such low soil water potentials are not detected by conventional soil water measuring device. At low predetermined soil water potential thresholds, the irrigated soil volume decreases and the water potential gradients from the irrigation point to the edges of the wetted zone are steeper. Consequently, a larger part of the root system is exposed to water stress (Bravdo and Proebsting, 1993). Both the volume of the irrigated soil and the average soil water potential formed due to any given irrigation threshold value control tree growth and regulate the vegetative to reproductive growth ratio (Bravdo *et al.* 1992a, Bravdo and Proebsting, 1993). It should be noted that contrary to drip, conventional irrigation with longer intervals and similar total amounts of water application results in deep root system with fewer small diameter rootlets developing in the deeper layers with their limited aeration (Levin *et al.* 1972, Assaf *et al.*1974). Such root systems are more dependent on the rate of water diffusion in the soil and increase the vulnerability of the plants to temporary atmospheric stresses. These stresses reduce leaf water potential, thus impairing various physiological functions (Flore and Lakso 1993, Kramer and Boyer 1995) and resulting in reduced fruit production and fruit size (Assaf *et al.*1974, 1978, 1982, 1985. Naor *et al.* 1995, 1996, 1997). Automated drip irrigation based on turning on and off the irrigation system at predetermined threshold soil water potential values is capable of changing the vegetative to reproductive growth ratio without reducing fruit size, although the adjustment process can take 1 to 3 years (Assaf *et al.* 1988, Bravdo *et al.*1992, 1992a, Richards, 1986). It may therefore be concluded that under such circumstances, reduction in fruit size is a consequence of water stress and or overcropping, whereas the physiological effect of root restriction obtained by drip irrigation or physical restriction by for example containers (Richards, 1986) or artificial fabric surrounding the root system (Williamson *et al.* 1992) results in reduced leaf area and crop level without reduction in fruit size.

4 Physiological parameters

Physiological parameters, such as stomatal resistance (or conductance), leaf water potential (Boyer, 1995), stem water potential (McCutchan and Shackel,1992, Naor *et al.* 1995, 1996, 1997), rate of fruit growth (Assaf et al, 1982, Fereres and Goldhamer, 1990), trunk shrinkage, (Huguet *et al.* 1992) etc. has been used with varying degrees of success to control or adjust irrigation scheduling.

Manual weekly measurements of fruit circumference are used in some commercial apple orchards in Israel. The circumference is converted to volume, which has been found to increase linearly throughout most of the season (Assaf *et al.* 1982). Although fruit growth is indeed linear under conditions of non- limiting water supply, the absolute weekly increment is dependent on crop load (Assaf *et al.* 1982, 1984, 1989). Stem water potential was found to better reflect the water status of various tree fruits than leaf water potential (McCutchan and Shackel, 1992, Naor *et al.* 1995, 1996, 1997).

None of these parameters, however is suitable for continuous control of an automated irrigation system for various reasons. Continuous measurements of stomatal conductance, or leaf or stem water potentials are too expensive and complicated for use by growers, whereas fruit and trunk growth and shrinkage react too slowly to changes in the plant's water potential and cannot be used for instantaneous irrigation commands to replenish water deficits in the immediate vicinity of the roots. The fastest- responding organ to plant water status is the leaf, which serves as a major barrier to water loss from the plant to the atmosphere at the site where the steepest water potential gradient of the SPAC system occurs. Leaf shrinkage is one of the first primary plant responses to water stress and is relatively easy to monitor. The shrinkage is a direct consequence of turgor loss which occurs concomitantly with reductions in water content and water potential. Shrinkage is expressed mainly as a reduction in leaf thickness, while leaf area is hardly affected (Levit and Ben Zaken, 1975, Heathcote *et al.* 1979). Concomitant measurements of various organs showed that twigs begin shrinking 20 min after the leaves whereas trunk shrinkage is delayed by ca. 2 h. (Fig. 5). These delays seem to be related to the buffering capacity of the cortex, which equilibrates with the xylem water potential via horizontal water flow due to a redistribution process (Molz and Klepper, 1972, 1973).

Leaf turgor was found to be linearly related to leaf thickness over a wide range of leaf water potentials (Fig. 6). The performance of an automated irrigation system consisting of an electronic leaf thickness sensor capable of continuous monitoring and feeding onto a data logger was studied for several years in citrus and avocado orchards (Sharon and Bravdo, 1996). The data logger was programmed to turn on a drip irrigation system whenever the relative leaf thickness dropped below a certain predetermined level. Figure 6 and 7 show that the relative leaf thickness, namely leaf turgor, of the sensor - irrigated treatment was kept higher than the control treatment which was irrigated once a day. Pulses of drip irrigation applied directly to concentration of active roots consisting of numerous rootlets with a very large surface area (Plate1), enable rapid water uptake, release of tension in the xylem, and turgor recovery. The mass water flush of the concentrated root zone removes any boundary layer resistance formed by water uptake lagging behind transpiration the later causing temporary plant water stresses even at high average soil water potentials.Young citrus plots (cv. Oroblanco) irrigated for 5 years by a computerized irrigation system exhibits higher yields, and lower water consumption than plots irrigated by drip irrigation using recommended class A pan evaporation coefficients (Sharon and Bravdo, 1996). This increase in water use efficiency was attributed to improved control of both stomatal conductance and water percolation to layers bellow the root zone.

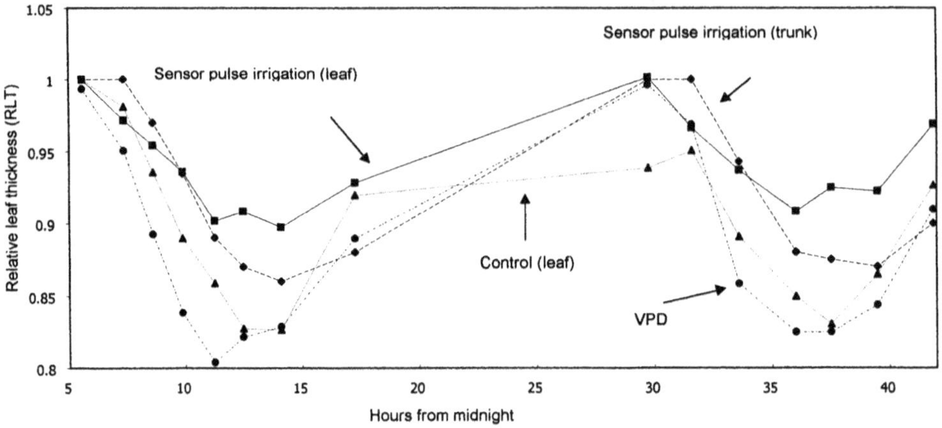

Figure 5. Diurnal variations in relative leaf thickness, trunk diameter and vapor pressure deficit of the air (VPD) of citrus trees. The sensor pulse irrigated trees were drip irrigated by an automated computerized system based on continuous monitoring of leaf thickness by sensor and turning on the irrigation at predetermined thresholds of leaf shrinkage. The control trees were irrigated once a day according to commercial recommendations. (Sharon and Bravdo, 1996).

4.1 Plant factors

All ET values, regardless of their source and of the parameters used to calculate the Kc factor, need additional corrections for the various crops. Even though leaf area index and canopy size are widely used as a basis for coefficient determinations, additional factors, such as canopy architecture, training system and row orientation are relevant. An important factor determining the water requirement for a given orchard is the crop load. The term "crop load" differs from crop level and expresses the ratio between the crop level and the vegetative growth rather than crop yield per land surface area. Since the vegetative growth of a tree is difficult to asses, various parameters, such as trunk cross - sectional area (Ferree, 1980, Erf and Proctor, 1987, Schecter *et al.* 1994) trunk cross - sectional area increment in apples (Assaf *et al.* 1985), pruning weight of winegrapes (Gal *et al.* 1997, Bravdo et al, 1984, 1985) and leaf area of winegrapes (Kliewer and Weaver, 1971). Leaf area to fruit weight ratio and fruit to pruning weight ratio was used as measure for crop load in winegrapes (Kliewer and Weaver, 1971, Bravdo *et al.* 1984, 1985, 1985a, Gal *et al.* 1977), whereas fruit weight to trunk cross sectional area were used for apples (Assaf *et al.* 1972, 1985). The above mentioned studies have suggested quantitative measures for assessing the crop load and its effect on

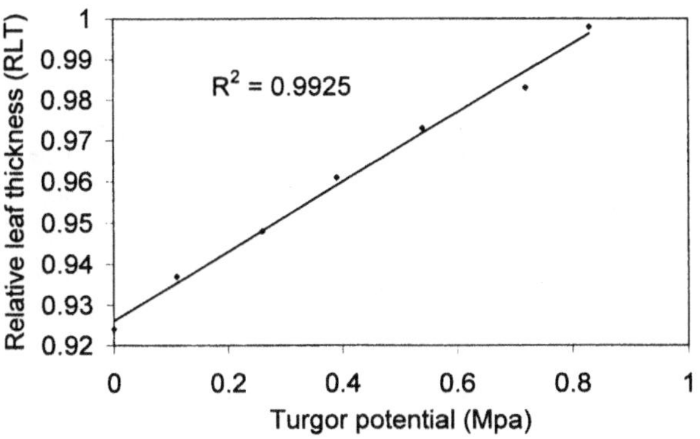

Figure 6. Relationship between relative leaf thickness and leaf turgor in Avocado. (Sharon and Bravdo, unpublished).

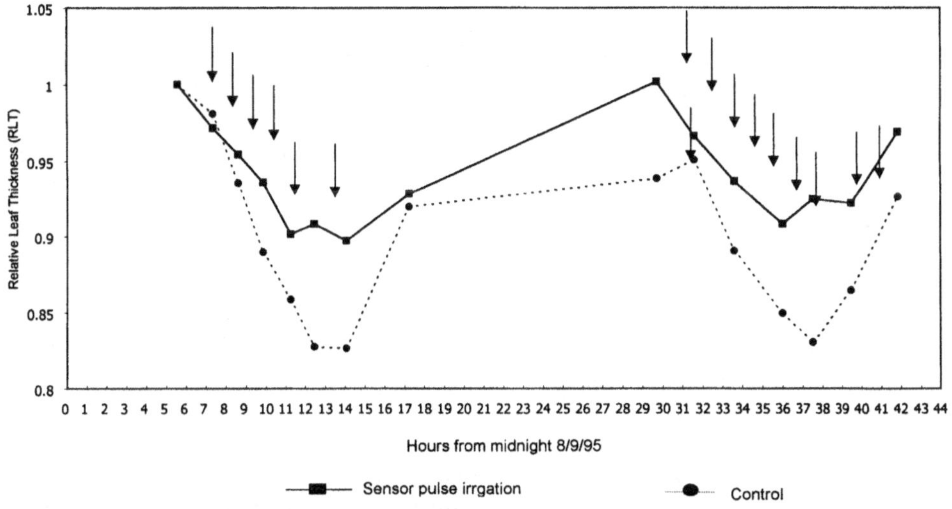

Figure 7. Citrus relative leaf thickness of trees irrigated by a computerized irrigation system based on leaf turgor sensor and control trees irrigated once a day. (Sharon and Bravdo, 1996).

production and quality. Analyses of various irrigation experiments show that the major irrigation effects on production and quality are indirect and are consequences of changes in vegetative and reproductive growth, while the most important direct effect is on turgor (Bravdo and Hepner, 1987, Bravdo and Proebsting 1993). Maintenance of high turgor enhances stomatal conductance, photosynthetic rate and vegetative growth. Micro irrigation methods based on partial wetting of the soil at high frequencies combined with fertilizer injection into the irrigation system can be used to efficiently control vegetative growth at various stages of growth and development and consequently production and fruit quality.

5 Soil water availability

5.1 General

Soil water availability is the main factor determining the rate of water uptake and consumption by plants and therefore also the frequency and amounts of water application by various irrigation methods. Since Viehmeyer and Hendrickson's (1950) postulation that soil water is readily available to plants throughout the entire range between field capacity and wilting point, many authors have advocated various threshold values for irrigation (Ritchie 1971), while others questioned the validity of the term "field capacity" from the physical point of view and proposed that only part of the water between field capacity and wilting point is equally available to plants (Ahuja and Nielsen, 1990). In Israel, the Ministry of Agriculture extension service recommended irrigating tree fruit when 50% of the available water had been depleted. This recommendation resulted in a shallow root system due to repeated irrigations of the upper soil layers while the lower layers remained continuously saturated with water leading to impaired root function due to lack of aeration (Assaf *et al.* 1974).

However, the major reason for the disagreement among researchers regarding the effect of water availability on plant performance appears to be insufficient understanding of the interactions between physiological and physical soil water processes. Moreover, the soil water availability data reported by various authors do not always relate to the same dynamic aspects. Soil water availability is, in fact, a dynamic rather than a static parameter because soil water potential at any given time is a function of the flow throughout the SPAC, which is a dynamic system.

It is difficult to determine a specific threshold value of available water under field conditions for many reasons: Water uptake is not uniform across the soil profile and neither is root distribution. Small diameter active rootlets tend to concentrate in the upper soil layers where organic matter and aeration are abundant (Assaf *et al.* 1974, Bravdo *et al.*, 1992). Therefore, most water uptake occurs first in the upper soil layers while uptake from the deeper soil layers proceeds at a slower rate, after water has been depleted from the upper layer (Levin *et al.* 1974, Assaf *et al.* 1974). Furthermore, irrigation affects root distribution and consequently soil water content across the soil profile, it is almost impossible to specify a single value which accurately represents the soil water content or availability for the entire root system along the soil profile. The problem of water availability determination is even more complicated under micro-irrigation regimes, particularly drip irrigation. Under daily drip irrigation regimes, the bulb- shaped irrigated soil volume remains almost constant and a gradient of soil water content from the point of water application to the edges of the wetted soil volume is formed. In this case, various parts of the root system are subjected to a wide range of soil water potentials. During each irrigation and right at its end, soil water potential varies from saturation under the dripper to wilting point at the edges of the wetted zone. It appears therefore, that under daily drip irrigation conditions some part of the root system is subjected to optimal soil water availability at any given time. Since water moves between roots according to water potential gradients (Taylor and Fenn, 1985, Baker and Van Bavel, 1988), water availability for the entire plant depends on the relative exposure of various parts of the root system to different soil water availability levels. It may be concluded that the conflicting data published on the optimal soil water availability to plants are due to the different irrigation methods and environmental conditions under which they were obtained. In the following paragraph, some major factors affecting water availability to plants will be reviewed.

5.2 Factors affecting soil water availability

a. Soil water conductivity

The transpiration process drives water uptake by the roots, which in turn initiates water potential gradients in the SPAC. Under conditions of increasing evaporative demand, the rates of water loss from the leaves and water uptake by the roots increase in parallel until water supply to the roots becomes limiting. The limitation is first apparent in the immediate vicinity of the roots where the soil first dries out (Dunham and Nye, 1973, Hasegawa and Sato, 1987, MacFall *et al.* 1990). Since the hydraulic conductivity decreases with decreasing soil water potential (Fig. 8) and, for short time intervals, water uptake is more dependent on water movement in the soil than on root growth (Kramer and Boyer, 1995), increased water potential gradients develop in the immediate vicinity of the roots. Such gradients are difficult to measure, because conventional soil water analyzers are capable of measuring average rather than point measurements of soil water content at specific locations, such as the very immediate vicinity of the roots. In reality, soil water potential varies as a function of distance from the root for any given average soil water potential (Fig. 9).

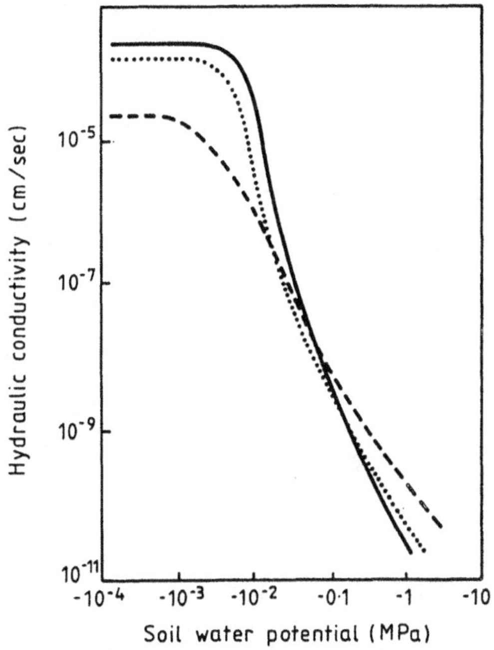

Figure 8. Relationship between hydraulic conductivity and soil water potential for indio loam (—), Pachappa sandy loam (······) and Chino clay (---). (after Gardner 1960).

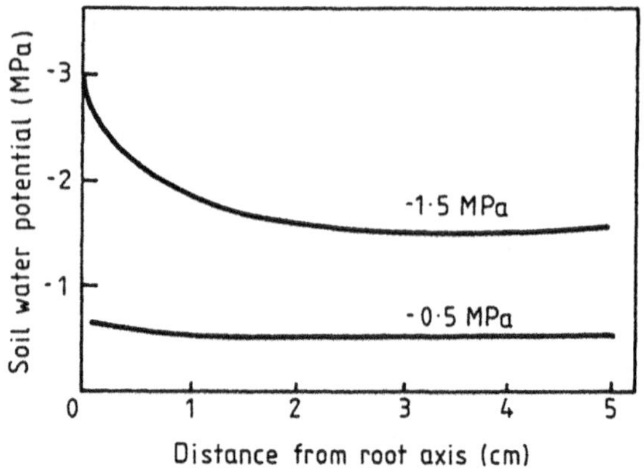

Figure 9. Profiles of soil water potential from soil mass to root surface, associated with soil mass values of soil water potential = -0.5 and -1.5 Mpa. Estimates refer to Pachappa sandy loam and an assumed water uptake rate of 0.1 cm^3/ cm root length. day. (after Gardner 1960).

b. *Evaporative demands*

The dynamics of water movement through the SPAC system is characterized by interactions between the system's various segments. Denmead and Shaw (1962) showed that transpiration rates were dramatically reduced at a soil water potential of -20 KPa and an ETp of 6.4 mm, whereas both the threshold values for the initial reduction in transpiration and the rate of reduction decrease with decreasing evaporative demand from 6.4 to 1.4 mm per day (Fig. 10). This further demonstrates that water availability is a dynamic rather than static situation which varies with evaporative demand and ET rate.

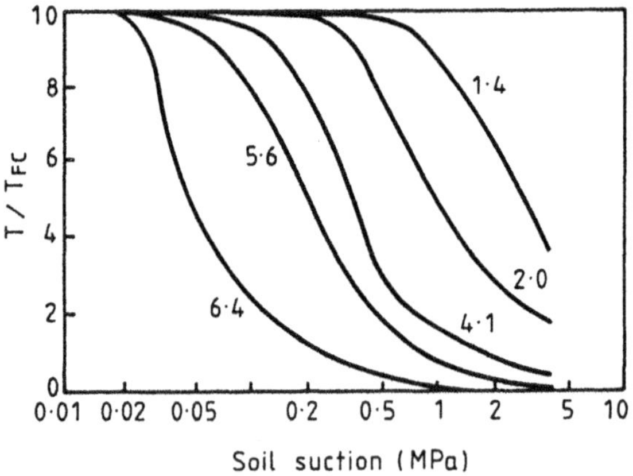

Figure 10. Relative transpiration rate (T/T_{FC}) as a function of soil suction in Colo silty clay loam for different potential transpiration conditions. The curves represent days on which the transpiration rates at field capacity were equal to the values shown in the body of the figure (after Denmead and Shaw 1962). T= actual transpiration rate, T_{FC} = transpiration with soil moisture at field capacity (mm/24h).

c. Efficiency of the root system

Root system efficiency is a function of root density, number, diameter and activity. Density is expressed as the number of roots per unit volume of soil. However, the rate of water and mineral uptake is dependent on a few physical and physiochemical properties of the roots. The calculated surface area of small - diameter rootlets is larger than that of large - diameter roots with the same accumulative cross sectional area. In addition, factors such as rate of growth and branching, as well as membrane permeability and hydraulic conductance, affect the efficiency of water and mineral uptake. Fig. 11 clearly demonstrates the effect of root density on soil water availability under constant environmental conditions. The fall in transpiration rate, indicating a limiting soil water availability threshold that occurs in the less dense root system at a relatively high soil water content, and this threshold value decreases with increasing root density. In this case, the interaction between soil hydraulic conductivity, root density and climatic conditions, greatly affects the threshold value as well as the rate of declining transpiration as expressed in the slope of the transpiration vs. soil water content graph. The low-density root system shown in Figure 11(1), represents sprinkler or flood irrigation, whereas the other extreme (4) a very high root density, more closely resembles a drip irrigation situation. In drip irrigation systems, water is applied at frequent intervals to a dense concentration of roots and consequently, water uptake is less dependent on soil water diffusion. However, a rapid depletion of available water irrigation occurs a while after irrigation is stopped.

It may be concluded that although water availability is a major factor affecting plant growth and agricultural production, its control is rather complicate and its dynamics and interactions with root growth are not fully understood. The introduction of modern irrigation and fertigation methods based on partial wetting of the orchard soil opens up new possibilities for controlling the root environment. A brief discussion of drip irrigation characteristics follows.

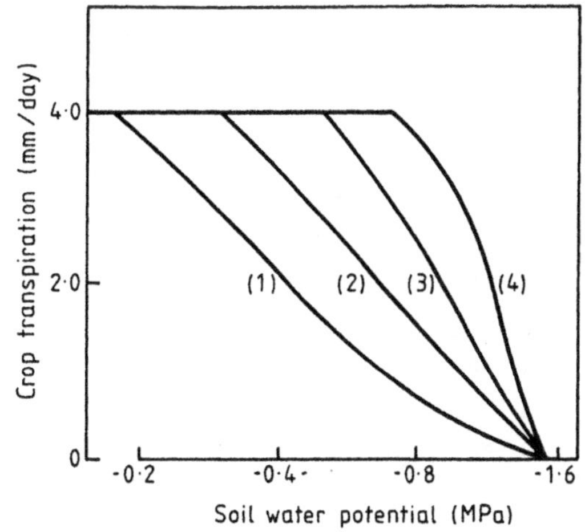

Figure 11. Effect of root density on the expected relationship between daily crop transpiration and soil water potential. Curves 1 to 3 refer to root densities of 8, 4 and 2 cm^3 soil/cm root length, and curve 4 to very dense rooting (after Cowan 1965).

6 Characteristics of drip irrigation

6.1 Water availability

Unlike conventional irrigation where uniformity of irrigation water distribution is a major objective, drip irrigation is characterized by non-uniform water distribution within the bulb – shaped, irrigated soil volume. The shape of the irrigated soil volume under the dripper varies with the hydraulic characteristics of the soil. The horizontal distribution of the water under the dripper depends on the hydraulic conductivity of the soil, whereas the vertical dimension is determined by both the hydraulic conductivity and gravitational force. Therefore, the vertical dimension is always longer than the horizontal one, and heavy soils characterized by high hydraulic conductivity have a higher horizontal, to vertical dimension ratio.

The soil water potential in the irrigated soil volume is never uniform; gradients exist between the point source of water application and the margins of the wetted soil volume. The roots in the irrigated soil volume are subjected to varying soil water potentials, from saturation to complete dryness. It is therefore practically impossible to determine a single representative value of water availability for the entire root system or to measure the total amount of water in the irrigated soil volume. On the other hand, it is well known that plants grow and produce well under conditions of non - uniform soil water potential. It seems, therefore, that plants can efficiently utilize water, even when only part of the root system is subjected to optimal soil water availability. The exact proportion of the root system needed for optimal utilization of the irrigation water by various tree fruits is not known. Moreover, the phenomenon of water transfer between roots in various plant species (Fig. 12) further complicates our ability to answer this question (Bravdo *et al.* 1992, 1992a Bravdo and Proebsting, 1993).

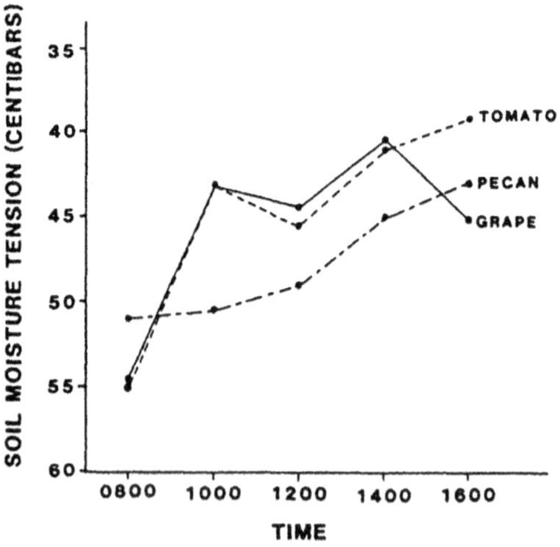

Figure 12. Average daytime reduction in soil moisture tension in the dry compartments of split root pecan, grape and tomato plants (each point is an average of 10 replicates; after Taylor and Fenn 1985).

6.2 Water transfer between roots

Water can move between roots of the same plant as well as between the soil and roots in both directions, according to water potential gradients (Rosene,1944, Jensen *et al.* 1961, Thorup,1969,, Taylor and Fenn, 1985 Baker and Van Bavel, 1988). Such water movement has been found in natural plant communities and this phenomenon is termed "hydraulic lift" (Corak *et al.* 1987, Richards and Caldwell, 1987, Dawson, 1993, Kramer and Boyer, 1995). The phenomenon is related to horizontal (Volk, 1947) or vertical (Tan *et al.* 1981) movement of water resulting in water, transfer through roots from wet to dry soil layers. Richards and Caldwell (1987) stated that if water is lost from the roots to drier soil, the root system can form a bridge for water transport between soil layers. Simulation models indicate that movement of water via such a bridge would be much more rapid than either liquid - or vapor-phase movement of water in the soil itself (Campbell, 1985).

Perhaps the most pertinent data concerning tree fruit were published by Taylor and Fenn (1985) who irrigated one compartment of split - root pecan, grape and tomato (Fig. 12) and found that water was transported from the irrigated to the non-irrigated compartment through the roots and even excreted to the surrounding soil, as evidenced by a reduction in tensiometer readings. Root excavations of vineyards and apple orchards in desert areas which had been drip irrigated since planting showed large root systems between the rows, even though these soil strips were never irrigated by drip or by rain (Bravdo, unpublished).

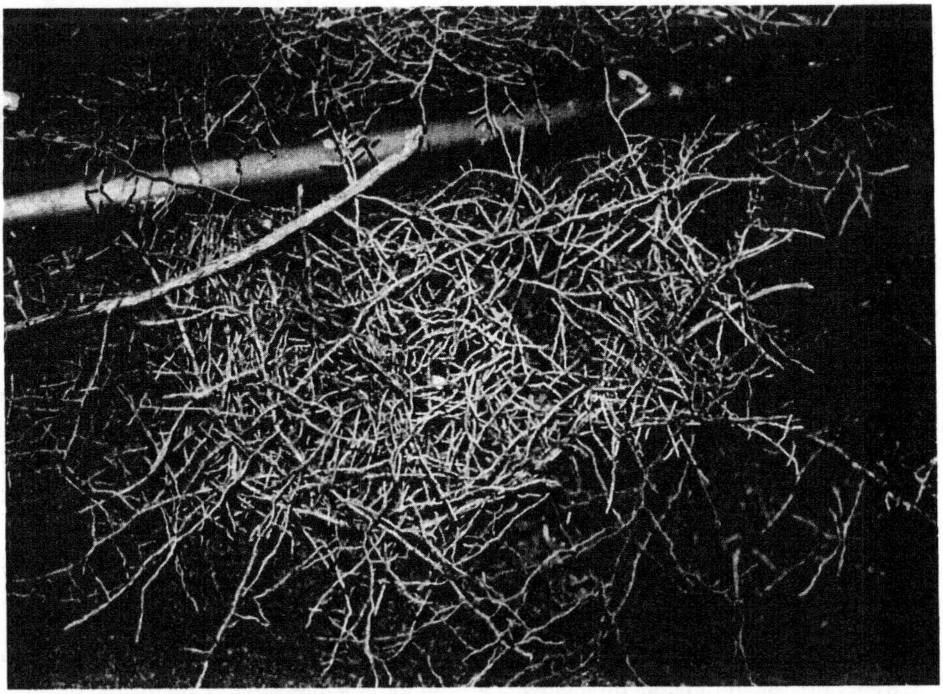

Plate 1. Citrus rootlets distribution under a dripper. The 5-10 cm soil in the vicinity of the pressure compensated inline dripper was washed by a stream of water (Bravdo unpublished).

It might therefore be concluded that gradients of soil water potential within the bulb shaped irrigated soil volume enable optimal water availability to a large part of the root system within the bulb due to water transfer between roots. Furthermore, the survival as well as growth of roots external to the irrigated soil volume throughout rainy and dry seasons may contribute mechanical support to the trees.

6.3 Physiological effects of root restriction

Plants grown in small containers that restrict their root system to a small volume respond by altering their vegetative to reproductive growth ratio. Richards (1986) reported a strong positive correlation between soil volume, canopy volume and butt circumference in container - grown peaches, and a negative correlation between soil volume and number of flowers per 100 cm lateral branches. Similar effects were observed in field - grown peaches where root growth was confined by porous fabric (Williamson *et al.* 1992). Precocity, a dwarfing effect and increased reproductive growth were observed in container-grown citrus (Lenz, 1967, Salomon, 1978), as well as increased reproductive growth of field-grown citrus and apples irrigated by drip were reported (Bravdo *et al.* 1992a, Bravdo and Proebsting, 1993)

Although the mechanism responsible for this effect is not fully understood, a few observations indicate the possible involvement of growth regulators. The importance of roots as growth regulator and signal producers rather than as water and mineral - absorbing organs, was reviewed by Davies and Zhang (1991). The major growth regulators produced are cytokinins, gibberellins and abscisic acid. Richards (1986) summarized several arguments supporting the involvement of growth regulators in the shoot - root relationship:

1. Cytokinins are produced in the root tips at various developmental stages.
2. The amount of cytokinins produced and transported in the xylem sap is proportional to the vigor of the roots.
3. There is some evidence that cytokinins produced in the roots promote auxin production in shoots, which in turn promotes root production.
4. In tissue culture cytokinins promote shoot growth, while auxin promotes root initiation.

Apple and citrus trees grown in containers develop a root system consisting of numerous tiny small - diameter rootlets with a very fast turnover rate, rapid branching and consequently, numerous tips. The number of root tips per unit soil volume in apples increased with the reduction in container size and was positively correlated with the amounts and concentrations of cytokinins found in the xylem sap (Bravdo, unpublished). The numerous small - diameter rootlets found under each dripper in citrus (Plate 1) under field conditions serve for water and mineral absorption, whereas the large roots scattered throughout the entire soil profile provide mechanical support for the trees. The ability to produce different types of roots is probably an evolutionary, genetic trait aimed at performing a few major root functions: increasing the efficiency of water and mineral uptake by the tiny rootlets, production of growth regulators by root tips and provision of mechanical support by the large - diameter strong roots. The development of an intensive rootlet system under the drippers does interfere with the concomitant production of the large roots which provide the mechanical support. However, experience shows that drip irrigated orchards in various parts of the world are not more sensitive to overturning by winds than orchards irrigated by conventional methods (where the entire soil volume is irrigated). Similarly, orchards irrigated by conventional methods or converted from sprinkler or flood irrigation to drip have a similar supporting root system as orchards which has been drip irrigated since their planting date. Conversion of a 25-year - old citrus orchard from sprinkler to drip irrigation resulted in the development of an intensive rootlet system right under the drippers within 2 to 3 weeks of the conversion date, and a dramatic increase in the number of fruits per tree within 1 year (Bravdo *et al.* 1992, 1992a). The increased number of fruits per tree was followed by stunted vegetative growth, resulted in overcropping and consequently small fruit size over the first 2 seasons. Efficient water and mineral application via the drip system resulted in large fruit size in the third season (Bravdo *et al.*,1992, 1992a). A similar 3-years equilibration period as observed in a peach root - restriction experiment (Richards, 1986). Changing the irrigated volume of soil in an apple orchard required a similar period of recovery for crop load expressed as fruit yield to trunk cross sectional increment area and fruit size ratio (Assaf *et al.*, 1978, 1985,).

6.4 Aeration of the root system

Aeration and soil water availability are generally conflicting parameters: at high soil water content or potential, air is pushed out of the soil pore spaces by the water. Moreover, root respiration further depletes oxygen and increases CO_2 concentration in both the soil and the root. This impairs water uptake (Jackson *et al.*, 1991, Kramer and Boyer, 1995) as well as the uptake of elements such as iron and nitrate, resulting in iron - induced chlorosis, reduced cytokinin production, leaf yellowing, death in extreme cases (Burrows and Carr, 1969), as well as increased ABA production, and stomatal closure (Kramer and Boyer, 1995). In drip irrigated orchards, there is no conflict between water availability and root aeration. Since the volume of the irrigated soil under the drippers is relatively small, efficient application of irrigation water requires frequent intervals and therefore almost continuous maintenance of high soil water potential under the drippers. Nevertheless, aeration deficiencies are very rare in drip irrigated orchards regardless of soil type or irrigation practice history. Moreover, there are many instances where conversion from flood or sprinkler irrigation to drip reduced or completely eliminated lack of aeration phenomena such as lime induced chlorosis (Levy, 1984, Korcak, 1987). A possible explanation is that the soil water potential gradients, ranging from saturation at the irrigation point to complete dryness at the margins of the wetted irrigated soil, are inversely correlated to gradients of air volume. Thus, various parts of the root system are subjected to various volumes of air and oxygen concentrations which can be transferred between the roots. Of course, the roots at the margins of the irrigated soil volume are exposed to maximal air and oxygen volumes. Gases can be transported from shoots to roots and between roots via large intercellular spaces and in dissolved form with the water (Esau, 1965, Zimmerman *et al.*, 1992, Kramer and Boyer, 1995). It can therefore be concluded that in typical drip irrigation regimes where irrigation is applied at frequent intervals, concomitant maintenance of water availability and aeration prevails due to opposite gradients of soil oxygen and water content. Frequent fresh - water applications, normally equilibrated with the surrounding air helps renewing soil oxygen levels as well.

6.5 Root development and spacing of drippers

Root development under drip irrigation conditions is highly dependent on water and mineral distribution in the restricted volume of irrigated soil. Data from a drip irrigation experiment (Levin *et al.*, 1979) in which three irrigation treatments were applied showed that root density is by far better correlated with their distance from the drippers than from the trees in all irrigation treatments studied (Levin *et al.*, 1979). It may therefore be concluded that drippers along an irrigation line should be spaced according to soil characteristics rather than to distance from the tree. Appropriate distances between drippers enables an overlapping of the bulb - shaped irrigated soil volumes and thereby provides means of maximizing water application without percolation to soil layers bellow the root zone.

Information is still lacking on the quantitative relationships between the irrigated volume of soil, tree vigor and fruitfulness, nevertheless, it seems obvious that relationships of that kind would vary with rootstock x scion combinations as well as with environmental conditions, making any set "recipes" impossible to formulate. The principle however, is clear: the smaller the irrigated soil volume, the lower the vegetative to reproductive ratio. It seems, therefore, that spacing aimed at creating a continuous strip of irrigated soil beneath the drip irrigation laterals would ensure sufficient irrigated soil volume to accommodate the roots and would provide means of controlling the vegetative to reproductive growth by varying the rate of emission through the drippers, frequency of irrigation or number of laterals per row.

Irrigation with one lateral per row can provide enough water for large canopy trees, even under high evaporative conditions and a wide enough range of irrigated soil volumes (Assaf *et al.* 1978, Bravdo *et al.* 1992). Under the extreme conditions of very shallow or sandy soils, a combination of two laterals per row, dense spacing of drippers, and pulse and low rate emitters may be advantageous due to their being less prone to water percolation and water stresses caused by low soil water conductivity.

6.6 Irrigation scheduling

The effect of the total amount of water applied per season or per any particular stage of growth and development is greatly dependent on the irrigation frequency. Since frequent irrigation by drip does not interfere with aeration (Assaf *et al.*, 1978, Bravdo *et al.*, 1992, Bravdo and Proebsting, 1993), the increase in frequency is associated with a reduction in the volume of the irrigated soil and therefore, in the size of the active root system. Short pulses, as well as slow water application by a large number of slow - rate emitters, approach a state of water application according to the rate of consumptive use, thereby making minimal use of the soil as a water reservoir. In contrast, decreasing the irrigation frequency is associated with increasing the use of the soil volume as a water reservoir, and consequently increasing the size of the active root system as well as the fluctuations in water availability (Levin *et al.*, 1973, Assaf *et al.*, 1988, Bravdo and Proebsting, 1993).

The normal irrigation frequencies used by drip systems in hot climates are once every 1-3 days and in extreme hot weather and sandy soils, a few times a day. Long irrigation intervals may necessitate prolonged applications of water which may result in water percolation to layers below the root zone.

6.7 Interaction between irrigation and fertilization

The use of micro irrigation enables to control the root environment with regard to water, minerals and aeration under field conditions. Injection of fertilizers into the irrigation water enables to control the concentration of the minerals in the root zone (Bravdo and Hepner 1987a). Since the uptake of any mineral element is a function of its concentration in the root media, controlling the mineral concentrations at various stages of growth and development provides means for regulating the uptake rate of minerals at any given time. Unlike the traditional approach of applying fertilizers on the basis of weight per unit land area *i.e..* t/ha, growers can now adopt concepts based on mineral concentrations, an approach similar to that of hydroponics. Such an approach is essential in order to optimize the use of drip irrigation since it increases the efficiency of mineral uptake, decreases the risk of clogging the irrigation system due to chemical precipitation, eliminates water and mineral percolation to soil layers below the root zone and (Hepner *et al.* 1985, Bravdo and Hepner, 1987, Bravdo and Proebsting, 1993). Due to the high efficiency of soil occupation by the roots, under drip irrigation regime, application of fertilizers through the irrigation water is essential in order to prevent mineral depletion. The fast mineral depletion sounds as a shortcoming since it increases the dependency of the plant and may result in mineral deficiency but in fact, it enables the grower to better control the composition and concentration of the minerals in the root zone (Bravdo and Proebsting, 1993, Hepner and Bravdo, 1985).

7 Conclusions

Microirrigation of tree fruit under warm climate conditions provides means for controlling water, mineral and aeration availability of the root zone. Irrigating part of the root system in the tree fruits has many advantages and has greatly contributed to improving orchard productivity grown in warm climate. The major advantages are:

1. Efficient control of the vegetative to reproductive growth thereby providing means for regulating fruit production and quality.
2. Prevention of incipient wilting under high evaporative demands due to frequent irrigation
3. Control of the composition and concentration of minerals as well as soil water potentials in the immediate vicinity of the roots, thereby increasing the efficiency of mineral uptake and the mineral balance in the plant tissue

4. Suitability of using computerized irrigation systems based on continuous monitoring of either soil water potential or plant parameters such as leaf turgor.
5. Capability of concomitant maintenance high water availability and root aeration.

8 References

Ahuja, L.R. and Nielsen, D.R. (1990). Field soil water relations In *Irrigation of Agricultural crops* (eds. Stewart, E.B. and Nielsen, D.) ASA monograph No 30, 143-190

Assaf, R., Bravdo, B. and Levin, I. (1974). Effects of irrigation according to water deficit in two different soil layers on yield and growth of apple trees. *J. Hort. Sci.* 49, 53-64.

Assaf, R., Bravdo, B. and Levin, I. (1978). An intensive hedgerow, drip irrigation system for apples as developed in Israel. *Compact Fruit Trees*. 11, 37-42.

Assaf, R., Bravdo, B. and. Levin, I. (1982). Apple growth as a measure for irrigation control. *Hort. Science* 17, 59-61.

Assaf, R., Levin, I. and Bravdo, B. (1985). Effect of drip irrigation on yield and quality of Golden Delicious and Jonathan apples. *J. Hort. Sci.* 59(4), 493-500.

Assaf, R., Levin, I. and Bravdo, B. (1988). Optimization of water for fruit trees by a computerized irrigation system. *Agronomie* 9, 1-6.

Baker, J.M. and Van Bavel, C.H.M. (1988). Water transfer through cotton plants connecting soil regions of differing water potential. *Agron. J.* 80, 993-997.

Bosman, H.H. (1987). The influence of installation practices on evaporation from Symon's tank and American class A – pan evaporimeters. *Agric. Meteor.* 41, 307-323

Boyer, J.S. (1995). *Measuring the Water Status of Plants and Soils*. Academic Press, N.Y.

Bravdo, B., Hepner, Y., Loinger, C., Cohen, S. and Tabacman, H. (1984). Effect of crop level on growth, yield and wine quality of a high yielding Carignane vineyard. *Am. J. Enol. Vitic.* 35, 247-252.

Bravdo, B., Hepner, Y., Loinger, C., Cohen, S. and Tabacman, H. (1985). Effect of crop level and crop load on growth, yield, must and wine composition and quality of Cabernet Sauvignon. *Am. J. Enol. Vitic.* 36, 125-131.

Bravdo, B., Hepner, Y., Loinger, C., Cohen, S. and Tabacman, H. (1985a). Effect of irrigation and crop level on growth, yield and wine quality of Cabernet Sauvignon. *Am. J. Enol. Vitic.* 36, 132-139.

Bravdo, B. and. Hepner.Y. (1987). Water management and effect on fruit quality in grapevines. In *Proc. of the Sixth Australian Industry Tech. Conf*, (ed. Lee T.) 150-158.

Bravdo, B. and Hepner, Y. (1987a). Irrigation management and fertigation to optimize grape composition and vine performance. *Acta Horticulturae* 206, 49-67.

Bravdo, B., Levin, L. and Assaf, R. (1992). Control of root size and root environment of fruit trees for optimal fruit production. *J. Plant Nutr.* 15, 699-712.

Bravdo, B., Salomon, E., Erner, Y., Oren, Y. and Shufman, E. (1992a). Effect of drip and microsprinkler fertigation on citrus yield and quality. *Proc. Inter. Soc. of Citriculture* 2, 646-649.

Bravdo, B. (1993) Effect of low volume irrigation on productivity of fruit trees. *Proc. 6th Inter. Conf. on Irrigation*. Agritech, Tel Aviv, Israel.,pp 97-106.

Bravdo, B. and Proebsting, E.L. (1993). Use of drip irrigation in orchards. *HortTechnology* 3, 44-49.

Bravdo, B. and Naor, A. (1996). Effect of water regime on productivity and quality of fruit and wine. *Acta Horticulturae* 427, 15-26 .

Behboudian, M.H. and Mills, T.M. (1997). Deficit irrigation in deciduous orchards. *Hort. Rev.* 21, 105-131.

Burrows, W.J. and Carr, D.J. (1969). Effect of flooding the root system of sunflower plants on the cytokinin content in the xylem sap. *Physiol. Plant.* 22, 1105 - 1112.

Campbell, G.S. (1985). Soil Physics with BASIC: *Transport Models for Soil Plant Systems.* Elsevier, New York, 150 p.

Campbell, G.S., and Mulla, D.J. (1990) Measurement of soil water content. In *Irrigation of Agricultural crops.* (eds. Stewart, E.B. and Nielsen D.) ASA Monograph No 30, 128-141

Chalmers, D.J. (1989). A physiological examination of regulated deficit irrigation. *New Zealand J. Agr. Sci.* 23, 44-48

Corak, S.J., Belvins, D.G. and Pallardy, S.G. (1987). Water transfer in an alfalfa – maize association : Survival of maize during drought. *Plant Physiol.* 84, 582-586.

Cowan, I.R. (1965). Transport of water in the soil – plant-atmosphere system. *J. App. Ecol.* 2, 221-239.

Davies, W.J. and Zhang, J. (1991). Root signals and the regulation of growth and development of plants in drying soil. *Annu. Rev. Plant. Physol. Plant Mol. Biol.* 42, 55-76.

Dawson, T.E. (1993). Hydraulic lift and water use by plants: Implications for water balance performance and plant - plant interactions. *Ecologia* 95, 565-574.

Dejong, T.M. (1982). Leaf nitrogen content and CO_2 assimilation capacity in peach. *J. Amer. Soc. Hort. Sci.* 107, 955-959.

Dejong, T.M (1983). CO_2 assimilation characteristics of five Prunus tree fruit species. (1983). *J. Amer. Soc. Hort. Sci.* 108, 305-307.

Dejong, T.M. (1986). Fruit effects on photosynthesis in Prunus persica. *Physiol. Plant.* 66, 149-153.

Dejong, T.M. (1986a). Effects of reproductive and vegetative sink activity on leaf conductance and water potential in prunus persica (L) Batsch. *Scientia Hort.* 29, 131-137.

Denmead, O.T. and Shaw, R.H. (1962). Availability of soil water to plants as affected by soil moisture content and meteorological conditions. *Agron. J.* 54, 385-390.

Dunham, R.J. and Nye, P.H. (1973). Soil water content gradients near the plane of onion roots. *J. Appl. Ecol.* 10, 585-598.

Erf, A.J. and Proctor, J.T.A. (1987). Changes in apple leaf water status and vegetative growth as influenced by crop load. *J. Amer. Soc. Hort. Sci.* 112, 617-620.

Esau, K. (1965) *Plant anatomy,* 2nd Edition. Wiley, New York.

Fereres, A., and D.A. Goldhamer, (1990) Deciduous fruit and nut trees. In *Irrigation of Agricultural Crops.* (eds. Stewart E.B. and Nielsen D.) ASA monograph No 30, 435-474.

Ferree, D.C. (1980). Canopy development and yield efficiency of Golden delicious apple trees in four orchard management systems. *J. Amer. Soc. Hort. Sci.* 105, 376-380.

Flore, J.A. (1989) Environmental and physiological regulation of photosynthesis in fruit trees. *Hort. Rev.* 11, 111-149

Gal, Y., Naor, A. and Bravdo, B. (1997). Effect of shoot density, crop level and crop load on fruit and wine quality of Sauvignon Blanc. *Acta Horticulturae* 427, 151-160

Gardner, W.R. (1960). Dynamic aspects of water availability to plants. *Soil Sci.* 89: 63-73

Geiger, R. (1961). *The Climate Near the Ground.* Cambridge M.A. Harvard, University Press.

Gradmann, H. (1928). Untersuchungen uber the wasser verhaltnisse des bodens als grundlage des pflanzenwachstums. *Jahrb. Wiss. Bot.* 69, 1-100.

Hansen, P. (1971). The effect of fruiting upon transpiration rate and stomatal opening in apple leaves. *Physiol. Plant.* 25, 181-183

Hasegawa, S. and Sato, T. (1987). Water uptake in roots in cracks and water movement in clayey subsoil. *Soil. Sci.* 143, 381-386.

Hatfield, J.L. (1990) Methods of estimating evapotranspiration. In *Irrigation of Agricultural crops.* (eds.

Stewart E.B. and Nielsen D.) ASA monograph No 30, 987-1017

Heathcote, D.G., Etherington, J.R. and Woodward, F.I. (1979) An instrument for non - destructive measurement of pressure potential (turgor) of leaf cells. *J. Exp. Bot.* 30, 811-816.

Hepner, Y., Bravdo, B., Loinger, C., Cohen, S. and Tabacman, H. (1985) Effect of drip irrigation schedules on growth, yield, composition and wine quality of Cabernet Sauvignon. *Am. J. Enol. Vitic.* 36, 77-85.

Hepner, Y. and Bravdo, B. (1985) Effect of crop level and drip irrigation scheduling on the potassium status of Cabernet Sauvignon and Carignane vines and its influence on must and wine composition and quality. A*m. J. Enol. Vitic.* 36, 140-147.

Honert, T.H., van den (1948) *Water transport in plants as a catenary process*. Disc. Faraday Soc. 3, 146- 153.

Howell, T.A. (1990) Relationships between crop production and transpiration, evapotranspiration and irrigation. In *Irrigation of Agricultural crops*. (eds. Stewart E.B. and Nielsen D.) ASA monograph No 30, 392-429.

Huber, B. (1924) Die beurtailung die des wasserhaushaltes der pflanze. *Jahrbb. Wiss. Bot.* 64, 1-20.

Huguet, J.G., Li, H. Lorendeau, J.Y. and Pelloux, G. (1992) Specific micromorphometric reactions of fruit trees to water stress and irrigation scheduling automation. *J. Hortic. Sci.* 67, 631-640.

Jackson, M.B., Davis, D.D. and Lambers, H. (1991) *Plant Life Under Oxygen Deprivation*. SPB Publ. Co., The Hague.

Jackson, D.I. and Lombard, P. (1993) Environmental and management practices affecting grape composition and wine quality – a review. *Am. J. Enol.Vitic.* 44, 409-430.

Jensen, M.E. (1973) Consumptive use of water and irrigation water requirements. *Am. Soc. Civ. Eng.* New York

Jensen, M.E. and Haise, H.R. (1963). Estimating evapotranspiration from solar radiation. *J. Irrig. Drain. Div. Am. Soc. Civ. Eng.* 89, 15-41.

Jensen, R.D., Taylor, S.A. and Wieb, H.H. (1961). Negative transport and resistance to flow through plants. *Plant Physiol.* 36, 633-638.

Kliewer, W.M. and Weaver, R.J. (1971). Effect of crop level and leaf area on growth, composition and coloration of "Tokay" grapes *Am. J. Enol. Vitic.* 22, 172-177.

Korcack, R.K. (1987). Iron deficiency chlorosis *Hort Rev.* 9, 133-186.

Kramer P.J. and Boyer, J.S. (1995). *Water Relations of Plants and Soils.* Academic Press.

Lenz, F. (1967). Relationship between vegetative and reproductive growth of Washington navel orange cuttings. *J. Hort. Sci.* 42, 31-39.

Lenz, F. (1986). Fruit effects on transpiration and dry matter production in apples, In *The Regulation of Photosynthesis in Fruit Trees.* (eds. Lenz F. and Lakso A.N.) N.Y. State Agr. Exp. Sta., Geneva, NY. 101-104.

Levin, I., Assaf, R. and Bravdo, B. (1972). Effect of irrigation treatments for apple trees on water uptake from different soil layers. *J. Amer. Soc. Hort. Sci.* 97, 521-526.

Levin, I., Bravdo, B. and Assaf, R. (1973). Relation between root distribution and soil water extraction in different irrigation regimes. In: *Ecological studies Analysis and synthesis.* (eds. Jacobs J. *et al.*) Springer Verlag, Berlin. Vol: 4, 351-359.

Levin, I., Assaf, R. and Bravdo, B. (1974). Soil moisture distribution and depletion in an apple orchard irrigated by tricklers. *Proc. 2nd International Drip Irrigation Congress* San Diego. Univ. of Calif. Press. 252-258.

Levin, I., Assaf, R. and Bravdo, B. (1979). Soil moisture and root distribution in an apple orchard irrigated by tricklers, *Plant and Soil* 52, 31-40.

Levit, J. and Ben Zaken, R. (1975). Effects of small water stresses on cell turgor and intercellular space.

Plant Physiol. 34, 273-279.

Levy, Y. (1984). The effect of sprinkler and drip irrigation on lime - induced chlorosis of citrus. *Scientia Hortic.* 22, 249-255.

MacFall, J.S., Johnson, G.A. and Kramer, P.J. (1990). Observation of a water-depletion region surrounding loblolly pine roots by magnetic resonance imaging. *Proc. Natl. Acad. Sci.* USA 87, 1203-1207.

McCutchan, H. and Shackel, K.A. (1992). Stem-water potential as a sensitive indicator of water stress in prune trees (Prunus domestica L. cv. French). *J. Amer. Soc. Hort. Sci.* 117, 607-611.

Molz, F.Z. and Klepper, B. (1972). Radial propagation of water potential in stems. *Agron. J.* 64, 469-473.

Molz, F.Z. and Klepper, B. (1973). On the mechanism of water stress induced stem deformation *Agron. J.* 65, 304-306.

Monteith, J.L. (1965). Evaporation and environment. *Symp. Soc. Exp. Biol.* 19, 205-234.

Monteith, J.L. (1985). Evaporation from Land Surfaces: Progress in Analysis and Prediction Since 1948 in *Advances in Evapotranspiration* ASAE. St. Josef MI. 1-12.

Naor, A., Bravdo, B. and Hepner, Y. (1993). Effect of post veraison irrigation level on yield, quality and water relations of Sauvignon blanc grapevines. *S. Afr. J. Enol. Vitic.* 14, 19-25.

Naor, A., Bravdo, B. and Gelobter, J. (1994). Gas exchange and water relations of field grown Sauvignon blanc grapevines. *Am. J. Enol. Vitic.* 45, 423-428.

Naor, A., Klein, I. and Doron, I. (1995). Stem water potential and apple fruit size. *J. Amer. Soc. Hort. Sci.* 120, 577-582.

Naor, A., Klein, I., Bravdo, B., Doron, I., Gal, Y and Ben David, Z. (1996). Irrigation and crop level interactions in relations to apples yield and fruit size distribution. *J. Hort. Sci*, 122, 411-414.

Naor, A., Klein, I., Doron, I., Gal, Y., Ben-David, Z. and Bravdo, B. (1997). The effect of irrigation and crop load on stem water potential and apple fruit size. *J. Hort. Sci.* 72, 765-771.

Naor, A., Gal, Y. and Bravdo, B. (1998). Crop level affects assimilation rate, stomatal conductance, stem water potential and water relations of field-grown Sauvignon blanc *J. Exp. Bot.* 48, 1675-1680

Passioura, J.B. (1982). Water in the soil plant atmosphere continuum. In *Encyclopedia of Plant Physiology.* (eds. Lang O.L., Nobel P.S., Osmond C.B. and Ziegler H.) Springer Verlag, New York. Vol 12B, PP 5-33.

Penman, H.L. (1948). Natural evaporation from water, bare soil and grass. *Proc. Royal Soc. London, Ser. A* 193, 120-146.

Penman, H.L. (1956). Evaporation: An Introductory Survey. *Neth. J. Agric. Sci.* 9-29.

Phene C.J. and Campbell, R.B. (1975). Automatic pan evaporation measurement with a float- activated electronic sensor. *Agric. Meteorol* . 15, 181-191.

Philip, J.R. (1957). The physical principles of soil water movement during the irrigation cycle. *Int. Com. Irrig. Drain 3rd Cong. Ques.*8:125-128

Pruitt, W.O. (1966). Empirical method of estimating evapotranspiration using primary evaporation pan. *Proc. Conf. on evapotranspiration*, 57-61. ASAE St. Joseph, MI.

Reich-Gelfat, S., Assaf, R., Bravdo, B. and Levin, I. (1974). The keeping quality of apples in storage as affected by different irrigation regimes. *J. Hort. Sci.* 49, 217-225.

Richards, D. (1986). Tree growth and productivity – the role of roots. *Acta Horticulturae.* 175, 27-36.

Richards, J.H. and Caldwell, M.M. (1987). Hydraulic lift: Substantial nocturnal water transport between soil layers by Artemisia tridentana roots. *Ecologia* 73, 486-489.

Ritchie, J.T. (1971). Dryland evaporative flux in subhumid climate: iii. Soil water influence. *Agron. J.* 64, 168-173

Rosene, H.F. (1944). Control of water transport in local root regions of attached and isolated roots by means of osmotic pressure of the external solution. *Am. J. Bot.* 28, 402 - 410.

Safran, B., Bravdo, B. and Bernstein, Z. (1975) Irrigation de la vigne par goutte a goutte. *Bull. O.I.V.* 48-31, 405-429.

Salisbury, F.B. and Ross, C.W. (1984). *Plant Physilogy*, Belmont California, USA.

Salomon, E. (1978). Induction of dwarfing and early cropping through root treatment in citrus. *Acta Horti*culturae 65, 147-148.

Sharon, Y. and Bravdo, B. (1996). Irrigation control of citrus according to the diurnal cycling of leaf thickness. *Proc. 6th Inter. Conf. on Irrigation*. Agritech, Tel Aviv, Israel, 97-106.

Schechter, I., Proctor, J.T.E. and Elving, D.C. (1994). Carbon exchange rate and accumulation in limbs of fruiting and non - fruiting apple trees. *J. Amer. Soc. Hort. Sci.* 119, 150-156.

Szeicz, G., C., Van Bavel, H.M and Takami, S. (1973). Stomatal factors in the water use and dry matter production by sorghum. *Agric. Meteor.* 12, 361-389.

Tan, C.S., Cornelisse, A. and Buttery, B.R. (1981) Transpiration, stomatal conductance and photosynthesis of tomato plants with various proportions of root system supplied with water. *J. Amer. Soc. Hort. Sci.* 106, 147-151.

Taylor, R.M. and Fenn, L.B. (1985) Translocation of water within root systems of pecan, grape and tomato. *Hortscience* 1, 104-105.

Thorup, R.M. (1969) Root development and phosphorus uptake by tomato plants under controlled soil moisture conditions. *Agron. J.* 61, 808-811.

Tunssuwan, T. and Buneman, G. (1973) Behavior of stomates on leaves of apple trees with and without fruits. *Gartenbauwisssenschaft* 38, 109- 115.

Van Bavel, C.H.M. (1966) Potential evaporation: The combination concept and its experimental verification. *Water Resour. Res.* 2, 455- 467.

Viehmeyer, F.J. and Hendrickson, A.H. (1950). Soil moisture in relation to plant growth *Annu. Rev. Plant Physiol.* 1, 285-304.

Volk, G.M. (1947). Significance of moisture translocation from soil zones of high moisture tension by plant roots. *Agron. J.* 39, 93-106.

Williamson, J.G., Coston, D. and Cornell, J.A. (1992). Root restriction effects on shoot development of peach in a high density orchard *J. Amer. Soc. Hort. Sci.* 117, 362-367.

Zimmerman, U., Rygol, J., Balling, A., Klock, G., Metzler, A. and Hasse, A. (1992). Radial turgor and osmotic profiles in intact and excised roots of Aster tripolium. *Plant Physiol.* 99, 186-196.

4 FERTILIZATION OF TEMPERATE-ZONE FRUIT TREES IN WARM AND DRY CLIMATES

ISAAC KLEIN [1] *and* STEVEN A. WEINBAUM [2]

[1] *Institute of Horticulture, Agricultural Research Organization, The Volcani Center, Bet Dagan 50250, Israel*
[2] *Department of Pomology, University of California, Davis, CA 95616, USA*

1 Introduction

It is impossible to discuss the unique aspects of fruit tree fertilization in warm and dry climate without outlining the basic principles of tree nutrition which are applicable everywhere. Temperate-zone fruits referred to in the present book include mainly deciduous crops, such as stone and pome fruits, various nut crops and some small fruits. In the discussion, however, references are made to other crops, such as the olive and the avocado, which are native or widely grown in warm and dry climates. This is justified because some of the topics which need to be addressed when discussing fertilization include soil variables and various agrotechnical practices which influence nutrient fixation, movement and availability to all tree crops.

Most fruit tree nutritionists emphasize the plant, rather than the soil, when considering diagnostic work. However, when corrective measures are taken the soil texture and chemistry need to be considered. Nutrient fixation and availability, dependent largely on soil genealogy, the Cation Exchange Capacity (CEC) and soil pH, determine the exact course of corrective measures to be taken to meet plant needs. Rainfall patterns (quantity and frequency) and irrigation practices have major consequences on nutrition and crop production and, in this respect, warm and dry climates are unique, as plant growth and development can be controlled more precisely by irrigation and fertilization. Shortage of water resources under dry and warm climates was a primary incentive for the development of irrigation technology and the concept of fertigation which will be discussed.

Finally it should be emphasized that fertilization, like any other production parameter, is governed by the well known rule of a limiting factor. Higher intensity or balanced nutrition can improve production and fruit quality only as long as it is the limiting factor. This should be kept in mind both by extension and farm personnel who are constantly encouraged by salesmen to apply more fertilizers as well as by researcher when experimenting under non-deficient conditions.

2 Effect of soil pH on fertilization requirements

Plant nutrition is profoundly affected by soil pH. With the exception of calcifuges which prefer a soil pH of 4.2-5.5, the optimum soil pH for nutrient availability and

Amnon Erez (ed.), Temperate Fruit Crops in Warm Climates, 77-100.
© 2000 *Kluwer Academic Publishers. Printed in the Netherlands.*

agricultural production, including fruit tree cultivation, is around 6.5 and decreases toward the extremes of the pH scale (Mengel and Kirkby 1987). In areas of high precipitation rates, H^+ ions replace mono and divalent cations and low pH soils are prevalent. Low pH soils are characterized by high solubility of microelements, low calcium content and low phosphate availability. Liming is recommended to raise low soil pH, to alleviate aluminum and manganese toxicities and to correct phosphate deficiency. Fertilizers containing nitrate nitrogen have a basic reaction in the soil and are preferred for acid soils. In high pH soils microelement availability is low and phosphate is moderately soluble. The presence of calcium carbonate in high pH soils accentuates microelement deficiencies. Soils of arid and warm climates are usually neutral or basic (Marschner 1986), although low pH soils can also be encountered (Williams and Vlamis 1978). Acidifying nitrogen fertilizers are usually preferred for basic soils. Fertigation with acidifying fertilizers can not decrease soil pH when CEC is high (Klein and Spieler 1987) and as long as calcium carbonate is present in the soil. However, soil acidification can occur if neutral pH, low CEC soils (without calcium carbonate) are fertigated with an acidifying fertilizer (Haynes 1990, Nielsen *et al.*, 1994, Zasoski *et al.* 1992).

3 Fertilization methods

3.1 Conventional soil application

Traditional fertilization methods, such as broadcasting and trenching, are being practiced under dry farming (Klein and Lavee 1977) or conventional (flood and sprinkle) irrigation methods (Rolston *et al.*, 1979). Potentially high yields may be obtained by a single or a split broadcast application of a fertilizer. Correction of K deficiencies by broadcast application is not feasible, particularly in heavy soils. Under flood or sprinkler irrigation growers often resort to trenching and foliar application of KNO_3 prior to harvest (Aldrich *et al.*, 1978) to satisfy the high demand for K during the final stage of stone fruit development (Weinbaum *et al.*, 1994b).

3.2 Foliar

Foliar feeding is an integral part of the fertilization program of tree crops in temperate as well as warm and dry climates. Discussion of principles involved and references to previous work can be found elsewhere (Weinbaum 1988). The potential benefits of foliar nutrition may be predicted if the following information is available:
a. The nutrient concentration which can be applied without phytotoxicity,
b. volume of solution (nutrient) retained on leaf surfaces and,
c. foliar uptake as a percent of nutrient retained on the leaf surface. Table 1 demonstrates the comparative benefits of a single foliar urea application to various crops, based on the above criteria.

Table 1. Leaf urea retention and nitrogen uptake from a single foliar urea spray

	Apple, at				
Parameter	*Fruit Set*[1]	*Leaf Abscission*[2]	*Almond*[1]	*Avocado*[1]	*Olive*[1]
Urea concentration (%w/v)	0.6	4.0	0.5	2	3
Retention (μl/cm²)	6.0	6.0	1.0	2.5	1.9
Uptake (%)	90	90	90	85	75
N-Uptake (μg/cm²)	14.9	99.4	2.1	19.6	19.7

[1] Sprayed at the threshold of phytotoxcity.
[2] Phytotoxicity is of limited concern just prior to leaf abscission.

A single fall application to apple, close to leaf abscission when phytotoxicity is of no concern, results in fifty fold more uptake per unit leaf area than an application to almond. Measurements of solution retention showed that wetting of abaxial leaf surfaces is critical for realizing the maximal benefit from foliar application since solution retention is 3-5 greater on the abaxial as compared to the adaxial leaf surface (Klein and Zilkah 1986).

Deficiencies of Mg, Ca, Zn, Mn, B and Cu are corrected exclusively by foliar application in the high pH soils prevailing in warm and dry climates. Nitrogen (urea) and K are also applied by foliar sprays under certain circumstances. Prevalent deficiencies and methods of fertilization are discussed in section 4.4.

3.3 Fertigation

a. Influence on tree development

The concept of fertigation evolved with the introduction of restricted soil volume (drip and microjet or microsprinkler) irrigation systems. The water applied from a point source in drip irrigation is subject to a relatively large gravitational force causing percolation and leaching, and a smaller capillary suction force causing lateral distribution. As a result, a two dimensional gradient of water and nutrient is formed within the soil profile (Bresler 1975, Bar-Yosef and Sheikholslami 1976). Water and solute movement in microjet irrigation is one dimensional, although the intensity of irrigation along the radii, from the water source to the periphery of the wetted circle, declines at a certain rate characteristic of the design of the microjet. The discharge at a distance of half radii is usually equal approximately to the average discharge rate of the microjet. Thus, characterization of solute movement in the soil is considerably simpler under the microjet than under the dripper. The dripper wets approximately 30%-50% of soil volume, depending on soil type, discharge rate and the system design. Microjets with varying discharge rates and irrigation diameters are available to wet *ca.* 20%-60% of the soil surface. Both methods of irrigation restrict the soil volume available for root exploration and therefore require maintenance of high soil water content by high frequency irrigation to compensate for the buffering capacity of a larger soil volume. The consequences of the high frequency irrigation in a restricted soil volume and maintenance of high water content is the development of a shallow root system, usually not deeper than 1 m, in a highly leached soil profile. The maintenance of a low soil matric potential under restricted soil volume irrigation results in the development of a high density fibrous root system in the vicinity of the water source. Root restriction causes reduction in root and top growth (Bar-Yosef *et al.,* 1988). Increased nitrogen supply can not compensate for the soil volume restriction, although root nitrogen uptake flux is increased (Bar-Yosef *et al.,* 1988). The extent of root restriction under standard drip or microjet irrigation systems, is not sufficient to restrict total canopy growth (Proebsting *et al.,* 1977), provided soil depth is *ca.* 1 m or more. Therefore, it can be assumed that functional root length, surface area and number of root tips are not reduced under these methods of irrigation. In contrast, total root mass may be reduced, since the requirement for structural roots to support water extraction from a large soil volume is not required under these conditions. The extent and consequences of such a possible reduction in total root mass, and its possible impact on root and tree storage capacity for nutrients and carbohydrates, has received little attention.

b. Efficiency of fertilization

The confinement of root growth to a small volume of the soil under drip or microjet irrigation systems makes it feasible to change soil fertility levels in the root zone. The efficiency of fertilization can be increased considerably, if fertilization practices are adapted to the irrigation system. Fertilization efficiency can be increased either by higher rates of dry matter production per unit of nutrient applied or by rapid correction of a nutrient deficiency.

The rapid development of a highly fibrous and dense root system under drip irrigation probably contributes simultaneously to both higher N and irrigation efficiency in young trees. Fertigation increases nitrogen use efficiency, as measured by reduced nitrogen inputs and increased tree growth in newly planted and young orchards, under restricted soil volume irrigation, compared with conventional fertilization-irrigation practices (Bussi *et al.,* 1991). Microjet, compared with drip irrigation of newly planted trees may be slightly less efficient, both in water consumption and N use, although with time, as root density increases it may catch up with the drip system (Nielsen *et al.,* 1995) and even surpass it in shallow and stony soils (Klein *et al.,* 1989b). Some additional benefits, such as overcoming replant problems and promoting earlier and greater flowering was noted when phosphate was included in the fertigation of newly planted fruit trees (Nielsen and Yorston 1990).

Nitrogen use efficiency, as measured by reduced N inputs and maintenance of steady growth, yield and leaf N concentration, was also shown to increase in mature orchards, (Kenworthy 1979, Smith *et al.,* 1979, Worley *et al.,* 1995). ^{15}N fertilizer recovery in mature vines was three times greater under drip irrigation than under furrow irrigation (Williams 1991). Nitrogen inputs, however, are not always scaled down under restricted soil volume irrigation, under the pretense that a reduced root volume requires compensation by maintenance of high soil N concentrations, or as a result of an apparent need of N under excessive irrigation which cause nitrate leaching. Thus, the capacity for increased nitrogen use efficiency under fertigation relative to conventional fertilization-irrigation practices may not be realized. Excess N may cause salting effects, particularly in microjet irrigation when water percolation and leaching is controlled. In drip irrigation salts are leached to the wetted front, allowing normal functioning of the root system in the leached zone of the soil. Salinity levels of 5-6 ds/m at a distance of 50 cm from the dripper had no adverse effect in the orchard (Klein *et al.*, 1989b)

In restricted soil volume irrigation, dissolution of the fertilizer into the irrigation water is either discontinuous, in small doses at frequent intervals by means of fertilizer tanks, or continuous ('proportional' fertigation) by means of a fertilizer pump that injects a steady concentration of a fertilizer solution into the water. With the exception of coarse textured or very shallow soils, proportional fertigation has not proven to be more advantageous than frequent discontinuous application, since readily mobile nutrients which are not leached during an individual irrigation cycle, redistribute by mass diffusion within soil aggregates, and leaching from within soil aggregates is not instantaneous (Cameron and Haynes 1986). Thus, weekly fertilizer applications and daily or biweekly irrigations (Klein and Spieler 1987) do not accentuate leaching compared with proportional fertigation in medium and heavy textured soils under controlled irrigation. In sandy soils, the discharge of NH_4^+ in the last quarter of an irrigation cycle retarded its leaching (Fuller and Moolman 1992).

Under sprinkler irrigation, an application of more than 200 g/tree of Fe-EDDHA chelate may be required to correct iron deficiency, depending upon the intensity of chlorosis, size of the tree and soil type (Samish and Hoffman 1976). In deciduous crops irrigated by sprinklers the usual practice was to treat only trees showing deficiency symptoms. In drip or microjet irrigation only 5 -15 g/tree of the chelate is required, and the accepted practice is to apply the chemical through the irrigation system to all the trees in the orchard. It can be calculated from these figures that in drip or microjet irrigation a saving of the chelate is realized when more than 5-10% of the trees suffer from iron deficiency, although it is applied to all trees indiscriminately.

The movement of potassium into the soil profile is accelerated when applied through the drip irrigation system (Klein and Spieler 1987, Klein *et al.*, 1989b, Klein 1992, Uriu *et al.*, 1980), due to K saturation of the clay lattices beneath the drippers and movement along macropores (Cameron and Haynes 1986). Potassium deficiency can be corrected more readily by the continuous application of relatively small annual doses under restricted soil volume irrigation system than by the conventional treatment of trenching large quantities of potassium into the soil once every several years. However, potassium saving under drip irrigation on a long term basis is unlikely, in view of the continuous applications required to maintain elevated K concentrations in the soil solution and the need to compensate for its loss by leaching. The efficiency of potassium fertilization is therefore expressed mostly through the speed and ease by which the deficiency can be alleviated rather than by the amount of K saved.

Measurements of root distribution (Levin *et al.*, 1979) and random excavations in mature orchards indicated that roots traverse the distance between rows, particularly in the deeper parts of the soil, under drip irrigated high density plantation (3-4 m alleys between rows). Under wider row spacing (6-7 m between rows, *i.e.* in almonds), the roots of adjacent rows do not meet, particularly if winter rains are not sufficient to wet the soil profile. Fertigation ensures that fertilizers are placed within the main root system, rather than wasted in the alleys, between rows. Winter rains which wet the soil profile encourage root development and water utilization from regions of the soil between rows which are not wetted by the irrigation system. Root proliferation and water extraction between rows which contributes to water economy, does not require maintenance of soil fertility between rows since nutrient supply to part of the root system is sufficient to correct deficiencies and maintain productivity. Potassium deficiency of apple could be readily corrected by enrichment of the soil to a distance of only 60 cm from the dripper (Klein 1992). Assuming that the root system occupied only 50% of the soil volume to a depth of 1 m and that the root density close to the dripper was twice the average density of the whole root system, it can be calculated that *ca.* 50% of the root system was exposed to elevated soil K. This exposure is probably high compared with values encountered when K trenching is practiced.

c. *Nutrient movement in the soil*

The movement of fertilizer nutrients under microjet and drip irrigation have been investigated in soil columns and in field experiments (Bar-Yosef and Sheikholsami 1976, Bresler 1975, Goldberg *et al.,* 1971, Goode *et al.,* 1978, Killingmo 1966, Klein and Spieler 1987, Klein *et al.,* 1989a,b, O'Neil *et al.,* 1979, Rinot *et al.,* 1971, Uriu *et al.,* 1980, Zohar 1971). Fertilizers which are not fixed in the soil (urea, iron chelates, nitrate and chloride) may leach and accumulate at the wetted front if not intercepted by the root system. Leaching of soluble fertilizers and chloride can cause salinity at the wetted front. This is especially so following excessive application of fertilizers, the use of low quality water, saline soils and irrigation regimes that do not alter the position of the wetted front. Nitrates and chlorides which are not taken up by the plant during the summer are leached out by winter rain or accumulate permanently at the wetted front under extremely dry climates. Because nitrate is readily leached, it is preferable to apply it in small doses using frequent, discontinuous or proportional fertigation. A greater proportion of residual fertilizer ^{15}N was found in the soil (subject to leaching), and less in the tree, when a high nitrogen rate was applied (Feigenbaum *et al.,* 1987). It is questionable if leaching can be eliminated altogether even under low N inputs (Dasberg *et al.,* 1984), and particularly under drip irrigation where the soil water content of the column under the dripper is above field capacity for extended periods. The control of wetting depth and nutrient leaching under field conditions is easier in microjet irrigation where the soil infiltration profile is one dimensional. The high leaching profiles in drip irrigation, encourage higher rates of fertilizer inputs. High fertigation rates are being used indiscriminately with microjet irrigation as well, frequently causing salinity damage. Nitrate use efficiency is a function of nitrate input in relation to its interception. Nitrate interception is dependent on root density, rate and extent of water percolation and the sink (plant and soil microbial) demand. When nitrogen input exceeds sink demand, nitrate concentrations increase at depth (Klein *et al.,* 1989a) and at the wetted front. In contrast, when nitrogen input is equal or is less than the demand, plant interception of nitrate is high and its concentration decreases with distance from the emitter (Klein and Spieler 1987, Klein *et al.,* 1989a). Nitrogen interception was found to be slightly greater along the tree row than perpendicular to the tree row, probably reflecting differences in root distribution and density (Klein and Spieler 1987).

Fixation and ion exchange reactions under the dripper restrict the soil mobility of certain nutrients (*i.e.* ammonium, potassium, phosphrus). Soil tests and leachate analyses from lysimeters in shallow stony soil with high calcium carbonate, have indicated calcium ion displacement by ammonium and potassium ions (Klein *et al.,* 1989b). Potassium, in contrast to ammonium, was also leached readily through the lysimeters. Ammonium and potassium, having similar charge and size of hydrated radii, compete in their fixation to clay particles (Cameron and Haynes 1986). When applied at equal rates, 3 times more NH_4^+ is fixed at the soil surface than K^+ (Cameron and Haynes 1986)· The ammonium ion penetrated to a depth 70 cm in a sandy soil (Fuller and Moolman 1992) but only to a depth of 20 cm in a silty loam soil (Haynes 1990). The ammonium ion is practically immobile in high pH soils having a high CEC and calcium carbonate and is fixed very close to the soil surface under the dripper. Leaching of ammonium ions through 25-70 cm deep lysimeters, packed with a native 'soil' composed of 40-80% stones and high calcium carbonate, could not be detected during 6 years of daily fertigation with up to 62 ppm NH_4NO_3 (Klein *et al.,* 1989b). Therefore, in soils with high pH and calcium carbonate content it is impossible to regulate the NH_4/NO_3 ratios at the root surface under orchard conditions, and uptake is mainly in the form of nitrate following nitrification. Consequently, many sand culture studies of ammonium/nitrate ratio have little relevance to orchard conditions, particularly in arid and warm climates where high pH soils prevail.

Contrary to many fixation studies carried out on disturbed soils in packed columns, field studies have documented movement of phosphrus and potassium to depths of 60-90 cm and 50 cm from the drippers (Klein and Spieler 1987, Goldberg *et al.,* 1971, O'Neil *et al.,* 1979, Rauschkolb *et al.,* 1976, Rolston *et al.,* 1979, Uriu *et al.,* 1980). Nutrient movement in soils with normal aggregate structure results from percolation of water (up to 50%) and accompanying nutrients through macropores (Cameron and Haynes 1986, Bar-Yosef 1992). The concentration of K and P in the soil solution increased 50% and ten to twenty fold, respectively, as a result of weekly fertigation (Klein and Spieler 1987). Concentrations of these ions declined to pre-fertigation levels within 1-2 weeks after fertigation ceased. Before ammonium and potassium are permanently fixed in the inner layers of 2:1 clay lattices, these ions are maintained in a more readily available form (Cameron and Haynes 1986). Ion exchange reactions after fixation in the soil are rate limiting steps (Bar-Yoseph 1992). Thus, there is evidence that potassium and phosphate deficiency (see section 4.2) can be corrected by relatively small doses, without actually saturating the soil CEC, if they are supplied continuously. In this way P and K are available to plants before their stable fixation (Klein and Spieler 1987). Adherence to this important principle necessitates that continuous fertigation be practiced rather than the single application typical of conventional fertilization programs. Accordingly, K fertigation should be continuous and probably even intensified when demand for K is greatest and exceeds potential supply (Weinbaum *et al.,* 1994b). The higher mobility and availability of K increases its leaching potential, therefore, an annual maintenance dose of K fertigation is recommended under drip irrigation.

The movement of trace elements in high pH soils is very limited. A cumulative dose of 1500 kg/ha zinc sulfate over a period of five years, increased the concentration of zinc under the dripper to a depth of only 15 cm (Klein unpublished data).

d. Effect of system design and irrigation rate and method on nutrient mobility

Increasing lateral water distribution from a point source decreases the leaching potential of nutrient not fixed in the soil, thereby increasing nutrient use efficiency. System designs which increase lateral water and nutrient distribution include higher dripper discharge rates (Bresler 1975, Haynes 1990), use of more than one lateral and optimal (adjusted to soil type) spacing of drippers along the lateral. Financial considerations play an important role in systems designs and layouts, and usually efficiency is compromised somewhat by installation of a single lateral and low discharge rate drippers (2-4 l/h instead of 8 l/h). Water percolation, and presumably nutrient leaching, also decreases as irrigation frequency increases (Levin *et al.,* 1979). However, in the long run, the benefits of an extremely high frequency irrigation are offset by the reduced rooting volume which decreases interception paths and increases the chances of water stress following technical failure of the irrigation system.

The choice of an irrigation method, in relation to soil type and rooting depth, have considerable importance when corrections of certain deficiencies are concerned. Potassium deficiency could be corrected readily by drip irrigation in soils with high clay content because of its high mobility and increased CEC saturation (Klein 1992, Uriu *et al.,* 1980). However the very same principle which makes drip fertigation of K so efficient in heavy deep soils, makes it less efficient than microjet fertigation in sandy or shallow and stony soils. Measurement of free flowing lysimeter leachates in a stony and shallow soil fertigated by low or high intensity N+K in drip and microjets (Fig. 1), indicated that the soil was completely saturated with K under the intensively-fertigated drippers within 1.5 years while partial soil saturation was delayed for 3 years under intensively-fertigated microjets (Klein *et al.,* 1989b). Nevertheless, a higher leaf K concentration could be maintained by the standard low level K microjet fertigation than by the high intensity K drip fertigation, suggesting that the level of K available to the plant and the efficiency of K recovery by the plant was higher under the microjet fertigation than under the drip fertigation in this very coarse textured soil.

Irrigation rate influences nutrient recovery by the tree. Under a standard dosage of 1200-1400 kg K/ha fertigated continuously with microjets over a 6 year period, low and medium rate of irrigation was associated with a 150% increase in avocado leaf K concentration but only a 50% increase when the rate of irrigation exceeded consumptive water use (Fig. 2)

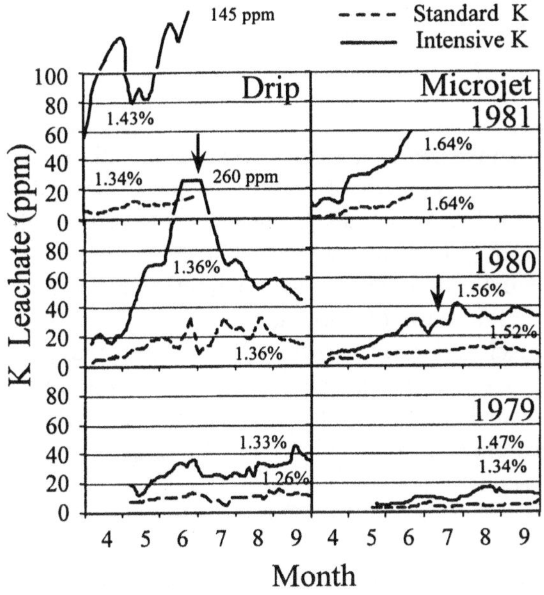

Figure 1. Leaching of K through shallow lysimeters (25-70 cm) packed with a stony soil (40-80% stones) and fertigated with standard and intensive levels of nitrogen and potassium (18 and 62 ppm N and 18 and 48 ppm K, respectively) under drip and microjet irrigation.

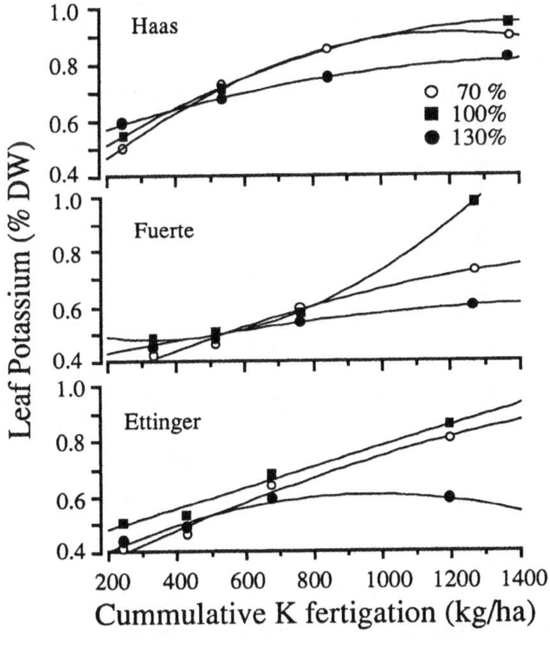

Figure 2. The effect of irrigation rate on potassium nutrition of avocado. Irrigation rates (70%, 100% and 130%) are based on class A open pan evaporation.

4 Macroelement nutrition

4.1 Nitrogen

Trees may be more dependent upon regular applications of fertilizer N in arid and warm climates than in temperate zones. Soil organic matter content is related inversely to mean annual soil temperatures (Stevenson, 1982), thus, the amount of potentially mineralizable N may be lower at high soil temperatures. Also, limited rainfall and the use of restricted soil volume irrigation systems may greatly reduce the soil volume available per tree for nutrient exploration.

When, how much and in what chemical form should nitrogen be applied? The answers to these questions are intimately related. Proper timing, *i.e.* when the sink demand is high, ensures better interception and less leaching. Consequently less N is required to achieve optimal yield and fruit quality.

a. Timing

Timing of nitrogen application should be optimized for maximizing uptake efficiency and minimize leaching, and for correcting transient or localized deficiency related to specific sink demands and processes, *i.e.* fruit set, root growth, etc.

Using isotopic N, Weinbaum *et al.*, (1978) showed that tree capacity for N uptake and distribution within young non-bearing prune trees varied seasonally in relation to sink demand. Uptake efficiency increased close to ten fold, once rapid shoot growth commenced in spring and declined, proportionally to leaf area, during leaf senescence and abscission in fall (Fig. 3).

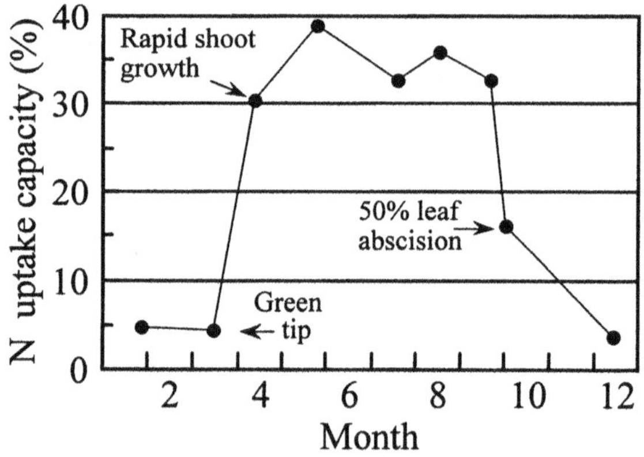

Figure 3. Nitrogen uptake capacity by prune. Redrawn after Weinbaum *et al.* (1978).

Nitrogen uptake remained fairly constant during the growing season, but its upward translocation and distribution was related to patterns of sink demand within the tree. Rapid shoot extension and leaf abscission of deciduous crops are easily distinguishable morphological markers to follow, for high N efficiency application. Interestingly, uptake efficiency of immature prune trees at the bud break stage, a few days prior to the start of rapid shoot growth was still low, indicating that pre-bud growth N fertilization was not absorbed efficiently. Furthermore, it has been shown that under non-N-deficient conditions winter or early spring fertilization does not contribute significant amounts of nitrogen to the initial flush of growth, flowering and fruit set. There is ample evidence indicating that early N demand by deciduous and evergreen fruit trees is met from internal N pools, stored within the tree (Titus and Kang 1982, Millard 1995). Apricot flowers of mature trees acquired up to 16% of their N from the previous fall nitrogen application compared to <1% from winter application (Weinbaum *et al.,* 1980). Similar results were found in grapes (Conradie 1992). Traces of isotopically labeled nitrogen uptake in almond could be detected within one week after application in the spring but the bulk of N used for initial growth and flowering originated from internal pools enriched during the previous year (Weinbaum *et al.,* 1984). Therefore, under sound fertilization practices, it is not necessary and it is rather impossible to 'prepare' the tree for a better start in spring by pre-bud break N application. Nitrogen storage and re-cycling is governed by the same principles in evergreens and deciduous tree species, with the added advantage of having the leaf as an over-wintering alternate N storage pool.

A pre-budding spring application of N may be advisable when N must be positioned within a deep root system, as may be the case under sprinkle or flood irrigation. Under microjet or drip irrigation, the restricted root system can be enriched with N at will and nitrogen application can be postponed until its uptake is highly efficient. A so called 'technical' fertigation of N applied close to bud break, for the purpose of fertilization rather than irrigation, delivering 10-20 kg/ha N in 4-8 mm irrigation, can effectively charge the root system with nitrogen following its leaching by winter rains. Late winter - early spring application of nitrogen should also be practiced when an early deficiency is diagnosed, although efficiency of uptake may be still low. Early diagnosis and correction of N deficiency under low uptake efficiency is feasible, as an example, in high elevations of arid and dry climates, when low chilling requiring cultivars of peach foliate but fail to grow until favorable temperatures set in.

Isotope labeling and nutrient analyses of whole trees, excavated sequentially, were used to determine temporal patterns of nutrient uptake in mature trees (Weinbaum *et al.,* 1994b, Rosecrance 1996). These studies suggest that nutrient uptake does not occur uniformly over the season, and the seasonal periodicity of uptake is conditioned by the kinetics of sink demand for nutrients. Three distinct sinks could be identified in woody species: vegetative (including shoot, leaf, root and cambial growth), reproductive, and N storage in perennial tree parts. When the rate of soil uptake lags behind demand, the demand is satisfied by remobilization from the perennial storage pool (*i.e.* during early season growth, Titus and Kang 1982), or by resorption from transitory pools, such as fruit pericarp (Weinbaum and Muraoka 1986) and leaves (Weinbaum *et al.,* 1994b, Rosecrance 1996). Remobilization of N is associated with net phloem (Titus and Kang 1982) and leaf (Vessey *et al.,* 1990) protein degradation and the transport of phloem mobile amino acids to the vegetative and reproductive sinks (Weinbaum *et al.,* 1994a). According to a model proposed by Cooper and Clarkson (1989), there is a linkage between tree demand for phloem-mobile macroelements and tree capacity for uptake of these nutrients.

Phloem-mobile macroelements cycle throughout the tree in the xylem and phloem saps (Weinbaum *et al.,* 1994a). A pulse of labeled N may circulate in the tree for years (Weinbaum *et al.,* 1987). During periods of high N demand, phloem mobile amino acids are removed from vascular circulation, and Cooper and Clarkson (1989) propose that this somehow signals the plant to resume uptake.

Sequential tree excavations and measurements of N uptake in mature pistachio trees at the end of the spring growth flush, the nut fill period and the post harvest period showed that despite differences in crop load and fruit N removal in alternate bearing trees, the annual N usage did not vary between heavily cropping 'on' and lightly cropping 'off' pistachio trees (Table 2).

Table 2. Nitrogen availablility (kg N/ha) from tree storage and from soil uptake and its partition into reproductive and storage sinks in 'on' and 'off' years in 'Kerman' pistachio trees[1]

		Uptake from Soil[2]						
			Nut fill					
Cropping Status	Available from storage[3]	Spring flush	Fruit	Storage	Post Harvest storage	Total	Resorption from leaf to storage	Total perennial storage
On	112	50	93	0.3	1.0	145	13.7	15
Off	15	65	9	69.4	0	144	42.6	112

[1] Data from Rosecrance 1996.

[2] Tree N storage pool size and soil N uptake was determined by difference in tree N content following sequential tree excavations on 24 May, 8 Sept. and 13 Dec. Consequently, soil uptake periods include 10 Dec.- 24 May, 24 May - 8 Sept. and 8 Sept.- 13 Dec., for spring flush, nut fill and post harvest storage, respectively.

[3] N available to 'on' year trees accumulated the previous, *i.e..* 'off' year and vice versa (see total perennial storage column at right side).

145 kg N/ha was taken up annually, both by 'on' and 'off' year pistachio trees, at identical timings. The N taken up, however, was diverted to either the reproductive sink in the 'on' tree or the permanent storage pool in the 'off' tree. Thus, despite the extreme bienniality of the pistachio, fertilizer application need not be adjusted to fruit load on an annual basis but rather on a perennial basis. Table 2 also demonstrates that in pistachio the major uptake period, including the recharge of the perennial storage pool in 'off' years, was the nut filling period (June-September).

Timing of N application is important for correction of transient or localized deficiency within the tree. It has been shown under temperate climates that autumn foliar application of urea to apple improved fruit set more than a spring soil application (Oland 1963). Recent experiments indicate that supplemental foliar application of urea and a limited amount of soil-applied N may be preferable to complete dependence on soil N application to minimize nitrate leaching, improve fruit color and control vigor of semi-dwarfing rootstocks used in warm climates (Klein, unpublished data). Supplemental Fall or Spring, but not Summer, foliar urea sprays could maintain productivity of N deficient trees, indicating that a specific N demand for fruit set could be met, although the tree as a whole was N-deficient. Nitrogen absorbed by the spur, from Fall or Spring applications was presumably available for fruit set while N absorbed in Summer was translocated throughout the tree and was not available during fruit set. In avocado, a mineral deficiency restricting root growth and reducing yield was diagnosed by application of an N-P fertilizer in the fall, to trees which were diagnosed by foliar analysis as non-deficient. Taking into account the shallow rooting of avocado, the use of restricted soil volume irrigation with high rates of N leaching and the prevailing tendency of N for limited basipatal movement (Klein *et al.,* 1984), a localized N and P deficiency in the root system can develop, if provisions are not made for fertilizer application at times when root growth is expected. Root growth of fruit trees is asynchronous with top growth and in general is thought to precede shoot growth in the spring, followed in many cases by a second wave of growth in autumn.

There is no evidence that crops other than avocado suffer from localized root N or P deficiency. Nevertheless, 'Fall' nitrogen application to many tree crops is a sound practice, not just to prevent a localized N deficiency of the root system under restricted soil volume irrigation characterized by high leaching rates, but also for the buildup of N reserves needed for early spring root and top growth. The buildup of N reserves within the tree permanent structure has the lowest sink priority, when compared to vegetative and reproductive sinks, and its timing is probably adjusted ecologically to the annual developmental cycle and physiology of each tree species and cultivar. The buildup in deciduous crops takes place normally after harvest and growth cessation, whether it is in mid-summer in 'off' pistachio trees (Rosecrance 1996), late-Summer in pears (Sanchez *et al.,* 1991) or Autumn in apple (Mason and Whitfield 1960). The buildup in grapes was found to be environmentally affected, occurring after harvest in a warm climate and before harvest in a temperate climate, where maturity is delayed (see discussion by Conradie 1993, and references therein).

Numerous references have been made in the literature to the adverse effect of summer N on fruit 'quality'. This term has been applied to an array of fruit attributes, including fruit size, color, texture, chemical composition, physiological disorders and storage properties. Unfortunately, none of the available investigations delineated clearly between effects of N timing and N dose. This should not be surprising since fruit trees store nitrogen, which renders many timing investigations of N in mature trees inconclusive. We have demonstrated that N available from within tree storage may contribute to poor fruit color and excessive tree vigor in 'Starking Delicuous' apple, although soil N was absent during a 10 week period of fruit enlargement, prior to harvest. Thus, adverse effects of excess N on fruit quality may occur independent of summer N application.

b. Dose

Nitrogen requirement of fruit trees has been estimated by measurement of nitrogen removed from the tree, from dose response experiments and by measuring annual atom % depletion (Weinbaum *et al.,* 1987) and total tree N content. Measurement of nutrient usage by whole tree excavations (Williams 1991, Weinbaum *et al.,* 1994, Rosecrance 1996) is difficult because of large tree size and tree variability. Estimation of N demand by N removed in fruit and prunings underestimate actual tree uptake because it does not take into account losses by flower and fruit abortion, leaf litter and root turnover, and the yearly increment into tree framework. Furthermore, to be accurate, N removal data should be based on long term measurements of high yielding orchards fertilized with optimal levels of available nitrogen, since N removed in fruit and prunings may increase considerably at high N application rates. More fruits as well as higher concentrations in the fruit contribute to a greater N removal. The N concentration of almond embryos increased from *ca.* 3% to 4% with increasing N fertilization rates (Uriu, personal communication). Fruit N concentration is sometimes more sensitive and contributes more to total N removal than fruit number. In peaches, fruit N concentration increased 175% while fruit number only 122% when a nitrogen deficiency was corrected (Table 3).

Table 3. Fertilization increases leaf and fruit N concentrations, yield and total N removal in peach[1]

Rate of N Fertilization	*Leaf N (% DW)*	*Fruit N (% DW)*	*Yield DW (kg/ha)*	*Total crop N removed (kg/ha)*
Zero	2.52	0.40	5616	22.5
Half dose	3.15	0.69	6875	47.4
Full dose	3.25	0.71	6866	48.5
		% increase (relative to zero treatment)		
Half dose	25	72	22	110
Full dose	29	76	22	115

[1] Data from Saenz, J.L. (1996) M.S. dissertation in preparation. University of California, Davis.

Estimation of N requirement from field trials requires measurement of all nitrogen transformations (changes in organic matter content of the soil, and losses from volatilization, denitrification and leaching) and utilization by tree and competing weeds, to carry out an accurate balance. An accurate balance is almost impossible to carry out and it has rarely been attempted. When such attempts were made, large quantities of fertilizer N could not be accounted for (Dasberg *et al.*, 1984). Therefore, dose response experiments are site specific and represent tree response rather than the net nutrient demand of trees.

Cost limits the use of isotopically labeled N in field trials with mature trees. Dilution of the isotope in the soil and the tree complicate interpretation of ^{15}N uptake results. In addition, interpretation of ^{15}N dilution within the tree, following pulse labeling, requires knowledge of the total N pool. In almond, a high nitrogen consuming fruit, a 50% annual dilution of ^{15}N was measured (Weinbaum *et al.*, 1987) while in grape, a low N consumer, the corresponding value was 90% (Conradie 1983). Measurements in grape indicated a 70% annual total N removal by crop and pruning (Conradie 1980). These values indicate that a very high turnover rate of a small N pool, *i.e.* in the grape, can lead to erroneous conclusions of high demand.

Nevertheless, in spite of these shortcomings, an attempt to estimate 'net N demand' of a crop has an important implications for estimating how efficiently fertilizer N is being used in the orchard. 'Net N demand' can be defined as the amount of nitrogen utilized by the tree itself, disregarding soil nitrogen losses, organic matter changes, weed utilization, etc. This value, if it could be measured accurately, is the true reference value for estimation of nitrogen use efficiency. It should be kept in mind that in contrast to the limited number of experiments that can be carried out, there are large numbers of orchards throughout the fruit industry with specific case histories of fertilizer use. Any measured value of 'net N demand' which we come up with, is being constantly validated by the industry since the occurrence of sustainably high yielding orchards planted in soils with low organic matter content and using less nitrogen than our reference value, may indicate that our estimate was wrong.

The apparent nitrogen requirement to maintain high yield is greater under sod cultivation and conditions favoring nitrogen losses by leaching or denitrification. Changing from sod to herbicide strip, and eventually to complete herbicide cultivation, reduced nitrogen requirement of apples in the Netherlands from 400 kg/ha to 50 kg/ha (Kipp 1992). In dry and arid climates, particularly where water is scarce, orchards are clean cultivated and therefore does not require additional nitrogen to compensate for weed growth. In drip and microjet irrigation technologies only 30-50% of soil volume is irrigated and maintained at a high available soil water content. Consequently, nitrate leaching and/or denitrification may be considerable and an apparent high N demand can be demonstrated if irrigation is excessive and proper fertigation practices are not maintained.

The apparent nitrogen requirement may be lower where nitrate-polluted canal or ground water is used for irrigation and the concentration of N in the water is neglected. A concentration of 10 ppm nitrate nitrogen in the irrigation water contributes 30-80 kg/ha N when the annual irrigation is 300-800 mm. Nitrogen requirement may be underestimated also from short term field trials, since productivity and growth can be maintained for limited periods by N derived from pools in the soil and the tree. It is frequently necessary to withhold nitrogen for 2-3 years, before yield and/or growth is reduced in the orchard. The soil nitrogen pool size is dependent on its clay and organic matter content and its micro biological activity affecting rate of organic matter transformations. Following the fate of ^{15}N depleted nitrogen in mature almond trees, it has been found that the maximal enrichment of leaves, flowers and spurs was maintained for one and two consecutive years, in light and heavy soils, respectively, indicating longer persistence of applied N in soils with higher clay content (Weinbaum *et al.*, 1984a, Weinbaum *et al.*, 1987). Extrapolated depletion functions from 5 consecutive years of measurement showed that traces of an ^{15}N depleted nitrogen pulse persisted in the soil-tree system ten times longer in a heavy soil as compared to a light soil (Weinbaum *et al.*, 1987). Soil derived N can contribute significant quantities of nitrogen, which is commonly neglected in field trials. The annual contribution of soil derived N to apple nutrition in a tuffic soil was estimated to be 100 kg/ha, an amount which exceeded the measured removal rate (Klein *et al.*, 1989a).

Young trees are dependent mainly on external supply of nitrogen for growth (Taylor and May 1967). For early production and rapid return on investment, the common practice, under warm climates, is to maximize canopy development by water and nutrient input. The low root density of young trees and high rates of nitrate mobility and leaching in the soil require the use of relatively high water and N doses to maintain maximal vegetative growth until the tree canopy is fully developed. In shallow and stony soils, an increase of N concentration in the daily fertigation from 18 ppm to 62 ppm, increased vegetative growth of apples and stone fruits (peach nectarine and prunes) by 12% and 35%, respectively (Klein *et al.*, 1989b). Because of low root density and poor nitrogen interception, an expression of nutrient concentration, rather than total annual dose may be advantageous when dealing with young trees immediately after plantation. Gradually, as root density, N interception, and the storage capacity within the tree permanent structure increase, N application based on solution concentration becomes irrelevant and often leads to N overfertilization. Rather, N inputs should equal demand to improve N use efficiency and minimize nitrate leaching and ground water pollution in the mature orchard (Dasberg *et al.*, 1984, Klein *et al.*, 1989b).

Nitrogen removal by fruit and prevailing input values in various berries and tree crops is summarized in Table 4. These values are presented to emphasize relative crop use as compared to prevailing fertilization practices, rather than a true 'net N demand' value. Values in Table 4 represent average N analyses of many fruit samples, average and above average yields and a wide range of nitrogen inputs.

Table 4. Nitrogen removal in the harvested crop and prevailing range of N application rates

Classification	*Crop*	*DW*[1] (%)	*N FW*[1] (%)	*Yield* (t/ha)	*N removed by fruit* (kg/ ha)	*Fertilizer N input* (kg/ha)
Berries	Blueberry	15.4	0.1072	10- 18	11-24	3-113
	Raspberries	31.4	0.1456	4- 12	6-18	28-112
	Blackberry	14.4	0.1152	10- 17	12-20	56-112
	Strawberry	8.4	0.0976	50-100	49-98	100-200
	Table Grape	11.2	0.0800	30- 50	24-40	50-150
	Wine Grape	18.0	0.1300	10- 20	13-26	50-100
Pome fruits						50 - 150
	Apple	16.1	0.0304	50-100	15-30	
	Pears	16.2	0.0624	40- 70	25-44	
Stone Fruits						150 - 200
	Apricot	13.7	0.2240	20- 50	45-112	
	Cherries	19.2	0.1920	5- 20	10- 38	
	Nectarine	13.7	0.1504	20- 60	30- 90	
	Peach	12.3	0.1504	30- 60	45- 90	
	Plum	14.8	0.1264	20- 50	25- 63	
	Prune	96.0	0.5920	4- 8	14- 28	
Nuts						200 - 250
	Almond	95	2.757	2.4-4.0	88-146	
	walnut	92	1.669	4.5-6.7	89-134	
	Pistachio	92	2.397	3.4-5.6	97-160	
	Pecan		0.917	3.0-5.0	37- 55	56 - 112

[1] With the exceptions of wine grapes, apples, peaches and the nuts, FW values are based on data of raw fruit analysis, from Gelbhart *et al.* (1982). Composition of foods. Agricultural Handbook No. 8-9, USDA, Human Nutrition Service. Nut % N in FW refer to In Shell values. N removed by fruit include the hull (data from analysis by S.A.Weinbaum).

Among the small berry fruits, the strawberry is the greatest consumer of N, on account of its heavy yield. High yielding grapes, pome fruits and the cherry remove approximately 30-45 kg N/ha. The remaining stone fruits remove approximately 60-112 kg/ha. With the exception of the pecan, which has a relatively low fruit N content, the dry nuts remove approximately 134-160 kg N/ha. Consumption in Table 4.4 does not take into account N removal by pruning, which can account in some of the berry crops to a significant amount.

The prevailing nitrogen input practices, although reflecting to some extent the overall consumption by fruit removal, vary considerably among crops and within each crop. The variation probably reflects a variable N use efficiency rather than an actual adjustment of dose to N consumption. Table 4 also demonstrates that N use efficiency is generally greater in crops which consume more nitrogen. In small berries, N input is up to five times the estimated quantity removed by the fruit, while in the strawberry and the dry nuts it is only *ca.* twice the removal rate.

4.2 Phosphate

It has been known for many years that fruit trees do not respond to phosphate application. The lack of response has been attributed to soil fixation and the resultant inability to raise P level at the root surface. This explanation is not valid under drip irrigation since the P concentration in the soil solution can be raised effectively around a significant portion of the root system (Klein and Spieler 1987, O'Neil *et al.*, 1979, Rauschkolb *et al.*, 1976). Tree responses to P application under drip irrigation have been reported recently in young and mature trees (Nielsen and Yorston 1990). In Israel, the phosphate level of mature apple trees could not be increased (Klein and Spieler 1987), while that of grape (Bravdo and Hepner 1987) and plums (Klein, unpublished) were raised effectively by drip fertigation. Thus, phosphate availability is no longer a rate limiting step and the need for its application can now be evaluated with greater confidence. Correction of P deficiency can be affected by soil application of soluble phosphate fertilizers, such as monoammonium phosphate and phosphoric acid, which are compatible with fertigation (Bar-Yoseph 1992). It has been claimed that soil P is depleted under drip irrigation. Close monitoring of petiole or leaf P concentration of grapes and fruit trees in Israel have not substantiated those claims so far, perhaps because crop removal of P, unlike that of N and K, is low.

4.3 Potassium

Large quantities of potassium are removed in fleshy fruits, resulting in low leaf K concentration during years of heavy fruit load. The demand for daily K uptake is greatest in certain crops (*i.e.* prunus, grape) at the final stage of fruit development, when uptake frequently does not keep up with demand (Weinbaum *et al.*, 1994). Since K is consumed mainly by the fruit, leaf analysis in newly planted orchards can not predict a potential deficiency in later years (Klein 1992). In grapes, petiole analysis performed at flowering does not reflect K status accurately, because the demand for K develops subsequently, *i.e.* between veraison and harvest (Hepner and Bravdo 1985). Fruit K responds readily to potassium supply, with measurable effects on composition, taste and storage properties related to K/Ca ratios of the fruit. Release of potassium from clay minerals and low ion exchange rates are the limiting steps in K supply (Bar-Yoseph 1992). Potassium deficiency can be widespread in various crops and soil types, and K application is probably second in importance only to nitrogen. Under restricted soil volume irrigation systems potassium mobility and leaching are accelerated (Klein and Spieler 1987, Uriu *et al.*, 1980). Unless an exceptionally high K is demonstrated by soil or leaf analysis, an annual maintenance dose of 200-300 kg/ha K is recommended for drip and microjet irrigation. Two to three times higher rates are recommended under K deficiency, depending upon severity and soil clay characteristics. Continuous fertigation enables the correction of the deficiency within the first year of application and is the method of choice.

4.4 Calcium

Calcium deficiency has been described as a deficiency of reproductive organs, because of its low phloem mobility. Translocation of Ca to the fruit is severely diminished after completion of cell division, as fruit supply of both water and nutrients gradually becomes phloem dependent. The fact that in arid and warm climates Ca is the predominant ion in the soil solution, and fruit trees are frequently grown in soils with very high calcium carbonate content, does not help to eliminate fruit-related problems of Ca nutrition. Nevertheless, calcium-related problems are probably less severe under dry and warm climates, compared with temperate zones, since high Ca in the soil solution antagonizes K uptake, and fruit K/Ca may be lower. While correcting K deficiency in a heavy soil, it was found that fertigating a moderate dose of potassium (600 kg/ha K) was sufficient to raise leaf K of 3 apple cultivars from 0.8% to 1.3% (Klein 1992). A higher dose (900 kg/ha) was required to improve yield significantly but leaf K was not increased above 1.3%. The additional K applied was consumed by more fruits, rather than enriching the leaves, raising also fruit K concentration and the K/Ca ratio. The change in Ca concentration and the K/Ca ratio were not sufficiently high to cause fruit calcium related disorders. Calcium-related physiological disorders and nutrition have been reviewed previously (Faust 1989).

4.5 Magnesium

Magnesium deficiency can be found in crops which are low Mg accumulators, *i.e.* the apple. Magnesium concentration in the apple leaf rarely exceeds 0.5% in DW, in contrast to much higher levels in stone fruits (*ca.* 1%), and the grape blade (*ca.* 1.5%) or petiole (*ca.* 2%). Magnesium, unlike Ca, is considered to be phloem mobile and will not accumulate to high levels as the leaf ages when its supply is limited. In the apple, 0.30% Mg in leaf DW is usually considered to be the critical level. Magnesium deficiency can be readily corrected by 2-3 foliar applications of 1% magnesium nitrate in the Spring. In grapes a cluster stem necrosis disorder which was initially attributed to low Mg, has been recently associated with high N and NH_4^+ levels (Christensen *et al.*, 1991).

5 Microelement nutrition

5.1 Iron

Iron chlorosis is the most common microelement deficiency, affecting all fruit tree crops. High soil pH, lime, water logging and certain rootstocks with sparse (Myroblan) or a nutritionally inefficient root system (Nemaguard) intensify iron chlorosis. The well known lime-induced chlorosis is probably associated with lack of aeration, since high lime concentrations in aerated stony soils do not intensify chlorosis. Leaf analysis is not a useful tool in the estimation of iron deficiency since bulk leaf iron concentration is only slightly reduced in chlorotic leaves. The autonomous nature of individual scaffolds and limbs of the tree, feeding sectorially from certain parts of the root system, frequently cause chlorotic patches within tree canopies which necessitates visual estimation when correcting iron deficiency. Fortunately, correction can be easily and accurately evaluated visually.

Iron is phloem immobile and its deficiency is always evident in the new leaves near the shoot tip. In warm and dry climates young trees which continue to grow 180-200 days every year, require continuous application of iron chelates, as long as a deficiency of iron is evident. In mature trees which frequently cease growth 6-10 weeks after bud break, correction measures are required as long as shoot growth continues. Iron is taken up at the root tip when the soil is relatively warm (*ca.* 12-15°C or above, Mengel 1987). In heavy, wet soils and under fluctuating air temperatures which often cause early bud break in dry and warm climates, canopy development starts before soil temperature is sufficiently high for root growth and iron uptake. Under these conditions the response to application of iron chelates in the Spring may be erratic and sluggish. When needed, Fall application of iron chelates can effectively eliminate iron chlorosis in the following Spring (Klein and Bareski 1974).

Iron deficiency is corrected by soil application of iron chelates. The effectiveness of a chelate is measured by the quantity required to correct the chlorosis and the speed of its action. The FeEDTA chelate is completely ineffective in high pH soils. Using fertigation, iron chlorosis can be corrected by small doses of FeEDDHA within 7-10 days. It is, however, difficult to 'guess' the optimal dose required since the severity of chlorosis (if not corrected) can vary with time, and the exact dose required is dependent upon changes in severity after the application (Klein and Bareski 1974). The evolution of chlorosis severity depends on undefined soil conditions (including overirrigation or lack of aeration), and rate of vegetative growth. A prescribed dose may be insufficient if soil conditions worsen or during a sudden surge of vegetative growth, and excessive if soil conditions improve. Multiple applications of small doses of highly effective chelates are usually the best strategy, both agronomically and economically, to overcome this problem.

5.2 Zinc

Symptoms of zinc deficiency may be sporadic within the tree and the orchard, and it is sometime difficult to determine the severity and extent of the deficiency. Deficiencies in apples (Hoffman and Samish 1966), peach and nut crops are often sufficiently severe to warrant prophylactic correction by Spring or Fall foliar sprays. A variety of zinc formulations have been used, in most cases with questionable success. Low biuret urea is used for improving penetration into apple leaves (Hoffman and Samish 1966). The uptake and translocation of Zn are major limiting steps. Zn analysis of tissue samples following foliar application is complicated by the fact that it is adsorbed irreversibly to the wax in the leaf cuticle.

5.3 Boron

Boron is required for cell division, and is involved in fruit set and fruit quality. Boron deficiencies have been identified in stone fruits and nut crops. The failure to recognize by foliar analysis a B deficiency associated with the reproductive process (Callan *et al.*, 1978), raises questions about the adequacy of present boron diagnostic indices and the status of B deficiencies in fruit trees. Boron uptake by the leaf is rapid (Pichioni *et al.*, 1995). Resorption of foliar applied B in species translocating sorbitol (*i.e.* pome fruits, stone fruits and almond) or mannitol (olive) via the phloem is quantitative and can be completed within 2-4 weeks (Brown and Hu 1996). In contrast, phloem mobility of B in sucrose-translocating species is limited. In these species B accumulates in the leaves when B toxicity occurs and after foliar application. Therefore, appropriate strategies to correct B deficiency by foliar application are crop specific. Species translocating sucrose, require a green tip application of 500 ppm B spray, while in sorbitol or mannitol translocating species either a green tip or fall application can be used. In arid and warm climates boron toxicity is not uncommon, and frequently occurs in association with elevated Na in the deeper soil layers.

5.4 Manganese

Manganese is phloem mobile and deficiency symptoms of it are noticeable in the older leaves of the shoot. Iron and manganese are antagonistic ions and their contents in leaves are negatively correlated over a wide range of concentrations. Manganese deficiency in fruit trees often results from an overdose of iron chelate. The consequences of manganese deficiency on vegetative and reproductive processes in fruit trees are not known, probably because severe deficiencies are rare. Manganese toxicity in low pH soils has a considerably greater impact on fruit tree culture than deficiencies in high pH soils.

5.5 Copper

Copper deficiencies of fruit trees are rare, although typical concentrations of copper in tissues are very low. Sprays of certain insecticides and pesticides which contain copper may significantly augment the leaf concentration. With the exception of the 'monkey face' symptoms in the fruit, Cu deficiency resembles B deficiency. Symptoms of copper deficiency, including severe fruit drop, shoot die back with 'witches broom' proliferation growth, and severe bark cracking have been found in the Golan Height in apples grown in a tuffic soil (Klein, unpublished data).

6 References

Aldrich, T.M., Carlson, R.M., Catlin, P.B., Fitch, L.B., Henderson, D.W., Ramos, D.E., Rauschkolb, R.S., Schulbach, H., Sibbet, G.S. and Uriu, K. (1978). *Irrigation fertilization and soil management of prune orchards. Leaflet 21016.* Division of Agri. Sci. University of California.

Bar-Yosef, B. and Sheikholshami, M. R. (1976). Distribution of water and ions in soils irrigated and fertilized from a trickle source. *Soil. Sci. Soc. Amer. J.* 40, 575-82.

Bar-Yosef, B., Schwartz, S., Markovitch, T., Lucas, B. and Assaf, R. (1988). Effect of root volume and nitrate solution concentration on growth, fruit yield and temporal N and water uptake rates by apple trees. *Plant and Soil* 107, 49-56.

Bar-Yosef, B. (1992). Fertilization under drip irrigation. *Fert. Sci. Technol. Ser.* Marcel Dekker. New York, N.Y.

Bravdo, B. and Hepner, Y. (1987). Irrigation management and fertigation to optimize grape composition and vine performance. *Acta Hortic.* 206, 49-67.

Bresler, E. (1975). Two-dimensional transport of solutes during nonsteady infiltration from a trickle source. *Soil Sci. Soc. Amer. Proc.* 39, 604-13.

Brown, P.H. and Hu, H. (1996). Phloem mobility of boron is species dependent: Evidence for phloem mobility in Sorbitol-rich species *Annal. Bot.* 77, 497-505.

Brown, P.H., Weinbaum, S.A. and Picchioni, G.A. (1994). Alternate bearing influences annual nutrient consumption and the total nutrient content of mature pistachio trees. *Trees.* 9, 158-64.

Bussi, C., Huguet, J.G. and Defrance, H. (1991). fertilization scheduling in peach orchards under trickle irrigation. *J. Hort. Sci.* 66, 487-93.

Callan, N.W., Thompson, M.W., Chaplin, M.H., Stebbins, R.L. and Westwood, M.N. (1978). Fruit set in 'Italian' prune following fall foliar and spring boron sprays. *J. Amer. Soc. Hort. Sci.* 103, 253-7.

Cameron, K.C. and Haynes, R.C. (1986). Retention and movement of nitrogen in soils. (ed R.J. Haynes) In: *Mineral nitrogen in the plant-soil system., Physiological ecology series* (ed T.T. Kozlowski), Academic Press, London. pp. 166-220.

Christensen, P., Boggero, J. and Adams, D.O. (1991). The relationship of nitrogen and other nutritional elements to the bunch stem necrosis disorder 'waterberry'. *International symposium on nitrogen in grapes and wine.* The Amer. Soc. Enol. and Viticulture. (ed JoAnne M. Rantz), Seatle Washington. pp. 108-9.

Conradie, W. (1980). Seasonal uptake of nutrients by Chenin Blanc in sand culture: I. Nitrogen. *S. Afr. J. Enol. Vitic.* 1, 59-65.

Conradie, W. (1983). The uptake and distribution of ^{15}N enriched nitrate by three rootstock cultivars grafted on Chenin Blanc. *S. Afr. J. Enol. Vitic.* 4, 45-7.

Conradie, W.J. (1991). Translocation and storage of nitrogen by grapevines as affected by time of application. *International symposium on nitrogen in grapes and wine.* Amer. Soc. Enol. and Viticulture. (ed JoAnne M. Rantz), Seatle Washington. pp. 32-42.

Conradie, W.J. (1992). Partitioning of nitrogen in grapevines during autumn and the utilization of nitrogen reserves during the following growing season. *S. Afr. J. Enol. Vitic.* 13, 45-51.

Cooper, H.D. and Clarkson, D.T. (1989). Cycling of amino-nitrogen and other nutrients between shoots and roots in cereals - a possible mechanism integrating shoot and root regulation of nutrient uptake.

J. Exp. Bot. 40, 756-62.

Dasberg, S., Erner, Y. and Bielorai, H. (1984). Nitrogen balance in a citrus orchard. *J. Environ. Qual.* 13, 353-6

Faust, M. (1989). *Physiology of temperate zone fruit trees.* John Wiley and sons, New York

Feigenbaum, S., Bielorai, H., Erner, Y. and Dasberg, S. (1987). The fate of ^{15}N labeled nitrogen applied to mature citrus trees. *Plant and Soil.* 97, 179-87.

Fuller, K.D. and Moolman, J.H. (1992). Fertigation: effect of timing on the redistribution of applied ammonium nitrate in a sandy orchard soil. *Deciduous Fruit grower* 42 (10), 363-7.

Gebhart, S. E., Cuttrufelli, R. and Matthews, R. (1982). Composition of foods. *Agricultural Handbook. No.* 8-9, USDA Human Nutrition Service.

Goldberg, D., Gornat, B. and Bar, Y. (1971). The distribution of roots, water and minerals as a result of trickle irrigation. *J. Amer. Soc. Hort. Sci.* 96, 645-8.

Goode, J.E., Higgs, K.H. and Hyrycz, K.J. (1978). Trickle irrigation of apple manuring. *J. Hort. Sci.* 53, 307-16.

Haynes, R.J. (1990). Movement and transformation of fertigated nitrogen below trickle emitters and their effects on pH in the wetted soil volume. *Fert. Res.* 23, 105-12.

Hepner, Y. and Bravdo, B. (1985). Effect of crop level and drip irrigation scheduling on the potassium status of Cabernet Sauvignon and Crignane vines and its influence on must and wine composition and quality. *Am. J. Enol. Vitic.* 36, 140-7.

Hoffman, M. and Samish R.M. (1966). The control of zinc deficiency in apple. *Israel J. Agr. Res.* 16, 105-114.

Kenworthy, A.L. (1979). Applying nitrogen to fruit trees through trickle irrigation systems. *Act. Hortic.* 89, 106-8.

Kilingmo, O.H. (1966). Studies on plant nutrition on soils. 1. Distribution of some nutritious elements in a soil fertilized by a drip irrigation equipment. *Acta Ag. Scand.* 16, 155-62.

Kipp, J.A. (1992). Thirty years fertilization and irrigation in Dutch apple orchards: a review. *Fertil. Res.* 32, 149-56.

Klein, I. and Lavee, S. (1977). The effect of nitrogen and potassium fertilizers on olive production. In: *Fertilizer Use and Production of Carbohydrates and Lipids.* Proc. 13th colloquium of International Potash Institute. pp. 295-304.

Klein, I. and Bareski, A. (1974). Correction of iron deficiency in deciduous fruits - second year experiments using Kelapher. *Alon Hanotea* 29, 71-4. (In Hebrew).

Klein, I. and Weinbaum, S.A. (1984). Foliar application of urea to olive. Translocation of urea nitrogen as influenced by sink demand and nitrogen deficiency. *J. Amer. Soc. Hort. Sci.* 109, 356-60.

Klein, I. and Weinbaum, S.A. (1985). Foliar application of urea to almond and olive: Leaf retention and kinetics of uptake. *J. Plant Nutr.* 8, 117-29.

Klein, I. and Spieler, G. (1987). Fertigation of apples with nitrate or ammonium nitrogen under drip irrigation II. Nutrient distribution in the soil. *Comm. Soil Sci. Plant Anal.* 18, 323-39.

Klein, I., Levin, I., Bar-Yosef, B., Assaf, R. and Berkovitz, A. (1989a). Drip nitrogen fertigation of 'Starking' Delicious' Apple trees. *Plant and soil. 119*, 305-14.

Klein, I., Bravdo, B., Levin, I. and Assaf, R. (1989b). Fruit tree growth in a shallow and stony soil. *Alon Hanotea.* 44, 95-104 and annual (1977-1986) research reports for the Bet Nekofa experiment. (in Hebrew).

Klein, I. (1992). The effect of potassium applied by drip irrigation on growth, fruit quality and storage of apples on a soil with marginal Mg content. *Potash Rev.* Intern. Potash Instit. No. 1/1992, 9/5, 7 pp.

Klein, I. and Zilkah, S. (1986). Urea retention and uptake by avocado and apple leaves. *J. Plant Nutr.*

9, 1415-25.

Levin, I., Assaf, R. and Bravdo, B. (1979). Soil moisture and root distribution in an apple orchard irrigated by tricklers. *Plant and Soil* 52, 31-40.

Marschner, H. (1986). *Mineral nutrition of higher plants*. Academic Press. London.

Mason, A.C. and Whitfield, A.B. (1960). Seasonal changes in the uptake and distribution of mineral elements in apple trees. *J. Hort. Sci.* 35, 34-55.

Mengel, K. and Kirkby, E.A. (1987). *Principles of plant nutrition*. International Potash Institute. Bern, Switzerland.

Millard, P. (1995). Internal cycling of nitrogen in trees. *Acta Hort.* 383, 3-14.

Neilsen, G.H., Hogue, E.J. and Yorston, J. (1990). Response of fruit trees to phosphorus fertilization. *Acta Hortic.* 274, 347-59.

Neilsen, G.H., Parchomchuk, P., Hogue, E.J., Wolk, W.D. and Lau, O.L. (1994). Response of apple trees to fertigation-induced soil acidification. *Can. J. Plant Sci. 74*, 347-51.

Neilsen, G.H., Parchomchuk, P., Nielsen, D., Berard, R. and Hogue, E.J. (1995). Leaf nutrition and soil nutrients are affected by irrigation frequency and method for NP-fertigated 'Gala' apple. *J. Amer. Soc. Hort. Sci.* 120, 971-6.

Oland, K. (1963). Responses of cropping apple trees to post-harvest urea sprays. *Nature* (London) *198*, 1282-3

O'Neil, M.K., Gardner, B.R. and Roth, R.L. (1979). Orthophosphoric acid as a phosphorus fertilizer in trickle irrigation. *J. Soil Sci. Soc. Amer. Proc.* 43, 283-6.

Picchioni, G.A., Weinbaum, S.A. and Brown, P.H. (1995. Retention and kinetics of uptake and export of foliage-applied, labeled boron by apple, pear, prune and sweet cherry leaves. *J. Am. Soc. Hortic. Sci.* 120, 28-35.

Proebsting, E.L., Middleton, J.E. and Roberts, S. (1977). Altered fruiting and growth characteristic of 'Delicious' apple associated with irrigation method. *HortScience* 12, 349-50.

Rauschkolb, R.S., Rolston, D.E., Miller, R.J., Carlton, A.B. and Burau, R.G. (1976). Phosphorus fertilization with drip irrigation. *Soil Sci. Soc. Am. J.* 40, 68-72.

Rinot, M., Goldberg, S.D. and Karu, N. (1971). Trickle irrigation in the vineyard. *Proc. Conf. on Drip (Trickle) and Automated Irrig. in Israel*, Herzlya-on-sea. pp. 132-8.

Rolston, D.E., Rauschkolb, R.S., Phene, C.H., Miller, R.J., Uriu, K., Carlson, R.M. and Henderson, D.W. (1979). *Applying nutrients and chemicals to trickle- irrigated crops,* Bul. *1893*. University of California, Div. of Agr. Sci. Davis. 14 pp.

Rosecrance, R.C. (1966). The effect of alternate bearing on nutrient uptake, storage and root growth in mature pistachio (*Pistachio vera L.*) trees. Ph.D. dissertation, University of California, Davis.

Samish, M.R. and Hoffman, M. (1976). Correction of deficiencies of available iron. *Abs. of Res. Activity, Dept. of Deciduous Trees, Vines and Olives for the years 1962-1965*. Volcani Institute of Agric. Res., Bet Dagan, Israel (in Hebrew).

Smith, M.W., Kenworthy, A.L. and Bedford, C.L. (1979). The response of fruit trees to injections of nitrogen through a trickle irrigation system. *J. Amer. Soc. Hort. Sci.* 104, 311-3.

Stevenson, F.J. (ed) (1982). *Nitrogen in agricultural soils*. Agron. Monogr. *22*, Amer. Soc. Agron., Madison, Wis.

Taylor, B.K. and May, L.H. (1967). The nitrogen nutrition of the peach tree. II. Storage and mobilization of nitrogen in young trees. *Austral. J. biol. Sci.* 20, 389-411.

Uriu, K., Carlson, R.M., Henderson, D.W., Schulbach, H. and Aldrich, T.M. (1980). Potassium fertilization of prune trees under drip irrigation. *J. Amer. Soc. Hort. Sci.* 105, 508-10.

Vessey, J.K., Raper, Jr C.D. and Henry, L.T. (1990). Cyclic variations in nitrogen uptake rate in soybean plants: uptake during reproductive growth. *J.Exp. Bot.* 141, 1579- 84.

Weinbaum, S.A., Merwin, M. and Muraoka, T. (1978). Seasonal variation in nitrate uptake efficiency and distribution of absorbed nitrogen in non-bearing prune trees. *J. Amer. Soc. Hort. Sci.* 103, 516-9.

Weinbaum, S.A., Uriu, K. and Muraoka, T. (1980). Relationship between $K^{15}NO_3$ application period and ^{15}N enrichment of apricot blossoms and developing fruit. *J. Plant Nutr.* 2, 699-706.

Weinbaum, S., Klein, I., Broadbent, F. and Micke, W. (1984a). Effect of time of nitrogen application and soil texture on the availability of isotopically labeled fertilizer nitrogen to reproductive and vegetative tissue of mature almond trees. *J. Amer. Soc. Hort. Sci.* 109, 339-43.

Weinbaum, S., I. Klein, F. Broadbent and W. Micke. (1984b). Use of isotope nitrogen to demonstrate dependence of mature almond trees on annual uptake of soil nitrogen. *J. Plant Nutr.* 7, 975-90.

Weinbaum, S.A. and Muraoka, T.T. (1986). Nitrogen redistribution from almond foliage and pericarp to the almond embryo. *J. Amer. Soc. Hort. Sci.* 111, 224-8.

Weinbaum, S., Klein, I. and Muraoka, T. (1987). Use of nitrogen isotopes and a light textured soil to assess annual contribution of nitrogen from soil and storage pools in mature almond trees. *J. Amer. Soc. Hort. Sci.* 112, 526-9.

Weinbaum, S.A., Muraoka, T. and Plant, R.E. (1994a). Intracanopy variation in nitrogen cycling through leaves is influenced by irradiance and proximity to developing fruit in mature walnut trees. *Trees* 9, 6-11.

Weinbaum, S.A., Neiderholzer, F.J.A, Ponchner, S., Rosecrance, R.C., Carlson, R.M., Whittlesey, A.C. and Muraoka, T.T. (1994b). Nutrient uptake by cropping and defruited field-grown 'French' prune trees. *J. Amer. Soc. Hort. Sci.* 119, 925-30.

Weinbaum, S.A. (1988). Foliar nutrition of fruit trees. In: *Plant growth and leaf-applied chemicals*, (ed P.M. Newmann), CRC press, Boca Raton, Fla. pp. 81-100.

Williams, D.E. and Vlamis, J. (1978). Acid soil infertility. In: *Soil and plant-tissue testing in California.* (ed M.H. Reisenauer), University of California, Div. of Agric. Sci., Bul. *1879.* pp. 46-48.

Williams, L.E. (1991). Vine nitrogen requirements - utilization of nitrogen sources from soils, fertilizers and reserves. *International symposium on nitrogen in grapes and wine.* Amer. Soc. Enol. and Viticulture. (ed JoAnne M.Rantz), Seatle Washington. pp. 62-6.

Winer, L., Reuveni, B., Bar, Y., Haas, J.H. and Zveibil, A. (1994). Influence of autumn fertilization with nitrogen and phosphorus on root activity and development in avocado. *Alon Hanotea* 48, 450-6 (In Hebrew).

Worley, R.E., Daniel, J.W., Dutcher, J.D. and Harrison, K.A. (1995). A long term comparison of broadcast application versus drip fertigation of nitrogen for mature pecan trees. *HortTechnology* 5, 43-7.

Zasoski, R.J., Meyer, R.D., Schulbach, H. and Edstrom, J.P. (1992). Effects of urea applications on bulk and rhizosphere soil properties in drip irrigated almonds. *J. Plant Nutr.* 15, 977-89.

Zohar, Y. (1971). Field observation plots and experiments in the use of trickle irrigation in orchards and vines. *Proc. Conf. on Drip (Trickle) and Automated Irrig. in Israel*, Herzlya-on-sea. pp. 161-74.

5 FLOWERING, FRUIT SET AND DEVELOPMENT UNDER WARM CONDITIONS

FRANK G. DENNIS, JR.

Department of Horticulture, Michigan State University, East Lansing, MI 48824-1325

1 Introduction

Flowering, fruit set, and development under tropical and subtropical conditions parallel these processes as they occur in temperate zones (for reviews see Buttrose, 1974; Monselise, 1986; Mullins. *et al.*, 1992; Sedgley, 1990; Srinavasan and Mullins, 1981); however, higher temperatures during the vegetative period and the lack of sufficient chilling pose problems not observed under most conditions in the temperate regions. Retardation or stoppage of growth favors flower induction, particularly in pome fruits that form terminal flower buds, but growth rate may be excessive under high temperatures. Marginal chilling results in erratic bloom, with flowers opening over an extended period of time. This leads to problems in pollination, in timing of pesticide and thinning treatments, and variable size and maturity of the fruits on the same tree. Thus cultural practices and choice of site often differ from those in colder regions. In this chapter I will discuss some of the environmental factors and cultural practices that affect flowering, fruit set, and fruit development in warm climates. Sherman and Rodriguez-A. (1996) recently reviewed temperature effects on growth and fruiting of peach under such conditions.

2 Flowering

The process of flowering can be separated into several phases - induction, initiation, maturation, and anthesis. Induction is the process by which a meristem becomes "programmed" to flower, although no macroscopic or microscopic changes are evident. Initiation begins when the change becomes microscopically evident as the vegetative meristem flattens, and continues as sepals, petals, stamens and pistils appear. Maturation is the completion of development of the floral organs, and anthesis is the opening of the flower.

The first three phases normally occur in one growth cycle, the last in the next cycle. (The final phases of maturation may occur in the second cycle, depending upon species and climate.) In regions with cold winters, chilling temperatures are required to break bud dormancy before flowers can mature and open, but in some areas in the tropics buds will break without chilling provided special treatments (*e.g.*, pruning, defoliation, drought or root pruning followed by irrigation) are applied (Janick, 1974; Saure, 1971, 1974). For example, in southern India (Bammi and Randhawa, 1968) and in Venezuela (Corzo,

Amnon Erez (ed.), Temperate Fruit Crops in Warm Climates, 101-122.
© 2000 *Kluwer Academic Publishers. Printed in the Netherlands.*

1987) pruning grapevines shortly after harvest induces a new flush of growth and a second flowering with no intervening rest period.

2.1 Induction and initiation

All seedling plants exhibit juvenility, a primary feature of the juvenile period being inability to flower. This phenomenon has been reviewed extensively elsewhere (*e.g.*, Hackett, 1985; Zimmerman, *et al.*, 1985), and as mature tissues are used to propagate most fruit plants, therefore this subject will not be discussed here.

a. Flowering habit

Flower initiation occurs only in lateral buds in grapes (not "true" lateral buds because of sympodial growth) and in stone fruits. Initiation can occur in both lateral and terminal buds in apple and pear, the latter being more common. Tropical conditions favor rapid growth with limited development of lateral buds. In apple cultivars with high chilling requirements, relatively few spurs are formed and few terminal buds are therefore available for flower formation (Giesberger, 1972). Apple cultivars that flower only on spurs or short shoots are generally not recommended for the tropics. Cultivars, such as 'Rome Beauty', that are "terminal bearers", initiating flowers at the tips of long shoots, are preferred. Although grapes are rapidly growing vines, flower initiation occurs in the lateral (morphologically terminal - see above) buds despite continuing growth of the terminals (Edwards, 1990; Lavee, *et al.*, 1967).

b. Time of initiation

Few data are available on the timing of flower initiation under tropical/subtropical conditions. In subtropical areas, where flowering occurs only once per year, induction apparently takes place at the "usual" time, *i.e.*, approximately 3-4 weeks after bloom. However, Ramirez and Saavedra (1990) reported that a mutant of 'Golden Delicious' apple initiated flowers a month earlier than the normal strain in Northeast Mexico, and that anthesis occurred 6 weeks earlier (early March *vs.* mid-April). Varietal differences in time of flower initiation at the same or similar latitudes have also been reported for plum in Mexico (Avitia-Garcia and Castillo-Gonzalez, 1991 - cited by Rumayor-Rodriguez, 1995), and for Asian pear in Taiwan (Lin, *et al.*, 1987) and Ceylon (Zeller, 1973), but temperature differences were probably responsible for the effects in pear.

In Taiwan, scions from high quality, high-chill pear trees at high elevations are grafted every year into poorer quality, low-chill cultivars at low elevations to obtain early cropping. Liaw (1987) tested the effects of taking scions from previous year's grafts at low elevations and chilling them prior to re-grafting on low-chill trees. Scions were taken from 5 October until 7 December and compared with scions from trees growing in the highlands. Percent flowering was highest in scions taken in November (83-85% *vs.* 35-40% for other times). Early collection did not allow time for differentiation, whereas late collection did not permit sufficent time for chilling to break dormancy.

In the tropics, induction in apple can occur at any time of year, with flowers opening at intervals during the course of fruit development. Yield data from Malaysia reflect this; Ko (1990) reported harvests of 'Rome Beauty' apple of 5.1, 1.8, 3.1, 8.3, and 7.6 kg/tree for 6-or 7-month intervals between January 1986 and January 1988, with intervals of no fruiting interspersed. Flower initiation was complete within 6 to 7 weeks of terminal bud formation in 'Rome Beauty' apple in Java (Edwards, 1990; Notodimedjo, *et al.*, 1981).

Lloyd and Couvillon (1974) manually defoliated peach trees in Georgia, U.S.A., at 2-week intervals from mid-July to late September. Terminal buds were also removed, growth having ceased. The number of flower buds forced increased, while the percentage of vegetative buds breaking declined, as defoliation was delayed (Fig. 1). Most flowers developing after early treatments were abnormal, but almost all were normal following defoliation in September. Fruit development was prevented by freezing temperatures.

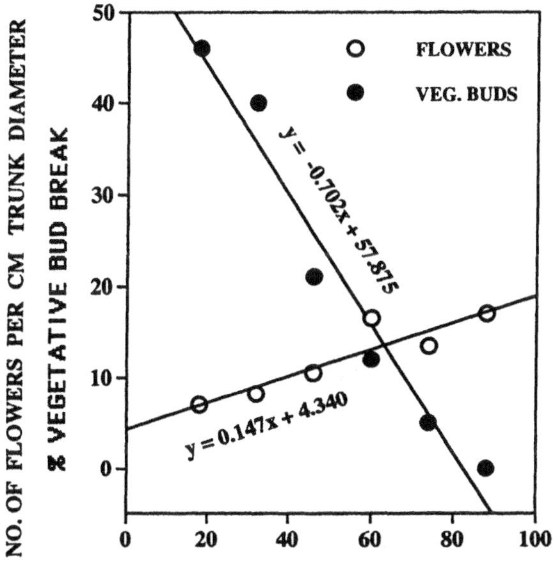

Figure 1. Effects of time of defoliation and apex removal upon breaking of floral and vegetative buds of > Washington = peach in Georgia, U.S.A. (Adapted from Lloyd and Couvillon, 1974. Used by permission.)

c. Climatic effects - photoperiod and light intensity

In some species, *e.g.*, black currant (*Ribes nigrum* L.) and many cultivars of strawberry (*Fragaria* X *ananassa* Duch.) short days (=long nights) are required for flower induction. In general, photoperiod *per se* has little effect on flowering of tree fruits. Buttrose (1970) demonstrated quantitative photoperiodic responses in several grape cultivars, more inflorescences being formed on long than on short photoperiods, even when the same total radiant energy was supplied (Table 1).

Flowering is often reduced when light intensity is low, as in the interior of trees. Baldwin (1964) observed a high correlation between hours of bright sunshine and percentage fruitful buds in 'Sultana' grape. In growth chambers, flowering of grapes increased as light intensity rose from 900 to 3600 f-c (Buttrose, 1969a, 1970), but no flowering occurred in 'Muscat of Alexandria' at 20°C, regardless of light intensity (Buttrose, 1969a) (Table 1).

Table 1. Effects of temperature and light intensity on flowering of grape (*Vitis vinifera L.*). Number of primordia per bud following 13 weeks of exposure to conditions shown[z]

		Light intensity (f-c)		
Temp.(° C)[y]	*Cultivar*	*900*	*1800*	*3600*
20/15	Reisling	-	-	0.70
	Gordo	-	-	0
	Sultana	-	-	0
	Muscat	0	0	0
25/20	Reisling	0.40	0.80	1.60
	Gordo	0.05	0.70	1.05
	Sultana	0	0	0.05
	Muscat	0.05	0.60	0.90
30/25	Reisling	1.90		
	Gordo	1.40		
	Sultana	0.10		
	Muscat	0.50	0.95	1.33
18 plus				
30 for: 1 hr.				0.05
4 hr.				1.23

[z] Buttrose (1969a,b), used by permission. Data approximate, based upon graphed data.
[y] Day (16 hr)/night (8 hr)

d. Climatic effects - temperature

Although daylength is not critical for induction in most deciduous tree fruits, temperature can have a marked effect. At high temperatures, such as those experienced at low elevations in the tropics (Edwards, 1987, 1990), induction does not occur in peach or apple, probably because growth is rapid and continuous, and induction can occur only when growth slows. [In contrast, Lichou and Fournier (1981) observed that some peach cultivars failed to flower at *high* elevations (see below).] Diaz, *et al.* (1986) reported that apple trees failed to differentiate flowers in Western Mexico until autumn; they attibuted this to high summer temperatures. Failure to form axillary flower or leaf buds in peach has been associated with high temperature during rapid summer growth (Richards, *et al.*, 1994). The incidence of such "blind" nodes declines with elevation in Mexico, and varies with genotype.

The effects of temperature on flower induction in grape vines varies with species and cultivar. Alleweldt (1963) reported a positive correlation between temperature in mid-June to mid-July in Germany and the number of inflorescences produced by *Vitis vinifera* L. vines the following year. Baldwin (1964) noted a similar correlation for the percentage of fruitful buds in 'Sultana' (also *V. vinifera*) *vs.* daily mean temperature. Buttrose (1969a,b, 1970) demonstrated that cultivars differed in response to temperature. 'Muscat of Alexandria' formed no inflorescences at 20°C, but did so at 25 and 35°C. In contrast, 'Reisling' and 'Shiraz', (both *V. vinifera)*, and 'Delaware' (*V. labrusca L.)* initiated inflorescences at day temperatures of 20°C or above, but not at 15°C (Table 1) (Buttrose, 1970). Flowering also occurred in vines grown at 18°C, when exposed to daily 4-hour periods at 30°C (Buttrose 1969b). Relatively brief periods ("pulses") at high temperature also induced flowering in 'Muscat Gordo Blanco' (Buttrose, 1969b). 'Delaware' grapes can initiate flowers at 10-21°C (Kobayashi, *et al.*, 1965).

e. Cultural practices

Several other treatments are used commercially for stopping growth, such as restricting water supply (Bederski, 1987, 1988; Black, 1952 - cited by Giesberger, 1972), branch bending (Liaw, 1987), light pruning, breaking shoots, notching (Hill and Campbell, 1949, Scholz, 1964), and treatment with growth retardants. Water stress can either reduce or enhance flower bud formation. In California, flowering of peach was increased by post-harvest water stress (Larson, *et al.*, 1988).

Most of the conditions that affect flower induction in the temperate zone (leaf/fruit ratio, solar radiation, adequate control of insects and diseases to maintain sufficient leaf surface, etc.) also apply in the tropics/subtropics. Partial removal of fruit from heavily cropping trees (fruit thinning) is essential, not only for optimum fruit size (see below), but also to prevent biennial bearing in apples. Some of the apple cultivars used at low latitudes, *e.g.*, 'Rome Beauty', are annual bearers, others, *e.g.*, 'Winter Banana', may become biennial if the fruits are not thinned.

Bending can stimulate flowering of lateral buds and/or force them into growth, thus providing more sites for flower bud formation. In many peach cultivars, very vigorous shoots often branch profusely; although flowers may be initiated in some lateral buds, others that might have formed flowers become vegetative shoots.

The use of growth retarding chemicals has been suggested as a means of slowing growth and inducing flowering (Edwards, 1990), as they are effective in the temperate zone (see Miller, 1988). In the tropics/subtropics they have been effective on pear in Taiwan (Liaw, 1987), peach in Israel (Erez, 1984) and nectarine in Australia (George and Nisson, 1987), although in nectarine only flower bud weight, rather than percent of the buds flowering, was evaluated. Notodimedjo, *et al.* (1981) observed that the growth retardant daminozide hastened terminal bud and flower formation in apple, although it had little effect on final percentage of buds flowering. In relatively cool lowland sites in the Philippines, growth ceased earlier in paclobutrazol-treated trees, and flower bud initiation occurred on lateral shoots that stopped growing (Llanes, *et al.*, 1987). In a factorial experiment, Liaw (1987) demonstrated that branch bending of Asian pear reduced shoot growth, and stimulated both flowering and fruit set. Treatments with paclobutrazol, CCC, or ethephon had similar effects on branches that were not bent but did not increase the effect of bending. Use of paclobutrazol by commercial growers in Australia has resulted in some detrimental effects, such as complete suppression of development of laterals (Lloyd, 1992).

f. Summary

In tropical and subtropical climates, the processes of flower induction and initiation can differ considerably from those in the temperate zone. Terminal bearing cultivars may be preferable in pome fruits, given the tendency for lateral buds to remain dormant. Timing of flower initiation is noticeably different, occurring once in the temperate zone *vs.* more or less continuously in the tropics. High temperature can also inhibit flowering in tree fruits, while favoring flowering in grapes. Light intensity is a critical factor regardless of climate, as flowering is inhibited by shading.

Many cultural practices favor flowering, among the most important being thinning of pome fruits when crops are heavy, and techniques to limit excessive growth, such as withholding of water, limb bending, and application of growth retardants.

2.2 Flower maturation

Once induction and early phases of initiation have occurred, many factors may influence subsequent flower development A healthy leaf surface is essential for development of "strong" flowers, *i.e.*, ones that will be capable of setting fruit (Goldwin, 1981; May, 1970). Favorable light and temperature conditions are important. Yuda (1981 - cited by Verheij, 1990) ascribed a poor second crop of apples in Java to overcast, wet weather during the maturation of the ovules and anthers. Such conditions may have limited the supply of carbohydrates required for proper flower maturation.

a. Temperature and /or water stress

Rumayor-Rodriguez (1995) used multiple regression modeling to evaluate the effects of environmental factors on yields of peach over eight seasons. He reported a negative relationship between temperature at bloom *vs.* yield in three plum cultivars (Ozark, Premier, and Burbank) in Zacatecas, Mexico. In two additional cultivars (Frontier and Santa Rosa) yield was optimum at moderate temperatures, lower yields being obtained at higher or lower temperatures. In 'Shiro', temperatures during flower initiation the previous summer interacted with temperature at bloom, a high summer temperature reducing yield. Rumayor-Rodriguez (1995) attributed both effects of high temperature to reduced ovule longevity.

Lam-Yam and Parisot (1990), working at an elevation of less than 325 m on Reunion Island, reported no flowering in 'Flordared' peach following a mild winter. The flower primordia stopped growing and the buds abscised, suggesting that the "slow" stages of bud development require cool temperatures. This raises some interesting questions, as "bud drop" often occurs in marginal areas of the temperate zone when chilling is insufficient, but does not appear to occur in the tropical lowlands, suggesting that partial chilling may predispose buds to abscission. Considerable variability exists among stone fruit genotypes in the degree of bud abscission that occurs (W.B. Sherman, pers. comm.)

Nakasu, *et al.* (1995) also reported flower bud abortion in Asian pear (*Pyrus pyrifolia* x *P. ussuriensis*) in the highlands of southern Brazil. In 'Nijisseiki' abortion was negatively correlated with the number of chilling hours experienced, but this was not the case with other cultivars. Nakasu suggested that large fluctuations in temperature might be responsible. Zeller (1973) also observed flower abortion in apple and pear in Ceylon. Brown (1952) noted flower abortion in pear in California when chilling was insufficient.

Few sweet cherries are grown in the tropics/subtropics, primarily because the chilling requirement (800-1700 h < 7°C) is too high; breeders are attempting to obtain low chill cultivars by using chance seedlings and interspecific crosses (Richards, *et al.,* 1995). An additional problem is the formation of double pistils or "spurring" in areas with high summer temperatures. Philp (1933), for example, reported severe problems with doubling of sweet cherries in the central valley of California one year, whereas little or no doubling occurred in cooler coastal areas. Tucker (1934, 1935) reported high rates of doubling in Idaho, U.S.A, in 1932 and 1935, and attributed this to higher maximum daily temperatures in late May in 1931 and 1934. The percentage of doubling varied with the location of the spurs - 61% on those on the top of the limbs, 34% on the sides, and 13% on the undersides - suggesting that exposure to the sun raised tissue temperature. Micke, *et al.* (1983) compared 19 cultivars growing at Parlier, Calif., over a period of 8 years. The percentage of double fruits, averaged over years, varied from 0.5 to 34.4 among cultivars, with 'Chinook' producing the highest percentage and 'Jubilee' the lowest. Cultivars were classified into three groups according to the percentage of doubles. Year-to-year variations were also evident, with 16 *vs.* 60% in 'Chinook' in 1977 and 1979, respectively. Thompson (1996) listed 'Angela', 'Sweet Ann', and 'Utah Giant' as cultivars that form few doubles in Utah, whereas 'Bing' produces many.

Southwick, *et al.* (1991) reduced doubling and suturing (abnormally deep sutures) in 'Bing' cherry by evaporative cooling with overtree sprinklers. Controls were drip-irrigated. The percentage of normal fruits was increased from 17 (control) to 37% in 1989, and from 51 to 79% or 71 to 77% (2 locations) in 1990. The authors were uncertain as to whether temperature or irrigation differences were responsible for the reduction in abnormal fruit.

In peach trees, *cv.* Regina, drought stress significantly increased the percentage of doubles from 2% (control) to 23% (severe stress) when averaged over a 3-year period (Johnson, *et al.,* 1992). The effect of stress was less dramatic in 1985, when average maximum temperatures in August were lower, than in 1984 or 1986. Johnson, *et al.* (1992) suggested that water stress may reduce transpiration, leading to a rise in tissue temperature, which in turn affects pistil morphology. Sherman (personal comm.)

believes that water deficit, rather than temperature, *per se* is responsible for doubling in peach. Labib (1990) reported that hand thinning was required to eliminate double fruits of 'Flordaprince' peach in the western desert region of Egypt, where summer highs ranged from 35 to 40°C. Water stress could obviously have been a factor here, as well. Diaz, *et al.* (1985) associated high rates of doubling in Western Mexico with reduced irrigation following harvest.

Lichou and Fournier (1981) observed cultivar differences in flower initiation at various elevations in the tropics. Some cultivars failed to flower at high elevations, others flowered, but many of the flowers had multiple ovaries.

Doubling can be a problem with some cultivars of plum. 'Stanley' plum forms numerous doubles in the relatively cool climate of Michigan (personal observation). Johnson, *et al.* (1994) reported doubling in about 2% of the fruits of the early-maturing cultivar Red Beaut in the central valley of California. In this case, however, drought stress did not exacerbate the problem.

Withholding irrigation during the fruit development period reportedly increases flower bud differentiation and/or fruit set the following year in several species, including apple and pear (Aldrich and Work, 1934; Magness, 1952), and regulated deficit irrigation (RDI - applying water when fruits are growing rapidly, but not when shoots are most active, see chapter 3) has been proposed as a means of increasing fruit size, blossom density, and fruit set in pear and peach (Chalmers, *et al.*, 1981; Mitchell, *et al.*, 1984,1989). Although no studies are known in which RDI has been used to control flower initiation or fruit development in tropical/subtropical production of deciduous fruits, Johnson, *et al.* (1992) increased flowering and/or fruit set of peach by restricting irrigation after harvest. Most of the additional fruits, however, were removed in thinning.

b. Summary

Flower maturation is important in determining flower "quality", or the ability to set and mature a good-sized fruit. Such quality is dependent upon sufficient, well-exposed, leaf surface to provide needed nutrients, as well as adequate solar radiation. Both high temperature and water stress can reduce ovule longevity, resulting in low fruit set. Irrigation, especially regulated deficit irrigation, has been proposed as a means of increasing fruit size, but its effects upon flower quality are not well documented. Lack of winter chilling may cause bud abortion, whereas high temperatures during flower bud development can lead to doubling of pistils in stone fruits.

2.3 Anthesis

Defoliation by pathogens or insects or by hand can induce flower buds to open prematurely. Apple and peach buds can be forced throughout the year in central Florida by such treatments (W. B. Sherman, pers. comm.), whereas timing is more critical with peaches in Georgia (Lloyd and Couvillon, 1974, see above). Defoliation can be used in the tropics to prevent the onset of dormancy and induce budbreak (see chapter 2). If performed too early, flowering is reduced; if performed too late, bud break is limited (Edwards, 1987b). Optimum timing for apple in Indonesia was within 4 weeks following harvest (Edwards, 1987b; Saure, 1971). Bending or shoot tip removal alone had less effect than defoliation alone, but bending increased response to defoliation, particularly during the dry season (Edwards and Notodimedjo, 1987). Spraying with caustic chemicals (*e.g.*, copper sulfate, urea, or sodium chlorate) also induces leaf abscission (Edwards, 1987b). Magnesium chlorate and hydrogen cyanamide were effective defoliants on a range of deciduous fruit trees (Erez, 1985).

In self-unfruitful cultivars bloom periods must overlap with those of pollinizer cultivars in order to maximize fruit set. This is a problem even in the temperate zone, but in warmer climates the differences in bloom periods among cultivars are accentuated. Growers can hasten bloom of a cultivar by rest-breaking treatments with chemicals such as dinito-o-cresol (DNOC) to synchronize flowering with the pollinizer (see Chapter 2). Thiourea has been used on 'Spadona' pear (Erez and Lavee, 1973) and hydrogen cyanamide on sweet cherry (Snir and Erez, 1988) for this purpose. In Florida, bloom of 'Ein Shemer' apple can be hastened to coincide with that of 'Anna' (W. B. Sherman, pers. comm.), which is self-unfruitful in Florida (Crocker and Sherman, 1977).

Rest-breaking treatments also concentrate blossom opening so that flowering occurs within a relatively short interval. This reduces variability in the time of fruit maturation/harvest. Early flowering can also improve fruit set (see below), for temperatures are lower early in the season.

3 Fruit set

The next critical phase in cropping is fruit set - the growth of the ovary, or ovary and closely related parts, to become a fruit.

3.1 Pollination

Fruit species differ as to the need for pollination. No major temperate zone tree fruits, with the exception of some cultivars of persimmon *(Diospyros kaki* L.) are parthenocarpic under normal conditions. Several grape cultivars produce seedless fruits, some (*e.g.*, 'Black Corinth') being parthenocarpic, others (*e.g.*, 'Sultana') stenospermocarpic - pollination is necessary, but the seeds abort.

Verheij (1990) suggested that cross-pollination was not necessary in several apple cultivars, including 'Red Delicious' and 'Manalugi', in Java. Fruit size was similar in these cultivars with or without pollination. In contrast, 'Anna' required cross-pollination, as previously reported (Crocker and Sherman, 1977; Crocker, *et al.,* 1979; Diaz, *et al.,* 1986). 'Dorsett Golden' is recommended as a pollinizer for 'Anna' (Crocker, *et al.,* 1979; Lichou and Thiery. 1981), and, in some climates at least, is self-fruitful (Diaz, *et al.,* 1986). Note that 'Anna' is also self-fruitful under some conditions (Lichou and Thiery, 1981). Yuda (1981) reported that 'Rome Beauty' flowers developed into fruits even when the flowers were emasculated and bagged. However, pollen tubes were present in the styles, suggesting that cleistogamy had occurred. Thus fruit set may have resulted from self-pollination. Saure (pers. communication) cites evidence that 'Rome Beauty' is self-fertile at relatively warm sites in Israel, but requires cross-pollination at cooler sites.

3.2 Flower quality

Lack of chilling results in poor set because the flowers are poorly developed (Erez, 1987; Giesberger, 1972; Weinberger, 1950). Peach anthers fail to dehisce, pollen production is poor, and the styles and stigmas may fail to develop (Weinberger, 1950). High temperature shortly before anthesis has been associated with abnormal meiosis and therefore poor pollen germinability in apple (Heilborn, 1930 - cited by Gur, 1985). In apple, flowers appear before the leaves when chilling is marginal; with adequate chilling leaves and flowers develop simultaneously (Sherman and Sharpe, 1971). Assuming that photosynthates are required for fruit set, inadequate chilling would delay foliation, thus limiting set. However, fruit set in peach can occur in the absence of leaves following early autumn defoliation (George, *et al.*, 1988). W. B. Sherman (pers. comm.) offers another explanation for poor set. Temperatures rise as the season progresses, and set of peach is reduced when temperatures exceed 14°C (see below).

3.3 Temperature

Llanes *et al.* (1987) observed that several peach cultivars set fruit at highland sites in the Philippines, but none did so at low elevations. In apple, however, fruit set was apparently similar at both highland and lowland sites. After comparing temperatures at these and other sites, Edwards (1987a) suggested that high temperatures during bloom (mean monthly minima of 19-20°C) at the lowland site limited set of peach in comparison with highland sites (*ca.* 15-15.5°C). By contrast, mean monthly minima in areas of commercial peach production elsewhere are 6-8°C. In the Philippines, however, fruits set at Baguio in November (mean monthly minimum 14.9°C), but not in April (15.5°C); this suggests that a small difference in temperature can have a marked effect on set.

George, *et al.* (1988) and Sherman and Topp (1990) have observed that fruit set in peach is greatly reduced when night temperature rises above 14°C. For this reason, cultivars that bloom after March in north Florida are less productive than those that bloom earlier. Williams and Menegazzo (1988) reported 3 waves of bloom in apples grown in Guatemala. The first was killed by frost, surviving flowers producing large fruits, the second presumably set normal fruit, and the third produced small fruit of poor quality. Zeller (1973) observed the development of weak lateral flowers in pear buds in which the primordia had been killed. Such flowers may have been responsible for the "third wave" observed by Williams and Menegazzo (1988).

3.4 Water stress

In California, restricting irrigation following harvest increased fruit set the following spring in peach (Larson, *et al.*, 1988).

3.5 Shoot length

Lin and Chang (1988) evaluated set in 'Kyoho' grape in Taiwan as a function of shoot length at bloom. Shoots less than 30 cm. long produced short clusters that required more hand thinning; those longer than 60 cm. had clusters with few berries and more seedless fruit. Optimum shoot length was 30-60 cm.

3.6 Rest-breaking treatments

Chemical treatments that hasten bloom may have variable effects on fruit set, in part because of variability in toxicity (see Chapter 2). For example, application of hydrogen cyanamide to advance bud break and fruit maturity of nectarine reduced set and yield (George and Nissen, 1988). In contrast, set of apple in South Africa was increased when dinitro-o-cresol (DNOC) was used to hasten bloom (Terblanche, *et al.*, 1979). Yield of 'Golden Muscat' grapes was increased 5-6—fold following treatment with hydrogen cyanamide to stimulate budbreak (Lin, *et al.*, 1985), primarily because of greater numbers of clusters per vine. Cluster and berry size were reduced, as might be expected with the greater yield.

3.7 Summary

The need for cross-pollination in apple appears to vary, depending upon location. In some areas, for example, the apple cultivar Anna requires cross-pollination; in other areas it does not. Set may be reduced when chilling is inadequate or when temperatures exceed an optimum, or when rest-breaking treatments cause excessive injury.

4 Fruit growth

Fruit growth is dependent upon foliage and solar radiation to supply photosynthates, water, nutrients, and suitable temperatures; competition among fruits often limits size. Effects of these factors on fruit development in the temperate zone have been reviewed elsewhere (*e.g.*, Monselise, 1986, and chapters therein).

4.1 Temperature

The rate of fruit growth increases with temperature. Boynton (1959) compared the performance of 'Delicious' apple, 'Kieffer' pear, and 'Yellow' plum at five elevations, all between 18 and 19°N latitude, in Mexico. At the highest elevation (2800 m), chilling was adequate, but fruit size was limited by cool temperatures. Darrow (1953) made similar observations, but noted that some cultivars developed good flavor even at low temperatures.

Limited chilling leads to prolonged flowering, which in turn results in lack of uniformity in fruit size (Maggs, 1975). Late opening flowers characteristically produce small fruits, as noted by Williams and Menegazzo (1988) for apple. Topp and Sherman

(1989b) evaluated the relationship between monthly mean temperatures and several fruit characteristics of low-chill (100-550 chill units) peaches grown in Australia. Twenty-two cultivars were compared at 13 different sites representing a range of climates. The results indicate that fruit diameter declined with temperature.

Hale and Buttrose (1974) tested the effects of three temperature regimes (18/13°, 20/15°, and 35/30°C, day/night) on 'Cabernet Sauvignon' berry growth. During Stage I of fruit development, growth rate was proportional to temperature; in Stage II, growth was more rapid at moderate than at low, or, especially, high temperature. In Stage III, temperature had relatively little effect on size.

4.2 Chemical thinning

Minimum diameter of apples required by the markets is 65 mm and larger, or from 130 to 150 g per fruit, depending upon cultivar. Chemical thinning has been used commercially for some 40 years to reduce crop load and increase fruit size. Ebert and Kreuz (1988) used chemicals to thin 'Fuji' - a difficult-to-thin cultivar of apple - in Brazil. Although fruit density was reduced, increases in size were disappointing, the best treatments producing fruits with weights only slightly above 100 g *vs.* 90 to 100 g for the controls. Hand thinning likewise had little effect on size. In a second report (Ebert, 1988), weight of 'Gala' fruits was increased from 82 g (control) to a maximum of 112 g - still small by market standards. None of the treatments increased profit per hectare. However, thinning often increases flowering the following year, and such treatments could increase income over a two-year interval. Several other reports are available on the effects of chemical thinners under subtropical conditions (*e.g.*, Pereira, 1988; Tiscornia, 1988). Klein and Cohen (1995) reported that higher concentrations of monocarbamide dihydrogensulfate (trade name, 'Wilthin') were required for blossom thinning of peach in Israel than in the U.S.A. (see Myers, *et al.*, 1993). They also noted that concentrations of NAA of 40 ppm are recommended in Israel for apple fruit thinning (Ben-Arie, 1992), whereas much lower concentrations (10-20 ppm) are recommended in the USA (Williams, 1979). These differences suggest that climate may affect response to such chemicals. Myers worked in South Carolina, where conditions are warm and moist, Williams in Washington State, where they are relatively cool and dry, whereas Israel is warm and dry. High concentrations are also required in South Africa, where conditions are similar to those in Israel (Erez, pers. comm.).

4.3 Time of bud break and fruit ripening

Correlation was apparent between shoot length - a possible indication of time of bud break - and fruit size among a series of cultivars of peach tested in Thailand (Subhadrabandhu, 1987). The fact that some early-ripening cultivars of apple, *e.g.*, 'Paulared', can attain sizes similar to those of later-ripening ones, *e.g.*, 'Granny Smith', despite similar timing of budbreak in Michigan (personal observation), suggests that the time needed for maturation, *per se*, is not a good indicator of fruit size among cultivars.

Evaporative cooling in warm climates may increase grape berry size; this is most effective prior to veraison (Kliewer and Schultz, 1973). Bammi and Randhawa (1968) observed that the length of the fruit development period of grapes in India was inversely proportional to temperature. In north India temperature is higher, and fruits mature earlier, than in south India, where temperatures are more moderate.

4.4 Apical dominance

Grapes grown in the tropics produce small bunches, with fewer berries per bunch and clusters per vine than those grown in temperate zones (Chapman, 1990). Basal buds are often unfruitful. Chapman (1990) attributed poor flower and fruit development in the lower shoots to apical dominance; when the first shoots were removed, fruit size was improved.

4.5 Fruit morphology

The shape of fruits is often affected by climatic factors. Two of the best documented effects in apple are the greater length/diameter (L/D) ratio and the prominence of the calyx lobes in 'Delicious' ("typiness"), induced by low night temperatures during early fruit development. In contrast, fruits grown in warmer areas have low L/D ratios - so much so that a cultivar may not be recognizable (Ticho, 1970). Grapes are affected similarly. Berry shape is round or elongated depending upon whether night temperature is high or low (Kobayashi, *et al.*, 1965). As noted above, apple fruits developing from late bloom are often small and have short pedicels. The L/D ratio of 'Anna' apple declines as seed number increases (Diaz, *et al.*, 1986).

The prominence of the stylar tip in peach has been associated, not with growing season temperature, but with mean temperature during the coldest month of the year, suggesting that this characteristic may be determined prior to flowering (Topp and Sherman, 1989b). The authors note, however, that response was cultivar-dependent; some cultivars did not form prominent tips at any site. In Texas, prominent apices and elongated fruit shape in peaches were associated with high growing season temperatures (Bowen, 1971; Rouse, 1988). George, *et al.* (1990) attributed prominent sutures and tips in peaches grown in Australia to insufficient chilling.

4.6 Summary

As with other physiological processes, the rate of fruit growth increases with temperature up to an optimum. However, high temperature can result in early maturation, thus abbreviating both growth period and final size. Other factors being equal, the longer the period of development, the larger the fruit. Thus rest-breaking sprays may increase fruit size by providing a longer period of development. In grape, the effect of temperature on growth varies with the stage of development; little effect is evident in the third stage. Differential rates of growth in various portions of the fruit can affect fruit shape; this phenomenon is related to night temperature, seed number, and other factors.

5 Fruit maturation

5.1 Time of bloom

Oppenheimer (1962), based upon his apple breeding work, reported that time of fruit maturation was independent of the time of bud break; nevertheless, late leafing genotypes could not be among the earliest fruit to ripen because of their short growing period.

George, *et al.* (1990) used 'Dormex' (hydrogen cyanamide) to induce early flowering of 'Flordaprince' peaches in New South Wales, Australia; some treatments increased the length of the fruit development period by as much as 40 days. They reported a significant positive correlation between days from budbreak to harvest and mean fruit weight, and suggested that early flowering may be preferable where frosts are not a problem.

Lin, *et al.* (1982) defoliated peach trees in Taiwan by spraying with pentachlorophenol. Bud break occurred with no chilling and the trees bloomed in October/November, and the fruits were harvested in February/March - 5 months early. Although fruit size was satisfactory, many fruits abscised, resulting in reduced yield, and the harvested fruits were "slightly bitter".

5.2 Temperature

In contrast with the observations of Oppenheimer (1962, see above), Blake (1930) reported that the earlier the bloom, the more time was required for fruit development in peach. He attributed this to low temperatures in April and May in New Jersey; as bloom was delayed, average temperature rose. Weinberger (1948) quantified the response for 'Elberta' peach at Fort Valley, Ga., U.S.A. He compared times from bloom to maturity over 10 years at Fort Valley with mean maximum temperatures from bloom until 30 to 70 days after bloom. Correlation coefficients (r) increased with time, and then stabilized at 50 days ($r = -0.93$, significant at $P < 0.01$). Based upon the regression curve (Fig. 2), a 1°F rise in average maximum temperature will reduce the fruit development period about 3.5 days (= 6.3 days per 1°C).

Topp and Sherman (1989 a,b) used a similar method in comparing growing season temperatures for 22 cultivars of low-chill peach at 13 different sites in Australia. The results (Table 2, Fig. 2) parallel those of Weinberger (1948), except that the slope of the regression line was 2.8 days per 1°F (=5 days per 1°C) change in temperature. Similar relationships apply in apricot and prune (Baker and Brooks, 1944). Munoz, *et al.* (1986) evaluated base temperatures, as well as heat units required for five low-chill peach and nectarine cultivars, finding that optimum base temperature varied from 2.2 to 4.5°C among genotypes.

Grape berries also ripen sooner in hot climates; under controlled conditions, temperature during Stage II was the primary determinant of days to maturity (Hale and Buttrose, 1974).

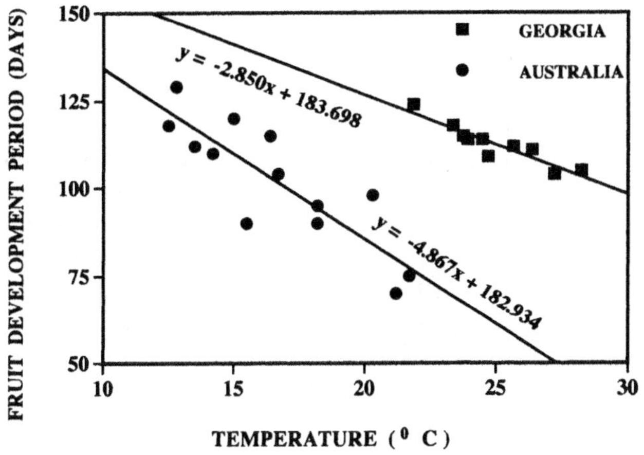

Figure 2. Regression of peach fruit development period on temperature for: (a) 'Elberta' at Fort Valley, Ga., U.S.A, (Weinberger 1948) and (b) 22 cultivars at 13 locations in Australia (Topp and Sherman, 1989a). Data from Georgia are for mean maximum temperature for 50 days following full bloom, those for Australia are mean temperatures for Aug.- Nov. (Adapted from originals; used by permission).

Table 2. Correlations between mean monthly temperature during fruit development and several fruit characteristics of 22 low-chill peach cultivars at 13 sites in Australia (Topp and Sherman, 1989b)

Characteristic	*R value*
Diameter	-0.83**
Firmness	0.62*
Flavor	0.72**
Prominence of stylar tip	-0.65*z

z Value for entire fruit development period not significant at $P < 0.05$; value shown is for mean July (coldest month) temperature

In 'Ribier' in Bali and 'Carolina Black Rose' grapes in Queensland, Australia, maturation varied among berries within a cluster (Chapman, 1990). Chapman suggested that this problem might be reduced by using plastic bunch covers or pruning to reduce shading, the use of ethephon as a ripening agent, and/or a reduction in vine vigor. Others (Sherman, pers. comm.) associate the variability in ripening with high night temperature during the last month prior to maturation.

5.3 Cultivar

Peaches of Spanish origin have been grown in some areas of Latin America for 400 years, and are propagated by seed. Time from bloom to harvest varies from 120-150 days. The clingstone fruits are small to medium-sized and firm-fleshed (Sharpe, 1969; Sherman and Lyrene, 1984). New cultivars developed by breeders at the University of Florida and elsewhere ripen earlier (60-90 days), have melting flesh and shorter shelf life, and are unsuitable for non-refrigerated transport (Diaz and Alvarez, 1984; Sharpe, 1969; Sherman, *et al.,* 1979; Sherman and Lyrene, 1984). Despite their early ripening, these fruits are considerably larger than the traditional seedling peaches.

5.4 Summary

Time of bloom and time of fruit maturation are independent across cultivars. However, within cultivars early bud break, such as that induced by rest-breaking sprays, can lead to early ripening, even though the total period from bloom to maturation may be longer. Maturation is hastened by high growing season temperatures, while surface color is reduced.

6 Market quality

Many of the low-chill fruit varieties are inferior to high chill cultivars in terms of consumer acceptance (Lloyd, 1992), probably because of their early ripening.

6.1 Temperature

Although he provides no data, Chandler (1928) observed that 'Bartlett' pears of the best flavor and keeping quality were grown in areas with summers too hot and dry for production of flavorful apples, and that sweet cherries were most flavorful in areas with cool summers. He also noted differences in climatic adaptability among apple cultivars, some being more flavorful at lower, others at higher, latitudes. Maxie and Claypool (1957) observed internal browning in prunes exposed to temperatures exceeding 40°C. Evaporative cooling increased firmness and reduced fruit drop in one study (Alani, 1976). Baker and Brooks (1944) reported that temperatures greater than 40°C delayed ripening - not surprising, given the optimum temperature ranges of most enzymatic reactions.

Although peach fruit size declines as growing season temperature increases (see above), firmness and flavor increase with temperature (Topp and Sherman, 1989a,b).

Surface color is an important selling point for some fruits and in some markets, as witnessed by the popularity of redder strains of apples. Red color development in many cultivars of apple is dependent upon cool temperatures during the 4-6 weeks prior to maturation; fruits grown at low elevations in the tropics and subtropics often are less highly colored than the same cultivars grown at higher elevations or in the temperate zone (Diaz and Alvarez, 1984; Diaz, *et al.,* 1986; Ticho, 1970). Similarly, low night temperatures improve color in grape (Buttrose, *et al.*, 1971; Kliewer and Torres, 1972; Naito, 1964).

7 Summary

Flowering and fruit development of temperate zone fruits in the tropics and subtropics resemble the same processes in the temperate zone. Notable differences include limited chilling that delays/prolongs flowering and subsequent fruit development, the possibility of multiple cropping with only brief periods of "dormancy", apical dominance that inhibits spur formation and therefore fruit production on such spurs, and the existence of zones where high temperatures limit flowering. High temperatures may also reduce fruit set, color, and quality, and affect fruit morphology and ripening pattern. Methods for compensating for many of these effects have been developed, but numerous problems remain to be solved. The introduction of new and better adapted cultivars will alleviate some of these problems, as genotypes differ greatly in their responses to environmental conditions.

8 References

Alani, K. (1976). Die Wasserbespruhung der Obstbaume vor der Ernte und ihre Wirkung auf Haltrekraft und Fruchtfestigkeit bei Steinobst. *Erwerbobstbau*, 18, 188-90, 19, 7-9.

Aldrich, W.W. and Work, R.A. (1934) Effect of leaf-fruit ratio and available soil moisture in heavy clay soil upon amount of bloom of pear trees. *Proceedings of the American Society for Horticultural Science*, 31, 57-74.

Alleweldt, G. (1963) Einfluss von Klimafaktoren auf die Zahl der Infloreszenzen bei Reben. Wein. Wiss., 18, 61-70.

Avitia-Garcia, E. and Castillo-Gonzalez, A.M. (1991) Diferenciación floral en *Prunus*. Nat'l. Congress of the Sociedad Mexicana de Ciencias. *Horticolas IV*: 304 (Abstract). (Cited by Rumayor-Rodriguez, 1995.)

Baker, G.A. and Brooks, R.M. (1944) Climate in relation to deciduous fruit production in California. III. Effect of temperature on number of days from full bloom to harvest of apricot and prune fruits. *Proceedings of the American Society for Horticultural Science*. 45,95-104.

Baldwin, J.G. (1964). The relation between weather and fruitfulness of the Sultana vine. *Australian Journal of Agricultural Research*. 15, 920-8.

Bammi, R.K. and Randhawa, G.S. (1968).Viticulture in the tropical regions of India. *Vitis,* 7, 124-9.

Bederski, K. (1987). Selection and dormancy management of temperate zone deciduous fruit tree cultivars in coastal valleys of Peru. *Acta Horticulturae*, 199, 33-38.

Bederski, K.A. (1988). Apple growing in the coastal areas of Peru. *Acta Horticulturae*, 232, 51-55.

Ben-Arie, Z. (ed.) (1992). Spray recommendations of deciduous orchards. (Hebrew) Israel Ministry of Agriculture (pamphlet).

Black, M.W. (1952) The problem of prolonged rest in deciduous fruit trees. *Rept. 13th Intern. Hort. Congress*, London, V. II:1122-31.

Blake, M.A. (1930). Length of the fruit development period of the Elberta and some other varieties of peaches. *New Jersey Agr. Expt. Sta. Bul.* 511.

Bowen, H.H. (1971). Breeding peaches for warm climates. *HortScience* 6, 153.

Boynton, D. (1959). Observations on the temperature limitation of the apple in tropical America. *Proceedings Caribbean Region, American Society for Horticultural Science* VII, 69-88.

Brown, D.S. (1952). Relation of irrigation practice to the differentiation and development of apricot flower buds. *Botanical Gazette* 114, 95-102.

Buttrose, M.S. (1969a). Fruitfulness in grape-vines: effects of light intensity and temperature. *Botanical Gazette* 130, 166-73.

Buttrose, M.S. (1969b). Fruitfulness in grape-vines: effects of changes in temperature and light regimes. *Botanical Gazette* 130, 173-9.

Buttrose, M.S. (1970). Fruitfulness in grape-vines: the response of different cultivars to light, temperature and day length. *Vitis* 9,121-5.

Buttrose, M.S. (1974). Fruitfulness in grape-vines: effect of water stress. *Vitis* 12, 299-305.

Buttrose, M.S., Hale, C.R. and W.M. Kliewer. (1971) Effect of temperature on the composition of 'Cabernet Sauvignon' berries. *American Journal of Enology and Viticulture* 22, 71-5.

Chalmers, D.J., Mitchell, P.D. and Van Heek, L. (1981) Control of peach tree growth and productivity by regulated water supply, tree density and summer pruning. *Journal of Horticultural Science* 69, 1035-41.

Chandler, W.H. (1928) *North American Orchards: Their crops and some of their problems.* Lea and Febiger, Philadelphia, Pa.

Chapman, K.R. (1990) Table grapes in the tropics - areas for research and market opportunities. *Acta Horticulturae* 279, 91-6.

Corzo, P. (1987). Tropical viticulture in Venezuela. *Acta Horticulturae* 199, 27-9.

Crocker, T.E. and Sherman, W.B. (1977) Self-fruitfulness in 'Anna' apple. *HortScience* 12, 421 (Abstract).

Crocker, T.E., Sherman, W.B. and Knight, R.J., Jr. (1979). Self-unfruitfulness of 'Anna' apple. *Fruit Variety Journal* 33(2), 65-6.

Darrow, G.M. (1953) Deciduous fruits in central America, Colombia and Ecuador. *Ceiba* 4, 69-80.

Diaz, D.H. and Alvarez, A. (1984) (cited by Diaz, 1992).

Diaz, D.H., Martinez, J.J. and Sherman, W.B. (1986) Apple and peach production in warm climates of northwest Mexico. *Fruit Variety Journal* 40, 121-5.

Ebert, A. (1988). Hand thinning in the apple cultivars Gala, Golden Delicious and Fuji in the state of Santa Catarina. *Acta Horticulturae* 232, 147-58.

Ebert, A. and Kreuz, C.L. (1988) Technical and economic practicability of chemical thinning of Fuji apples in the state of Santa Catarina, South Brazil. *Acta Horticulturae* 232, 127-36.

Ebert, A. and Kreuz, C.L. (1989) [Use of growth regulators for fruit thinning in the the apple cultivar Gala.] (Portuguese). *Revue Brasileira de Fisiologia Vegetal* 1,55-62.

Edwards, G.R. (1987a) Temperature in relation to peach culture in the tropics. *Acta Horticulturae*, 199, 61-2.

Edwards, G.R. (1987b) Producing temperate-zone fruit at low latitudes: avoiding rest and the chilling requirement. *HortScience* 22, 1236-40.

Edwards, G.R. (1990) Ten year's experience with temperate fruits in the tropics. *Acta Horticulturae* 279, 47-51.

Edwards, G.R. and Notodimedjo, S. (1987) Defoliation, bending and tip pruning of apple under tropical conditions. *Acta Horticulturae* 199, 125-7.

Erez, A. (1984) Dwarfing peaches by pruning and by paclobutrazol. *Acta Horticulturae*, 146, 235-41.

Erez, A. (1985) Defoliation of deciduous fruit trees with magnesium chlorate and cyanamide. *HortScience*, 20, 452-3.

Erez, A. and Lavee, S. (1973) How to achieve earlier ripening of fruit and increased fertility in fruit trees

(Hebrew). *Hassedeh* 54, 251-3. (Cited by Ruck, 1975).

George, A.P. and Nissen, R.J. (1987). Effects of paclobutrazol (PP-333) on vegetative and floral development of Fla 6-3 and Sunred nectarines. *Acta Horticulturae* 199, 173-8.

George, A.P., Nissen, R.J. , Lloyd, J. and Richens, K. (1990) Factors affecting fruit quality of low-chill stone fruit in subtropical Australia. *Acta Horticulturae* 279, 559-70.

George, A.P., Nissen, R.J. and Sherman, W.B. (1988) Overlapping double and early single cropping of low-chill peaches in Australia. *Fruit Varieties Journal* 42, 91-5.

Giesberger, G. (1972) Climatic problems in growing deciduous fruit trees in the tropics and subtropics. *Tropical Abstracts* 27, 1-8.

Goldwin, G.K. (1981) Hormone-induced setting of Cox apple, *Malus pumila*, as affected by time of application and flower type. *Journal of Horticultural Science* 56, 345-52.

Gur, A. (1985) Rosaceae, in *Handbook of Flowering*, Vol. .1, (ed. A.H. Halevy), CRC Press, Inc., Boca Raton, Fla. pp. 355-89.

Hackett, W.P. (1985) Juvenility, maturation and rejuvenation in woody plants. *Horticultural Reviews* 7, 109-55.

Hale, C. R. and Buttrose, M.S. (1974) Effect of temperature on ontogeny of berries of *Vitis vinifera* L., cv. Cabernet Sauvignon. *Journal of the American Society for Horticultural Science* 99, 390-4.

Heilborn, O. (1930) Temperature und Chromosomenkonjugation. Svensk Botanisk Tidskrift 24, 12-24. (Cited by Gur, 1985)

Hill, A.G.G. and Campbell, G.K.G. (1949) Prolonged dormancy of deciduous fruit trees in warm climates. *Empire Journal of Experimental Agriculture* 17(68), 259-64.

Janick, J. (1974) The apple in Java. *HortScience* 9,13-5.

Johnson, R.S., Handley, D.F. and Day, K.R. (1994) Postharvest water stress of an early maturing plum. *Journal of Horticultural Science* 69, 1035-41.

Johnson, R.S., Handley, D.F. and DeJong, T.M. (1992). Long-term response of early maturing peach trees to postharvst water deficits. *Journal of the American Society for Horticultural Science* 117,881-6.

Klein, J.D. and Cohen, S. (1995) High concentrations of monocarbamide dihydrogensulfate are needed to thin nectarine blooms in Israel. *HortScience* 30, 628.

Kliewer, W.M. and Schulz, H.B. (1973) Effect of sprinkler cooling of grapevines on fruit growth and composition. *American Journal of Enology and Viticulture*, 24, 17-26.

Kliewer, W.M. and Torres, R.E. (1972) Effect of controlled day and night temperature on grape coloration. *American Journal of Enology and Viticulture* 22, 71-7.

Ko, W.W. (1990) Apple culture in the highlands of Malaysia. *Acta Horticulturae* 279, 97-108.

Kobayashi, A., Yukinaga, H. and Itano, T. (1965a) Studies on the thermal conditions of grapes. III. Effects of night temperatures at the ripening stage on the fruit maturity and quality of Delaware grapes. *Journal of the Japanese Society for Horticultural Science* 34, 26-34.

Kobayashi, A., Yukinaga, H. and Matsunaga, E. (1965b) Studies on the thermal conditions of grapes. V. Berry growth, yield and quality of Muscat of Alexandria as affected by night temperature. *Journal of the Japanese Society for Horticultural Science* 34, 7-14.

Labib, S. (1990) Peach culture in Egypt. (Unpublished manuscript).

Lam-Yam, L. and Parisot, E. (1990). Preliminary study on peach floral differentiation in mild winter areas. *Acta Horticulturae*, 279, 231-7.

Larson, K.D., DeJong, T.M. and Johnson, R.S. (1988) Physiological and growth responses of mature peach trees to postharvest water stress. *Journal of the American Society for Horticultural Science* 113, 296-300.

Lavee, S., Regev, V. and Samish, R.M. (1967) The determination of induction and differentiation in grape vines. *Vitis* 6, 1-13.

Liaw, W.J. (1987) The reutilization of shoots from top-grafted scions of pear in Taiwan. (Chinese with English summary), in *Forcing Culture of Horticultural Crops*. (ed L.R. Chang), Spec. Publ. No. 10 of Taichung District Station, Changhua, Taiwan. p. 175-93.

Lichou, J. and Fournier, P. (1981) Le conduite du pêcher. Troisième partie: Le comportement en zone tropical d'altitude. *Fruits* 36, 43-6.

Lichou, J. and Thiery, D. (1981) Comportement de quelques variétés de pommiers a l'île de la Reunion. *Fruits* 36, 317-26.

Lin, J.H. and Chang, L.R. (1988) Effect of shoot growth on fruit setting and quality of grape, in Symposium on Grape Industry Research and Development, (eds C.C. Tu and C.H. Hsiao).Taiwan Agr. Res. Inst., Taichung, Taiwan. pp. 34-41.

Lin, C.H., Chang, L.R., Lin, H.S. and Lee, C.H. (1987) Seasonal observations on the flower bud differentiation in two cultivars of *Pyrus serotina* Rehd. in Taiwan. *Gartenbauwissenschaft* 52, 200-4.

Lin, J.S., Liau, W.J. and Lin, H.S. (1982) The regulation of harvest time of peach (*Pyrus persica* sib. et zucc.) in *Taiwan. Bull. Taichung District Agr. Improvement Station* N.S. 6, pp. 88-96. [Chinese with English summary].

Lin, C.H., Lin, J.H., Chang, L.R. and Lin, H.S. (1985) The regulation of the Golden Muscat grape production season in Taiwan. *American Journal of Enology and Viticulture* 36, 114-7.

Llanes, R., De Las Alas, R.A., Otomo, E. and Edwards, G.R. (1987) Apples and peaches in tropical lowlands of the Philippines. *Acta Horticulturae* 199, 64.

Lloyd, D.A. II and Couvillon, G.A. (1974) Effects of date of defoliation on flower and leaf bud development in the peach [*Prunus persica* (L.) Batsch.]. *Journal of the American Society for Horticultural Science* 99, 514-7.

Lloyd, J. (1992) Temperate zone fruit production in the tropics and subtropics: Oceania. *Acta Horticulturae,* 296, 219-29.

Maggs, D. H. (1975) Influence of a prolonged flowering period on the uniformity of an apple crop. *Australian Journal of Experimental Agriculture, Animal Husbandry* 15, 709-20.

Magness, J. R. (1952) Soil moisture in relation to fruit tree functioning. *Report 13th International Horticultural Congress, London* 1, 230-9.

Maxie, E.C. and Claypool, L.L. (1957) Heat injury in prunes. *Proceedings of the American Society for Horticultural Science* 69, 116-21.

May, P. (1970) Physiological and horticultural aspects of flowering and fruit-set. *Proceedings 18th International Horticultural Congress* 4, 211-21.

Micke, W.C., Doyle, J.F. and Yeager, J.T. (1983) Doubling potential of sweet cherry cultivars. *California Agriculture*, Mar/Apr., 24-5.

Miller, S.S. (1988) Plant bioregulators in apple and pear culture. *Horticultural Reviews* 10, 309-401.

Mitchell, P.D., Jerie, P.H. and Chalmers, D.J. (1984) The effects of regulated water deficits on pear tree growth, flowering, fruit growth and yield. *Journal of the American Society for Horticultural Science* 109, 604-6.

Mitchell, P.D., Van den Ende, B., Jerie, P.H. and Chalmers, D.J. (1989). Responses of 'Barlett' pear to withholding irrigation, regulated deficit irrigation and tree spacing. *Journal of the American Society for Horticultural Science* 114, 15-19 .

Monselise, S. (ed) (1986) *Handbook of Fruit Set and Development*. CRC Press, Boca Raton, Fla.

Mullins, M.G., Bouquest, A. and Williams, L.E. (1992) *Biology of the Grapevine*. Cambridge Univ. Press,

Cambridge, U.K.

Munoz, C., Sepulveda, G., Garcia-Huidobro, J. and Sherman, W.B. (1986) Determining thermal time and base temperature required for fruit development in low-chilling peaches. *HortScience* 21, 520-22.

Myers, S.C., King, A. and Savelle, A.T. (1993) Bloom thinning of 'Winblo' peach and 'Fantasia' nectarine with monocarbamide dihydrogensulfate. *HortScience* 28, 616-7.

Naito, R. (1964) Studies on coloration of grapes. V. Influence of light intensity on the coloration and pigmentation of some black and red grapes. *Journal of the Japanese Society of Horticultural Science* 33, 213-20.

Nakasu, B.H., Herter, F.G., Leite, D.L. and Raseira, M.C.B. (1995) Pear flower bud abortion in southern Brazil. *Acta Horticulturae* 395, 185-92.

Notodimedjo, S., Danoesastro, H., Sastriosumarto, S. and Edwards, G.R. (1981) Shoot growth, flower initiation and dormancy of apple in the tropics. *Acta Horticulturae* 120, 179-186.

Oppenheimer, C. (1962). Breeding of apples for a subtropical climate. *Proceedings 16th International Horticultural Congress* 3, 18-24.

Pereira, A.J. (1988). Effect of timing and concentration of naphthaleneacetic acid (NAA) and carbaryl, with and without mineral oil, on the thinning of apples, cv. Gala. *Acta Horticulturae* 232, 141-6.

Philp, G.L. (1933) Abnormality in sweet cherry blossoms and fruit. *Botanical Gazette* 94, 815-820.

Ramirez, H. and Saavedra, L.L. (1990) A low chilling requirement Golden Delicious apple mutant from Northeast Mexico. *Acta Horticulturae* 279, 67-3.

Richards, G.D., Kantharajah, K., Porter, G.W. and Sherman, W.B. (1995) Progress with low-chill cherry breeding in Australia and USA. *Acta Horticulturae* 403, 179-81.

Richards, G.D., Porter, G.W., Rodriguez-A., J. and Sherman, W.B. (1994) Incidence of blind nodes in low-chill peach and nectarine germplasm. *Fruit Variety Journal*, 48, 199-202.

Rouse, R. E. (1988) Tropical adapted peach cultivars for commercial production in South Texas. *Journal of the Rio Grande Valley Horticultural Society*, 41, 41-5.

Ruck, H. C. (1975). *Deciduous fruit tree cultivars for tropical and sub-tropical regions.* Horticultural Review No. 3, Commonwealth Bureau of Horticultural. and Plantation Crops, E. Malling, Kent. 91 pp.

Rumayor-Rodriguez, A. (1995) Multiple regression models for the analysis of potential cultivation areas for Japanese plums. *HortScience* 30, 605-10.

Saure, M. (1971) Beobachtungen uber den Apfelanbau im tropischen Indonesie als Beitrag zur Frage der Blütenbildung, des Kältebedurfnisses und des Alterungsprozesses unserer Apfelbaume. *Gartenbauwissenschaft* 36, 71-86.

Saure, M. (1973) Successful apple growing in the Tropical Indonesia. *Fruit Variety Journal* 27, 44-5.

Sedgley, M. (1990) Flowering of deciduous perennial fruit crops. *Horticultural Reviews* 12, 223-64.

Scholz , H. (1964) Der Apfelenbau in Angola. *Afrika Heute* 16, 206-8. (Cited by Giesberger, 1972)

Sharpe, R.H. (1969) Sub-tropical peaches and nectarines. *Proceedings of the Florida State Horticultural Society* 82, 302-6.

Sherman, W.B. and Lyrene, P.M. (1984) Biannual peaches in the tropics. *Fruit Variety Journal* 38(2), 37-9.

Sherman, W.B., Lyrene, P.M. and Andrews, C.P. (1979) Development of peach cultivars from the Florida breeding programs. *Proceedings of the American Society for Horticultural Science, Tropical Region,* 23, 25-8.

Sherman, W.B. and Rodriguez-A., J. (1996) Peach tree growth and fruiting as affected by temperature. *Proceedings of the Interamerican Society for Tropical Horticulture* 38, 122-6.

Sherman, W.B. and Sharpe, R.H. (1971) Subtropical apples. *Proceedings of the Florida State Horticultural Society* 84, 337-8.

Snir, I. and Erez, A. (1988) Bloom advancement in sweet cherry by hydrogen cyanamide. *Fruit Varieties Journal* 42, 120-1.

Southwick, S.M., Shackel, K.A., Yeager, J.T., Asai, W.K. and Katacich, M., Jr. (1991) Over-tree sprinkling reduces abnormal shapes in 'Bing' sweet cherries. *California Agriculture* 45(4), 24-6.

Srinivasan, C. and Mullins, M.G. (1981) Physiology of flowering in the grapevine — a review. *American Journal of Enology and Viticulture* 32, 47-63.

Subhadrabandhu, S. (1987) Some characteristics of peach varieties grown in the highlands of northern Thailand. *Acta Horticulturae* 199, 83-9.

Terblanche, J.H., Stassen, P.J.C., Hesebeck, I. and Strydom, D.K. (1979) Effects of autumn nitrogen nutrition and winter rest-breaking spray on the growth, development and chemical composition of young 'Golden Delicious' apple trees grown in sand culture. *Scientia Horticulturae* 10, 37-48.

Thompson, M. (1996) Flowering, pollination and fruit set, in *Cherries: Crop Physiology, Production and Uses* (eds A.D. Webster and N.E. Looney), CAB International, Wallingford, Oxon, UK. Ch. 8, pp. 223-41.

Ticho, R.J. (1970) Low chilling temperate zone fruits in Israel. *Proceedings of the Florida State Horticultural Society* 83, 332-96.

Tiscornia, J.R. (1988) Chemical thinning of apples in Rio Grande do Sul. *Acta Horticulturae* 232, 137-40.

Topp, B.L. and Sherman, W.B. (1989a) The relationship between temperature and bloom-to-ripening period in low-chill peach. *Fruit Variety Journal* 43, 155-8.

Topp, B.L. and Sherman, W.B. (1989b) Location influences on fruit traits of low-chill peaches in Australia. *Proceedings of the Florida State Horticultural Society*, 102, 195-9.

Tucker, L.R. (1934) Notes on sweet cherry doubling. *Proceedings of the American Society for Horticultural Science*, 31, 300-2.

Tucker, L.R. (1935) Additional notes on sweet cherry doubling. *Proceedings of the American Society for Horticultural Science* 33, 237-9.

Verheij, E.W.M. (1985) Apple growing in East Java: the scope for improvements. *Acta Horticulturae* 158, 47-51.

Verheij, E.W.M. (1990) Adaptations of apple to tropical growing conditions. *Acta Horticulturae* 279, 83-90.

Weinberger, J.H. (1948) Influence of temperature following bloom on fruit development period of Elberta peach. *Proceedings of the American Society for Horticultural Science* 51, 175-8.

Weinberger, J.H. (1950) Chilling requirements of peach varieties. *Proceedings of the American Society for Horticultural Science* 56, 123-33.

Williams, M.W. (1979) Chemical thinning of apples. *Horticultural Reviews* 1, 270-300.

Williams, W.T. and Menegazzo, G. (1988) Apple culture in the highlands of Guatemala. *Acta Horticulturae*, 232, 57-66.

Yuda, E. (1981) Tree fruit growing in Java. *Ecological studies via the utilization of cultivated plants in rural communities of Southeast Asia and Oceania with special reference to fruit trees and production* (mimeographed report) (ed T. Ogo). College of Agriculture, Okayama University, Japan, 144 pp. (Cited by Verheij, 1990).

Zeller, O. (1973) Blühhrhythmik von Apfel und Birne im tropischen Hochland von Ceylon. *Gartenbauwissenschaft* 38, 327-42.

Zimmerman, R.H., Hackett, W.P. and Pharis, R.P. (1985) Hormonal aspects of phase change and precocious flowering. Hormonal Regulation of Development III. *Encyclopedia of Plant Physiology* (n.s.) 11 79-115.

6 EFFECTS OF HIGH TEMPERATURES AT THE ROOT ZONE AND THE GRAFT UNION ON THE DEVELOPMENT OF TEMPERATE FRUIT TREES

ARYE GUR

The Hebrew University of Jerusalem, Faculty of Agriculture, Rehovot, Israel

1 Introduction

Among the detrimental effects of warm climates on temperate fruit crops, supraoptimal root temperatures may be of considerable importance, the optimal temperatures varying according to the species involved. Even if we disregard the temperature in the upper 5 cm of the soil, which may reach extremely high levels in subtropical countries, soil temperatures exceeding 30°C are to be found in areas growing deciduous fruit trees as in parts of the U.S.A, Japan, Israel, Australia, Turkey and Greece (Ashbel *et al.,* 1965). Mulching and frequent drip irrigation reduces temperatures in the root zone, but still the damage caused by supraoptimal root temperature has to be considered.

2 Supraoptimal root temperatures

2.1 Growth and flowering responses

The optimal root temperature for various fruit tree species and fruit rootstock clones has been established by subjecting their root system to controlled temperatures, without changing the temperature of their tops and measuring the rates of root and shoot growth, as well as other parameters connected with bud break, flowering and fruiting. The root temperature optimum depends on the season. With dormant apple trees chilling the root system at 4°C resulted in earlier bud break and increased shoot growth compared to non chilled trees. However with peaches chilling of the rootstock delayed bud break and retarded shoot growth (Young and Werner, 1984 a,b).

In another experiment, chilling the rootstock did not influence the date of full bloom in apple and peach trees, but caused somewhat earlier blooming in cherries (Hammond and Seeley, 1978). The contradicting results of both authors regarding apples may be caused by the application of chilling in the latter investigation after dormancy was already broken. In both investigations the rootstocks included both roots and a small stem piece and it is not clear whether the reported temperature effects are due to the roots or the stem pieces.

After dormancy was broken a root temperature of 20°C was more efficient in inducing bud break than 10°C, particularly if combined with 20°C shoot temperature

Amnon Erez (ed.), Temperate Fruit Crops in Warm Climates, 123-135.
© 2000 *Kluwer Academic Publishers. Printed in the Netherlands.*

instead of 10°C (Belding and Young, 1987). With persimmons bud break was earlier at 13°C root temperature, compared to 7°C and 23°C (Mowat, 1994).

During the main growing season the optimal root temperature has been determined for several temperate fruit tree species. Many and frequently conflicting results have been reported for apple trees. One reason for such conflicting results may be the use of different rootstock clones varying in their susceptibility towards supraoptimal root temperature.

Another problem is the disregard to differences between the controlled root temperatures to the non - controlled air temperatures at various locations. Comparisons between responses of various apple rootstock clones have been provided by several authors. A summary of the optimal root temperatures for various deciduous fruit species, as to their effect on root and shoot growth is presented in Table 1.

Table 1. Optimal root temperatures for root and shoot growth of various deciduous fruit species and apple rootstock clones

Species/rootstock	*Optimal root temperature (°C)*	*Source*
Apple		
M 1[a]	13	Nelson & Tukey, 1956
MM 104	13	Carlson, 1965
MM 106	16 - 18	Carlson, 1965
M 7[a]	16	Nelson & Tukey, 1956
MM 111	16 - 18	Carlson, 1965
M 26	16 - 18	Tromp, 1978
M 16[a]	25	Nelson & Tukey, 1956
Seedlings[a]	25	Nelson & Tukey, 1956
M 9[b]	25	Gur *et al.*, 1976 a
M 9	28 - 30	Tromp, 1984, 1992
MM 106	25	Tromp, 1978
M 2	22	Gur *et al.*, 1976 a
M 7	30	Gur *et al.*, 1976 b
Khashabi	30	Gur *et al.*, 1976 b
Peach	24	Proebsting, Young, 1980
Grapevine	27 - 28	Kubota & Shimamura, 1984 Nakamura & Anima, 1970a
Black walnut[c]	17 - 19	Kuhns *et al.*, 1985
Pecan[c]	30	Woodroof & Woodroof, 1934
Blueberry (highbush)[b]	32	Bailey & Jones, 1941

[a] cultured in nutrient solution

[b] refers to total shoot growth only

[c] refers to root growth only

The considerable discrepancy between the optima for the M7 apple rootstock -16°C, according to Nelson and Tukey (1956), and 30°C according to Gur *et al.* (1976), may be due to the fact that the lower optimum was obtained in water cultures, whereas the higher one is based on sand cultures. As will be detailed later the damage of supraoptimal root temperature stems partly from anaerobic conditions in the root system, which may be aggravated in water cultures.

Another method for the evaluation of apple rootstock clones according to their resistance towards supraoptimal root temperatures was the establishment of the relative weight reduction of plants grown at 35°C root temperature (a definitely harmful temperature) as compared to plants of the same clone grown at 25°C root temperature (Gur *et al.,* 1976b) (Table 2). Among the most resistant clones, according to this method, emerged M 25 and M 7. Very susceptible to damage was M 2. Early decline of apple trees grafted on this rootstock in warm countries has been reported from Israel (Samish *et al.,* 1975) and Zimbabwe (Foster, 1963).

Table 2. The fresh weight of roots of apple rootstocks grown in sand cultures at a root temperature of 25 and 35°C from May 17th to July 5th and the elongation of excised root tip cultures at 37 relative to 32°C (Reproduced from Gur *et al.*, 1976b)

Rootstock clone	*Root Temperatures*			*Elongation of root Tips at 37°C as Proportion of 32°C*
	25°C	*35°C*		
	g/plant	*g/plant*	*% of weight at at 25°C*	
M25	50	41	82	87
M7	57	42	74	39
MM 109	57	34	60	38
M9	28	12	43	42
M2	62	23	37	35
Italian Doucin	27	10	36	48
Khashabi	72	22	31	34

Significant F values for root weight: for rootstock = 5.29^{++}
for temperature = 72.50^{+++}
for rootstock x temperature = 1.90^{+}

Still another method for establishing the response of apple rootstocks to root temperature was the cultivation of their root tips *in vitro* at various temperatures (Muromtsew, 1962; Gur *et al.,* 1976b). However the results obtained with this method were not identical, as to the optimal root temperature and the respective resistance of various apple rootstock clones towards supraoptimal root temperature, with results obtained with intact plants (Gur *et al.,* 1976b). Apparently the plant tops and not only the roots are important with respect to these parameters, a fact which may be also recognized by the differences in the resistance of apple cultivars grafted on the same rootstock, *e.g.* the cultivar 'Orleans' being relatively resistant and 'Starking Delicious' susceptible with respect to the reduction in the dry weight of their roots at 35 as against 25°C (Gur *et al.*,1976b). When two cultivars on the M 9 rootstock were grown at constant root temperatures for 5 months the total shoot growth and particularly the total growth of laterals increased with rising root temperature in the range of 7 to 28°C and so did their number, but not their average length. Lateral shoots tended to occur further down the main stem with increasing temperature (Tromp, 1992). Total leaf weight of the plants was reduced by supraoptimal root temperature in short-term experiments (6 weeks), but in long-term ones (1 yr. excepting the winter months) this was largely compensated by the formation of many spur-type branchlets bearing a large number of small leaves, both at 29° and at 36°C (Gur *et al.*,1976a). The number of well developed flower clusters (with at least four flowers per cluster) increased as well with rising root temperature at least up to 28°C. Due to this fact and the increase in the number of laterals fruit production may be expected to be earlier with rising root temperature (Tromp, 1992).

An indirect effect of supraoptimal root temperature on the cultivation of apple trees was found to be the effect of lowering the resistance of their roots towards the attack of the fungus Sclerotium rolfsii (Sacc.) This resistance is caused by oxidizable polyphenols, and their content in apple plants is considerably reduced by high root temperatures (Lavee, 1963). Clonal rootstock differences with regard to this resistance are also known (Lavee and Samish, 1960).

The importance of the choice of a proper rootstock under conditions of high root temperature has been demonstrated with apricots. Grafted on Prunus domestica or Prunus cerasifera rootstocks, apricot growth is restricted at high soil temperatures compared to the growth on apricot seedlings. On the Prunus insitita rootstock winter dormancy of apricots is prolonged in warm soils (Rom, 1991).

Typical responses of several other fruit species to supraoptimal root temperatures have been reported. With peaches and pears growth stopped almost entirely at 35°C root temperature. Many pear trees died at this temperature (Proebsting, 1943; Young, 1980). At 38°C root temperature pecan growth almost stopped and many shoot tips died (Woodroof and Woodroof, 1934).

For grape vines the optimal root temperature for bud break was 27°C for Muscat of Alexandria on "Hybrid Frank" rootstock (Kubota and Shimamura, 1984), but 35°C for ungrafted Cabernet Sauvignon (Kliewer, 1975). As a rule 35°C was found to be deleterious to grape vines according to several criteria like the average shoot length, the leaf number and area, the stem and cluster weight (Kliewer, 1975). The maximal percent of fruit set was at 30°C (Woodham and Alexander, 1966). A distinct influence of root temperature on grape quality has been established for Delaware vines, 28°C being most favourable for high sugar and low acid content. The strongest fruit color at maturation was obtained by 35°C root temperature, though somewhat later no difference was found between this temperature and 28°C. At 21°C a particularly high fructose/glucose ratio was found (Nakamura and Anima, 1970b).

For kiwi plants the optimal root temperature in nutrient solution cultures for root and shoot growth was 20°C, but even at 30°C, the highest temperature tested, no visual damage occurred (Smith *et al.*, 1989).

2.2 Morphological and anatomical features

Morphological and antomical changes due to supraoptimal root temperature have been described in detail for apples and peaches by Nightingale (1935). He found a reduced root diameter at such temperatures in both species, combined with less mechanical strength. Roots were also more branched. A similar decrease in root diameter at supraoptimal root temperatures was also noted with grape vines (Kubota *et al.*, 1987). The cortex of apple and peach roots was ruptured and sloughed off at such temperatures (Nightingale, 1935; Nelson and Tukey, 1965). The stele was encircled by strongly developed cork cambium, producing external rapidly suberizing cells. When shifting from 16 to 35°C root temperature the active pericycle became strongly suberized and pericycle fibers appeared between four to five days later (Nightingale, 1935). At a root temperature of 35°C apple roots were suberized almost to the tip (Gur *et al.*,1976b). The meristem of the root tips was very limited and the root cambium layer was only one to two cells thick at 29°C, and even less at 32°C. Much of the primary phloem remained at this temperature intact for a long time without secondary development. Secondary xylem was also limited in amount at these temperatures. Vessels were strongly lignified and they were somewhat more abundant and larger at 29°C than at 32°C (Nightingale, 1935). At 35°C many apple roots died. Leaves showed interveinal necrosis and shedding (Gur *et al.*, 1976 b). Supraoptimal root temperatures also reduced apical dominance in apple, increasing the number of laterals but not their length (Tromp, 1992).

2.3 Effect on respiration, photosynthesis and their products

Root respiration rates rise with rising temperature, though with apple roots maximal rates are reached at 35°C, declining at higher temperature (Gur *et al.*, 1972). With grape vines rates increased up to 34°C (the maximum tested) (Kubota *et al.*, 1987).

Due probably to rising CO_2 concentrations the pH of meristematic apple and peach root tissue decreased with rising root temperature (Nightingale, 1935). Increased respiration rates at high root temperatures may produce anaerobic conditions within the root tissue. Thus in apple roots at high temperatures (particularly at 35°C) products of anaerobic respiration like acetaldehyde and ethanol were accumulated (Gur *et al.*, 1972). Both substances are translocated to the leaves, and as feeding apple plants with ethanol proved, can reduce their photosynthetic activity. Such a reduction was actually found in apple plants grown at 36°C root temperature (Nightingale, 1935; Gur *et al.*, 1976c). Partial closure of stomata at such root temperature may also contribute to reduction in photosynthetic rates (Gur *et al.*, 1972). Ethanol content of the roots of four apple rootstock clones at 35°C root temperature as compared to their content at 25°C was correlated with their susceptibility to the damage of supraoptimal root temperatures (Gur *et al.*, 1972).

The high respiration and the low photosynthetic activity at supraoptimal root temperature causes a certain depletion of carbohydrates in the roots. The fate of individual carbohydrates depended on the species. 29 to 35°C root temperature resulted in high levels of starch accumulation in peaches, probably due to the less efficient enzymatic starch decomposition at such relative high temperature, but levels in apple roots were very low, rising only at 32°C (Nightingale, 1935). Acids of the TCA cycle like malic, succinic and fumaric acids, were reduced in apple roots at 35°C root temperature. The decrease in the malic acid being more pronounced in susceptible rootstock clones towards supraoptimal root temperature than in resistant ones (Gur *et al.,* 1972). The possible explanation for this fact may be that pyruvic acid obtained by glycolysis may under anaerobic conditions turn either to potentially harmful acetaldehyde and ethanol or by dark CO_2 fixation into harmless malic acid (Mazelis and Vennesland, 1957).

2.4 Effect on leaf chlorophyll

The chlorophyll content of apple leaves decreased at supraoptimal root temperature (Nightingale and Blake, 1934). The chlorophyll content of several apple rootstock clones was smaller at 35°C root temperature than at 25°C, however the content at 35°C was proportional to the content of leaves of the tested clones at 25°C. The reduction in the chlorophyll content at supraoptimal root temperatures does not seem to be a major factor in the harmful effects of such temperatures (Gur *et al.,* 1976b).

2.5 Effect on water supply

With regard to the effect of supraoptimal root temperature on the water household of trees, a short and a long term effect should be distinguished. Transfering plants from 24°C to 35°C did not affect the foliage immediately due to the well developed system of young roots formed prior to the transfer (Nightingale, 1935). But when apples and peaches were grown at 32 or 35°C root temperature, and occasionally even at 29°C for an extended period slight wilting of the leaves occurred, probably due the increased shoot/root ratio at such temperatures (Gur *et al.,* 1976a). Transpiration rates were maximal at a root temperature of 30°C as compared to 25 and 35°C, however shifting the root temperature for only 12 hours from 25 to 40°C increased transpiration considerably (Gur *et al.,* 1972), as leaf temperature rises with rising root temperature and high leaf temperature decreases the leaf resistance towards the passage of water vapour (Stalfelt,1962). In addition the root permeability to water increases with rising root temperature (Brouwer, 1965), explaining the initial increase in the transpiration rates even at supraoptimal root temperatures. However a prolonged exposure to such temperatures increases the leaf water potential, closes the stomata and reduces transpiration rates (Gur *et al.,* 1972; Gur *et al.,* 1976c). The gradually declining shoot/root ratio under such conditions (Carlson, 1965) also contributes to this effect.

The total water use of M 26 and MM 109 apple clones increased with rising root temperatures up to 30°C, but with the M 9 clone only up to 18°C. Growth increased only up to 25°C at most. Hence water deficits could not generally be regarded as the cause for growth reduction at root temperatures exceeding 25°C (Tromp, 1978).

2.6 Effect on mineral nutrition

The effect of supraoptimal root temperature on the mineral nutrition of deciduous fruit trees has been studied with respect to apples and peaches.

Leaf nitrogen in peaches was found to be high at 30°C root temperature, which was above the optimum for root and shoot growth (Young, 1980). However Nightingale (1935) found low levels of organic nitrogen in apple and peach plants at supraoptimal root temperature. Leaf nitrogen in apples was higher at 35°C root tempera ture than at 30°C (Gur *et al.*, 1979). However according to Nightingale (1935) the main effect of root temperature on the nitrogen metabolism of apple trees was not on the nitrate uptake, but on the nitrate reduction, which increased initially with rising temperature, but dropped with time due to the increase in dead root cortex and inert xylem tissue at high root temperatures.

The total potassium content of apple plants as related to root temperature followed a similar trend to the growth of the trees, reaching maximal levels between 18 to 24°C (Tromp, 1978). The leaf potassium content of apples reached maximal levels at 25 or 30°C root temperature, depending on the rootstock clone. The susceptible M 2 and M 9 clones preferred the lower temperature, the more resistant M 7 and M 13 - the higher one for maximal K content in the leaves (Gur *et al.*, 1979). The deterioration in the potassium uptake at supraoptimal root temperature is possibly related to the reduction in the percentage of fine roots found at such temperatures (Gur *et al.*, 1976b). However the potassium levels of leaves declined more at 36°C than that of roots or shoots (Gur *et al.*, 1979). K Fertilization of apple plants grown in sandy loam or clay, and particularly the use of foliar potassium nitrate sprays minimized the deleterious effect of 29°C root temperature on the growth of the plants, but with 36°C potassium application was without effect (Gur *et al.*, 1976a). Application of potassium fertilizer also reduced the deleterious effect of 35°C root temperature on the chlorophyll content, and the photosynthetic and transpiratory rates of apple leaves (Gur *et al.*, 1976c).

Under conditions of potassium deficiency putrescine accumulates in apple leaves, this being an indication for K deficiency (Hoffman and Samish, 1972), however at 30 and 35°C root temperature no such accumulation occurred (Gur and Shulman, 1972). The accumulation of putrescine under conditions of potassium deficiency seems to be related to the accumulation of certain organic acids, which are instrumental in the conversion of arginine into putrescine (Smith and Sinclair 1967). In potassium deficient apple plants acetic acid accumulated particularly in the roots at optimal root temperature. At supraoptimal root temperatures this accumulation was only minimal. Thus the accumulation of acetic acid and putrescine in potassium deficient apple plants is probably linked (Gur and Shulman, 1972).

The Ca uptake and translocation of apples was less affected by high root temperature than the K uptake and was not related to shoot growth as was the K uptake. However in the M 9 rootstock clone the Ca uptake was reduced at 30°C as compared to 24°C root temperature (Tromp, 1978). The Ca content of the leaves was maximal at 30°C, however at very high K supply this maximum dropped to 25°C probably due to a synergistic interaction between K and Ca at 25°C as against an antagonistic effect at higher temperatures (Gur and Shulman, 1972). Both root and leaf Mg and Na content in apples reached a maximum at 36°C root temperature, the differences between 22 to 29° being small, except with root Na, which increased sharply between 22 to 29°C (Gur *et al.*, 1979). Increased Na uptake as against decreased K uptake is characteristic of stress conditions, like root hypoxia (Labanauskas *et al.*, 1965).

The involvement of root hypoxia in the effects of supraoptimal root temperatures has been already discussed.

A very marked effect of supraoptimal root temperature of considerable practical importance is the reduction in the leaf Zn content. With certain apple rootstock clones this reduction occured between 25 to 30°C, with others above 30°C root temperature. Certain rootstock clones showing normal leaf zinc levels at 30°C were definitely deficient at 35°C. The leaf zinc content was in certain rootstock clones (M 7, M 9) 72% of the content at 25°C, in other clones (Khashabi) it was as low as 43%. (Samish *et al.,* 1968) (Table 3). As the source of Zn in plants is mainly the upper soil layer (Hibbard, 1940), which heats most, the high frequency of Zn deficiency in apples growing in warm zones is of considerable importance. The deleterious effect of high root temperatures on the Zn levels was not limited to the leaves but occured also in stems and shoots, the effect was particularly marked in roots (Gur *et al.,* 1979). With respect to iron, data for peaches indicate that at 30°C root temperature, which exceeds the growth optimum, leaf Fe was much reduced compared to 20°C (Young, 1980)

Table 3. The effect of high temperature upon the zinc content of apple rootstock plants (Reproduced from Samish *et al.*, 1968)

	Zinc in leaf dry matter		
Rootstock Cultivar	*at 25°C (ppm)*	*at 35°C (ppm)*	*at 35°C (% of Concentration of 25°C)*
M 7	49	36	73
M 9	62	45	73
M 2	54	39	72
MM 102	56	38	68
M 1	67	43	65
Italian Doucin	76	43	57
M 25	98	47	48
Khashabi	93	40	43

Significance according to F-test: for temperature = 40.3^{+++}
for rootstock = 2.7^{+}
for interaction = 2.2^{+}

2.7 Effect on growth regulators

Several responses of apple plants grown at supraoptimal root temperatures point towards the involvement of cytokinins, produced in the roots and transported to the tops. With root temperatures of 29°C and above the number of nodes formed decreased, whereas their length remained constant (Gur *et al.,* 1976a). In peas for example the number of nodes was positively related to the cytokinin levels in the plants (Sachs and Thimann, 1967). The number of apple shoot laterals and flowers increased with rising temperature up to 28°C (Tromp, 1992). The effect of cytokinins on the reduction of apical dominance (resulting in an increase in the number of laterals) (Phillips, 1975) and on increasing the number of apple flowers (McLaughlin and Greene, 1991) has been demonstrated previously.

The reduction in the apple leaf chlorophyll content at 40°C root temperature could be prevented by benzyladenine or kinetin (Gur *et al.*, 1972), pointing also towards a lack of endogenous cytokinins in the leaves at this temperature.

Finally it has been shown directly that the cytokinin content of apple roots was lower in apple plants grown at 35°C than in plants grown at 25°C root temperature (Gur *et al.*, 1972). The maximal zeatin + zeatin riboside content in apple roots was found to occur at 18-20°C root temperature (Skogerbo and Mage, 1992). These results do not correspond well with the root temperature optima for the formation of flowers and laterals.

3 Effects of high root temperature on rootstock-scion incompatibility

3.1 Field observations

A high temperature at the graft union seems to be an aggravating factor in certain cases of rootstock-scion incompatbility, though direct evidence is rather scarce. A comparison of two fruit growing areas in France regarding the incompatbility between apricots and certain Prunus domestica rootstock clones proved that symptoms of graft incompatbility were much more severe in the warmer area (Gard province as against Gironde) (Duquesne, 1970). Many pear cultivars are grown successfully on quince rootstocks in central Italy, but failed totally in warmer Sicily (Carrante, 1941). The list of pear cultivars which succeed in the warm climate of Israel on quince rootstoks is very limited and many cultivars which are compatible on quince in temperate zones are clearly incompatible in Israel. Examples are 'Beurre Superfin', 'Beurre d'Amanlis', 'Beurre Diel' (Gur, 1957). To some degree was this incompatbility alleviated by soil mounds around the trunk covering the graft union. These mounds, which were originally installed for inducing scion rooting, prevented the decline of the trees due to graft incompatbility to some degree, even in cases when no scion rooting occured (Gur, unpublished), probably due to their cooling effect.

3.2 Mechanisms involved

The temperature effect on the incompatbility between quince rootstocks and pear scions finds its explanation in the theory linking this case of incompatibility with the enzymatic decomposition of the cyanogenic glycoside prunasin found in quinces, when ascending into the pear tissues, where it is not found normally (Gur *et al.*, 1968). The glycoside is decomposed in a narrow zone close to the graft union, setting free toxic hydocyanic acid, which causes the death of tissues in this zone and prevents the free passage of assimilates through the graft union. The resulting sugar depletion in the rootstock causes the decomposition of the glycoside prunasin in the rootstock tissues, and by setting free hydrocyanic acid causes the gradual death of the rootstock bark. The evolution of hydrocyanic acid from quince bark slices was found to depend much on temperature, rising steeply at temperatures exceeding 30°C and reaching a maximum at 47°C. Particularly marked was the temperature effect if the quince bark slices were obtained from quince rootstocks beneath incompatible pear cultivars, deprived of sugars ('Beurre Superfin', 'Bartlett'). The increased respiration rates at high temperatures apparently contributed to the sugar depletion in the rootstocks. With ungrafted quince the Q_{10} of the reaction was 2.2, which is common for enzymatic reactions. With quince bark beneath a compatible pear cultivar the Q_{10} was 5.0, whereas with bark from an incompatible combination the Q_{10} was 6.6 (Gur *et al.*, 1968) (Fig. 1)

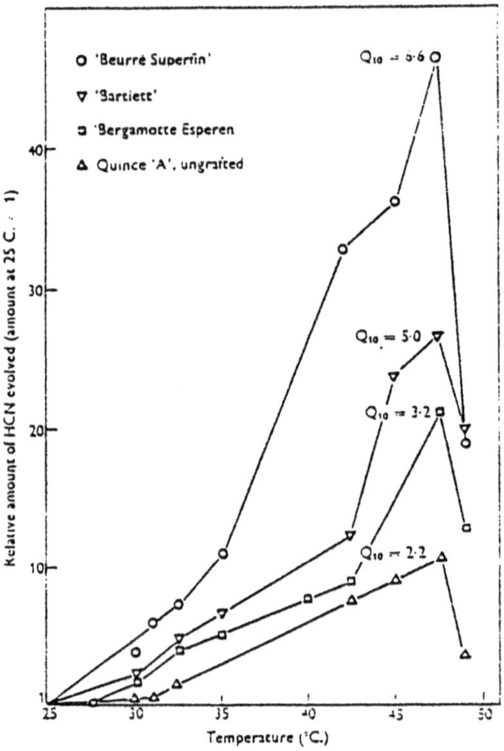

Figure 1. The influence of temperature on the evolution of HCN from stem bark of quince 'A' rootstocks grafted with various pear cultivars. (5-year old trees, reaction time-3hrs). Redrawn from Gur *et al.* 1968.

A confirmation regarding the importance of the temperature factor was obtained by *in vitro* studies, dealing with the elongation of pear and quince shoots in media containing the cyanogenic glycoside amygdalin. Growth and survival of the shoots was much affected by amygdalin, but only at temperatures exceeding 27°C (Vasilakakis, 1991). For certain reasons this author envisages still other unknown quince constituents which may be toxic to pear tissues besides prunasin.

3 References

Ashbel, D., Eviatar, A., Doron, E., Ganor, E. and Agmon, Y. (1965). Soil Temperature in Different Latitudes and Different Climates. *The Hebrew University of Jerusalem,* Israel.

Bailey, J.S. and Jones, L.H. (1941). The effect of soil temperature on the growth of cultivated blueberry bushes. *Proc. Amer. Soc. Hort. Sci.* 38, 462-464.

Belding, D. and Young, E. (1987). Shoot and root temperature effect on carbohydrate levels during budbreak in young apple trees. *J. Amer. Soc. Hort. Sci.* 112, 596-599.

Brouwer, R. (1965). Water movement across the roots, in *The State and Movement of Water in Living Organisms.* S.E.B. Symposia XIX, (ed. G.E. Fogg) Cambridge University Press, pp.131-149.

Carlson, R.F. (1965). Responses of Malling Merton clones and Delicious seedlings to different root temperatures. *Proc. Amer. Soc. Hort. Sci.* 86, 41-54.

Carrante, V. (1941). Per il miglioramento della frutticoltura in Sicilia con particolare riguardo alla Zona Etnea. *Ann. Rep. Staz. Frutt., Agrum. Aci reale* 16, 27-102.

Duquesne, J. (1970). Etude de la compatibilité de greffe de quelques cultivars de Prunus armeniaca (Koehne) sur divers types de Prunus. II. Compatibilité des cultivars "Canino" et "Rouge de Roussillon" sur divers clones de Prunus domestica. *Ann. Amel. Plantes* 20, 453-67.

Foster, L.Y. (1963). The performance of clonal apple rootstocks in Nyassaland. *Rhodesian J. Agric. Res.* 1, 39-46.

Gur, A. (1957). The compatibility of the pear with quince rootstocks. *Agricultural Research Station, Rehovot, Special Bulletin* No. 10 (Hebrew, English Summary).

Gur, A., Samish, R.M. and Lifshitz, E. (1968). The role of the cyanogenic glycoside of the quince in the incompatibility between pear cultivars and quince rootstocks. *Hortic. Res.* 8, 113-134.

Gur, A. and Shulman, Y. (1972). The influence of high root temperature on the potassium nutrition and on certain organic constituents of apple plants, in *Recent Advances in Plant Nutrition* (ed. R.M. Samish), Gordon and Breach Sci. Publ., New York, London, Paris, Vol. 2, 643-656.

Gur, A., Bravdo, B. and Mizrahi, Y. (1972). Physiological responses of apple trees to supraoptimal root temperature. *Physiol. Plant.* 27, 130-138.

Gur, A., Hepner, J. and Mizrahi, Y. (1976 a). The influence of root temperature on apple trees. I. Growth responses related to the application of potassium fertilizer. *J. Hortic. Sci.* 51, 181-193.

Gur, A., Mizrahi, Y. and Samish, R.M. (1976 b). The influence of root temperature on apple trees. II. Clonal differences in susceptibility to damage caused by supraoptimal root temperature. *J. Hortic. Sci.* 51, 195-202.

Gur, A., Bravdo, B. and Hepner, J. (1976 c). The influence of root temperature on apple trees. III. The effect on photosynthesis and water balance. *J. Hortic. Sci.* 51, 203-210.

Gur, A., Hepner, J. and Shulman, Y. (1979). The influence of root temperature on apple trees. IV. The effect on the mineral nutrition of the tree. *J. Hortic.Sci.* 54, 313-321.

Hammond, M.W. and Seeley, S.D. (1978). Spring bud development of Malus and Prunus species. *J. Amer. Soc. Hortic. Sci.* 103, 655-657.

Hibbard, P.L. (1940). Accumulation of zinc in soil under long persistent vegetation. *Soil Sci.* 50, 53-55.

Hoffman, M. and Samish, R.M. (1972). Free amine content in fruit tree organs as an indicator of the nutritional status with respect to potassium, in *Recent Advances in Plant Nutrition* (ed. R.M. Samish) Gordon and Breach Science Publishers, New York, London, Paris, Vol. 1 pp. 189-206.

Kliewer, W.M. (1975). Effect of root temperature on budbreak, shoot growth and fruit set of 'Cabernet Sauvignon' grapevines. *Amer. J. Enol. Vitic.* 26, 82-89.

Kubota, N. and Shimamura, K. (1984). The effect of root temperature under forcing conditions on bud break, shoot growth and flower cluster development in 'Muscat of Alexandria' grape vines. *J. Jap. Soc. Hortic. Sci.* 53, 242-250.

Kubota, N., Yanagisawa, G. and Shimamura, K. (1987). The effect of root temperature on budbreak, shoot growth and development of flower clusters in fruiting cv. Muscat of Alexandria grape vines. *J. Jap. Soc. Hortic. Sci.* 56, 16-23.

Kuhns, M.R., Garret, H.E., Teskey, R.O. and Hinckley, T.M. (1985). Root growth of black walnut trees related to soil temperature, soil water potential and leaf water potential. *Forest Sci.* 31, 617-629.

Labanauskas, C.K., Letey, J., Stolzy, L.H. and Valoras, N. (1965). Effects of soil oxygen and irrigation on the accumulation of macro - and microelements in citrus seedlings. *Soil Sci.* 101, 378-84.

Lavee, S. (1963). The effect of mulch on the resistance of EM II apple stock to Sclerotium rolfsii (Sacc.).*Proc. Amer. Soc. Hortic. Sci.* 82, 25-35.

Lavee, S. and Samish, R.M. (1960). Resistance of apple rootstocks to Sclerotium rolfsii (Sacc.). *Ktavim* 10, 5-13.

Mazelis, M. and Vennesland, B.(1957) Carbon dioxide fixation into oxalacetate in higher plants. *Plant Physiol.* 32, 591-600.

McLaughlin, J.M. and Greene, D.W. (1991). Fruit and hormones influence flowering of apple. II. Effects of hormones. *J. Amer. Soc. Hortic. Sci.* 116, 450-53.

Mowat, A.D. (1994). The effect of root temperature on the bud dormancy in persimmon, *in Handbook of Environmental Physiolgy of Fruit Crops*, (ed. B. Schaffer and P.C. Andersen) CRC Press, Boca Raton, Fl. Vol. 1, p. 217.

Muromtsev, I.A. (1962). Temperature and root growth. *Soviet Plant Physiol.* 9, 334-37.

Nakamura, R. and Anima, H. (1970). Effect of soil temperature on the growth of Delaware vines. *Sci. Rep. Fac. Agric. Okayama,* No. 35, pp. 45-55.

Nakamura, R. and Arima, H. (1970). Effects of soil temperature on the quality of the berries of Delaware vines. *Sci. Rep. Fac. Agric. Okayama, No.* 35, pp. 57-71.

Nelson, S.H. and Tukey, H.B. (1956). Effects of controlled root temperatures on the growth of East Malling rootstocks in water culture. *J. Hortic Sci.* 31, 55-63.

Nightingale, G.T. (1935). Effects of temperature on growth anatomy and metabolism of apple and peach roots. *Bot. Gaz.* 96, 581-639.

Nightingale, G.T. and Blake, M.A. (1934). Effect of temperature on the growth and composition of Stayman and Baldwin apple trees. *N. Jersey Agric. Exp. Sta., Bull.* 566.

Phillips, I.D.J. (1975). Apical dominance. *Ann. Rev. Plant Physiology* 26, 341-67.

Proebsting, E.I (1943). Root distribution of some deciduous fruit trees in a California orchard. *Proc. Amer. Soc. Hortic. Sci.* 43, 1-4.

Rom, R.C. (1991). Apricot rootstocks: Perspective, utilization and outlook. *Acta Hortic.* 293, 345-353.

Sachs, T. and Thimann, K.V. (1967). The role of auxins and cytokinins in the release of buds from dominance. *Amer. J. Bot.* 54, 136-44.

Samish, R.M., Gur, A. and Shulman, Y. (1968). The effect of root temperature on the mineral composition of the apple leaf. *Control de la fertilization de las plantas cultivadas*, (*II coloquio Europeo y Mediterrano*) pp. 675-85.

Samish, R.M., Gur, A. and Arzi, M. (1975). Apple rootstock trial at Kevuzat HaSharon - Ramat David (Hebrew, English summary). *Agricultural Research Organization and Fruit Growers' Association,* Bet Dagan, Israel. Pamphlet No. 148, pp. 62.

Skogerbo, G. and Mage, F. (1992). Xylem cytokinin content of apple (Malus domestica Borkh.) as affected by season, soil management and root temperature. *Norwegian J. Agric. Sci.* 6, 485-497.

Smith, G.S., Buwalda, J.G., Green, T.G.A. and Clark, C.J. (1989). Effect of oxygen supply and temperature at the root zone on the physiolgy of kiwifruit. *New Phytol.* 113, 431-37.

Smith, T.A. and Sinclair, L.C. (1967). The effect of acid feeding on amine formation in barley. *Ann. Bot.* 31, 103-11.

Stalfelt, M.G. (1962). The effect of temperature on opening of the stomata cells. *Physiol. Plant.* 15, 772-79.

Tromp, J. (1978). The effect of root temperature on the absorption and distribution of K, Ca and Mg in three rootstock clones of apple budded with Cox's Orange Pippin. *Gartenbauwiss.* 43, 49-54.

Tromp, J. (1984). Flower - bud formation in apple as affected by air and root temperature, air humidity, light intensity and day length. *Acta Hortic.* 149, 39-47.

Tromp, J. (1992). The effect of soil temperature on lateral shoot formation and flower - bud formation in apple in the first year after budding. *J. Hortic. Sci.* 67, 787-93.

Vasilakakis, M.D. (1991). An *in vitro* study of graft - incompatibility between pear and quince its expression in relation to rootstock, scion, cultivar and temperature. *Advances in Hortic. Sci.* 5, 51-54.

Woodham, R.C. and Alexander D. McE. (1966). The effect of root temperature on development of small fruiting Sultana vines. *Vitis* 5, 345-50.

Woodrof, J.G. and Woodrof, N.C. (1934). Pecan root growth and development. *J. Agric. Research* 49,511-30.

Young, E. (1980). Response of seedling rootstocks of peach to soil temperature. *HortScience* 15, 294-96.

Young, E. and Werner, D.J. (1984). Rootstock and scion chilling effects on apple and peach bud break. *Compact Fruit Trees* 17, 32-36.

Young, E. and Werner, D.J. (1984). Effects of rootstock and scion chilling during rest on resumption of growth in apple and peach. *J. Amer. Soc. Hortic. Sci.* 109, 548-51.

7 PHYSIOLOGICAL CONSIDERATIONS FOR GROWING TEMPERATE-ZONE FRUIT CROPS IN WARM CLIMATES

MIKLOS FAUST*

Fruit Laboratory, Beltsville Agricultural Research Center, Beltsville, MD, USA

1 Introduction

Temperate fruit trees evolved in climates prevalent between the 34th and 48th northern parallels. Various species of fruit trees are native to China, Kazakhstan, the Caucasus, and Europe. Temperate-zone fruit trees are adapted to a winter - summer climate where the typical winter is relatively cold and the summer is moderately mild. Evolution forced them to produce their crop, form flower primordia during the summer and fall and complete flower development just before bloom during next spring, store a certain amount of reserves to survive the winter and support the early spring activities from reserves before the leaves are old enough to produce photosynthates again. At the beginning of this century, trees started to be planted in large quantities in milder than traditional climates. This introduced new considerations. The trees did not receive enough cold in southern California (Chandler 1925) or in southern Georgia (Weinberger 1950), but also did not need the reserves for the harsh winters. The high summer heat encountered in the new areas also caused new problems especially in the southeastern United States where high night temperatures increased respiratory losses and this limited productivity. Beginning in the 1940s trees were planted in even more tropical climates. This magnified both types of difficulties: those caused by the lack of winter and those created by high summer heat. Problems related to lack of winter cold are discussed in the various chapters of this book. This chapter is limited to a discussion of those problems of tree physiology that are related to the high temperatures of the subtropical areas.

2 Climatic considerations

Aspects of tropical climate that influences productivity of temperate-zone fruit trees the most are those that influence budbreak and bloom. This focuses attention on climatic factors occurring during the dormant period. In warm climates, if the bloom is relatively normal then the climatic factors that affect the tree during the growing period become of paramount importance. In general we are concerned with four climates that influence the general physiology of the trees. These are:

1. Climates where temperate-zone trees are well adapted.
2. Climates that provide enough chilling but the growing period is warmer than temperate-zone trees are accustomed to.

Amnon Erez (ed.), Temperate Fruit Crops in Warm Climates, 137-156.
© 2000 *Kluwer Academic Publishers. Printed in the Netherlands.*

* Deceased June 1998

3. Climates that are insufficient to fully chill the trees but there is some cold period and trees need help for overcoming dormancy. Summer temperatures are troublesome for the trees.
4. Climates that are continually warm, trees receive no chilling at all and the climate is warm enough that the trees be forced to regrow immediately after harvest. To understand the difficulties temperate-zone fruit trees face in tropical and subtropical climates we have to further analyze the classical climates of temperate-zone trees that occur during the growing period. This can be categorized into four groups:
    1. The climate is characterized by cool days and cool nights. Prime examples are England, Holland, Germany or northern France and the mountain areas such as the Bolzano Valley of Italy and the Bodensee area of Germany. Because the climate is generally cool, growth of trees is moderate. Apple and pear trees in this climate are relatively small, they can be grown on dwarfing rootstocks without much difficulty. This climate is generally too cool for peaches and apricots and they are not adapted for at these locations. Response to pruning is moderate thus heavy pruning can be practiced. Spurs of apples age and renewal spur pruning may be required.
    2. In this climate warm days and cool nights are prevalent. It occurs mostly in arid climates such as Washington State, central Italy and oceanic climates such as New Zealand or Tasmania, Australia. Trees in these climates are very productive retaining enough photosynthates during the cool night to be able to form flower buds abundantly. It is an ideal climate for high productivity in apples, pears, peaches, plums, and apricots.
    3. This is a clear variation of climate 2 in areas where hot days and moderate night temperatures occur, such as around the Mediterranean Sea, in California, southern Australia and central Chile. Apple and pear trees in both areas 2 and 3 are moderate in size, having abundant amounts of spurs and peach trees are large with upright branching. Response of the tree to pruning is moderate because growth takes place during the night and the night being cool thus regrowth, triggered by pruning, is limited.
    4. This climate is known for its warm days and warm nights and occurs typically in the Mid Atlantic area of the United States, southern France, Japan, and China. In this climate the productivity of trees is the lowest among all three climates. Trees typically loose 40 to 50% of the carbohydrates made during the day through respiration during the night. Shoot growth in apple is relatively long because night temperatures are high and shoot growth is not limited by temperature. Shoots are thin, spurs are relatively weak, and productivity is much lower than in climate group no. 2. Peaches have a branch structure that is more prostrated than in climate 2. Typical day and night temperatures, respectively, for the four climates are presented in Table 1.

Rain is about 800, 400, 400 and 1000 mm, respectively, in the four climatic regions. Because of the rain distribution, irrigation is essential in area types no. 2 and 3.

Table 1. General temperature regimes during growing period in various climates

Climate number	*Day °C*	*Night °C*
1	20	15
2	30	10-17
3	30	20
4	30	24

(Adapted from Climate and man, 1941)

In tropical areas trees often are grown at high elevations to assure the minimum amount of cold required for breaking dormancy. The mid growth-season temperatures in temperate climate no 4 (30 and 24°C for day and night, respectively) are not greatly different from that experienced by trees in a tropical climate. The climate at the tropics at higher elevations, for example in Kenya at 2000 m where 27-30: 12-15°C day:night temperature occur, is comparable to climate type no. 2. However, the difference is still very important. In the temperate zone around 30-35 days of such temperatures occur during the 160 days growing season of apples. In tropical climates this period is much longer and may extend the entire year. The course of monthly mean temperatures are illustrated in Fig. 1 with increasing latitudes and altitudes. It is obvious from this illustration that near the equator or at low altitudes even at 10° N temperatures never drop below the 20°C range. In order to have night temperatures in the 15-17°C range that is normal for temperate fruits, one must be at the 1000-1500 m altitude or higher.

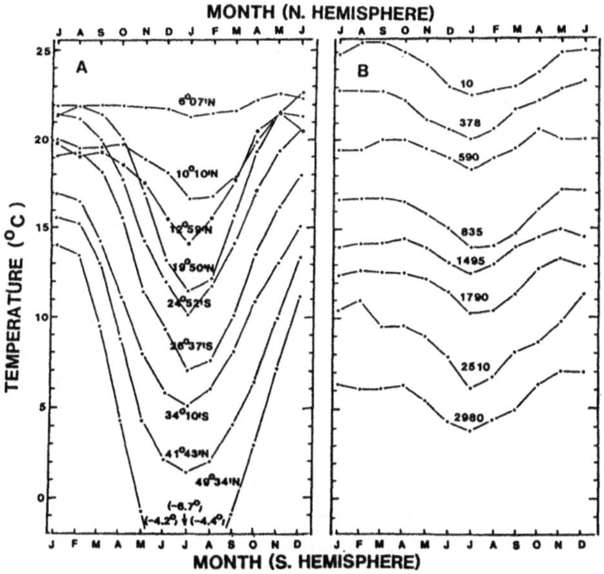

Figure 1. Monthly mean minimum temperatures (°C) recorded at fruit growing areas of differing latitudes (**A**) and elevations (**B**) in meters. All locations in (B) in Venezuela. From Edwards (1987c).

At other tropical locations, usually at lower elevations, where temperatures are always above 20°C year around, continuous cropping is practiced. Trees, after producing their crops, are defoliated and the prevailing high temperatures allow the start of a new growth cycle. This continuous production, with no reserve accumulation, introduces an entirely new physiological demand on the tree of which little is known about. Overall climatic conditions encountered by several authors working with temperate-zone fruit trees in tropical areas are listed in Table 2.

Table 2. Environmental data of selected tropical areas where temperate-zone fruits are grown

Location	*Latitude Degree*	*Elevation m*	*Precip. mm*	*Max °C*	*Min[x] °C*	*References*
Brazil						
Santa Catarina	27	800 -1000	1600	27	17	Denardi et al. 1988
Ecuador	0	2000 -3000	1000	24.4	8.8	Niegel, 1988
Guatemala	15	1400 -2000				Williams & Menegazzo, 1988
Venezuela						
Peach growing	10	1200 -1750	400	30	20	Ramirez, 1987
Apple growing	8	2000 -2510	1100			Colmenares, 1988;
Peru						Edwards, 1987
Below cloud layer	13	100 - 400		30	18	Bederski, 1987
Above cloud layer	17	780 -1400		30	12	Bederski, 1987
Mexico						
Durango	24			24-26	6-12	Diaz et al., 1987
Puebla	19	1800		28	17	Ignazio & Laborde, 1987
Indonesia						
Batu	7	1100	1348	26.7	18.4	Edwards, 1987
Kenya						
Kitale	1	1500		23.3	11.5	Edwards, 1987
Philippines						
Lowland	13	200	1849	29.7	20.1	Edwards, 1987
Highland	17	1500	3422	23.3	15.1	Edwards, 1987
Taiwan						
Lowland	23		3000	27		Edwards, 1987
Pakistan						
Murree	34		1350			Ahmad 1987
Peshawar	34		400	27	15	Ahmad, 1987
Reunion Island	21	800 -1500	1280	22.6	13.6	Aubert & Bertin, 1987
Thailand						
Highlands	19	1000	1544	24-26	13-16	Subhadrabandhu & Punsri 1987

[x] Minimum temperatures were reported for the year and they do not represent the minima during the growing period. Nevertheless they can serve as descriptors of the specific climate.

3 Distribution and utilization of carbohydrates

Producing a crop in a fruit tree is a complicated process. One of the crucial step in the process is allocating the photosynthates for various processes and development of organs. Respiration, a process greatly dependent on temperature, is important in utilization of photosynthates because the rate of respiration determines how much carbohydrates the tree can allocate for structures that determine productivity. The yearly respiratory loss is very high even in temperate climates. Hansen (1977) in Denmark and Lakso (1994) in New York and New Zealand estimated that about 40-45% and 45-50% of fixed carbon was used for respiration in apple, respectively. Evenari *et al.* (1977) estimated a similar respiratory loss of 45-50% in apricot from the desert areas of Israel. From the physiological point of view not all the respired carbon is lost. Some of the respiration is used for maintenance and a considerable amount of energy generated by respiration is used for growth. Growth takes place during the night because this is the time when turgor pressure can force the enlargement of the cells. In temperate climates cool nights limit growth. In tropical climates, especially at low latitudes, where nights are very warm, growth is abundant and respiration is high. From the production point of view excessive growth is an unwanted activity because it consumes carbohydrates on the expense of fruiting (Forshey and Elfving, 1989). Thus limiting shoot growth is the only effective way to reduce respiratory losses and make additional carbohydrates available for activities of the tree that are essential for fruiting.

Temperate-zone fruit trees distribute most of the available carbohydrates to their fruit, considerably less for leaves and wood and relatively limited amount for the root. Considering the fact that nearly half of the photosynthates are used for respiration, the proportion of total photosynthates utilized for fruit ranges from an estimated 26% in peach (Grossman and DeJong 1994) and 33.5% in apple (Lenz, 1986) to as low as 1.7 to 11.8% in apricot (Schulze *et al.*, 1977). These values were obtained in traditional temperate-zone climates and adjusted for respiration which in those conditions rarely if ever exceed more than half of the total photosynthate made. Naturally respiration depends on prevailing temperatures. In general, respiration responds to temperature logarithmically over the range of temperatures from 0 to 42°C (Lakso, 1994). Grossman and DeJong (1994) estimated Q_{10} values near 2 for peach leaf respiration when they compared R_{20} (20-30°C) with R_{30} (30-40°C), with somewhat lower values (1.61) for stems. Similar Q_{10} values were reported for peach fruit between 20 and 30°C (DeJong *et al.*, 1987; Pavel and DeJong, 1993). Somewhat higher values (Q_{10} 2.32) were reported for apple (Butler and Landsberg, 1981). If the existing information is projected to tropical conditions one must specify the elevation where respiratory losses are considered. While at low elevations where temperatures are considerably higher during the night, one must conclude that respiratory losses are considerably higher and consequently carbohydrates left for use for fruit production or especially for root growth must be much lower. In contrast, at higher elevation where night temperatures are comparable to those occurring in the temperate zone, respiratory losses are probably similar to those reported above.

There are three periods when carbohydrate availability and distribution is crucial within the tree. These are:

1. before harvest when shoot and fruit growth occur and they compete for carbohydrates with root growth;
2 after harvest when reserves are accumulated; and
3 as the growth cycle starts before leaves are fully expanded bloom and fruit set solely depend on reserves.

In the temperate zone, after the fruits are harvested, carbohydrates are deposited as reserves to be used for the next growth cycle activities (Faust, 1989). Flowering and fruit set normally occur before leaf buds break and growth starts. If shoot growth occurs before flowering, reserve-carbohydrates are not sufficient for a high rate of fruit set and also satisfying shoot growth. Fruit set and early shoot growth require stored carbohydrates. In a tropical double cropping system, when trees are forced into the second crop soon after the first crop, stored carbohydrates may not be available for the burst of activities initiated by forcing the second crop. This may be the reason why trees are not defoliated immediately after harvest. During this 3-4 weeks "rest" period, needed carbohydrates can accumulate in the wood to support the ensuing bloom. However, if the climate is warm and leaf buds break at the same time or before bloom occurs, the available carbon is not sufficient to support both activities. At this time the shoots are the stronger sinks and no fruit set occurs. In tropical conditions, especially after hand defoliation budbreak of leaf buds precede and exceed flower budbreak (Llanes *et al.,* 1987). Young apple shoots, for 15 to 25 days after budbreak, depend on stored carbohydrates because they do not generate enough carbohydrates to be self sufficient (Johnson and Lakso, 1986). This can be devastating for fruit set after defoliation.

Edwards (1987a) reported that in the Philippines when temperatures are high (at or above 15°C) no fruit set occurred in peaches. He attributed this to a temperature effect. However, at high temperatures leaf buds often break at the same time flower buds open and there is a direct competition for the meager supply of stored carbohydrates. This results in poor or no fruit set. In Guatemala, because of the lack of deep dormancy, apple trees flower and fruit simultaneously and they tend to have three waves of flowering. The first wave often suffers from frost, but the surviving fruits are large and of high quality. Fruit produced by the third wave of bloom, when obvious shoot competition exists, are small and of poor quality (Williams and Menegazzo, 1988).

The level of photosynthetic activity and the distribution of carbohydrates are crucial in obtaining high productivity from the trees. In lowland areas of the tropics where night temperatures are high, respiratory losses are also expected to be high and fruit yield low. This is obvious when comparing apple yields of east and west coast of the United States to those of New Zealand. Using the productivity of the east coast as the base, the west coast is twice as productive, while in New Zealand the productivity is four fold of that attainable on the East Coast. The differences in productivity are directly related to prevailing night temperature of the area. Mean night temperatures for mid growing season in the above areas are 24, 17 and 10°C, respectively. In the highland areas of the tropics where night temperatures are lower we must expect comparable respiratory losses to trees in the temperate-zone. However, trees located in tropical highlands often loose their crop due to freezes (Williams and Menegazzo, 1988) and the resulting light crop results in lessened competition for carbohydrates and greatly increased shoot growth.

Budbreak of leaves prior to bloom may occur in both low-land and highland tropical conditions. Early leafing sets up competition for fruit set and results in a fruitless state. Thus, in recording budbreak, differentiating between opening of leaf and flower buds is necessary. We have to consider factors such as carbohydrate utilization in warm-night climates where losses are likely to be very high or conditions when the trees are defoliated soon after harvest and have no opportunity to accumulate sufficient reserves to start their new growth cycle. Naturally, carbohydrate distribution will influence many other activities of the tree, some of which are mentioned in this chapter.

4 Root growth

Root growth and root function is an integral part of tree productivity (Faust, 1989). Schulze *et al.* (1977) reported in a detailed study with apricots that 47.4% of the net CO_2 fixation was respired during the night, 42.5% was used for the permanent structures, leaves, fruit and twigs (of this only 11.8% was used for fruit), and 10.1% was transported to the roots. His test trees were located in an arid climate where night temperatures were relatively low. His work may indicate that in a relatively warm arid climate 10% of carbohydrates transported to the roots is sufficient to satisfy the requirements for relatively high productivity of the tree. In the temperate-zone, in general, roots start to grow at low temperatures before bloom. After its initiation, root growth follows an irregular course with period of active growth alternating with less active periods. The periodicity and extent of root growth greatly depends on shoot growth and fruit load of the tree. The end of the initial peak in root growth in the temperate zone usually corresponds with the beginning of active shoot growth and the second peak starts after shoot growth ceases (Head 1967, Rogers and Head, 1969). Quinlan (1965) suggested that the bimodal periodicity for root growth is due to competition between the shoots and roots for carbohydrate reserves. Several independent studies corroborate the hypothesis that photosynthates are translocated to the roots only after the main period of growth (Priestley *et al.,* 1976). This competition between shoots and roots is even more important where shoot growth is extended and is very vigorous. Such vigorous shoot growth create a higher level of competition for root growth than normally would be expected in the temperate zone. Cropping of the tree further reduces root growth (Atkinson, 1977, 1980). Although the crop load in the tropics is usually limited, it still adds to the total competition considerably.

In tropical areas, the soil occasionally is very warm. Diaz and Romo (1988) reported soil temperatures of 37 and 30°C in the top 20 cm layer of the soil and 35 and 30°C in the 20 to 40 cm layer when air temperature was 45 and 34°C, respectively (Fig. 2). Gur *et al.* (1972; 1976) reported that temperatures above 30°C caused a marked reduction in root growth with severe effects on leaf growth at temperatures above 35°C. There are limited numbers of studies on the effect of soil temperature on root and shoot growth of fruit producing plants. Spiers (1995) investigated the effects of soil temperatures of 16, 27, and 38 °C on root and shoot growth of blueberries. Both shoot and root growth decreased by 1/3 when soil temperature increased from 16 to 27°C and again by 1/3 when temperature reached 38°C. One reason for the decrease in vigor is that at high soil temperatures root and leaf cytokinins are reduced (Gur, 1972). Usually when root growth is reduced, a reduction in shoot growth follows. We have no data on root growth in tropical soils. If root growth is reduced because above ground competition for carbohydrates and/or high soil temperatures one would expect shoot growth also to be reduced. However this has not been observed in orchards located in tropical conditions.

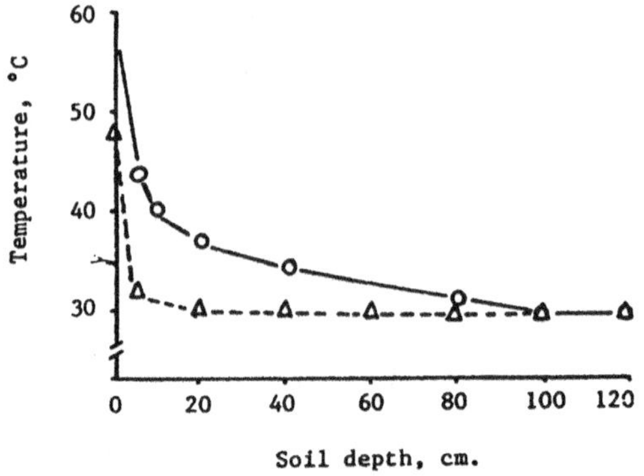

Figure 2. Soil temperatures at various depths, taken under midday sun during 45°C (circles) or 34° C (triangles) air temperatures in Sonora, Mexico during August (From Diaz and Romo, 1988).

Root distribution in warm Mexican soils in not sufficiently different from those found in the temperate zone. Roots of 'Anna' on MM106 rootstock extended to 120 cm from the trunk and reached the depth of 120 cm, whereas seedling roots extended to 160 cm and reached the 120 cm depth (Fig. 3). Periodicity of root growth greatly depends on shoot growth and fruit load on the tree. We have to consider the partially opposing

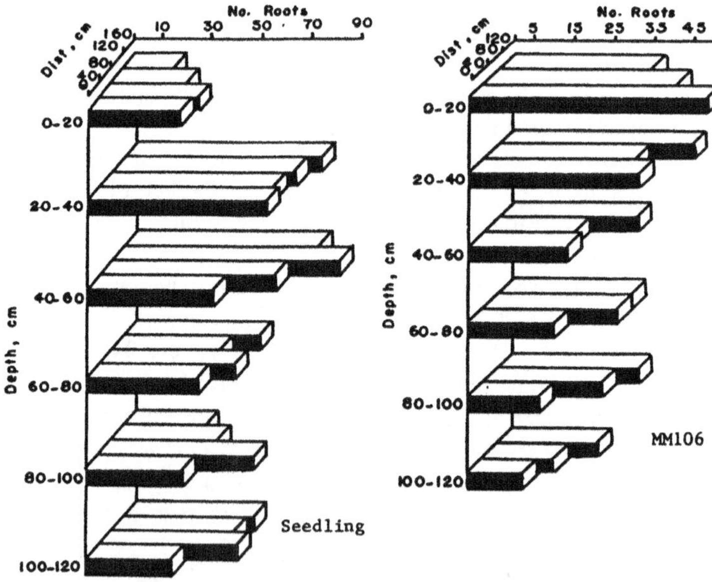

Figure 3. Root distribution of 5-years old 'Anna' apple on MM.106 or seedling rootstocks, at various depths and distance from their trunk in Sonora, Mexico (From Diaz and Romo, 1988).

effects of carbohydrate distribution in shoots of trees in tropical orchards where growth is strong and long in duration which limits translocation of carbohydrates into the root, and at the same time the crop of the tree is light allowing an increased translocation to the roots. Judging from the limited data presented above the two effects apparently cancel each other and translocation to the root and root growth in the tropics is similar to that in temperate regions.

5 Nutrient requirement

In fruit trees nutrient levels are most often established by leaf analysis of mid-summer leaves. This time was chosen for the analysis because nutrient levels of the leaves are relatively stable during this period (Faust, 1989) and the stable period is around 90 days after full bloom. Diaz and Romo (1988) reported seasonal changes of foliar nutrients in 'Anna' apples from Mexico. Their data does not show the plateaus of nutrient content that are commonly seen in the temperate zone (Fig. 4). All major nutrients show a steep increase or decrease 90 days after bloom and even with a long growing season of 270 days, satisfactory plateaus are not seen. Reasons for differences in changes in nutrient composition between trees located in the temperate and tropical zone is not known.

Basso and Wilms (1988a,b) published nutrient concentration in apples in Santa Catarina, Parana, and Rio Grande do Sul states of Brazil. For N they found 74%, 75%, and 75% of samples, in the three states respectively, above the level that is established as normal (2-2.5%) for temperate zones. In K, over fertilization was much less with samples being over normal with 10%, 24%, and 27% in the tree states respectively. Concentrations of P, Ca, Mg, and minor elements were close to those considered normal in temperate-zone conditions. Especially in N one can not always apply the values determined in studies of temperate-zone locations. For example normal N values are around 2% in Michigan, yet 0.5% higher values (2.5%) are required in the slightly warmer North Carolina. Therefore, additional information is needed for optimal values of nutrients in tropical climates. To adjust N to the needed level and not higher is crucial. Later in this chapter I discuss high shoot growth rates that occur in tropical conditions. High shoot growth rates compete with flower bud formation. Therefore, adjusting N is essential to decrease an overly high shoot growth rate and to set the trees to fruit.

Bernardi (1988) reported flattened fruit as a problem in Santa Catarina, Brazil. Flattened and flattened-lopsided apple fruit are two different phenomena. In general, fruit enlargement apparently depends on the availability of gibberellins in the fruit (Williams and Stahly, 1969). Treatment with gibberellin biosynthesis inhibitors results in small fruit and/or flattened fruit (Webster and Crowe, 1969). In warm night areas, such as on the East Coast of the United States, apples are always round (flattened) whereas in cool night conditions, such as the Washington State area, fruit is elongated. Thus in warm-night tropical areas one can expect round fruit. In addition to gibberellins governing apple fruit shape, an unequal or lopsided condition of flattened fruit is often associated with insufficient levels of B and in turn with low levels of IAA in the tree (Faust, 1989). This seem to be accentuated in warm conditions especially with sensitive cultivars. Lopsided flat fruit of 'Fuji' is often found in California and lopsided flat fruit of 'York Imperial' in Maryland,

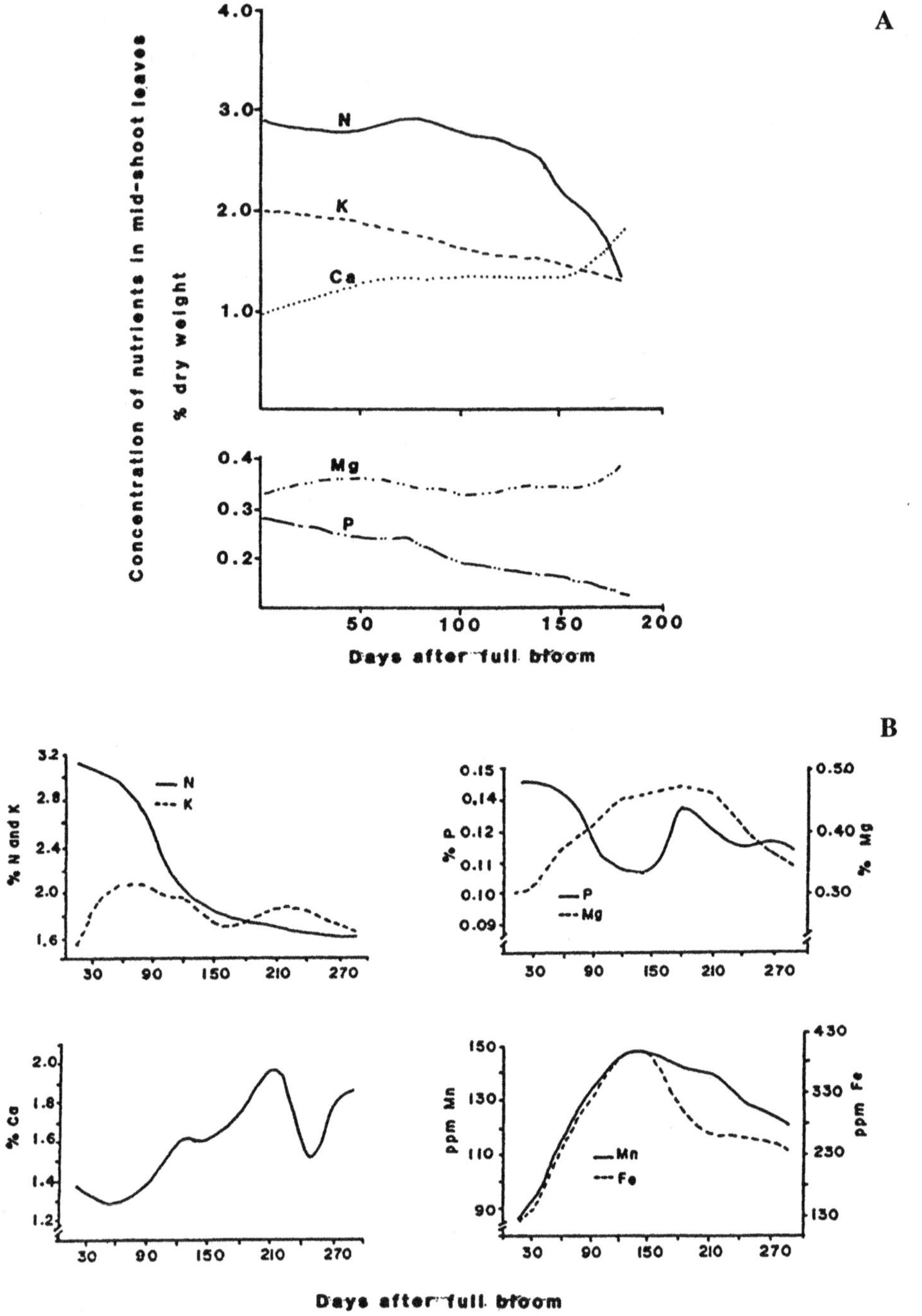

Figure. 4. Seasonal changes of foliar N, K, Ca, Mg, and P (**A**) in 'Delicious' apples in the temperate-zone in Wenatchee WA. (Full bloom Apr. 20) and N, K, Ca, Mg, P, Mn, and Fe (**B**) in 'Anna' apples in a warm-zone, in Sonora, Mexico (Full bloom Feb. 10). Note the much longer growing period at Sonora, Mexico and the lack of a plateau around 100 days after full bloom, the time nutrient levels are normally determined in the temperate-zone (From Rogers and Batjer, 1953 and Diaz and Romo, 1988).

whereas lopsided flat fruit is rarely seen in cooler areas. To break dormancy, in tropical areas the trees are trained with horizontal branches. IAA movement in horizontal branches is low. Boron deficiency increases IAA oxidation thus reducing IAA levels even further. In Beltsville we have found that less than 30 ppm of B content in the mid July leaves on a dry weight basis will produce lopsided flat fruit in 'York Imperial'. To avoid lopsided flat fruit B levels of 36-40 ppm are needed. Brown (1996) was able to show that in several fruit species B applied as spray to the leaves was immediately transported to the fruit. Thus correcting lopsided fruit is relatively easy with foliar B applications. However, B level of 70 ppm in the leaf will induce fruit drop. Therefore, growers must be careful what level of B they apply and in general they should not apply more than one spray of 100 ppm of B in early spring.

Boron is also essential for fruit set. Insufficient fertilization caused by low B usually occurs in cool soil conditions where B uptake is lagging and B requirement of bloom must be met completely from B reserves. If B reserves of the tree are insufficient fertilization and fruit set will be poor (Faust, 1989). This may not be a problem in tropical conditions where soil is warm and B uptake is likely to be sufficient. Also the B requirement for fruit set may be less important in warm tropical conditions because of the tendency to set fruit parthenocarpically. In California good correlation exist between parthenocarpic fruit set in solid blocks of 'Bartlett' pears and hours over 14°C during a 10 day bloom period. If more than 80 hours over 14°C occur during the ten days bloom period, good to heavy crops are set parthenocarpically, but crop set is poor if less than 80 hours over 14°C is received by the trees during bloom (Childers *et al.,* 1995). Since in tropical conditions apple bloom occur under warm conditions, the B requirement for pollen tube growth and fertilization may be less important.

Potassium nutrition is important in fruit production in any area (Faust, 1989). It, however, may have more importance in tropical conditions where fruit quality tends to be lower. Higher application of K increased sugar levels of peach cultivars in subtropical Australia (George *et al.,* 1990). Low K also tends to produce small, poorly colored, brownish-red fruit. Such fruit is quite common in the warm areas of China where K concentration in the leaf on a dry weight basis is 0.8-1.0%. In warm climates where red coloration is poor because lack of cool temperatures, K at 1.5% or higher is essential.

In peach, in South Australia, K effects on fruit quality are evident within four months of imposing different fertilization regimes. Higher application rates increase sugar levels, as well as increasing consumer acceptability in terms of both flavor and texture (Table 3). Increasing K also increases the fruit size of peach (George *et al.,* 1990). George *et al.* (1990) did not report leaf K levels of peach, but the clear difference seems to be between none and 32.5 kg K applied per hectare which indicates that peaches do respond to K fertilization.

Peaches in warm climates in the northern hemisphere exhibit a pointed structure typified as '*Southern Chinese Peach with beaked fruit*' (Faust and Timon, 1995). Similar fruit exhibiting sutures and prominent tips were reported from subtropical Australia (George *et al.,* 1990). Potassium fertilization, because of its effect on fruit size often minimize the prominence of such tips. As it mentioned above, high soil temperatures may be detrimental for root growth. High soil temperatures above 25°C, that reduce uptake of N, P, K, and Zn (Carlson, 1964; Gur *et al.,* 1979), can be detrimental for production of temperate-zone fruits in warm-soil type tropical conditions.

Table 3. Effect of potassium fertilization on fruit quality attributes of peach

Potassium applied (kg per hectare) July	September	*Mean Fruit Weight (g)*	*Sugar Brix (°)*	*Taste (0-20) scale*	*Texture (0-20) scale*
0	0.0	65a	11.3a	9.0a	9.2a
10	22.5	69b	11.8b	10.8b	10.0b
20	45.0	67b	11.7b	10.0b	10.2b
40	90.0	69b	12.6c	10.2b	10.6b

(Reproduced from George *et al.*, 1990)

6 Water requirement

In general, there are two areas in warm climates where temperate-zone fruits are produced:
1. high rainfall areas that are wet and humid, and
2. low rainfall areas that are arid. Producing temperate-zone fruits under high rainfall conditions has negative effects mainly due to enhanced fungal and bacterial diseases. Temperate-zone fruit production in arid areas is productive but in these areas water requirements need to be supplied by irrigation.

The need of the tree for water is determined by two factors: the radiation leaves are receiving and the water pressure deficit (WPD) between the air and the leaf. These factors and the availability of water in the soil create internal forces that control the use of water in the tree and are generally referred to as the water potential of the tree, which also is greatly influenced by its crop load. For discussions on general physiological control of water status in temperate-zone fruit trees readers are directed to Jones *et al.* (1985), Flore and Lakso (1989), Flore (1994) and Lakso (1994). For tropical conditions there are few reports in the literature.

Utilization of water in tropical orchards of temperate fruits is important in arid tropical areas.

Water pressure deficits between the leaf and air increase with increased temperature (Flore, 1994). Vapor pressure deficits greater than 1.5 kPa have been shown to cause stomatal closure in sour cherry (Beckman *et al.,* 1992). Often under field conditions, gas exchange in *Prunus,* judged from photosynthetic activity, is still 50-70% of maximum when temperatures are between 35 and 38 C (Flore, 1994) which means that transpiration still proceeds at least at half the capacity at these temperatures.

Transpiration for cooling the parts of the tree is very important but the most crucial is for the fruit. As little as 2°C increase from 29.6 to 31.6 caused 7.2°C increase in fruit temperature which greatly decreased the quality of the fruit (Unrath, 1972). Bark, leaf, and fruit temperatures in a sprinkling-cooled orchard with overtree irrigation, compared with undertree irrigated and non irrigated orchards are given in Table 4.

Table 4. Temperatures of the part of an apple tree at various air temperatures

Irrigation type	*Mean Temperature °C*			
	Air	Bark	Leaf	Fruit
Overtree	29.6	25.6	26.1	27.6
Undertree	30.4	29.8	30.9	32.6
Unirrigated	31.6	31.9	31.2	34.8

In warm conditions to produce high fruit quality sufficient water need to be supplied to maintain a transpiratory rate that keeps the fruit temperature to the minimum. The problem in warm climates is in maintaining a high rate of irrigation that usually also increases shoot growth that is already high in subtropical or tropical areas. Therefore irrigation to maintain low fruit temperatures should be carefully balanced with irrigation that minimizes shoot growth.

Chalmers *et al.* (1981, 1982) have proposed that timing the amount of irrigation may control vegetative growth on long growing period peaches in Australia without having a detrimental effect on the fruits. The method is termed as "Controlled deficit irrigation" and involves reduced water supply during stage II period of the growth curve in stone fruits (Mitchell and Chalmers, 1982). However, many of the low-chilling-requiring cultivars have either short or medium growth seasons where shoot growth occurs at the same time as fruit growth. Therefore with these cultivars deficit irrigation only can be practiced after harvest. Because the growing period is long under tropical conditions this seems to be a reasonable solution. Larson *et al.* (1988) evaluated postharvest irrigation under California conditions by irrigating trees

1. at 3-week intervals after harvest,
2. one irrigation during the postharvest period, and
3. no irrigation at al. Trees receiving treatment 3 were smaller in trunk diameter, had lower weight of dormant prunings, return bloom was greater and fruit set was increased over either treatment 1 or 2. After thinning to comparable levels there were no differences in fruit yield or size.

This is in contrast with apricot, postharvest water stress is detrimental to flower bud development and subsequent fruiting (Uriu, 1964). If stress was imposed from July to the end of the growing period, little effect on vegetative growth occurred (Veiheimer, 1975). In warm conditions double fruit seems to be a problem. Johnson *et al.* (1992) reported that early maturing peaches developed higher flower and fruit numbers and an increased occurrence of double fruits when postharvest water deficits were practiced. After 4 years of postharvest deficit irrigation, vegetative growth was reduced, but after normal hand thinning yield and fruit size were not affected.

Deficit irrigation in apples was evaluated by Ebel *et al.* (1995). Early season regulated deficit irrigation suppressed fruit growth rate compared to drip and furrow irrigation. When full irrigation was restored to regulated deficit irrigated trees, after terminal buds set, fruit growth rate increased. Fruit weight at harvest was a result of interaction of irrigation treatment and crop load. They recommended ending regulated deficit irrigation before fruit growth rate declines below the standard curve.

It is unquestionable that drought promotes flower bud formation. This may be the reason why in Peru a dry period of 20 days is imposed between crops (Bederski, 1988).

7 Flower bud formation and fruiting

Flower bud formation can be a problem especially at low altitudes where vegetative overgrowth often occurs. This is a problem throughout the world regardless whether the trees are in temperate or tropical climate. An inverse relationship between vegetative growth and fruiting in apples is well established (Forshey and Elfving, 1989) and, in essence, vegetative growth needs to be controlled for a high rate of productivity. This may be difficult under tropical conditions. Vegetative growth is highly enhanced by availability of nitrogen, and high night temperatures. In lowland tropical conditions night temperatures are high, in some upland tropical conditions rainfall is excessive. Neither one can be controlled. However, close spacing to encourage competition, tipping and bending to reduce shoot growth and encourage spur development (Edwards, 1987d; Edwards and Notodimedjo, 1987), and controlling nitrogen all could lead to decreased shoot growth. In Taiwan oriental pear cultivars tend to have excessive upright growth and limited flower bud formation (Lin *et al.*, 1990). Growth restriction, topping, twisting of shoots and possibly growth inhibitor applications limiting growth are required to promote flower buds. Without flower bud development growers have to get buds from higher elevations and graft flower buds on strong shoots. This is a labor intensive and an unsatisfactory method (Lin *et al.*, 1990). In Reunion Island peach growth is excessive and flower buds do not form without restricting growth (Lam-Yam and Parisot, 1990). Too vigorous growth of peaches is also a problem in subtropical regions of Australia (Lloyd, 1992) and in Mexico (Edwards, 1987a). In contrast, in Venezuela, locally selected peach clones do not grow excessively and form flower buds and fruit profusely (Ramirez, 1987; Edwards, 1987b). These peaches may have a different genetic origin. Diaz (1974) considers them to be similar to the evergreen 'Tetela' peach grown in localized areas south of Mexico City. In Java, under tropical conditions, 'Rome Beauty' apples form a few vigorously elongating shoots rather than many spurs. Tying such shoots down to a horizontal position stimulates lateral growth, increases the number of spurs, and promotes terminal bud formation. Providing that regrowth does not occur, flower development is completed within 6 weeks of terminal bud formation.

The time of defoliation in a continuous cropping system is also an important means to promote flower bud formation. Flowers must reach a certain stage of development to be functional. Forcing budbreak too early resulted in poor fruit set in Israel (Erez, 1987). When plants are defoliated and budbreak is forced, flowering may occur within 1 to 4 weeks (Edwards, 1990). This is a long time interval and must reflect the stage of development of the buds at the time of defoliation. However, detailed data are not available on the effect of bud development on the time required between defoliation and budbreak. In both peach and apple in the temperate zone, flowers enter into the winter fairly well developed, but preparation for flowering occurs only during the spring. Microsporogenesis, elongation of stigma and stamen filaments, occur only during the spring about a month before bloom. The process can be remarkably speeded up. In New Jersey an apple orchard that was defoliated in August bloomed 4 weeks later.

Forcing buds to grow too early may produce abnormal flowers. Lloyd and Couvillon (1974) defoliated peaches at 2 week intervals from 7/8 to 9/26. The first two defoliations produced a high percentage of abnormal flowers, then the number of abnormal flowers decreased and during the final 2 defoliations there were no abnormal flowers. The vegetative budbreak decreased with each successive defoliation date.

Early fall defoliation of low-chill 'Flordaprince' peaches produced early leafing and bloom, but fruit size was reduced by early relative to late defoliation. In contrast to 'Flordaprince', vegetative and floral budbreak of 'Flordagold' were delayed by early defoliation. These results indicate that early defoliation affects depth of dormancy and growth ability of buds, but that the extent to which these factors affect bud development depends on cultivar (Lloyd and Firth, 1990). Chandler and Tufts (1933) were able to show that in peach a period of arrested development was able to cause developmental changes in 2 weeks similar to those that occurred only after 3 months when development was continuous (Fig. 5). Although in the study of Chandler and Tufts the arrested period was created by placing the plants into cold storage (0°C); it is possible that defoliation also causes similar effects of arrested development after which bud development is greatly speeded up resulting in bloom within 1-4 weeks after defoliation as mentioned before.

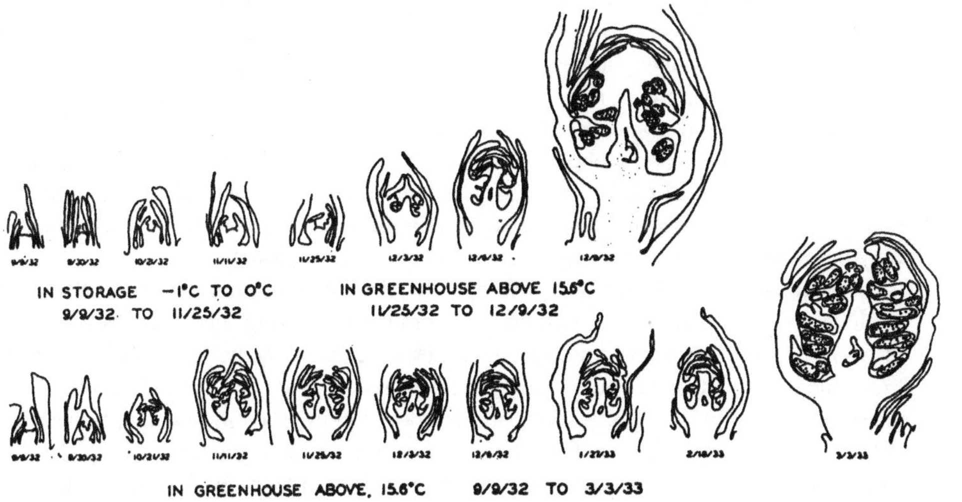

Figure 5. Influence of the rest period on the development of fruit buds of the 'Lovell' peach: Lower group, continuously warm, development slow; upper group, held in cold until November 25 and moved to warm greenhouse. After cold treatment there was as much development during 14 days in the warm greenhouse as in buds in 133 days of continuously warm conditions (From Chandler and Tufts, 1934).

Drought, because of its effect on decreasing growth may speed up flower bud formation. In double cropping systems drought is applied during a "rest" period between crops

(Hoffman, 1987). The rest must be applied to an intact tree. In Peru, trees must be foliated and the terminal leaves must not wilt to have an effect of drought (Bederski, 1987). Crops usually require 160 days cycle each and there is a 20 days of rest and drought between (Bederski, 1988). In Java, with 'Rome Beauty' apples, premature defoliation prevents flower initiation, whereas delayed defoliation reduces budburst (Edwards, 1990). This is not different from the behavior of apple in temperate zones. Premature leaf fall after harvest greatly decreases the rate of bloom next spring (Faust, 1989). Often the female part of the flowers develop after harvest. If this process is prevented by premature withdrawal of the carbohydrate supply (defoliation) then flowers do not complete their development. This may be the case in premature defoliation in tropical conditions.

In warm climates flower bud initiation can be a long drawn out process. In Thailand, at the Ang-Khang station, 'Gulfruby' plums form flower buds from August to December (Suriyapananont *et al.,* 1990). Thus, a viable leaf area must be maintained for the entire period to promote the flower development process. Peach flower induction proceeds over a long period in Reunion Island and proceeds with acropetal direction of twig morphogenesis (Lam-Yam and Parisot, 1990).

High temperature seems to adversely affect bud development. In Reunion Island, when temperatures are high during bud development, the resulting flowers are often abnormal. At low altitudes (below 325m) floral necrosis characterizes the 'Culemborg' peach cultivar. Although flower primordia remain alive in 'Flordared', 'Earlibelle', 'Flordabelle' and 'Sunred' peaches and nectarines, the primordial growth is arrested at this elevation when the temperatures are high which results in floral aberrations (Lam-Yam and Parisot, 1990).

High temperatures also severely limit fruit set. Temperature during bloom in temperate-zone climates ranges around 10°C. In the Philippines peach does not set fruit when temperature during bloom is above 15°C (Edwards, 1987a). Fruit set of peaches, apples, and grapes in Mexico also are all limited by temperatures above 30°C (Diaz 1992). Even when fruit set is satisfactory, high temperatures, especially at night, have been reported to cause fruit abscission (Dennis, 1979; Fukui *et al.,* 1984; Kondo and Takahashi, 1987) and poor fruit growth (Tukey, 1956; 1960). These phenomena may be related to excessive respiration as discussed previously.

In addition to the flower bud development changes that occur in tropical conditions, apple cultivars also appear to be more self compatible, setting seed with their own pollen. Fruits may grow parthenocarpically even in the absence of seeds. In an artificial pollination experiment with 'Rome Beauty' apples in East Java, fruit set was similar (about 60%) with emasculation only, hand pollination, or open pollination. The emasculated flowers formed 100% seedless fruit whereas the others had a poor seed formation of only 2 seeds per fruit (Verheij, 1990). Cleistogamy is more pronounced in high chilling than in low chilling cultivars such as 'Anna' (Verheij, 1990). In Peru, apple produces parthenocarpic fruit on quince rootstock but pollination is needed on apple rootstock for fruit set (Bederski, 1987). The rate of parthenocarpic fruit production in early blooming, low chilling-requiring apples, such as 'Anna', is excessive when on quince root. In contrast, late blooming, high chilling-requiring apples produce excessive growth when on seedling root and for good production they needed to be propagated on quince root. Pollinizers are not necessary when apples are grown on quince root but are necessary when the rootstock is seedling apple (Bederski 1987). This phenomenon is also observable in the temperate zone.

In California, in the relatively warm Lake Region 'Bartlett' pears set fruit parthenocarpically, but need pollinizers in the cooler area of Wenatchee Washington.

8 Conclusions

Growing temperate-zone fruit trees outside their adapted climatic range affects the physiology of the trees profoundly. High temperatures encountered in tropical or subtropical conditions increase respiration and growth of trees which in turn decreases fruiting and productivity. Physiological processes, such as leaf and flower budbreak, are well synchronized in temperate climates whereas the synchronization seems to be non existent in tropical or subtropical conditions. High soil temperatures encountered by the roots changes their metabolism and they can not fulfill their role expected from experiences gained from temperate conditions. New production techniques that are possible in tropical condition, such as double cropping, are taxing the reserves of the tree and must be thoroughly understood to be applied effectively. Double cropping forces the completion of flower bud formation in a short period of time which may produce incomplete products *i.e.* malformed flowers. Opposing the effects that cause difficulty, trees have a much longer period of productivity during the annual cycle which may compensate for the higher demands on carbohydrates (Erez 1986). These are only a few examples, but they point to our inadequate knowledge to understand the physiology of fruit trees brought about by a new regime of climates to which trees were not adapted to during the evolution. Nevertheless, orchards are planted and temperate-fruits are grown in tropical and subtropical climates and understanding the physiology under these conditions bound to improve their productivity in the future.

8 References

Ahmad, I. (1987) Cultivar selection and performance of stone fruits in Pakistan. *Acta Horticulturae* 199, 77-81.

Atkinson, D. (1977) Some observations on the root growth of young apple trees and their uptake of nutrients when grown in herbicide strips in grassed orchards. *Plant & Soil* 49, 459-471.

Atkinson, D. (1980) The distribution and effectiveness of the roots of tree crops. *Horticultural Reviews* 2, 424-490.

Aubert, B. and Bertin, Y. (1987) Aptitude of mild winter areas for growing peaches with low chilling requirements in Reunion Island. *Acta Hort.* 199, 51-60.

Basso, C. and Wilms, F.W.W. (1988a) Nutritional status of apple orchards in southern Brazil. *Acta Hort.* 232, 187-192.

Basso, C. and Wilms, F.W.W. (1988b) Soil amelioration for apple production in southern Brazil. *Acta Hort.* 232, 193-203.

Beckman, T.G., Perry, R.L. and Flore, J.A. (1992) Short-term flooding affects gas exchange characteristics in containerized sour cherry trees. *HortScience* 27, 1297.

Batjer, L.P. and Rogers, B.L. (1953) Blossom blast of pears: An incipient boron deficiency. *Proc. Amer. Soc. Hort. Sci.* 62, 119-120.

Bederski, K.A. (1987) Selection and dormancy management of temperate zone deciduous fruit tree cultivars in coastal valleys of Peru. *Acta Hort.* 199, 33-38.

Bederski, K.A. (1988) Apple growing in coastal areas of Peru. *Acta Hort.* 232, 51-55.

Bernardi, J. (1988) Behavior of some apple cultivars in the subtropical region of Santa Catarina, Brazil. *Acta Hort.* 232, 46-50.

Brown, P. (1996) Phloem mobility of boron is species dependent: evidence for phloem mobility in sorbitol-rich species. *Ann. Bot.* 77, 497-505.

Butler, D.R. and Landsberg., J.J. (1981) Respiration rates of apple trees, estimated by CO_2-efflux measurements. *Plan, Cell Environ.* 4, 153-159.

Carlson, R. (1964) Responses of Malling Merton clones and Delicious seedlings to different root temperatures. *Proc. Amer. Soc. Hort. Sci.* 86, 41-45.

Chalmers, D.J., Mitchell, P.D. and Van Heek, L. (1981) Control of peach tree growth and productivity by regulated water supply, tree density and summer pruning. *J. Amer. Soc. Hort. Sci.* 106, 307-312.

Chandler, W.H. (1957) *Deciduous orchards.* Lea & Febiger, Philadelphia.

Chandler, W.H. and Tufts, W.P. (1934) Influence of rest period on opening of buds of fruit trees in spring and on development of flower buds of peach trees. *Proc. Amer. Soc. Hort. Sci.* 30, 180-186.

Childers N.F., Morris, J.R. and Sibbett, C.S. (1995) *Modern Fruit Science* 10th revised edition. Horticult. Public. Gainsville.

Climate and Man. (1941) *Agriculture handbook. USDA.* Washington D.C.

Colmenares, S. (1988) Tropical apple production in Venezuela. *Acta Hort.* 232, 56.

DeJong, T.M., Doyle, J.F. and Day, K.R. (1987) Seasonal patterns of reproductive and vegetative sink activity in early and maturing peach (*Prunus pesica*)cultivars. *Physiol. Plant.* 71, 83-88.

Denardi, F., Hough, L.F. and da S. Bonetti, J.I. (1988) Low chilling and disease resistance as main objectives of apple breeding in Santa Catarina, Brazil. *Acta Hort.* 232, 15-21.

Dennis, F.G. Jr. (1979) Factors affecting yield in apple, with emphasis on 'Delicious'. *Hort. Rev.* 1, 395-422.

Diaz, D.H. (1974) Vegetative and reproductive growth habits of evergreen peaches in Mexico. Proc. XIX *Int. Hort. Cong. Warsaw* 1B, 525.

Diaz, D.H. (1992) Temperate zone fruits in the tropics/subtropics: Southern North America and Central America. *Acta Hort.* 296, 205-212.

Diaz, D.H., Alvarez, A. and Sandoval, J. (1987) Cultural and chemical practices to induce uniform bud break of peach and apple under warm climates in mexico. *Acta Hort.* 199, 129-133.

Diaz, D.H. and Romo, R. (1988) Nutritional status and root system of Anna apple under alcaline soil in warm dry climate of Mexico. *Acta Hort.* 232, 177-186.

Ebel, R.C., Proebsting, E.L. and Evans R.G. (1995) Deficit irrigation to control vegetative growth in apple and monitoring fruit growth to schedule irrigation. *HortScience* 30, 1229-1232.

Edwards, G.R. (1987a) Temperature in relation to peach culture in the tropics. *Acta Hort.*199, 61-62.

Edwards, G.R. (1987b) Potential for apple production in Venezuela. *Acta Hort.* 199, 19-25.

Edwards, G.R. (1987c) Producing temperate-zone fruit at low latitudes: avoiding rest and the chilling requirement. *HortScienc e* 22, 1236-1240.

Edwards, G.R. (1987d) Conditions of growth, dormancy and rest to produce temperate zone fruits under tropical conditions. *Acta Hort.* 199, 128.

Edwards, G.R. (1990) Ten year's experience with temperate fruits in the tropics. *Acta Hort.* 279, 47-51.

Edwards, G.R. and S. Notodimedjo. (1987) Defoliation, bending and tip pruning of apple under tropical conditions. *Acta Hort.* 199, 125-127.

Erez, A. (1986) The significance of the length of the leafless stage in deciduous fruit trees on fruit production potential. p. 23. In: *The Regulation of photosynthesis in fruit trees.* (Lakso A. and Lenz F. eds.) Symp. Proc. Publ. New York State Agr. Exp. Sta. Geneva, New York.

Erez, A. (1987) Use of rest avoidance technique in peaches in Israel. *Acta Hort.* 199, 137-149.

Evenari, M., Lange, O.L., Schulze, E.D., Kappen, L. and Buschbom, U. (1977) Net photosynthesis, dry matter production, and phenological development of apricot trees (*Prunus armeniaca* L.) Cultivated in the Negev highlands (Israel). *Flora* 166, 384-393.

Faust, M. (1989) *Physiology of temperate-zone fruit trees.* Wiley, New York.

Faust, M. and Timon, B. (1995) Origin and dissemination of peach. *Hort. Rev.* 17, 341-379.

Flore J.A. and Lakso, A.N. (1989) Environmental and physiological regulation of photosynthesis in fruit crops. *Hort. Rev.* 11, 111-157.

Flore, J.A. (1994) Stone fruit. In: *Handbook of enviromental v physiology of fruit crops.* Volume 1: Fruit crops. (Schaffer B. and Andersen P.C. eds) CRC Press, Boca Raton.

Forshey, C.G. and Elfving, D.C. (1989) The relationship between vegetative growth and fruiting of apple trees. *Hort. Rev.* 11, 229-288.

Fukui, H., Imakawa, S. and Tamura T. (1984) Relation between early drop of apple fruit, ethylene evolution and formation of abscission layer. *J. Japan Soc. Hort. Sci.* 53, 303-309.

George, A.P., Nissen, R.J., Lloyd, .J. and Richens, K. (1990) Factors affecting fruit quality of low chill stonefruits in subtropical Australia. *Acta Hort.* 279, 559-570.

Grossman, Y. and DeJong, T.M. (1994) Carbohydrate requirements for dark respiration by peach vegetative organs. *Tree physiol.* 14, 37-48.

Gur, A., Bravdo, B. and Mizrahi, Y. (1972) Physiological rsponses of apple trees to supraoptimal root temperatures. *Physiol. Plant.* 27, 130-138.

Gur, A. Hepner, J. and Shulman, Y. (1979) The influence of root temperature on apple trees. IV. The effect on the mineral nutrition of the tree. *J. Hort. Sci.* 54, 313-321.

Hansen, P. (1977) Carbohydrate allocations, In: *Environmental effects of crop physiology,* (Landsberg, J.J. and Cutting, C.V. eds.). Academic Press, London.

Head, G.C. (1967) Effects of seasonal changes in shoot growth on the amount of unsuberized root on apple and plum trees. *J. Hort. Sci.* 41, 197-206.

Johnson, R.S., Headley, D.F. and DeJong, T.M. (1992) Long-term response of early maturing peaches to postharvest water deficit. *J. Amer. Soc. Hort. Sci.* 117, 881-885.

Johnson, R.S. and Lakso, A.N. (1986) Carbon balance model of a growing apple shoot: I. Development of the model. *J. Amer. Soc. Hort. Sci.* 111, 160-164.

Jones, A.G., Lakso, A.N. and Syvertsen, J.P. (1985) Physiological control of water status in temperate and subtropical trees. *Hort. Rev.* 7, 301-344.

Kondo, S. and Takahashi. Y. (1987) Effects of high temperature in the nighttime and shading in the daytime on the drop of apple fruit 'Starking Delicious'. *J. Japan Soc. Hort. Sci.* 6, 142-146.

Lakso, A.N. (1994) Apple. In: *Handbook of environmental physiology of fruit crops.* Volume 1: Temperate Crops. (Schaffer B. and Andersen P.C. eds.) CRC Press, Boca Raton.

Lam-Yam, L. and. Parisot, E. (1990) Preliminary study of peach floral differentiation in mild winter areas. *Acta Hort.* 279, 231-237.

Larson, K.D., DeJong, T.M. and Johnson, R.S. (1988) Physiological and growth responses of mature peach trees to postharvest water stress. *J. Amer. Soc. Hort. Sci.* 113, 296-300.

Lenz, F. (1986) Fruit effects on transpiration and dry matter production in apples. In: *The Regulation of photosynthesis in fruit trees.* (Lakso N. and Lenz F. eds.) Symp. Proc. Publ. New York State Agr. Exp. Sta. Geneva, New York.

Lin, Hsin-Shan and Liaw, Wan-Jean (1990) Production of oriental pears at low latitudes. *Acta Hort.* 279, 75-82.

Llanes, R., Otomo, E., De La Alas, R.A. and Edwards, G.R. (1987) Apples and peaches in tropical lowlands of the Philippines. *Acta Hort.* 199, 64.

Lloyd, J. (1992) Temperate zone fruit production in the tropics and subtropics: Oceania. *Acta Hort.* 296, 219-229.

Lloyd, J. and Couvillon, G.A. (1974) Effect of date of defoliation on flower and leaf bud development in the peach *Prunus persica* (L) Batch. *J. Amer. Soc. Hort. Sci.* 9, 514-517.

Lloyd, J. and Firth, D. (1990) Effect of defoliation time on depth of dormancy and bloom time for low chill peaches. *HortScience.* 25, 157 5-1578.

Mitchell, P.P. and Chalmers, D.J. (1982) The effect of reduced water supply on peach tree growth and yield. *J. Amer. Soc. Hort. Sci.* 107, 853-856.

Niegel, W. (1988) History, actual advance and future of the Equatorian apple growing. *Acta Hort.* 232, 67-73.

Pavel, E.W. and DeJong, T.M. (1993) Seasonal CO_2 exchange patterns of developing peach (*Prunus persica*) fruit in response to temperature, light and CO_2 concentration. *Physiol Plant.* 88, 322-330.

Priestley, C.A., Catlin, P.B. and Olson, E.A. (1976) The distribution of ^{14}C labelled assimilates in young apple trees as influenced by doses of supplementary nitrogen. I. Total ^{14}C radioactivity in extracts. *Ann. Bot.* 40, 1163-1170.

Quinlan, J.D. (1965) The pattern of distribution of ^{14}Carbon in a potted apple rootstock following assimilation of ^{14}carbon dioxide by a single leaf. *Rpt. East Malling Res. Sta. For* 1964. P. 117-118.

Ramirez, R. (1987) Peach production in Venezuela. *Acta Hort.* 199, 26.

Rogers, B.L., and Batjer, L.P. (1953) Seasonal trend of several nutrient elements in Delicious apple leaves expressed on a per cent and unit area basis. *Proc. Amer. Soc. Hort. Sci.* 61, 1-5.

Rogers, W.S. and Head, G.C. (1969) Factors affecting the distribution and growth of roots of perennial woody species. In *Root growth* (Whittington, W.J. ed) Butterworths, London.

Schulze, E.D., Lange, O.L, Kappen, L. and Evenari, M. (1977) Stomatal responses to changes in humidity in plants growing in the desert. *Planta* 108, 259-270.

Spiers, J.M. (1995) Substrate temperatures influence root and shoot growth of southern highbush and rabbiteye blueberries. *HortScience* 30, 1029-1030.

Subadrabandhu, S. and. Punsri, P (1987) Deciduous fruit trees as an alternative to opium poppy in northern Thailand. *Acta Hort.* 199, 39-44.

Suriyapananont, S., Sariyapananonat V. and Tuntawiron, O. (1990) Flower bud development of plum variety 'Gulfruby' in Thailand. *Acta Hort.* 279, 253-257.

Tukey, L.D. (1956) Some effects of night temperatures on the growth of McIntosh apples. *I. Proc. Amer. Soc. Hort. Sci.* 68, 32-43.

Tukey, L.D. (1960) Some effects of night temperatures on the growth of McIntosh apples. *II. Proc. Amer. Soc. Hort. Sci.* 75, 39-46.

Unrath, C.R. (1972) The evaporative cooling effects of overtree sprinkler irrigation on 'Red Delicious' apples. *J. Amer. Soc. Hort. Sci.* 97, 55-58.

Uriu, K. (1964) Effect of postharvest soil moisture on subsequent yield of apricots. *Proc. Amer. Soc. Hort. Sci.* 84, 93-97.

Veihmeyer, F.J. (1975) The growth of fruit trees in response to different soil moisture conditions measured by width of annual rings and other means. *Soil Sci.* 119, 448-452.

Verheij, E.W.M. (1990) Adaptation of apple to tropical growing conditions. *Acta Hort.* 279, 83-90.

Webster, D.H. and Crowe, A.D. (1969) Effects of gibberellic acid, N-dimethylaminosuccinamic acid (Alar), ringing and thinning on McIntosh apple shape. *J. Amer. Soc. Hort. Sci.* 94, 308-310.

Weinberger, J.H. (1950) Chilling requirements of peach varieties. *Proc. Amer. Soc. Hort. Sci.* 56, 122-128.

Williams W.T. and Menegazzo, G. (1988) Apple culture in the highlands of Guatemala. *Acta Hort.* 232, 57-66.

8 STONE FRUIT GENETIC POOL AND ITS EXPLOITATION FOR GROWING UNDER WARM WINTER CONDITIONS

DAVID H. BYRNE [1], WAYNE B. SHERMAN [2]
and TERRY A. BACON [3]

[1,3]*Department of Horticultural Sciences, Texas A and M University, College Station, TX 77843-2133, USA*
[2] *Horticultural Science Department, University of Florida, Gainesville, FL 32611, USA*

1 Introduction

1.1 Stone fruit breeding efforts

Peaches, plums, apricots, and cherries are temperate fruit crops that generally require a cold winter to break dormancy and produce fruit on a yearly basis. As these stone fruit species have spread around the world, however, genotypes have been selected for adaptation to more marginal, and even extreme, local environments by farmers and researchers. In marginal environments such as in mild winter regions of the world, locally selected well-adapted cultivars often do not possess commercially acceptable fruit quality. Starting with these poor quality but adapted cultivars, fruit breeders have developed and are developing commercially acceptable stone fruit cultivars suited to a wide range of mild winter conditions.

1.2 Mild winter regions throughout the world

The Subtropical climate occupies the tropical margins of the middle latitudes and is transitional in climatic character between the Tropical (lacking winter) zone and the Temperate (very cold winters) zone. This 'Mild Winter Region' (MWR) contains some of the world's most important areas of stone fruit production. This is especially true for 'Mediterranean' type Subtropical climates typically situated along the western sides of continents in such locations as: the borderlands of the Mediterranean Sea, central and coastal southern California, central Chile, the southern tip of South Africa, and parts of southernmost Australia (Fig. 1). Environmental conditions in these areas are ideal for stone fruit production with hot, dry, sunny summers that favor tree growth and fruit development without excessive disease pressure, and mild-but-constant winter temperatures that usually, but not always, provide sufficient winter chill units.

The MWR, however also contains areas of marginal-to-extreme environmental conditions for stone fruit production including the Subtropical-Humid, and the Dry climate types (Fig. 1). Subtropical-Humid zones, typically located on the eastern sides of continents have hot, humid, rainy summers that favor disease development.

Amnon Erez (ed.), Temperate Fruit Crops in Warm Climates, 157-230.
© 2000 *Kluwer Academic Publishers. Printed in the Netherlands.*

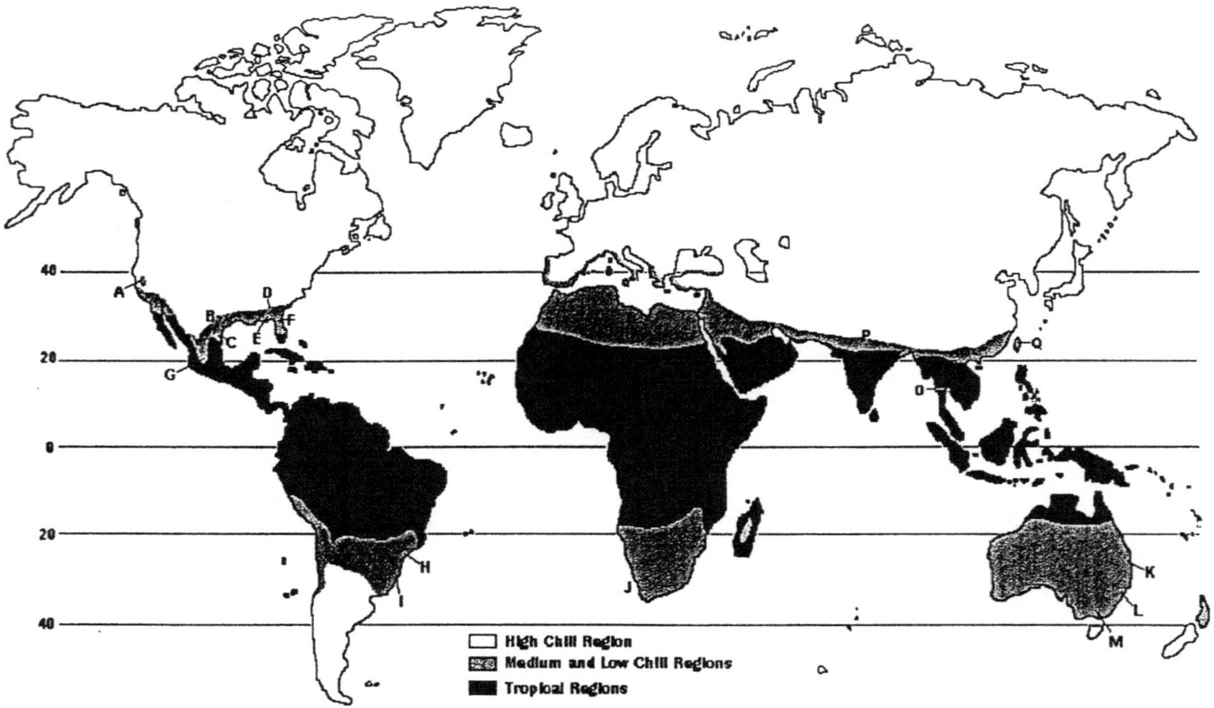

Figure 1. Medium and low chill regions of the world with locations of major stone fruit breeding programs.

UNITED STATES	MEXICO	SOUTH AFRICA	TAIWAN
A: California	G: Queretaro	J: Stellenbosch	Q: Taichung
B: College Station, TX	Chapingo		
C: Weslaco, TX		AUSTRALIA	THAILAND
D: Byron, GA	BRAZIL	K: Gatton, Queensland	O: Bangkok
E: Attapulgas, GA	H: Sao Paulo	L: Richmond, NSW	
F: Gainesille, FL	I: Pelotas	M: Tatura, Victoria	INDIA
			P: Solan, HP

Furthermore, inconsistent winter dormancy conditions, caused by conflicting air masses of tropical and polar origin, result in both insufficient chill-unit accumulation in some years, as well as perennial spring freeze danger during bloom. Dry climate zones, located in the central region of continents, do not have the disease pressure of humid zones, but insufficient chill-accumulation is still a frequent problem, and year-round dry conditions and lack of ground water limit production to areas near surface water sources.

Mild winter regions of the world have been classified, by chill-requirement ratings of cultivars that can be used commercially, into three major adaptation zones: the Transitional, Medium, and low-chill zones (Fig. 1). The Transitional chill zone includes regions in which most commercial cultivars have chill requirement (CR) of 650 to 950 chilling units (CU) but have years in which the higher chill varieties produce poorly due to mild winters. This is true of the southern San Joaquin valley in California on the west coast and central Georgia and the coastal plain of South Carolina on the east coast of the USA, the southern 25% of Australia, and the southern and higher altitude regions of South Africa. The medium-chill zone includes areas where cultivars requiring 400 to about 650 CU can be grown consistently. This would include areas such as southern Georgia, southern Louisiana, south central Texas, southern Spain, the highlands in Mexico, India, and other subtropical areas, and lower altitude areas of South Africa. The peach varieties that mark the higher limits would be 'June Gold', 'Juneprince', 'Springcrest', 'SpringBaby' and 'TexRoyal' and the lower limit would be 'Flordaking' and 'Flordacrest'. The low-chill zone is characterized by frequently receiving less than 400 CU. Common varieties grown in these areas include 'Flordaprince', 'Earligrande', 'TropicBeauty', and 'TropicSnow'.

Table 1. Peach and nectarine breeding programs in the world classified by activity and adaptation zone

Activity level	*Low-chill (<400 CU)*	*Medium-chill (400-700 CU)*	*Transitional (600-850 CU)*
Active	Florida - UFL - Gainesville Brazil - Pelotas, RS Brazil - Sao Paulo, Campinas	California - Zaiger Genetics Texas - TAES, College Station Mexico - Chapingo Mexico - Queretaro South Africa - Stellenbosch	California - USDA, Fresno California - Zaiger Genetics Georgia - USDA, Byron Australia - Tatura, Victoria
New	Texas - TAES, Weslaco Australia - Gatton, Queensland Australia - Richmond, NSW Taiwan - TARI, Taichung Thailand - Kasetsart Univ., Bangkok	Georgia - USDA, UGA and UFL - Attapulgus China - Zhengzhou, Henan	India - Y.S. Parmar Univ. of Hort. & Fores., Solan, HP
Inactive	California - SunWorld India - Saharanpur, Uttar Pradesh	California - SunWorld Louisiana - LAES, Idlewild	California - Armstrong California - Chaffey Jr. Coll. California - USDA, Palo Alto California - UC, Riverside South Carolina - Clemson University

The active breeding for new lower chill varieties of peach and nectarines was spurred by the appearance of an "obscure disease" in southern California in 1907 which was later identified as insufficient chilling (Horne, Weldon, and Babcock, 1926; Weldon, 1934). These early California programs (USDA at Palo Alto, Armstrong, Chaffey Junior College, and University of California at Riverside) were in the transitional chill zone (Table 1) and produced both high-chill and medium-chill germplasm. This material contributed substantially to the programs in the lower chill zones as a source of germplasm.

The majority of the work and programs emphasize the development of medium and low-chill peach and nectarines cultivars (Table 2 and 3). Many of the breeding efforts in plum, apricot, and cherry in the medium and low-chill regions are small efforts associated with larger peach breeding programs (Table 4 and 5). The largest efforts for all the stone fruits are in the transitional chill zone in the United States (California and Georgia), South Africa, and Australia

As compared to the programs in the transitional chill zones which served an existing industry, the medium and low-chill breeding efforts in stone fruit were creating cultivars for areas where no industry existed and thus have been small and poorly supported. Nevertheless, the dramatic advances of the Florida and other breeding programs in the development of low and medium-chill peach and nectarine cultivars have spurred the development of stone fruit production and additional breeding efforts in mild winter areas of the world (Tables 1-5).

2 History of breeding

2.1 Fresh market peaches

Institutional breeding for low and medium-chill peaches began in 1907 at the University of California (Riverside) and was subsequently extended by Weldon at Chaffey Junior College in Ontario, California. These programs released 'Babcock' that contributed substantially to following work. Later, the Armstrong program, used 'Babcock' and other cultivars to develop a range of nectarines and peaches. Important cultivars with respect to breeding are the peaches 'Flamingo', 'June Gold', 'Robin', and 'Springtime' and the nectarines 'Armking', 'Panamint', and 'Palomar'.

The first two programs (Riverside and Ontario) ended about the time other programs began in the USA (Florida, California, Georgia, Louisiana, and Texas), Brazil (Pelotas and Campinas), India (Saharanpur, Uttar Pradesh) and South Africa. The most active of these in the medium and low-chill regions are the programs in Florida, Pelotas, Campinas, and South Africa. The Texas program, although it officially began in 1935, was not very active until the mid seventies when more resources were devoted to it. The Louisiana program has also continued at a low level of activity throughout its lifetime. The programs in California (Zaiger's Genetics and USDA, Fresno) and the program in Georgia were active but with their main emphasis in the high-chill adaptation regions. Nevertheless, they were also releasing a few medium-chill cultivars. In particular, the USDA program in Georgia was instrumental in getting the Florida effort going by cooperating with them on some of the initial crosses (Sharpe, 1961; Sherman and

Table 2. Peach and nectarine breeding programs in medium and low-chill zones: USA

Location	Chill range (CU)	Fruit type[z]	Objectives	Releases	
California, Armstrong Nursery Ontario 1935 to 1982?	500 to 900	MT Fresh market Peach/nectarine Yellow/white	Home and commercial	Armking Armgold Flamingo Bobolink	June Gold Panamint Palomar Robin Springtime
California, Chaffey Junior Colleg G. P. Weldon Ontario 1919 to 1947?	500 to 900	MT Fresh market Peach White/yellow	Home and commercial Medium-chill	Early Babcock Fontana Weldon Honey Elberta Maywel	
California,USDA-ARS Fresno 1950 to present	500 to 1200	MT Fresh market Peach/nectarine Mainly yellow	Commercial shipping type Extend season Emphasis early ripening	Sunfre Goldcrest Mayfire	Crimson Baby Spring Baby
California, USDA and Stanford University Palo Alto 1920 to 1941?	700 to 1200	NM Processing peaches Peaches	Tolerant to delayed foliation	Andora Carson Corona Coronado	Cortez Farida Fortuna
California, University of California Riverside 1907 to 1961?	500 to 800	MT Fresh market Peach Yellow/white	Moderate chilling types Home garden and commercial	Babcock C.O. Smith Hermosa Ramona Ventura	

Table 2. continued

Location	*Chill range (CU)*	*Fruit type*[z]	*Objectives*	*Releases*	
California, Sunworld Bakersfield 1987 - preset	100 to 1000	MT Fresh market Peach/nectarine Yellow/white	Shipping type Early ripening Southern California	none	
California, Zaiger Genetics Modesto 1958 to present	200 to 1000	MT Fresh market Peach/nectarine	Shipping type Extend season, especially early Large size, high color, round Low and sub acid types	Mayglo Gulf Pride Gulf Queen	Aprilglo Earliglo Maypride
Florida, University of Florida Gainesville 1948 to present	100 to 500	NM/MT Fresh market Peach/nectarine Yellow/white	Early (FDP < 100 days) High color, firm, round Large size Resistance to bacterial leaf spot	Flordaking Flordaprince Flordacrest Flordaglo Flordario Florda series	TropicBeauty Tropic series with Texas
		Major emphasis on early ripening and more recently non melting types for fresh market		Sundollar Suncoast Sunmist Sunraycer Sun series	OroA Oro series with Mexico UFGold
Georgia, USDA, University of Georgia, and University of Florida Attapulgus 1986 to present	400 to 700	MT/NM Fresh market Peach/nectarine Yellow/white Changing emphasis to non-melting flesh.	Early season High color, firm, round Large size Resistance to bacterial leaf spot	Sunsplash	

Table 2. continued

Location	*Chill range (CU)*	*Fruit type*[z]	*Objectives*	*Releases*	
Georgia, USDA-ARS Byron 1936 to present	650 to 1000	MT Fresh market Peach/nectarine Yellow/white	Consistent yield Extend season Size (7 cm), firm, 70% blush Resistance to bacterial leaf spot and buttons	Springcrest Starlite Goldprince Juneprince	
Louisiana Louisiana State University Idlewild 1940 to present	450 to 750	MT Fresh market Peach Yellow/white	Large size, freestone Extend season Resistance to bacterial leaf spot	Idlewild La Feliciana La Festival	La Pecher Hawthorne Delta
Texas, Texas A&M University College Station 1935 to present	400 to 700	MT /NM Fresh market Peach/nectarine Yellow/white	Consistent yield and quality Early to mid season Large size, high blush, round Resistance to bacterial leaf spot	Sam Houston Texstar TexRoyal	
Texas, Texas A&M University Weslaco 1989 to present	0 to 250	MT Peach/nectarine Yellow/white	Consistent yield and quality Early season Large size, high blush, round	EarliGrande FlordaGrande TropicBeauty	TropicSweet TropicSnow ValleGrande

[z] MT = melting and NM = non melting flesh.

Table 3. Peach and nectarine breeding programs in medium and low-chill zones: Outside the USA

Location	*Chill range (CU)*	*Fruit type*[z]	*Objectives*	*Releases*	
Australia Tatura, Victoria 1972 to present	450 to 950	NM Processing Fresh market Yellow Peaches	Improved production and quality Lower chilling requirement Bright orange color Extend season Resistance to brown rot Resistance to peach leaf curl	Tatura 204 Tatura 207 Tatura 211 Tatura 212 Tatura 223	Golden Blaze Golden Charm Golden Delight Goldensun
1992 to present		NM-white canning	Flavor = Okubo	None, began 1992	
Australia Gatton, Queensland University of Queensland 1992 to present	0 to 500	NM/MT Fresh market Peach/nectarines Low acid and peento types	Domestic and export Early season, Sept. to Dec. Lower fruit set than FLA Resistance to bacterial leaf spot	None, new program	
Australia Richmond, NSW Univ. West Sidney 1992 to present	300 to 600	NM/MT Fresh market Peach/nectarine Low acid and peento	Domestic and export Early season, Sept. to Dec. Resistance to bacterial leaf spot	None, new program	
Brazil CPACT-EMBRAPA Pelotas, Rio Grande do Sul 1953 to present	150 to 500	MT-peach/nectarine Fresh market Yellow/white NM Processing and dual use Yellow	Extend season Late bloom, frost avoidance Increase size, firm, quality Dual purpose High SS, attractive skin color Resistance to bacterial leaf spot Resistance to brown rot	Nectarines, MT Linda Branca Peaches, MT Eldorado Vanguardia Chula Pilcha Sentinela	NM- Agata Bolinha Diamante Onix Precocinha Safira Turquesa

Table 3. continued

Location	*Chill range (CU)*	*Fruit type*[z]	*Objectives*	*Releases*	
Brazil IAC-Sao Paolo 1950 to present	0 to 150	MT Peaches/nectarines Fresh market Yellow/white NM Yellow peach Dual purpose Processing	Adaptation to very low-chill Extend harvest season especially towards earlier	Alo-Docura Biuti Bolao Canario Catita Cristal	Docura Natal Reliquia Talisma Tutu
China Zhengzhou Institute of Pomology Zhengzhou, Henan ? to present	Lower than 500	MT Peach/nectarine Fresh market White/yellow	FDP, 70-90 days Firm, round, sweet	Several 800 CU cultivars	
India Saharanpur 1957 to ?	Lower than 500	MT Peaches	Low-chill Adaptation to high temperatures Increased quality, non-browning Early ripening (60-70 FDP)	Local selections Sharbati Saharanpur 1 Saharanpur 6 Safeda Early Cream Kateroo	Saharanpur Prabhat
Dr. Y.S. Parmar Univ of Hort. & Forestry 1990 to present	500 to 900	MT Peaches	New program: medium-chill Late ripening		
Mexico Centro de Fruticultura Colegio de Postgraduados Chapingo,Mexico 1984	100 to 600	NM Dual purpose types Fresh and processing Yellow Peach	Extend season, FDP < 150 days Resistance to powdery mildew Resistance to blossom brown rot Low red coloration	OroMex OroAzteca (=Aztecgold) Diamante Mejorada Diamante Especial	

Table 3. continued

Location	*Chill range (CU)*	*Fruit type*[z]	*Objectives*	*Releases*	
Mexico	0 to 650	NM	FDP from 75 to 200 days	Seed propagated lines	Clonal releases
INIFAP-CENGUA	5 zones of	Peach	Medium vigor, less bushiness	Lupita	Amari
Queretaro, Qro	adaptation	Yellow	Resistance to powdery mildew	Productivo Agosto	Avega
1984 to present		Fresh/processing		Rendidor	Regio
					Toro
					Seguro
South Africa	400 to 750	NM	Dual purpose cultivars	MT- fresh market	NM-processing
Horticultural Research		Peach	Extension of season	DeWet	Kakamas
Institute, Pretoria		Processing and fresh	Sturdy upright growth	Albatros	Oom Sarel
FFTRI, Stellenbosch		Yellow /white	Storability	Clocolan	Neetling
1960? to present			Round shape	Imperani	Walgant
		MT	Late blooming	Safari	Drying
		Peach/nectarine	- frost avoidance	Ingwe	Bokkeveld
		Fresh and drying	Resistance to bacterial leaf spot	White fleshed	Hantam
		Yellow /white	Resistance to *Pseudomonas*	Orion	Waveren
				Culemborg	
Taiwan	0 to 200	MT	Large, low fuzz, high blush	In-ge taur	
TARI		Fresh market	Attractive and firm	Kuu Taur	
Taichung		Peach	Early ripening		
1992 to present		White	(FDP 60 to 120 days)		
		Low acid	Resistance to		
			Trunk gummosis		
			Bacterial leaf spot		
			Peach leaf curl		

[z] MT = melting flesh and NM= non melting flesh.

Rodriguez, 1987). The four most active programs (Florida, Pelotas, Campinas, and South Africa) have worked independently except for an occasional exchange of germplasm (Sherman and Rodriguez, 1987; Bruckner, 1987; Barbosa *et al.*, 1995; Smith, Personal Communication (PC) 1996)

In the 1980s the Armstrong program terminated and programs were initiated in California (SunWorld, Bakersfield), south Georgia (Attapulgus, cooperative program with the University of Georgia, University of Florida and USDA, Byron, GA), Texas (*low-chill* program in south Texas), and Mexico (Chapingo and Queretaro). The US programs have been building on the germplasm developed by Florida, Georgia (USDA, Byron, GA), and California (various programs).

Whereas the previous programs have emphasized the development of melting flesh types for the fresh market, the two Mexican programs work with nonmelting flesh. In Mexico, as in Spain, these types (pavia) are for the fresh and processing markets. The program in Chapingo has worked closely with Florida which is developing nonmelting, early ripening, low-chill peaches for the fresh market. The major difference is that the Mexican program selects for low to medium red skin color whereas the US based programs select for high red skin color. The Queretaro program has worked more with local selections and a few Brazilian materials (Perez, 1995; Rodriguez and Sherman, 1995).

At present, there appears to be a trend to develop nonmelting types for the fresh market, especially in the early season. This is evident in the medium to high-chill cultivars with the release of 'Maysun', 'Davidsun' (AgriSun Nurseries, CA), 'Crimson Lady', 'Crown Princess' (Bradford, California), 'Delta' (LAES, Louisiana), and 'Spring Baby' (USDA, Fresno), the Golden series (Tatura, Victoria, Australia), and the objective of developing dual purpose peaches in Brazil (Campinas and Pelotas) and in South Africa. The Florida program is also developing a series of nonmelting highly colored selections. 'UFGold' is the first release in this low-chilling nonmelting class (Sherman and Lyrene, 1997). The Florida germplasm is being used in two new programs in Australia (Gatton, Queensland and Richmond, NSW), in south Georgia (Beckman *et al.*, 1996), and in Texas. The objective of this work is to increase the firmness and storage qualities of the fruit so that it can be picked in a more mature and flavorful state than is possible with fruit that has melting type flesh. Discovery of a semifreestone nonmelting type creates greater opportunities to extend the use of the nonmelting flesh character in markets that traditionally prefer a freestone peach (Beckman and Sherman, 1996; Van Der Heyden *et al.*, 1997).

Several recently initiated programs in Asia (Taiwan and Zhengzhou, Henan, PRC) are concentrating on developing low-acid types for the fresh market. Although these are a side interest for many programs, these are the only programs that emphasize their development. The low-acid types are preferred by the Asian and the French palate and some breeders (especially Zaiger Genetics in California) are developing these types as specialty peaches.

2.2 Peaches for processing

The program in Palo Alto (CA, USDA/Stanford University) was the first to select for adaptation to medium-chill conditions (tolerance to delayed foliation) for processing peaches. The released cultivars require 700 to 800 chill units. This program ended in the 1940s and subsequently the breeding of processing peaches continued in the high-chill zone at the University of California, Davis location (Gradziel *et al.,* 1993).

Brazil (Pelotas) and South Africa initiated breeding efforts for processing peaches in the 1950s and 1960s because the cultivars introduced from other countries were not well adapted. Consequently the programs have based their breeding on locally adapted cultivars and worked to improve the production, yield consistency, and quality of what was available. Their releases are numerous and are being grown extensively in their country and also are being used in the breeding of a wide range of nonmelting flesh types.

The work in Tatura (Victoria, Australia) began in 1972 to develop a range of 'Golden Queen' types with greater production and a wider range of ripening for the local processing industry. This work developed a series of releases, many of which were lower chill than 'Golden Queen' (850 CR). The most outstanding is Tatura 204 that requires about 450 CU and is more productive and consistent than any other cultivar. This has become the new commercial standard and recent work emphasizes the development of lower chill selections like Tatura 204 (Richards, PC, 1996). A new objective is the development of white fleshed, nonmelting peaches with the low-acid flavor of 'Okubo' for the Asian market (Topp, 1993).

In the 1980s, the two programs in Mexico became active. These programs are developing dual purpose peaches because the peach traditionally grown in Mexico (low red skin color, nonmelting flesh) is equally acceptable for both the fresh and processing markets. Their work involves the selection of both seed-propagated and clonal varieties from locally grown peaches (criollos), and the use of germplasm from programs in Brazil, Florida, and South Africa (Perez, 1995; Rodriguez and Sherman, 1995).

2.3 Japanese plums

All the plum breeding in the medium and low-chill regions deals with the diploid Japanese-type plum (*P. salicina* and hybrids). This work was initiated at several locations in Brazil (Pelotas, RS and Campinas, SP), the United States (Florida, Georgia and Texas), and South Africa prior to 1975 (Ojima *et al.,* 1992; Okie, 1995; Smith, PC, 1996; Sherman and Lyrene, 1985; Topp and Sherman, 1990). The most cultivars have been released from the programs at Campinas, SP, Brazil (9 cv), Georgia, USA (6 cv), and South Africa (6 cv). The program in Texas, was first described in 1937 (Flory, 1937) but the level of effort was minimal until recently. In addition, new programs in Asia (Taiwan and Thailand), Australia (Queensland and New South Wales), Mexico (Chapingo), and Brazil (Santa Catarina, and Parana) were initiated (Ou, PC, 1996; Boonprakob, PC, 1996; Rodriguez, PC, 1996; Okie, 1995; Hauagge *et al.,* 1991).

The most common objective of these plum breeding programs beyond adaptation to the lower chill conditions is for better tree health, specifically resistance to bacterial leaf spot (*Xanthomonas pruni*), plum leaf scald (*Xylella fastidiosa*), bacterial canker (*Pseudomonas syringae*), and rust (*Tranzschelia pruni*) (Table 4). Among these, resistance to bacterial leaf spot is the most important, followed by resistance to plum leaf scald which is presently limited to the Americas. In the higher chill areas where Japanese plums are grown commercially such as California and parts of Italy, these diseases are not major problems due to a shorter growing season, a more severe winter, and drier weather conditions during the growing season. Consequently, much of the breeding focuses on intercrossing plums that are resistant to these diseases and/or have a low-chilling requirement with those which have better fruit quality.

2.4 Apricots

Again, as with plums, the most active breeding work is in the transitional chill zones of California, Georgia, South Africa, and South Australia where the adaptation problems are not as extreme as compared to the low-chill zones (Table 5). In all these areas, the major apricot cultivars from European or Californian breeding do not fruit consistently and therefore these programs have worked within the species *P. armeniaca* to develop more consistently fruiting genotypes (Smith, PC, 1996; Hough, 1989; Gathercole, PC, 1995; Okie, PC, 1997). Additional work is being done in the highlands of Mexico and India and the lowland areas of the medium (Texas) and low (Florida) chill zones of the USA (Sharma and Kumar, 1994; Perez, 1992; Sherman, 1987). In the lower chill areas, the use of *Prunus mume*, and various plum species are being explored as possible sources of low-chilling adaptation (Boonprakob and Byrne, 1990; Sherman *et al.*, 1984; Sherman, 1987)

2.5 Cherries

There is little done on the development of either sweet (*P. avium*) or sour (*P. cerasus*) cherries for the medium and low-chill zones. Both these crops grow best in areas that receive greater than 800 chilling hours and have cool growing conditions. Thus except for high altitude regions in the tropics, these crops are not grown. In the southern limits of cherry production, they exhibit poor fruit set after mild winters (Iezzoni *et al.*, 1990; Herter and Raseira, 1990) and frequently have problems with fruit doubling (twinning) if the summer temperatures are hot (Micke *et al.*, 1983; Southwick *et al.*, 1991). The degree of fruit doubling is dependent on the specific variety (Micke *et al.*, 1983; Tucker, 1934) and responds to negative selection. There are a few medium-chill (400 to 600 CR) sweet cherries such as 'Cristobalina' in Spain and others in Greece and the northern Mediterranean region. Among these are several self fertile types which although have a good flavor, tend to over crop and to produce small fruit.

Several cherry breeding efforts (Zaiger Genetics in Modesto and Marvin Nies in Lodi) in California are developing sweet cherries for the transitional chill zones by selecting within *P. avium*.. Small efforts in Australia (New South Wales), Florida (USA), and California (USA) to develop medium and low-chill sweet cherries have utilized related cherry species to obtain the lower chilling trait. There are no efforts to develop low-chill tart cherries (*P. Cerasus*).

Table 4. Japanese plum breeding programs in medium and low-chill zones

Location	*Chilling (CU)*	*Fruit type*	*Objectives*	*Releases*	*Parentage*	
Australia						
Nambour and Bundaberg Queensland 1994 to present	100 to 400	Fresh market	Low-chill adaptation	New program		
Richmond, NSW Univ. West Sidney 1992 to present	300 to 600	Fresh market	Low-chill adaptation	New program		
Brazil						
CPACT-EMBRAPA Pelotas, Rio Grande doSul 1953 to present	150 to 500	Fresh market	Resistance to Plum leaf scald Bacterial leaf spot	Pluma7 Pluma2	The First	Santa Rosa Satsuma
IAC- Campinas, Sao Paulo 1966 to present	0 to 150	Fresh market Red fleshed	Resistance to Rust	Carmesim Crancuore Golden Talisma Januaria Centenaria Rosa Paulista Gema de Ouro Rosa Mineira Kelsey-31	Santa Rita	Kelsey Satsuma
EPAGRI Urussanga, Santa Catarina	200 to 400	Fresh market	Resistance to Plum leaf scald Bacterial leaf spot	377290	Amarelinha	Kelsey Harry Pickstone

Table 4. continued

Location	*Chilling (CU)*	*Fruit type*	*Objectives*	*Releases*	*Parentage*	
EPAGRI Videira, Santa Catarina	200 to 400	Fresh market	Resistance to Bacterial leaf spot Plum leaf scald	None		
IAP - Curitiba, Parana	200 to 600	Fresh market	Resistance to Bacterial leaf spot Plum leaf scald	Irati	FS-89 Amarelinha	
Mexico						
Centro de Fruticultura Colegio de Postgraduados Chapingo,Mexico 1984	100 to 600	Fresh market Large size, firm	Low-chill	Corazon Rojo Cirena 1	Florida germplasm	Santa Rosa
South Africa						
Horticultural Research Institute, Pretoria FFTRI, Stellenbosch 1950 to present	500 to 750	Fresh market Size, 50-55 mm Round, no tip Firm but juicy TSS > 15% Yellow/ orange flesh Freestone Long storage life	Production, consistently >18 kg/tree Precocious Resistance to Bacterial leaf spot	Songold Harry Pickstone Laetitia Pioneer Red Gold Reubennel Sapphire Souvenir	Golden King Methley	Eldorado Gaviota Laroda Wickson
Taiwan						
TARI Taichung 1992 to present	0 to 200	Fresh Market Processed	Low-chill adaptation	No releases	Local plum cultivars	

Table 4. continued

Location	*Chilling (CU)*	*Fruit type*	*Objectives*	*Releases*	*Parentage*	
United States						
Florida University of Florida Gainesville 1969 to present	100 to 500	Fresh market Large, firm, productive	Resistance to Bacterial leaf spot Plum leaf scald Rust Low-chilling	Gulfruby[Z] Gulfgold[Z] Gulfbeauty Gulfblaze	Taiwan plum	Ozark Premier Burbank Methley Beauty Bruce
Georgia USDA-ARS Byron 1958 to present	650 to 1000	Fresh market Good quality, large, firm	Healthy long -lived tree Resistance to Bacterial leaf spot, *Pseudomonas*, Plum leaf scald Consistent yield	Robusto Explorer Segundo Byrongold RubySweet Black Ruby	Barstow Morris Ozark Premier *P. angustifolia*	Gaviota Mariposa Queen Ann Santa Rosa
Texas						
Texas A&M University College Station, 1937 to present	400 to 700	Fresh market Good quality Large, firm	Resistance to Bacterial leaf spot Plum leaf scald	Morris	Methley	Lower chilling is a new effort

[Z] 'Gulfruby' and 'Gulfgold' were not released by the University of Florida but rather unofficially by a nursery in Florida

Table 5. Apricot breeding programs in medium and low-chill zones

Location	*Chill range (CU)*	*Fruit type*	*Objectives*	*Releases*
Australia				
Nambour and Bundaberg Queensland QDPI 1994 to present	100 to 400	Fresh market	Low-chill adaptation	None, new program
Richmond, NSW Univ. West Sidney 1992 to present	300 to 600	Fresh market	Low-chill adaptation with good fruit set	None, new program
Loxton South Australia Primary Industries	600 to 900	Drying Fresh market	High soluble solids Firm fruit Good drying characteristics	Provisional release of two selections
India				
Dr. Y.S. Parmar Univ. of Hort. & Fores. Solan, Himachal Pradesh 1956 to present	500 to 900	Fresh market	Medium-chill	Chaubattia Alnkar Chaubattia Madhu Chaubattia Kesri
Israel				
The Volcani Center 1996	100 to 400	Fresh market	Low chill, early fruit ripening, high quality	None, new program
Mexico				
Univ. de Queretaro Queretaro 1984	400 to 700	Fresh market	Medium-chill Large, firm	Collecting local types
Centro de Fruticultura Colegio de Postgraduados Chapingo, 1984	100 to 600	Fresh market	Low enough chill to induce consistent fruit set Large size, firm	None, new program

Table 5. continued

Location	*Chill range (CU)*	*Fruit type*	*Objectives*	*Releases*	
South Africa					
Horticultural Research Institute, Pretoria FFTRI, Stellenbosch 1950 to present	500 to 750	Fresh market Drying Canning	Symmetric shape Firm and juicy TSS > 15% Freestone Good keeping quality Precocious, yield > 18kg Resistance to Bacterial leaf spot *Pseudomonas* canker	Drying types Ladisun Peeka Soldonne	Fresh Market Palsteyn Piet Cillie Super Gold
United States					
California Zaiger Genetics Modesto, CA 1962 to present		Fresh Canning Drying	Consistent production	Katy Gold Kist Improved Flaming Gold	
Florida University of Florida Gainesville 1969 to present	100 to 500	Fresh market	Resistance to Bacterial leaf spot Low-chilling Consistent fruiting	None	
Georgia USDA-ARS Byron 1964 to present	650 to 1000	Fresh market	Resistance to Bacterial leaf spot *Pseudomonas*, Healthy long-lived tree Good quality, large, firm Consistent yield	None	
Texas Texas A&M University College Station, TX 1990 to present	400 to 700	Fresh market	Resistance to Production consistency	None, new program Bacterial leaf spot	

3 Fruiting traits

3.1 Ripening period

Many of the medium and low-chill peach breeding programs have a major objective of extending the harvest period. Although there is a need to extend it both later and earlier, the emphasis, especially in areas which border higher chill production zones, is on earlier ripening cultivars. This is evident in the programs based in the California, Texas, Florida, Georgia, Australia (Queensland and NSW), and Taiwan where the emphasis is to develop lower chill cultivars which ripen before the higher chill production areas begin their harvest. In other areas, both extremes of the ripening periods are being more equally developed.

Very late ripening stone fruit (180 to 240 day fruit development period, FDP) can not be grown in most high-chill areas since their growing season is frequently less than 180 days. There is medium-chill peach germplasm (Mexican feral populations, 'Gaschina Novembre') with 190 to 220 days FDP (Perez, 1989; Byrne, PC, 1996) that could be used to develop these very late ripening cultivars.

In contrast to the wide range of FDP commercially available in peach and plum, commercially apricots and cherries are generally limited to relatively short seasons of 2 to 6 weeks. Nevertheless, there are reports of late ripening apricots that have FDP of up to 220 days (Mehlenbacher *et al.*, 1990).

3.2 Skin and flesh color.

Most programs that are developing peaches and nectarines are selecting for a bright (yellow, orange, or white, not green) ground color with an attractive red blush. The selection for a bright non-green ground color is particularly important in lower chill areas since many varieties when grown in lower chill regions tend to retain a greener ground color when ripe. For the traditional fresh fruit peach markets in Mexico and Spain the red blush is not required and in some growing areas of Mexico, red skin attracts birds which damage ripening fruit (Rodriguez, PC, 1996). Also in the peach processing market red coloration in the flesh turns brown upon processing and makes the final product unattractive. Most peach programs have produced yellow flesh varieties, but there is interest in white flesh and some new varieties are available. Red flesh is available in peach but there is little interest at present in varieties with blood red flesh, possibly because it stains badly. In all breeding programs, the selection away from the green flesh and skin ground color is essential.

Selection away from green color is also true for apricots but not necessarily for Japanese plums. In plums, there is a wide array of acceptable colors for both the flesh and skin. Skin color can be yellow, red, green, bronze, or black although the most common are red and black. Flesh color can range from blood red to orange, yellow, green or almost white. The most common colors in commerce would be in the red to yellow range although the other colors are not necessarily unacceptable. Blood red plums are an objective of Brazil and the NSW (Australia) programs and one has been released from Georgia (USA). In general, the blood plums have a less sour skin as in 'Mariposa' (USA) and 'Donsworth' (Australia).

3.3 Flesh types and pit adhesion

In the USA, many parts of Europe, Brazil, and Asia, melting flesh peaches and nectarines are used for the fresh market. Furthermore, the preference is for freestone peaches when these are available, but this trait is only expressed in genotypes with a FDP of greater than 90 days. Thus early ripening, genetically freestone, melting flesh peaches are physiologically clingstones (semicling). In contrast, this relationship does not seem to hold among apricots or sweet cherries. In these crops the freestone character is independent of the FDP. This freestone trait is highly desirable in varieties of sweet cherry and apricots but in plums, although freestone types exist, most varieties are clingstone.

Nonmelting flesh in peach is used for the processing market since it does not disintegrate when cooked. In addition, it is also commonly used for the fresh market in Mexico and Spain and used some in Australia ('Golden Queen'), Brazil ('Diamante' and others), Venezuela, Bolivia, Colombia, Peru (local varieties), and Japan. Recent releases of highly colored nonmelting flesh cultivars for the early market were successful in both the USA and European fresh fruit markets which traditionally use melting flesh. Comparison of some Florida nonmelting and melting genotypes found that the flavor profiles, soluble solids, pH, and titratable acidity did not differ between the two flesh types (Brovelli *et al.*, 1995). The advantages of nonmelting flesh are better firmness, enhanced storage qualities, and greater flavor at harvest as compared to melting flesh types (Beckman and Sherman, 1996; Robertson *et al.*, 1992). This flesh type is also found in firm sweet cherry cultivars and has been reported in an apricot land race from Pakistan (Thompson, 1993).

Firmness in peach has several components. Uniformity of firmness is essential and phenotypes that soften along the suture or on the tip are generally eliminated because they are difficult to handle. A good firmness at a yellow or orange ground color (all green gone) is another aspect of firmness that is essential. It is easy to select for very firm fruit at a green ground color but the market requires attractive fruit that does not have a green ground color. This is very important for lower chill zones since a given peach variety when grown in a lower chill as compared to a high-chill zone will exhibit a greener ground color. Finally, the fruit needs to maintain its firmness for as long a time as possible (extended storage life). Storage life is emphasized in the peach, plum, and apricot breeding in South Africa and is beginning to attract more active selection in various programs in the USA as well.

A perceived disadvantage of nonmelting flesh in peach was the lack of separation between the pit and the flesh. In early ripening (FDP < 90 days) melting flesh cultivars, this is not a disadvantage since these are all clingstones (semicling). In later ripening types, the preference for freestones is, in part, connected to the ease of preparing peaches for home canning but this activity has decreased and current very late season nectarine cultivars are clingstones. Perhaps the freestone trait in the fresh market trade is not essential even in markets that traditionally prefer these.

In peach, two types of freestone are seen in both melting and nonmelting flesh. One type expresses a layer(s) of cells adhering to the pit and in the other type the pit is dry. Suprisingly, some early ripening (70 day FDP) semi freestone, melting flesh cultivars (*i.e.* 'Flordaking') have a dry pit. While the number of genotypes have been limited in short FDP nonmelting types, we have observed the dry pit only in genotypes that have a FDP between 100-110 days.

In addition, semifreestone, nonmelting flesh peach genotypes were found among the Florida nonmelting breeding population (Beckman and Sherman, 1996; Van Der Heyden *et al.*, 1997) and nonmelting genotypes such as 'Precocinho' from Brazil have been reported to separate cleanly from the pit (Raseira, PC, 1996). This combination of traits, will also facilitate the incorporation of nonmelting fresh market cultivars into the traditionally melting flesh freestone market. This freestone flesh may also be important in canning peaches by reducing pit fragments clinging to the flesh after mechanical pitting.

Another peach flesh type called the stony hard type has some potential in the development of firmer peaches with tree ripe flavor, longer storage life, and with the freestone trait. This flesh has a reduced ethylene production compared to melting flesh peaches (Goffreda, PC, 1992; Yoshida, 1976) and thus has a longer storage life. The gene in US breeding comes from Japanese stock (7-28 selection, 'Okubo') which is white fleshed, high in soluble solids (16.3%) and low in acid (0.28%) (Robertson *et al.*, 1990). It appears to be inherited by a single gene that is recessive to melting flesh and thus is easily incorporated into a breeding population (Scorza and Sherman, 1996). Another trait possibly useful in extending storage life or the harvest season is the non ripening or slow ripening trait found in peach (Ramming, 1991; Brecht and Kader, 1984) and plum (Mowrey, PC, 1997; Yamaguchi and Kyotani, 1986).

3.4 Fruit size

In most markets large fruit sells more easily and at better prices than small fruit. In markets such as in the USA or Europe, fruit smaller than 6 cm can not be sold except in the very early season. This is not true in the traditional peach markets in Mexico or with the low-acid peaches in Taiwan where they can be picked early as "thinnings" and sold crunchy. Nevertheless, all programs select for large fruit size in the whole range of ripening times, although given the strong genetic correlation between the fruit development period and fruit size, the development of varieties with large fruit is more difficult for the early ripening as compared to the mid season and later ripening season (Souza, *et al.* 1998). A similar situation appears to be operative in plum, cherry, and apricots as well.

3.5 Fruit shape

The shape of peach fruit of a given cultivar becomes more elongated when grown in low-chill compared to higher chill region or after a milder winter (Bacon and Byrne, 1990; Byrne and Bacon, 1989; Topp and Sherman, 1989b). It is not clear whether this is due to the lower chilling received and/or to a higher temperature experienced during the early stages of fruit development. In either case, it is possible to select for genotypes that are rounder under low-chill conditions (Topp and Sherman, 1989b; Bacon and Byrne, 1990). The additional need in medium and low-chill areas is for cultivars that maintain their round shape in the range of chilling conditions where they are productive.

3.6 Fruit quality: acidity and sugar content

Fruit quality is a difficult concept to define since it changes with the region and nationality of the people. For example, in the USA the current preference is for medium to high acid, sweet peaches whereas in Asia the preference goes towards low-acid sweet peaches. Yet in all regions there is a range of acceptability in the flavor and texture components of fruit.

The acidity of a peach is controlled, mostly by a dominant major gene (honey = low acid) but most current varieties are high acid types (Monet, 1979). Given the continuous range of acidity observed among peaches (Byrne, Nikolic, and Burns, 1991), modifying genes also influence the final acidity of the homozygous recessive acidic genotypes. Among commercial peaches, the acidity ranged from low-acid (titratable acidity (TA) = 2.1 meq malic acid/mL) to highly acid (TA = 10.9 meq/mL) (Byrne, Nikolic, and Burns, 1991; Robertson *et al.*, 1988; 1990). Although most breeding programs concentrate on the development of high acid cultivars, some (Zaiger, Taiwan, Australia) work on low acid and subacid types appropriate for the Asian markets.

It should be noted that the amount of acidity does influence the perception of sweetness. A peach at a given level of soluble solids will taste sweeter as the acid level is decreased. This is important when developing very early ripening and very late ripening stone fruit since the sugar content is lower under the cooler temperatures at which these varieties ripen and therefore selection needs to be for lower acidity to maintain a good sugar to acid balance (Zaiger, PC, 1996).

A general objective in most breeding programs is higher sugar content, which is usually measured with soluble solids. It has been shown that selection for higher soluble solids is not highly effective (Hansche *et al.*, 1972). The low heritability estimates indicate that the environmental effects, which may include weather (cloudiness, rainfall, day/night differential temperatures), nitrogen availability, fruit development state, and the measurement procedure among other factors, affect the final level of soluble solids. Also the possibility of developing early ripening peaches with high soluble solids may be limited since there is a strong genetic correlation between fruit development period and soluble solids. Thus early ripening (<90 FDP) cultivars have lower soluble solids than mid-season ripening cultivars (Souza, *et al.* 1998). Early ripening cultivars had soluble solids contents between 9 and 12% brix as compared to 14 to 17% in the mid season cultivars (Byrne *et al.*, 1991; Malone and Snelling, 1997). There are efforts to develop high soluble solid peaches and nectarines. 'Arctic Rose', recently released white nectarine from the Zaiger Genetics breeding program has been reported with a brix of 18 to 21% (Ramming, PC, 1997; Zaiger, PC, 1997). Peaches and nectarines with this level of soluble solids may be useful for the dried fruit and baby food markets. There are four major sugars produced in the peach and although the relative amounts vary among cultivars, their individual contribution to sweetness is not well understood (Robertson *et al.*, 1988; Byrne *et al.*, 1991) nor are there inexpensive methods for measuring the levels of individual sugars. Much work needs to be done in this area.

Apricot germplasm has a greater range of soluble solids which can be over 30° brix in some of the drying apricot cultivars from Turkey and Pakistan (Thompson, 1993). Unfortunately, most of this germplasm is from the Central Asian group which requires high-chilling conditions and has small fruit, pale flesh and skin color and a tendency to crack under humid conditions (Mehlenbacher *et al.*, 1990; Gathercole, PC, 1995). The current breeding in South Australia is focused on incorporating this high soluble solids content into a lower chill background with better fruit characteristics (Gathercole, PC, 1995).

Commercial plums range from 10 to 16% brix (Byrne, PC, 1996; Dhatt *et al.*, 1992; Ramming, PC, 1997; Hauagge *et al.*, 1991; Sharma and Josan, 1993; Ved *et al.*, 1994; Topp and Sherman, 1990b; Wills *et al.*, 1983). Nevertheless, there are reports of plums with a TSS of greater than 20% such as the varieties 'Frank Ann' and 'Doris'. There appears to be germplasm available for higher soluble solids content in the higher chill germplasm of the Japanese plum although the soluble solids level obtainable may be limited in early ripening cultivars as is seen in the peach.

3.7 Peach versus nectarine

Although only one gene (G/g, pubescence/glabrous) difference exists between a peach and a nectarine, there are various pleiotropic effects on the fruit characters associated with this gene (Wen *et al.*, 1995a; 1995b). In general, nectarines were higher in sugars and acids, had redder skin color, had a rounder shape, were smaller in size, and had higher specific gravity (denser flesh) than peaches. Additive gene variability (heritability) for these quantitative traits is moderately high. Thus, in choosing parents that involve crossing between these two fruit types, one should consider what characters are likely to be expressed in a higher percentage of the progeny. For example, if a nectarine with a slightly pointed tip and marginal red skin color was hybridized with a peach, it is more likely to contribute less toward round shape and red skin color in resulting peach genotypes than is expressed in the nectarine parent. Similar trends may be used in reverse. For example, a peach heterozygous for nectarine, with pointed fruit, marginal red skin color, and moderate sugars is likely in crosses with nectarine to transmit to its nectarine progeny rounder fruit, more red skin color, and higher sugars than expressed in the peach parent. This difference in character expression between peach and nectarine makes it easier to breed these two fruit types separately.

4 Adaptation traits

4.1 Chilling and heat requirement for bloom

Chilling requirement is most commonly estimated by relative bloom time. This selection procedure does not distinguish between the effects of chilling requirement and heat requirement. Large differences in heat requirement among cultivars of similar chilling requirement, are detected by bloom reversal when comparing their relative bloom periods after winters in which little versus excessive chilling was received (Sherman,

PC, 1996). Generally, in commercial peach germplasm bloom reversal is not seen and the major determinant of bloom time is thought to be the chilling requirement. Nevertheless, late blooming peach genotypes have been pursued by several programs (South Africa, Mexico, and Brazil) as a route to frost tolerant cultivars (Van Rooyen, 1988; Raseira, PC, 1996; Perez, PC, 1996). There are reports of peaches that bloom later (bloom with high-chilling cultivars) than most and still fruit as heavily as the earlier blooming genotypes (Zegbe and Rumayor, 1995; Byrne, PC, 1996). Whether the high set of the later blooming genotypes is due to a greater heat requirement or better setting ability at the higher night temperatures of the later bloom period as compared to the other cultivars blooming at the same time is not known.

In contrast, the order of bloom changes from moderate chill to low-chill conditions for some varieties of plum and apricot. Thus, there is major variation in both chilling requirement and heat requirement for bloom in plum and apricot. In fact, most apricots in high chill climates are the first stone fruit to bloom indicating they possess a low heat requirement to break dormancy. Estimates in apricot indicate the heat requirement among apricots range about 4 fold, from less than 2,450 to about 8,000 growing degree hours with a modal heat requirement of 4000 to 6000 growing degree hours (Guerriero *et al.*, 1992). The heat requirement variability of dormancy in sweet cherry is unknown.

4.2 Tolerance to high temperatures during bloom

Peaches generally set poorly at night temperatures above 15°C (Edwards, 1987). Apricots appear to need a lower night temperature for fruit set than do peaches or plums.

Field observations indicate there are differences between peach genotypes in their ability to set under high bloom temperature, but thus far careful studies have not been done. This is one area that is important especially with respect to developing cultivars that are well adapted to double cropping or forced cropping systems in peach (George *et al.*, 1988a; Sherman and Lyrene, 1984). An ability to set under a wider range of temperature conditions would give the producer more flexibility in the timing of the harvest seasons. At present, the timing of bloom is limited by the temperature regime.

4.3 Blind nodes

Low-chill peaches are highly susceptible to blind nodes, but ample genetic variability exists to overcome this problem through breeding and selection (Boonprakob *et al.*, 1994; 1996; Richards *et al.*, 1994). The blind node problem is probably more evident in the low-chill genotypes because most ripen before summer. Trees without a crop load tend to flush growth in the summer and rapid growth during the hot weather is the condition under which blind nodes develop. Trees grown in cooler summer temperature, such as occur in higher elevations of the subtropics and in the coastal climates with cool summer onshore breezes, develop fewer blind nodes. In fact, genotypes selected in those climates (such as the highlands in Zacatecas, Mexico) often exhibit a high degree of blind nodes when grown in hot humid summer climates as in the deep south of the USA (latitude 29° to 36° N). The 'Armking' nectarine and the 'EarliGrande' peach are highly susceptible to blind nodes whereas the 'TropicBeauty' peach and the 'Suncoast' nectarine are highly resistant. Interestingly, in climates similar to Seville, Spain (central valley of California and Hermosillo, Mexico) summer temperatures are high enough to inhibit growth in July and August (autumn growth is good in low to medium-chill genotypes) and susceptible varieties exhibit few blind nodes.

4.4 Twin or double fruit

Water and heat stress during flower bud formation in mid summer greatly increases the formation of fruit twins or doubles in cherries (Southwick *et al.*, 1991) and peaches (Diaz *et al.*, 1986; Patten *et al.*, 1989; Zaiger, PC, 1995). This is why sweet cherry is grown east of the coastal mountain range break at San Francisco in the Central Valley as the cool oceanic air spills in here as the central valley warms up. Low to moderate chill peaches in the southern USA and NW Mexico tend to be subjected to twinning because little attention is given to irrigating vigorously growing trees during the long, post-harvest, dry summer months when the flower buds are forming. There is much genetic variability among sweet cherries (Micke *et al.*, 1983) and peach and nectarine genotypes for twinning. 'Flordagold' and 'Desertred' peaches are highly susceptible whereas 'TropicBeauty', 'EarliGrande', 'Sunhome' and several Florida selections are resistant (Diaz *et al.*, 1986; Table 6).

Table 6. Percent of fruit doubles in peach at College Station, Texas following drought stress

Variety or selection	*Full bloom date*	*% double fruit*
BY4-7127	8 Mar	64
BY5-938	8 Mar	41
BY8-3908	9 Mar	35
BY9-1041	7 Mar	29
Cherrygold	26 Feb	28
Desertred	18 Feb	64
Fla. 1-8	17 Feb	5
Fla. 9-6N	16 Feb	1
Fla. M2-7N	2 Mar	19
Goldcrest	4 Mar	44
Goldprince	2 Mar	26
Juneprince	2 Mar	32
Maylady	5 Mar	34
Redskin	12 Mar	40
Springcrest	05 Mar	23
Sunhome	18 Feb	16
Sunland	07 Mar	44

4.5 Bud Drop (Failure)

Spring flower bud drop prior to swelling is best known among apricots in the southeastern USA, but drop also occurs in some years in peach in the southern USA and the central valley of California. High minimum temperatures during mid dormancy (Dec-Jan) are correlated with flower bud drop at Fresno, California (Weinberger, 1967a). Winter temperatures fluctuate more often in low-chill areas, consequently low-chill genotypes are more subjected to bud failure conditions than are genotypes adapted to and grown in high-chill areas. These conditions occur in central Florida to South Georgia in most years and thus selection for a high bud set and low bud failure has resulted in genotypes that more consistently crop especially in years when a frost occurs during bloom since flower buds that have not opened may survive.

5 Breeding in the low-chill zone

5.1 Pollen germination, pollen tube growth and fruit set

For peach, pollen germination is generally excellent (above 85% of maximal level) between the temperatures of 15 and 25°C with the maximal germination at 20°C. Even at higher temperatures (28 to 30°C) pollen germination is still good (77 to 95% maximal germination), however, pollen tube growth decreases rapidly at or above 30°C (Medeiros, 1979; Weinbaum *et al.*, 1984). This agrees well with breeding experience that indicates that the best set from hand pollinations occurs during "short sleeve" weather (>20°C) and that if the temperature becomes too hot (>30°C) success in pollination decreases.

In addition, although many breeders emasculate and pollinate their peach and apricots the same day for convenience, the South African breeding program delays pollination two days after emasculation. They have increased fruit set from 5 to 24% by using this procedure (Smith, PC, 1996). In peach, pollination the day after emasculation resulted in greater set (45%) than either pollination the day of emasculation (35%) or the second day after pollination (20%) (El-Agamy and Sherman, 1987). This may be related to timing the pollination for the day on which the flower would normally open as this is when optimal set in peach occurs (Wang *et al.*, 1989).

5.2 Seed germination

Stonefruit seed germination is usually more difficult among the low-chill types because early ripening is more prevalent among low-chill cultivars. Seed germination in peach, plum, and cherry and to some extent, apricots decreases as the FDP shortens below 100 days. In fact, peach seed ripening less than 120 days will not tolerate drying without a reduction in percentage germination. Peach embryos from less than 100 day FDP are usually removed from the pit prior to stratification. Peach embryos from 75 to 90 day FDP may require embryo culture, and those below 75 day FDP generally require in ovule culture. Thus a 90 day FDP peach gives about 50% germination and seedling production. Above average temperature during FDP and heavy crop load increases germination in peach. This probably occurs because, the rate of seed development is more rapid under high temperatures than is the rate of fruit maturity (even though high temperature will reduce FDP). Braak (1978) showed that in sweet cherry low temperature during FDP had an adverse effect on embryo development. Glasshouse temperatures above 18°C during FDP increased percent germination. Heavy crop loads delay fruit ripening and allows seed to mature (fill out) to a larger extent. Thus seed germination of a 90 day peach is usually higher in the southern USA than in California.

Seedling rosetting at germination is a major problem in low-chill stone fruit breeding programs because many seedlings are from relatively immature seed that is either embryo cultured (<85 days FDP) or removed from the pit and stratified without drying (85 to 100 FDP). There are three factors that affect rosetting: the physiological state of the ovule at initiation of stratification, length of stratification, and high temperatures during germination.

The first factor can be reduced to near zero in genotypes ripening in over 120 days FDP by drying the seed prior to imbibing and stratification and in seed with more than 140 days of development without drying (Bacon and Byrne, 1995). Interestingly, seed dried prior to stratification required less time in stratification than seed put directly into stratification after removal from the endocarp.

For seed too immature to dry, rosetting is decreased by removing the endocarp immediately after harvest and stratifying at 3 to 5°C. Although the seed needs to be stratified longer at this temperature than similar seeds stratified at 7 to 10°C these seed will germinate with less rosetting. The rosetting can be further decreased by growing these seedlings at 15 to 20°C for the first month or so of growth.

Finally, although high-chill genotypes generally produce more rosettes than low-chill genotypes, even the lowest chill genotypes will rosette more as the temperature during germination is raised. This is a problem in low-chill areas since these genotypes fruit early and thus the seed germinates in stratification and requires planting in the greenhouse during the hot summer months. In the Florida program, rosettes are minimized by growing the sprouted stratified seeds (with newly emerged radicals in stratification) in a refrigerated chamber at about 20°C under continuous light or near the cooling pads of the greenhouse.

5.3 Autumn photoperiod

low-chill germplasm does not react to shortening autumn photoperiods for cessation of the terminal growth as readily as high-chill genotypes. Thus low-chill peach genotypes continue shoot growth into very late autumn or early winter. Furthermore, the total growing season is longer in low-chill than in high-chill areas. Trees at Gainesville, Florida and Weslaco, Texas often grow from bloom (late January) to the beginning of dormancy (first of December) or about 10 months in the year. This extra growth encourages the early fruiting of seedlings (2 years from seed to seed in the peach fruiting nursery) and budded trees (commercial crops in third leaf) (Rodriguez *et al.*, 1986; Sherman *et al.*, 1973; Sherman and Rodriguez, 1987). This permits more rapid selection and evaluation on budded trees thus speeding up the breeding program (only 3-5 years per generation) and release of cultivars. Some selections made the second year from seed have been used as pollen or seed parents (original seedling) in the third year from seed. In fact, several cultivars have been released in only 8 years from the original hybridization (fruited and selected the second year from seed, budded and evaluated 3-4 years with commercial crops on budded trees).

6 Host plant resistance

6.1 Favorable environments

The medium and low-chill production areas suffer from a spectrum of pests and disease problems that differ from those of the higher chill zones because the lower chill regions have milder winters and longer growing seasons. Although the milder winters make pathogens such as *Cytospora* canker, which depends on winter injury for infection sites, non existent in low-chill areas, this same condition also allows a greater winter survival of a range of other pests and diseases. The extended growing season creates a situation in which pests (mites, white peach scale, San Jose scale) and pathogens (bacterial leaf spot, rust, phony peach, plum leaf scald, and root-knot nematodes) have a longer time period each year to build up their populations. This combined with the mild winters through which a high percentage of these organisms can easily survive, creates a situation in which the pest and pathogen pressures may be higher. A good example of this is the development of peach rust in Texas and Florida. This pathogen first appears in the southern most regions of Texas and Florida in mid May and if no sprays are used, can defoliate the trees by June. This leaves weakened trees that have to survive the rest of the growing season (until the beginning of December) through the hot and dry but humid summer. Frequently these defoliated trees will begin to bloom in September once the autumn rains begin. This, of course, will lower the number of fruit buds available during their normal bloom period in late January or early February. Thus these trees are weakened by a lack of leaves during the long summer growing season and have a diminished number of flower buds which all results in a lower potential yield.

The range of environments in the medium and low-chill regions is highly varied such that only a few of the pests and diseases are present in any given location. For example, in areas with cool spring weather such as in the tropical highlands, southern areas of Spain, and coastal areas of southern California, problems with aphid infestations, powdery mildew, and peach leaf curl are common whereas in the warmer, moist summer areas, catfacing insects, bacterial leaf spot, bacterial canker, rust, and brown rot are much more important. In the desert production areas, there is little concern over bacterial leaf spot or brown rot which are major problems in the more humid areas but other pests such as mites can be a major problem.

Peach, plum, and apricot breeders in low and medium-chill areas most commonly direct their efforts to select for resistance to bacterial leaf spot (*Xanthomonas campestris pv. pruni* (Smith) Dye), plum leaf scald (*Xylella fastidiosa* Wells *et al.*), powdery mildew {*Sphaerotheca pannosa* (Wallr.:Fr.) Lev}, peach leaf curl {*Taphrina deformans* (Berk.) Tul.}, bacterial canker (*Pseudomonas syringae pv. syringae* van Hall), rust {*Tranzschelia discolor* (Fuckel) Tranzschel and Litv.} and brown rot {*Monilinia fructicola* (Wint.) Honey and *M. laxa* (Aderh. and Ruhl.) Honey}. Although many other diseases occur, these are the diseases that are most important economically or difficult to control by other means and/or most amenable to combat genetically because there are high levels of resistance available.

6.2 Bacterial canker, *pseudomonas syringae pv. syringae* van Hall

This bacteria mainly is found on limbs and trunks of plum, peach, nectarine, almond, and cherry in many of the major production areas of the world (Hattingh and Ross, 1995, Ramming and Cociu, 1990). Beyond the most common pathovar, two others have been identified: *P. syringae pv. morsprunorum* (Wormald) Young *et al.,* which attacks mainly cherry and plum and *P. syringae pv. persicae* which has been reported on peach and plum (Hattingh and Ross, 1995). *Pseudomonas syringae pv. syringae* has been associated with Peach Tree Short Life (PTSL) in the southeastern USA (Lownsbery *et al.*, 1977) and apoplexy of apricots in Europe (Mehlenbacher *et al.*, 1990). Limited work with peach has detected low resistance in Lovell, NA8 (Weaver *et al.*, 1979) and others (Scorza and Okie, 1990). The Guardian® rootstock which is resistant to PTSL may have some resistance to bacterial canker since its increased survival in PTSL sites is not due to lower populations of ring nematodes {*Criconemella xenoplax* (Raski) Luc and Raski} (Okie *et al.*, 1994a; 1994b). Some plums appear more susceptible to bacterial canker than do peaches. Early field evaluation by Flory (1941) indicated that 'Methley', 'Shiro', and 'Wickson' were susceptible to very susceptible, 'Kelsey', 'Burbank', and 'Satsuma', were moderately resistant, and 'Bruce', 'Golden Beauty', 'Green Gage' and 'Bruce Junior' were resistant. More recent work reports that releases from the Alabama program ('AU Roadside', 'AU Producer', 'Crimson', 'Purple'), the USDA at Byron, Georgia ('Segundo', 'Robusto', 'Black Ruby'), 'Starking Delicious', 'Myrobalan B', and 'Shiro' were moderately to highly resistant to bacterial canker (Ramming and Cociu, 1990; Thompson and Prince, 1982; Okie, 1994). 'Shiro' has been reported as both moderately resistant and very susceptible which either indicates that it is susceptible but escaped or that several pathogenic races of the bacteria exist. This disease has also been reported as an important problem on apricots in Georgia, France, and South Africa (Audergon, PC, 1995; Okie, PC, 1997; Smith, PC, 1996). A few sources of resistance have been reported (Table 7) but little active breeding has been done towards developing resistant materials.

Table 7. Susceptibility and resistance to bacterial canker (Pseudomonas) in plum and apricot

Resistance category		*Plums*		*Apricots*
Susceptible	Eagle	Morris	Shiro(MR)[Z]	Chuli
	Friar	Red Beaut?	Ternel	
	Methley	Roysum	Wickson	
		Santa Rosa?	Wild Goose	
Moderately susceptible to moderately resistant	America	Elephant Heart	Ozark Premier	
	AU Rosa	Gigalia	Robusto	
	Black Ruby	Gold	Satsuma	
	Burbank	Homeside	Wade	
	Cumberland	Kelsey		
Resistant	Amarelinha	Bruce	Poole's Pride	Ladaki (immune)
	AU Amber	Byrongold	Purple	Nari
	AU Cherry	Compass	Rubysweet	Kaisi
	AU Producer	Crimson	Sapa	EarliOrange
	AU Roadside	Golden Beauty	Segundo	
	AU Rubrum	Green Gage	Starking Delicious	
	Borrachuda	Munson	XV de Novembro	
	Bruce Junior	Myrobalan B		
		Opata		

Data from Audergon, PC, 1995; Cummins, 1991; Flory, 1941; Gupta and Kashyap, 1991; Hauagge, PC, 1996; Norton, 1972; 1986; 1992; Okie, 1995; Ramming and Cociu, 1990; Russell, 1993; Thompson, 1981; Thompson *et al.*, 1982a; 1982b; Topp *et al.*, 1989; Topp and Sherman, 1990b.

[Z] (MR) indicates that it was also reported as moderately resistant to bacterial canker.

6.3 Bacterial fruit and leaf spot, *Xanthomonas campestris pv. pruni* (Smith) Dye

This bacterial disease is widespread throughout the world on peach, plum, and apricot particularly in areas where these fruits are grown on sandy soils under humid and moist climatic conditions. Primary leaf, stem, and fruit infections are favored by wet weather from late bloom to several weeks after petal fall and secondary infections are readily spread by wind driven rain. In the dry summer growing conditions of western USA, southern Spain, Italy, and other areas, this disease is not a problem (Ritchie, 1995a)

Since chemical control efficacy is not always high, the general suggestion has been to use resistant cultivars in areas where this disease is prevalent. Consequently several breeding programs have routinely selected for resistance to bacterial spot in peaches (Texas, Louisiana, Georgia, Florida, North Carolina, New Jersey, South Africa, Brazil), plums (Georgia, Alabama, Florida, Texas, Australia, South Africa), and apricots (New Jersey, Georgia, South Africa).

The basis of screening for resistance has generally been field observations which are done over several years. These multi-year observations are essential given the variable infection levels from year to year and to the apparent lower susceptibility of younger peach seedlings as compared to older trees (Scorza and Sherman, 1996). Field inoculations can be done, but the results are still dependent on the prevailing climatic conditions at the time of inoculations. In field plots in Florida with overhead irrigation, plums were allowed to infect naturally or inoculated via the leaf dip method, and then evaluated using a number of different criteria. This study concluded that it was acceptable to rate just the 2 or 3 most severely affected leaves as compared to randomly chosen leaves, that using either the number or size of spots were equally useful, and that the most repeatable measure was measures of canker length. The genetic basis for spot formation, leaf spot lesion expansion, and stem canker growth appear different although on the average, the selection for one would raise the level of resistance to the others although not necessarily to the same degree (Topp, Sherman, and Stall, 1991). In plum, good correlations between greenhouse inoculation methods and field evaluations were found with leaf infiltration methods (compressed spray) (Martins and Raseira, 1996) and the leaf dip inoculation method (Topp *et al.*, 1993b).

In South Africa, they have determined that genotypes resistant to *Xanthomonas* induce the production of peroxidase (POD) more rapidly than do susceptible genotypes. The use of this assay is being evaluated as a possible pre screening technique to identify resistance to bacterial spot (Smith, PC, 1996).

In addition, a detached leaf bioassay has been described (Civerolo *et al.*, 1982; Randhawa and Civerolo, 1985), but it has been little used. *in vitro* selection of peach callus resistant to the toxin produced by the fungus has also been used with some success (Hammerschlag, 1984; 1988a, 1988b) but is limited by the lack of an efficient regeneration system for peach.

Cultivars of peach, plum, and apricot vary widely in their resistance to bacterial leaf spot (Tables 8 and 9). In general, with peach and plum, the more resistant cultivars were developed in humid areas such as South Africa, Texas, Georgia, Alabama, and Australia and most of the cultivars developed in the dry summer production areas of California are susceptible. Unfortunately their resistance may differ dramatically in different geographic regions. There are examples of this phenomenon for plum ('Amarelinha', 'Harry Pickstone', 'Frontier', 'Reubennel', and others, Table 8) and peach ('Loring', 'Okinawa', 'Surecrop' and others, Table 9). This appears to be due to climatic differences or unique pathogenic races of the bacteria in different geographic regions (du Plessis, 1988; Martins, 1996). This makes the development of stable resistance to bacterial spot more difficult and emphasizes the need for controlling the movement of this pathogen among geographic regions. Also, it indicates the need for cooperative programs for the development of plum and peach resistant to bacterial leaf spot.

Table 8. Bacterial leaf spot resistance in Japanese-type plum and apricot

Resistance categories	*Japanese plums*		*Apricots*
Susceptible	Amarelinha (R)[z]	Frontier (R)	Blenril
	Angeleno	Gulfruby	Castleton
	Beauty	Harry Pickstone (R)	EarliOrange(MS)
	Black Amber	Laroda	Earliril
	Burbank (R)	Midnightsun	Early Golden
	Carmesin	Redheart (MR)	Harcot
	Formosa	Reubennel (R)	Hargrand
	Freedom	Roysum (MR)	Harogem
	Friar	Sanguinea	Sundrop
	Golden Japan	Red Beaut	Sunglo
	Grandrosa	Wade (MR)	Rival
			Veecot
Moderately susceptible	Catalina	Santa Rita	Curtis(R)
	Ebonysun	Santa Rosa	Goldrich
	Eldorado	Shiro	Perfection
	Gigaglia	Songold	Riland
	Queen Rosa	Stirling	
Moderately resistant	Apple	Late Santa Rosa	Farmingdale
	AU Rosa	Methley	Golden Giant
	Black Ruby	Nubiana	Hungarian
	Burmosa	October Purple	Newcastle
	Byrongold	Ozark Premier	Superb
	Casselman	Queensland	Wenatchee Moorpark
	Gulfbeauty	Earlisweet	Wilson
	Gulfblaze	Rubysweet	
	Homeside	Satsuma	
	Kelsey	Shirley	
	Laetitia	Simka	
		Six Weeks	
Resistant	AU Amber	Morris	Alfred
	AU Producer	Purple	Haggith
	AU Roadside	Queensland Bellerosa	Kohl
	AU Rubrum	Redgold	Scout
	Borrachuda	Robusto	Stepynak
	Bruce	Segundo	Sunshine
	Crimson	The First	
	Explorer	Wickson	
	Mariposa	Wilson	
		XV de Novembro	

Data from Byrne and Bacon, 1994; Hauagge, PC, 1996; Keil and Fogle, 1974; Layne, 1966; Millikan and Hibbard, 1964; Nakasu and Castro, 1990; Ojima *et al.*, 1992; Okie, 1994; Ritchie, 1995a; Russell, 1993; Texas notes, 1996; Topp *et al.*, 1989; Topp and Sherman, 1990a; 1990b; Uppal *et al.*, 1993.

[z] (MS), (MR), or (R)= also reported as moderately susceptible, moderately resistant, or resistant by another author.

Table 9. Bacterial leaf spot resistance in peach and nectarine

Resistance categories	*Peach and nectarine genotypes*			
Susceptible	Desertred	Flavortop(MS)	Okinawa(R)	Southland
	Dixiland	Flordabelle(MR)	Queencrest	Springcrest(MR)
	Elberta	Flordagold(MS)	Red Ceylon	Springold(MS)
	Elegant Lady	Flordaprince(MS)	Redgold	Starlite
	Fantasia(MS)[z]	Flordared(MR)	RioGrande(MS)	Suncrest
	Fay Elberta(MS)	LaWhite	Royal April	Sundar
	Fayette(MS)	Maravilha(MR	Royal Gold	Sunfre(MS)
	Fireprince(MS)	Maycrest	Safari	Sunhigh
	Flamecrest(MS)	Maygold(MS)	Sam Houston	Surecrop(R)
	Flamekist(MR)	O'Henry	Scarlet Pearl	Tasty Gold
	Flavorcrest(MS)		(MS/MR)	Tropicsweet(MS)
Moderately susceptible	Armking	Flordacrest(MR	LaFeliciana(MR)	San Pedro(R)
	Babcock(MR)	Flordadawn(MR)	LaFestival(MR)	Sentinel(R)
	Belle of Georgia	Flordagrande(R)	Loring(MR)	Stoneyhard
	Bicentennial(MR)	Flordaking(MR)	Mayfire	Sunblaze(R)
	Bolivian Cling	Flordastar(R)	Maysun	Suncoast
	Carymac(MR)	Gaschina Novembre	Ouachita Gold(MR)	Sungem(MR)
	Cherry Gold(MR)	Harvester(MR)	Redglobe(MR/R)	Sunland(MR)
	Columbina(MR)	Hawthorne(MR)	Redskin(MR)	Sunprince
	Desertgold	June Gold(MR)	RioOsoGem	TropicBeauty
	Diamonte	Juneprince(MR)	Royal Giant	(MR)
	Earligrande(MR)	Kakamas		ValleGrande
	Fayette			(MR)
Moderately resistant	Agua 12-12	Flordaglo(R)	Idlewild(R)	Sunlite(R)
	Chinese Cling	Flordaglobe(R)	La Pecher	Sunred(R)
	Delta	Goldcrest	McRed	Sunripe(R)
	Dixired(R)	Goldprince	Newbelle	Sunsplash
	Flordagem	Hermosillo	Springtime	TexRoyal
				Texstar
Resistant	Bounty	Nectared 5	LaGold	PI65974, Italy
	Dwarf Mandarin	OroA	Le Grand	PI65977, Italy
	Early Amber	PI102521, Italy	PI131034, Poland	PI78544, Italy
	Flordasun	PI104287, Morroco	PI131209, England	PI95501, China
	Flordawon	PI106062,	PI133982,	Redglobe
	Frank	South Africa	South Africa	Salcaja
	Gaucho	PI125025, India	PI133987	TropicSnow
	Montclar	PI129675,	South Africa	
		South Africa	PI442378, Mexico	
		PI130980, Germany	PI442380, Mexico	

Data from Ritchie, 1995a; Sherman and Lyrene, 1981; Texas notes, 1996; and Werner *et al.*, 1986.

[z] (MS), (MR), or (R)= also reported as moderately susceptible, moderately resistant, or resistant by another author.

6.4 Botryosphaeria Gummosis, *Botryosphaeria dothidea* (Moug. ex Fr.) Ces and de Not.

This fungal disease is favored by warm temperatures (20-28°C) (Ko and Sun, 1992) and causes severe gum exudation following rain or irrigation, brittleness of the fruiting wood, and tree debilitation of peach. It does not appear to attack either plum, cherry, or apricot. It was first observed in the southern US in the 1960s (Pusey, *et al.*, 1995; Riley and Okie, 1982; Weaver, 1974). It has since been recognized as a potentially devastating disease in the southeastern United States, eastern China, Taiwan, Japan and Australia (Li *et al.*, 1995; Ko and Sun, 1992; Pusey, 1989; Pusey *et al.*, 1995; Hough, 1985; Sherman, PC, 1996). Little is known about this increasingly important disease. Most commercial peach germplasm appears very susceptible although high resistance was found in a few accessions such as Harbrite (Canadian release) and Eagle Beak (PI43289, a peach introduced from China). No information is available about its inheritance (Daniell and Chandler, 1982; Okie and Reilly, 1983) but genetic variability exists. At present, it can be managed by sanitation and frequent fungicidal summer applications but given its increasing importance, the development of resistant scion varieties and rootstocks needs to be explored.

6.5 Brown rot, *Monilinia fructicola* (Wint.) Honey; *Monilinia laxa* (Aderh. and Ruhl.) Honey

Brown rot of stone fruit is a serious problem throughout the world. In medium and low-chill zones where the weather is rainy during bloom and/or fruit ripening, this disease is a major problem that is currently managed primarily by fungicides. It is considered a serious fruit rot problem in areas of Brazil and the southeastern United States and other areas where rains are frequent during the ripening period (Raseira, PC, 1996; Byrne, PC, 1996). It can also cause serious blossom blight in areas where bloom occurs during the warm (20°C), moist (morning dew or rain) weather. An example of this would be areas in Mexico where the bloom is forced in September and October for a winter peach crop (Rodriguez, PC, 1996) or in occasional years in central Georgia (USA) and New South Wales (Australia).

Despite its importance, there has been relatively little work done on the development of brown rot resistant stone fruit cultivars because it has been reliably controlled by fungicides. There are numerous reports of resistance to fruit brown rot within peach (Feliciano *et al.*, 1987; Scorza and Okie, 1990), plum (Okie, 1995; Ramming and Cociu, 1990), and apricot (Mehlenbacher *et al.*, 1990). In general, the level of resistance reported is low to moderate and the screening techniques are not highly reliable. Most studies have focused on skin (epidermal skin layer) characteristics such as the cell wall thickness, cracking of the skin surface, and cuticle thickness (Biggs and Northover, 1989; Fogle and Faust, 1975; Adaskaveg *et al.*, 1991; Gradziel and Wang, 1993), as the major factors in the resistance. In fact, it appears that much of the California peach germplasm is highly susceptible to fruit brown rot due to skin cracking after rain. Recent work comparing resistance to brown rot upon surface inoculation versus injured fruit inoculation showed that a much wider range of resistance exists to surface inoculation in plum, apricot and peach (Pascal *et al.*, 1994). This, along with the fact that nectarines are more susceptible than peaches to brown rot infection, further emphasizes the important role that the epidermal layer plays in preventing brown rot infection.

The brown rot resistance breeding being done in Brazil (Pelotas, RS), and the USA (University of California, Davis, CA) is based on transferring the epidermal based resistance that is found in the Brazilian cultivar 'Bolinha' into a better commercial fruit type. Thus far, it has been observed that this resistance is readily passed on to its progeny but unfortunately along with brown rot resistance the parent also passes on its tendency to have a green ground color, to drop before harvest and to be highly bruise susceptible (flesh browning) (Gradziel, 1994; Raseira, PC, 1995). Another possible source of resistance to brown rot that has not been exploited is the feral Mexican peach populations that have been selected under rainy summer conditions (Scorza and Sherman, 1996).

6.6 Peach leaf curl, *Taphrina deformans* (Berk.) Tul

Peach leaf curl occurs in all peach growing areas of the world that have a cool moist climate at bud break. It has also been reported on cherry, almond, and apricot. This fungus typically initiates its yearly cycle by forcibly discharging ascospores in early spring during bud break although in some areas such as the Pacific northwest it may occur between the bud scales during the winter months (Pscheidt, 1995; Scorza and Okie, 1990). Later season cycles are continued via bud-conidia. It is worst in regions with extended periods of cool moist weather at bud break because the fungus growth is optimal at about 20°C and the budding of the bud-conidia is active above 95% relative humidity. Once the temperature rises above 26 to 30°C the fungus grows poorly and does not cause major problems (Ackerman, 1953; Pscheidt, 1995) thus late blooming genotypes are less prone to infection than early blooming ones. It is a yearly occurrence in many production areas of the Pacific northwest and the coastal areas of California in the USA, southern Spain, and the highland production areas of the tropics and subtropical zones. In contrast, in the eastern and southern production areas of the USA its occurrence is erratic from year to year since it tends to warm up quicker in the spring. Currently this disease is controlled with fungicide sprays at leaf fall or during early bud development but resistant varieties would be preferable.

Numerous reports indicate that none of the commercial peach and nectarine cultivars are immune to the disease and most are moderately to very susceptible (Ackerman, 1953; King *et al.*, 1988; Scorza, 1992; Scorza and Okie, 1990; Sharma and Badiyala, 1994; Simeone, 1985) (Table 10). As is the case with most field evaluations done under different environmental conditions (different years and locations), there are some inconsistencies in the susceptibility levels reported for specific cultivars (Table 10). Most likely this is due to differing disease pressure during rating which is confounded by the time of bloom in some areas (Scorza, 1992; Ackerman, 1953) and the time of evaluation (spring versus summer) in others. When King *et al.* (1988) evaluated peach cultivars for resistance to summer infection they reported that although the cultivars 'Velvet' and 'Starks Sweet Melody' appeared resistant to summer infection, they were moderately to severely affected by the early spring infection. Nevertheless, the existing data gives some guide to the range of susceptibility encountered within the cultivated germplasm.

Table 10. Susceptibility and resistance to peach leaf curl in Prunus persica

Resistance categories	*Peaches*			*Nectarines*
Susceptible	Aguas 6-4	Flordaqueen	Pi Tao	Armking
	Aguas 12-13	Flordared	Platycarpa	Earliblaze
	Angel	Flordasun	Redglobe	Fantasia
	Babcock	Flordawon	Redhaven	Flamekist
	Babygold 8	Golden Jubilee	(MS/MR)	Flaming Gold
	Belle of Georgia	Herholdt's Late Cling	Redskin	(MS)
	(MS)[Z]	Ingwe	Regina	Flavortop
	Bolivian Cling (MS)	June Gold	Robin	Goldmine (MS)
	Cary Mac	Kakamas	Rubidoux	Le Grand
	China Flat (MS)	Killiekrankie (MS)	Saharanpur2	Nectared 1-9
	Dixiland (MR)	Late LeGrand	San Pedro	Panamint
	Early Amber	Loring	Shalil (MS)	Quetta
	Early Sungrand	Marina	Sharbati	Starks Earliblaze
	Fay Elberta	Mexican Honey	Springcrest	Starks Redgold
	Ferganensis 02446	Nishiki	Springold	Sungold
	Flamecrest	Olinda	Starlite	Sunred
	Flavorcrest	P. mira	Sunhigh	Zeegold
	Flordahome		Sunrise	
			Yunnan	
Moderately susceptible	Andross	Garnet Beauty	Reliance	Bronte
	Blake	Gaschina	Rio Oso Gem	Harko
	Chinese Cling	Novembre	Southland (MR)	
	Dixon	Halehaven	Springbrite	
	Elberta (MR)	Hamlet	Surecrop	
	Fairtime	Harvester	Tos China (MR)	
		Inkoos		
		Lovell		
Moderately resistant	Candor	Erica Rudolph	Parade	
	Clayton	G.X.	Pekin	
	Correll	Golden Queen	Pollardi (R)	
	Dixiland (S-Italy)	Harbelle	Rogani Gow	
	Duke of York	Inkoos	Rubired	
	Elberta	Kirkman Gem	South Haven	
	Ellerbe	Palora	Yennoh(R)	
Resistant	Baladi 1	GF305	Rheingold	
	Early Crawford	PC408 (immune)	Royal George	
	Genovese	Proskauer	Tzim Pee Tao	

Data from Ackerman, 1953; Fideghelli *et al.,* 1983; Monet, 1985; Ritchie and Werner, 1981; Scorza, 1992; Scorza and Okie, 1990; Sharma and Badiyala, 1994; Simeone, 1985.

[Z] (MS), (MR), or (R)= also reported as moderately susceptible, moderately resistant, or resistant by another author.

Although most peaches are susceptible to peach leaf curl, a few materials such as GF305, 'Early Crawford', 'Baladi 1', 'Tzim Pee Tao', 'Rheingold', 'Proskauer', 'Royal George', 'Genovese', 'Yennoh', and 'Pollardi' were reported resistant (less than 10% leaves infected) (Ackerman, 1953; Monet, 1985; Scorza, 1992). Only two accessions, a chance peach seedling from Italy (PC408) and *Amygdalus petunnikowii* Litw. from Russia, have been reported immune to this fungus (Fideghelli *et al.*, 1983; Evreinoff, 1939). Although these resistant and immune genotypes represent a wide range of germplasm, they are unfortunately high-chilling and/or poor fruit quality types. The resistance is quantitatively inherited (Monet, 1985) and moderately heritable (Ritchie and Werner, 1981). The inheritance of the source of immunity has not been reported.

Peach cultivars with eglandular leaf glands have been associated with higher resistance or less susceptibility to leaf curl, but given that exceptions do occur and PC408, the immune peach from Italy, has reniform glands, this relationship is not useful as a screening procedure for peach leaf curl. The most common evaluation procedure used thus far relies on natural infection in the field.

6.7 Plum leaf scald and Peach Phony, *Xylella fastidiosa* Wells *et al.*

This leaf hopper transmitted, xylem-limited bacterium causes disease in peach (peach phony), plum (plum leaf scald), almond (almond leaf scorch), grape (Pierce's disease), and possibly citrus (citrus blight). Although the pathogen infects a wide range of plants, a given strain may be specific to a given group of species. For example, a similar strain may cause peach phony and plum leaf scald but a different one causes Pierce's disease on grapes and almond leaf scorch (Hopkins, 1989; Sherman *et al.*, 1989; Raju *et al.*, 1982). Control, although complicated because many weeds can harbor the bacterium and the vector without showing symptoms (Hopkins, 1989), is done by the early detection and elimination of infected trees and wild plums such as *Prunus angustifolia* which harbor the pathogen and vector (Wells, 1995; Ritchie, 1995b)

Plum leaf scald was first reported in Argentina (Fernandez and Bakarcic, 1954; Bakarcic and Santis, 1969) and has since been reported in Brazil, Paraguay, and the southeastern United States (Boyhan *et al.*, 1997; French and Kitajima, 1978; French and Feliciano, 1982; Latham and Norton, 1980; Sherman *et al.*, 1989; Okie, 1995). Peach phony was described about 1890 as a curiosity in an orchard in Georgia but later lead to the widespread decline of many peach orchards during two prolonged outbreaks (1915-1929 and 1971-1978) in the southeastern USA. These diseases are found in tropical and subtropical climates and have been observed to be worse after particularly warm winters (Wells, 1995; Hopkins, 1989; Sherman *et al.*, 1989) with symptom expression magnified after extremely stressful conditions such as very hot and dry summers and overcropping (Hopkins, 1989; Okie, PC; 1996)

Given the potential damage it can cause and the difficulty of control, the most promising approach for management is rogueing infected trees and choice of plum scion cultivars (Okie, 1995; Wells, 1995; Fernandez and Bakarcic, 1963). Thus far within peach there are no confirmed reports of good resistance (Sherman *et al.*, 1989) although there is some evidence that some rootstocks are less attractive to the vector (Wells, 1995). This is the most devastating disease of plums in the southeastern USA (Okie, 1995) and in southern Brazil (French and Feliciano, 1982; French and Kitajima, 1978). Thus the plum breeding programs in the southeast USA (Georgia, Alabama, Florida, and Texas) and in southern Brazil (Santa Catarina, Rio Grande do Sul, and Parana) have the development of plum scald resistance as an important objective. Most of the evaluation is based on natural field infection over 4 to 5 years although it is possible to screen after chip budding plums with infected buds (Boyhan *et al.*, 1997; Norton, PC, 1996). There is a range of resistance to this pathogen in plums (Table 11) reported from South and North America. Much more evaluation needs to be done in light of some inconsistencies in ratings and the observation that seedlings of resistant germplasm from Brazil ('Amarelinha', 'Sanguinea') do not appear healthy under Georgia conditions (Okie, 1995). Nevertheless, the work thus far has resulted in commercially acceptable plum cultivars with greater levels of resistance to plum leaf scald than were previously available.

Table 11. Resistance to Xylella fastidiosa in stone fruit

Resistance categories		*Cultivars*	
Susceptible	All peaches	Frontier	Ozark Premier(MS)
	All almonds	Giant Cherry	Ozark Premier F-1
	73 local plums from Argentina	Gigaglia	Ozark Premier F-2
		Gulfgold	Purple
	Burbank D-1	Methley(MS)[z]	Santa Rosa
	Crimson	Methley E-1	Wade
	Fla 87-6		
Moderately susceptible	AU Producer(R)	Bruce?	Robusto(MR-)
	AU Rosa(MR)		
Moderately resistant	AU Amber(S/R)	Black Ruby	Gulfruby
	AU Producer (R)	Byrongold	Morris(R)
	AU Roadside	CB-28	Segundo
	AU Rubrum(MS)	Gulfblaze	
Resistant	Abundancia temprana de	Colorada de Gogaglia	Homeside
	Valenti	Don Jorge	Mariposa
	Amarelinha	Dura	Methley A-21
	Bonora	Estrella Purpura	Monterrey
	Borrachuda	Gulfbeauty	Sanguinea
Immune	Piamontesa I	Remolacha Marquez	Vacarezza
	Remolacha de Banos	Soledad	Vinagrilla

Data from Bakarcic and de Santis, 1969; Boyhan *et al.*, 1997; Cummins, 1991; Hauagge, PC, 1996; Latham and Norton, 1980; Norton, 1986; Okie, 1994; Okie *et al.*, 1992; Sherman *et al.*, 1989; Topp and Sherman, 1990.

[z] (MS), (MR), or (R) = also reported as moderately susceptible, moderately resistant, or resistant by another author.

6.8 Powdery mildew, *Sphaerotheca pannosa* (Wallr.:Fr.) Lev on peach, apricot and plum; *Podosphaera* spp. on cherries and occasionally on other stone fruit

Powdery mildew infections on fruit and leaves of stone fruit are favored by cool, moist nights and warm days which are characteristic of the tropical highlands, coastal areas of the western USA, and the semi arid regions of California, Pacific northwest and southwestern Europe (France, Spain, Italy) (Grove, 1995; Perez *et al.*, 1993; Scorza and Okie, 1990; Weinhold, 1961). In the highlands of Mexico, it is a major limiting factor in the commercial production of peaches (Perez *et al.*, 1993), especially in areas such as Morelos where peaches are double cropped and are harvested between December and February (Rodriguez, 1988). Powdery mildew can cause a complete crop loss in cherries and up to a 50% crop loss in peaches, nectarines, plums and apricots. The disease problems are little in areas or periods when the temperature exceeds 28 to 30°C and the relative humidity is less than 70 to 75% (Grove, 1995).

A range of susceptibility and resistance to powdery mildew exists in peach, nectarine, plum, and apricot (Table 12). The development of peaches resistant to powdery mildew is an objective in peach breeding programs in Mexico (Perez *et al.*, 1993, Rodriguez *et al.,* 1992), Italy (Nicotra, PC, 1990; Roselli and Bellini, 1976), and France (Kervella, PC, 1995). Although good levels of resistance are found within peach (Perez *et al.*, 1993, Nicotra, PC, 1991), higher levels of resistance have been found in several related species: *Prunus davidiana, P. mira, P. kansuensis* (Kervella, PC, 1995, Pisani and Bellini, 1976) and perhaps *P. ferganensis* (D'Bov, 1983; Pukanova *et al.,* 1980; Tuz, 1958). It appears that *P. davidiana* and *P. mira* exhibit a hypersensitive resistance response to powdery mildew whereas *P. kansuensis* is immune to the fungus (Kervella, PC, 1995). At present, both *P. persica* and the related species, *P. davidiana* and *P. ferganensis*), are being used as a sources of resistance to powdery mildew (Dirlewanger *et al.*, 1996; Nicotra, PC, 1991; Perez *et al.*, 1993).

Resistance and susceptibility to powdery mildew in peach have been associated with leaf glands. Eglandular, serrated leafed genotypes were generally more susceptible to powdery mildew than were those with globose and reniform glands and crenate borders (Connors, 1922; Paunovic and Ogasanovic, 1975; Paunovic *et al.*, 1976; Roselli and Bellini, 1976; Rodriguez *et al.*, 1992) presumably due to the greater wetability of eglandular leaves (Hesse and Griggs, 1950). Nevertheless, there is overlap in susceptibility among these groups of leaves. Thus while the mean resistance between the two groups is different, it is possible to have eglandular, serrated leafed peach genotypes with the same level (although intermediate) of powdery mildew resistance as a reniform gland, crenate leafed peach genotype and to have genotypes with globose or reniform glands which are very susceptible to powdery mildew (Perez, PC, 1996; Rodriguez *et al.*, 1992). Consequently, this relationship has not been used as a screening tool. Screening is most commonly done in the field but also has been done in the greenhouse with inoculation (Perez *et al.*, 1993; Perez and Gonzalez, 1992; Pisani and Roselli, 1983; Rodriguez *et al.*, 1992; Kervella, PC, 1995; Simeone, 1985). Greenhouse screening with either spray inoculation or leaf rubbing inoculation are well correlated to field evaluations of resistance (Rodriguez *et al.*, 1992).

Table 12. Powdery mildew resistance in stone fruit

Resistance categories	*Peaches*		*Nectarines*	*Plums Apricots Cherries*
Susceptible	Babcock	O'Henry	Laurie Red	Gaviota
	CP-Pintado	Palora	Panamint	Kelsey
	Davidsun	Peak	Red Supreme	Wickson
	Elegant Lady	Red Lady	Sundollar	
	Flamecrest	Robin	Flaminggold	Blenhiem
	Flavorcest	Tetela (evergreen)	Sunfre	Perfection
	Flordacrest		Sunred	Rival
	Flordawon	TropicSnowe		Tilton
	Mexican criollos most peaches	Yakima Hale	common nectarines in western USA	
Moderately susceptible	Canario	Flordagold		
	Chiripa	Flordaprince		
	CNF1	Lucero		
	Earligrande	Kakamas		
		Neethling		
Moderately resistant	Aztecgold	Hermosillo	Angelis	
	Diamante	Magno	Armgold	
	Fla 9-26C	Nemaguard	Blazing Gold	
	Flordagold	Okinawa	Candor	
	Flordagrande	OroA	Goodcheer	
	Flordahome	Pantao	Johnson	
	Flordaking	Springcrest	LaPremier	
	Halford		Stuart	
			Walton	
Resistant	Tichivyi Pozdnii	*Prunus kansuensis*		Harcot
	Prunus davidiana	*Prunus mira*		Palummela
	Prunus ferganensi			Riland
				San Catrese
				Prunus serotina

Data from Byrne, PC, 1996; Grove, 1995; Kervella, PC, 1995; Knight, 1969; Perez *et al.*, 1993; Perez and Gonzalez, 1992; Pisani and Roselli, 1983; Ramming and Cociu, 1990; Rodriguez *et al.*, 1992; Roselli and Bellini, 1976; Simeone, 1985; Tuz, 1958; Pukanova *et al.*, 1980.

A two loci system has been reported to control resistance to powdery mildew in a selection of *P. ferganensis*. The first locus controlled high resistance and the second conditioned a medium level of resistance (Scorza and Sherman, 1996). Although this may be the case with *P. ferganensis*, the resistance from *P. davidiana* appears to be controlled by at least one major and several minor genes (Dirlewanger *et al.*, 1996). The inheritance of the lower levels of resistance found among *Prunus persica* is conditioned by multiple genes. A preliminary study in Mexico has estimated the heritability of powdery mildew resistance to be high ($h^2 = 0.87$) (Perez and Gonzalez, 1992).

Cherries are prone to a complete crop loss due to a powdery mildew attack (Grove, 1995) but there are few reports on sources of resistance to powdery mildew among either sweet (*P. avium*) or tart (*P. cerasus*) cherries (English, 1947; Hedrick, 1917). A possible source of resistance among medium-chill cherries might be *Prunus serotina*, the black cherry. This is native to the highlands of Mexico where the conditions for powdery mildew infection are favorable as evidenced by the severity of this disease on peaches. In spite of the perfect conditions for disease development, the native black cherries examined were disease-free. This species is tetraploid and thus may be useful in the development of medium-chill, powdery mildew resistant tetraploid cherry cultivars. At present, the crossability of this species with *P. cerasus* is being studied by Salvador Perez (Queretaro, Mexico) and Amy Iezzoni (MSU, East Lansing, Michigan, USA).

6.9 Rust, *Tranzschelia discolor* (Fuckel) Transchel and Litv. found world wide on cultivated stone fruit; *Tranzschelia pruni-spinosae* (pers) Dietel mainly in Europe on non cultivated Prunus species.

The most common species of rust found on the cultivated stone fruits is *T. discolor*. This fungus exists in several races that are specialized to attack specific *Prunus* species (Bolkan *et al.*, 1985; Kable *et al.*, 1986). Thus, peach trees adjacent to severely infected plum or prune trees are frequently not infected (Byrne, PC, 1996; Bertrand, 1995). This fungus mainly exists in the asexual state and although it is found throughout the world, the outbreaks usually are not severe except in *Prunus domestica* in which it has been recorded to cause a 33% yield loss. On peaches, it does not cause major problems in the medium and high-chill zones because it usually appears late in the season (late August or September) and does not have time to cause significant defoliation before natural defoliation commences (Bertrand, 1995; Byrne, PC, 1996). However, in low-chill zones that have long growing seasons, intermittent rains or high humidity, and early arrival of the fungus, rust can cause complete defoliation within 1 or 2 months following harvest, at a time when the tree normally would be growing for an additional 4 to 6 months. In areas such as these (Brazil, Mexico, south Texas) protective fungicide sprays are required to maintain the trees healthy (Byrne, PC, 1996; Barbosa *et al.*, 1994). In contrast, in central Florida with sandy soils and frequent afternoon rains from mid-June to mid-September, the application of fungicides would be too frequent and best leaf retention is obtained by keeping nitrogen levels high with summer applications in early June and August.

Among peach, it appears that no genotype is completely free of infection although a range in tolerance is evident (Table 13). Among the peach germplasm evaluated in Mexico and Florida, more tolerant genotypes tended to have a more compact growth habit (thicker shoots and larger leaves), to be later ripening, and more vigorous (Perez *et al.*, 1993; Sherman, PC, 1997). In Florida, the infected leaves of one resistant type failed to yellow and dehisce. This reaction may be related to lack of ethylene sensitivity. Most Brazilian, European and American peach and nectarines evaluated in field trials were susceptible to very susceptible (Barbosa *et al.*, 1994; Perez *et al.*, 1993; Simeone, 1985) although a few such as 'Diamante', 'Josefina', 'Regis', 'Flaming Gold', 'Sunred', 'Rubidoux', and 'Nectared 2' showed a moderate level of susceptibility/resistance. In contrast, most local cultivars from Mexico showed more tolerance to rust than the American varieties evaluated with them (Perez *et al.*, 1993).

Although rust is a major disease problem in the production of the European plum (*P. domestica*) it is not a consistent production problem with the Japanese plum (*P. salicina* and hybrids) in medium to high-chill regions (Bertrand, 1995; Ramming and Cociu, 1990). Nevertheless this fungus on plum is widespread in the southern United States and is considered an important problem in the low-chill areas of Brazil (Campo Dall'Orto *et al.*, 1980; Bertrand, 1995). On susceptible genotypes rust sori can develop in early summer and cause premature defoliation. The highest level of rust resistance in Florida was seen in genotypes best adapted (generally the most vigorous) to mild winters (Decker and Sherman, 1976). Other genotypes reported as resistant include several plums from Brazil ('Amarelinha', 'Sanguinea', 'Gema de Ouro', 'Golden Talisma'), 'George Wilson', 'Santa Rosa', and 'Shiro' (Table 13). In the lower chill zones, it would be wise to begin to select for resistance in the breeding efforts.

6.10 Arthropod pests

There has been very little done to assess the extent of the available host plant resistance to the common pests of stone fruit (Scorza and Okie, 1990). Most of the control strategies for the common pests such as scale insects, catfacing insects, and mites are dependent on pesticides although the trend is towards the use of a more integrated approach to minimize the use of pesticides.

The only reports on resistance of stone fruit to pest problems are for mite resistance in peach (Meagher and Werner, 1992), the lesser peach tree borer in peach and peach-almond germplasm (Puterka *et al.*, 1993), and aphid resistance in peach (Monet, 1985; Kfoury and Massonie, 1995; Nicotra, PC, 1993). It is only with aphid (*Myzus persicae* (Sulz.)) resistance that the work has progressed to the development of resistant peach genotypes. This resistance which is inherited by a dominant gene (Monet, 1985) has been found in *P. persica* ('Rubira', S2605, S2678, S2672; and several weepers from S-37 seed), *P. davidiana*, and *P. cerasifera* (Massonie *et al.*, 1982; Pisani and Roselli, 1983; Nicotra, PC, 1993). Work in France has shown that it has remained stable after four generations of breeding (Monet and Massonie, 1994).

Table 13. Susceptibility and resistance to rust in stone fruit

Resistance categories		*Peach and nectarines*		*Plum*
Susceptible	Armqueen	Flordagold	Panamint	Burmosa
	Aurora1	FlordaGrande	Robin	Carmesin
	Aurora2	Flordahome	Rosalina	Duncan
	Babcock	Flordaprince	Rubidoux	Ozark Premier
	Bronte	Flordaqueen	Springcrest	Rosa Paulista
	Centenaria	Flordawon	Springold	Taiwan
	Colora	Goldmine	Tropical	
	Delicioso Precoce	Joia1	TropicBeauty	
	Dixiland	Joia2	TropicSnow	
	Dourado1	Joia5	TropicSweet	
	Dourado2	Le Grand	ValleGrande	
	EarliGrande	Nectared 3-9		
	Fantasia	Nectared1	Most US	
	Flordaglo		cultivars	
Moderately susceptible	Coacalco	La Rica	Nectagold	Early Bruce
	Diamante	LaFayette	Pocahontas	Excelsior
	Evergreen	Lucero	Prisco Blanco Ver	Gigaglia
	Flaming Gold	M1	Regis	Grancuore
	Fuzalode	Merrill Aurora	S100	Terrill
	IAC 4685-45	San Castrese 1	S115	
	IAC N2680-	Stark Sunglo	S135	
	IV13ad	Sunking	S159	
	Josefina	Sunred	S163	
		T01-13	S82	
Moderately resistant	Nectared 2			Amarelinha
				Florida wild plums
				IAC 1-65
				IAC 1-9
				Laroda
				Sanguinea
Resistant	None			Gema de Ouro
				George Wilson
				Golden Talisma
				Gulfbeauty
				Gulfblaze
				Gulfruby
				Santa Rosa
				Shiro

Data from Barbosa *et al.*, 1994; Byrne, PC, 1996; Campo Dall'Orto *et al.*, 1979; 1980; Decker and Sherman, 1976; Hauagge, PC, 1996; Ojima *et al.*, 1992; Perez, 1989; Ramming and Cociu, 1990; Simeone, 1985.

7 Germplasm Resources

7.1 peach germplasm

a. Low-chill peach germplasm

The founding clones of low-chill germplasm differ dramatically between the programs of the USA (Florida and Texas, Weslaco) and Brazil (Pelotas and Sao Paulo) (Table 14). The Florida program began by intercrossing poor fruit quality, low-chilling South China peaches such as 'Hawaiian', 'Jewel' (derived from Peento peach), and 'Okinawa' with peach germplasm from the northeast United States (Sharpe, 1961; Sharpe, 1969; Sherman and Rodriguez, 1987). Initially 'Southland' was the high chill cultivar used in breeding. Later other materials from eastern ('Early Amber', 'Harken', 'Redglobe', 'RioGrande') and the west coast North American peach germplasm ('Early Sungrand', 'June Gold', 'Springtime', 'Sunnyside') were used to develop the present melting flesh peach cultivars. In the mid 50s the Florida program began to introgress the nectarine gene into the breeding population by using 'Panamint' and NJN21 ('Cardinal and 'Flaming Gold' parentage) and more recently 'Armking', 'Armqueen', 'Merrill Princess', 'Mayfire' and others in their breeding program (Sharpe and Aitken, 1971; Sherman, Rodriguez, and Miller, 1984).

The breeding in south Texas (Weslaco) has thus far only worked with materials derived directly from the Florida program and most of the releases have been joint releases from crosses made and selected in Florida but evaluated in south Texas which is a lower chilling zone (150-250 CU) than is Gainesville, Florida (300-400 CU) (Bowen, 1980; Rouse, 1989; Rouse and Byrne, 1990; Rouse and Sherman, 1987; Rouse *et al.*, 1985; Sherman and Rouse, 1989).

The Brazilian programs have developed their germplasm by combining local cultivars with breeding materials from the USA. The germplasm at Pelotas consists of 49% Brazilian germplasm whereas Sao Paulo germplasm has 33% of its background derived from local Brazilian cultivars (Table 14). The Brazilian cultivars used for each program are also different indicating that these two breeding projects worked independently. The lower chill program at Sao Paulo initially made crosses with low-chill South China peaches such as 'Jewel' (second generation selection of the 'Peento' peach), and 'Suber' and later introduced some Florida materials such as 'Sunred', 'Sunlite', 'Maravilha', and 'Columbina' into their breeding efforts. Both 'Maravilha' and 'Columbina' were initially selected in Florida and sent as selections for evaluation in Brazil where they were named (Sherman, Rodriguez, and Topp, 1992). The Pelotas program initially intercrossed local cultivars and introduced seed of the germplasm used by the Florida program ('Southland' x 'Hawaiian' derivatives) and selected several cultivars from that ('Cardeal', 'Premier', 'Princesa'). Subsequently, peach germplasm from New Jersey and several nectarines ('Sunred', 'Panamint', 'Goldmine') were used in their breeding. Thus although both programs have used germplasm from the USA, one-half to one-third of its germplasm is based on Brazilian local cultivars (Barbosa *et al.*, 1995; Bruckner, 1987; Nakasu *et al.,* 1981; Raseira *et al.*, 1992).

Table 14. Percent contribution of the founding clones and the cultivars used in peach breeding of four low-chill breeding programs for melting flesh fresh market cultivars

Breeding program		*Florida - Melting*[z]		*Texas - Weslaco*[z]		*Brazil - Pelotas*[z]		*Brazil - Sao Paulo*[z]
Low-chill germplasm	26%	Hawaiian Peento Okinawa	39%	Hawaiian Peento Okinawa	49%	Delicioso Precoce Rosado Admiravel 15 deNov.	36%	Peento Suber Okinawa Hawaiian Strawberry
	9%	Armking						
					14%	Hawaiian Peento Strawberry	33%	Rei da Conserva Perola de Itaque Taichi
High-chill germplasm	37%	J. H. Hale St. John July Elberta Rio Oso Gem Boston Nectarines Flaming gold Cardinal	36%	J. H. Hale St. John	26%	J. H. Hale Rio Oso Gem Boston St. John	15%	J. H. Hale Rio Oso Gem Boston St. John
							4%	Cardinal
							5%	Lake City
Clones used in breeding		EarliGrande Early Sungrand Harken June Gold NJN21 Mayfire Panamint Princess Redglobe RioGrande Southland Springtime Sunnyside		Early Amber Flordaprince Flordasun Maravilha Springtime		FV338-90 Goldmine Nectared 5 NJ238 Panamint Sunred		Jewel Maravilha Sunlite Sunred

[z] Based on commercial releases since 1976.

Over all breeding programs, the 'Peento' or Chinese flat peach group contributed the most to the genetic background of the *low-chill* peach germplasm (Table 15). The early peach breeding programs in California used a *low-chill* 'Peento' directly as a parent. This apparently was imported as seed into the USA from Australia in 1869 by P. J. Berckmans of Augusta, Georgia. This original tree was very *low-chill*ing and only adapted to south Florida. Its fruit was oblate, medium to large size, had thin, bitter, yellowish-green skin, yellowish-white firm, melting, juicy flesh, and *clingstone* (Hume, 1902). Presumably, this is the Peento that was used in the early work (1909-1910) in California (Lesley, 1939a; 1939b; Lesley and Winslow, 1953). Later programs incorporated this germplasm by using either 'Jewel' (a 2nd generation selection from 'Peento' named in Florida), 'Suber' ('Peento' progeny named in Florida) or 'Babcock' (a cross with 'Strawberry' developed in California) and its descendants ('Panamint', 'Maravilha', 'Sunred', 'Sunlite', or NJ238) as parents.

Both 'Hawaiian' and 'Okinawa' were incorporated into the low-chill germplasm base by the Florida program. 'Okinawa' was a source for overcoming blind nodes and bud failure (Sherman *et al.*, 1996) which was derived in higher chill parents from 'Springtime', 'Armking' (blind nodes) and 'Loring' (high bud failure). Since both sources of low-chilling have late ripening (FDP = 105 to 150 days) small, white, melting flesh fruit with poor quality it took 3 generations after the initial cross with a high-chill good quality fruit cultivar ('Southland') to recover commercially acceptable fruit type with a low-chilling adaptation (Sharpe, 1961). These sources of the low-chilling trait were incorporated into the Brazilian breeding via selections from Georgia (FV338-90 and others), Florida cultivars ('Sunlite', 'Sunred') or materials with these in their background ('Maravilha' and NJ238, a cross with 'Flordawon').

'Lukens Honey' was used directly by the early programs in California to develop varieties such as 'Springtime' (Armstrong) and 'Honey Elberta' (Chaffey Jr. College). Subsequently, 'Springtime' was used considerably in breeding resulting in cultivars such as 'June Gold' and 'Springcrest'.

The Brazilian work used germplasm from the USA efforts and local germplasm. Each program used different local cultivars in their work. The Pelotas work used mainly 4 cultivars ('Delicioso', 'Precoce Rosado', 'Admiravel', and '15 de Novembro') and the Sao Paulo program used 3 local cultivars ('Rei da Conserva', 'Perola de Itaque', and 'Taichi') extensively. 'Precoce Rosado' is a yellow fleshed, red blushed (80%), good flavored freestone peach with occasional pronounced tips (Raseira, PC, 1996). Both 'Delicioso' and '15 de Novembro' are local white melting peach selections requiring about 200 chill hours. 'Delicioso' which is firmer than '15 de Novembro' and a freestone was used in the development of the fresh market cultivars and is in the background of half of the releases since 1976 (Bruckner, 1987). In Sao Paulo, the local cultivars, 'Perola de Itaquera' and 'Rei da Conserva' were used because there are very productive and firm fruited although only of average flavor. Thus they were crossed with 'Suber' and 'Jewel', which were flavorful but soft to develop peaches with better flavor (Barbosa *et al.*, 1995). 'Rei da Conserva' is still used as a processing peach in the Sao Paulo region as well as seedling rootstock (Raseira, 1988). 'Taichi' was used for its local adaptation although it is a small to medium size peach with a tendency for a pronounced tip and suture and a sweet flavored, firm, greenish-white flesh. This germplasm probably arrived into Brazil via seed brought over by the Portuguese explorers.

Table 15. Percent contribution of low-chill founding clones and the cultivars used in breeding for six peach and nectarine breeding programs

Low-chill founding clone	*Florida (melting)*[z]	*Brazil Sao Paulo*[z]	*Brazil Pelotas*[z]	*California Armstrong*[y]	*California Chaffey*[y]	*California Riverside*[y]
Aldrighi			11%			
Delicioso			10%			
Hawaiian	11%	5%	3%			
Lukens Honey	2%			5%	11%	
Okinawa	6%	5%				
Peento	7%	18%	5%	8%	17%	33%
Perola de Itaque		10%				
Precoce Rosado			4%			
Rei da Conserva		15%				
Suber		6%				
Taichi		7%				
Clones used in breeding	Hawaiian Jewel Okinawa Panamint	Jewel Maravilha Perola de Itaque Rei da Conserva Sunred Sunlite Suber Taichi	Aldrighi Delicioso FV338-90 NJ238 Panamint Precoce Rosado Sunred	Babcock Lukens Honey Swatow	Babcock Lukens Honey Peento	Peento

[z] Based on releases since 1976.

[y] Based on all releases from which information was available.

b. Medium-chill peach germplasm

Medium-chill cultivars have been produced by several programs in the southern USA in both the east, central, and western sides of the country. Much of this work has used cultivars or selections developed by Florida as evidenced by the existence of 'Hawaiian' and 'Okinawa' as founding clones, and the early California programs as evidenced by 'Peento','Lukens Honey' (via 'Babcock'), and 'Armking' in their background (Table 16). The influence of the high quality elite germplasm that has been used in the breeding of the higher chill cultivars in the eastern USA is very prominent. In these medium-chill programs of the USA the elite cultivars such as 'J. H. Hale', 'Rio Oso Gem', 'St. John', 'July Elberta', 'Mayflower', 'Boston', and 'Fay Elberta' account for between 24 to 59% of their background. The lowest amount is in the Zaiger Genetics breeding program in California but most likely this is an underestimate given that 'Rubygold' (23% of the genetic background) is derived from Anderson varieties (probably 'LeGrand', 'Early Sungrand', or 'RedGrand') which are mainly derived from 'J. H. Hale' and 'July Elberta'. Thus a better estimate would be at least 40%. The highest would be for the now inactive program in southern Louisiana which had 59% of its base trace back to 'St. John' and 'J. H. Hale', but if you consider that 'Southern Glow' appears to be a seedling from 'J. H. Hale', this is actually higher. This trend is also evident in the breeding done in South Africa where 22% of the background is from 'Rio Oso Gem' and 'Chinese Cling'. Thus except for the programs at College Station, Texas, Fresno, California, and South Africa, the proportion is greater than 40%.

In the lower chill germplasm, these cultivars represent about 35% of the background of the US programs and about 20% for the programs in Brazil (Table 15). Thus as Scorza *et al.* (1985) reported with Eastern USA fresh market peaches, a few founding clones have played a major role in the development of the medium and low-chill germplasm, although to a lesser extent than seen in the higher chill germplasm.

c. Nonmelting peach germplasm

More recently the Florida program has worked to develop low-chilling, nonmelting, early ripening peaches for the fresh market. Beyond the low-chilling and early ripening objectives of this program, this work differs from previous nonmelting breeding for processing types or breeding/selection for fresh market nonmelting flesh types in regions such as Mexico and Spain in the requirements for appearance. The Florida program is selecting highly colored (bright red blush over a yellow or orange ground color) types. This type of fruit, given the firmer characteristics of its flesh, allows harvest at the more mature stage with high flavor and long shelf life (Sherman and Rodriguez, 1987; Sherman, Topp, and Lyrene, 1990). One concern in nonmelting genotypes is the tendency of some to develop off flavors (especially fermented flavors) if left to ripen too long on the tree or stored too long. This is a trait of extreme importance and appears to vary among genotypes. Observations indicate that types that develop off flavors are common among germplasm derived from processing and Babygold cultivars. Some cultivars not showing the off flavor to much extent are 'Springbaby', 'Delta', 'Big Top', 'Crimson Lady', and 'UF Gold' (Sherman and Lyrene, 1997). They have

Table 16. Percent contribution of the founding clones and the major cultivars used in breeding for seven fresh market medium and transitional chill peach breeding programs

Breeding program	*California - Zaiger Genetics*[z]	*Texas - College Station*[y]	*Louisiana Idlewild*[z]	*California - USDA Fresno*[z]	*Georgia - USDA Byron*[z]	*South Africa*[x]	*California - Armstrong*[w]
Founding clones	15% Unknown low-chill 23% Rubygold 24% J. H. Hale Fay Elberta July Elb. 7% Quetta	16% Hawaiian Peento Lukens Honey 26% J. H. Hale St. John July Elberta Rio Oso Gem 4% Earligal 4% Armking	17% Southern Glow 53% St. John 17% Alexander 6% J. H. Hale	13% Lukens Honey Okinawa Peento Strawberry 19% J. H. Hale 12% July Elb. Mayflow. 5% Hiley 5% Levi 30% Armking	13% Lukens Honey Hiley 24% J. H. Hale 13% St. John 20% Rio Oso Gem July Elb. Mayflow. 7% Chinese Cling	14% RP745 7% Moreira Jubilee 4% RP1-62 19% Early Rivers 12% Rio Oso Gem 10% Chinese Cling 9% 5-6-1965	22% Peento Strawberry Lukens Honey 20% Rio Oso Gem 23% July Elb. Mayflow. J. H. Hale
Clones used in breeding	Desertgold Fayette Juneglo Maygrand RedGrand Springtime Sunred	Flamecrest Flavorcrest Flordacrest Flordaking Flordastar Goldprince Juneprince Mayfire Springold Suncoast Sunraycer TropicBeauty	Dixigem	Bonanza Fla8B27 LeGrand Springcrest and sister sdlg Springtime SunGrand	FV89-14 June Gold Loring Springtime	Early Dawn Fairtime Flamekist Kakamas Oom Sarel Rhodes RP745 Sunred	Babcock Goldmine Rio Oso Gem

z Based on medium-chill (<700 CR) releases since 1976.

y Based on the best selections in last four years.

x Based all releases since 1976.

w Based on all releases from which information was available.

developed nonmelting germplasm from three sources: feral Mexican source (Fla. 9-26C, 'Aztec Gold'), Brazilian germplasm ('Diamante' seed from Brazil from which 'OroA' was selected), and northeast USA (Fla. 9-20C via a North Carolina selection) source (Beckman and Sherman, 1996; Sherman, Topp, and Lyrene, 1990) (Table 17).

The Mexican source for nonmelting flesh in the Florida program goes back to seed collected from a late ripening (August, FDP = 150 days), large (2.5"), nonmelting criollo peach by Jim Pittman (Gerber Co.) in 1969 from a seedling orchard in Aguascalientes, Mexico. He sent this to Ralph Sharpe who planted it. Although it was too high-chilling (650 CR) to fruit well in Gainesville, in 1973 its pollen was used in a cross with 'Sunred'. This after several generations has yielded selections with lower chilling requirement (<300 CU), earlier ripening (FDP = 78 to 115 days), and good size and quality (Sherman, PC, 1996; Beckman and Sherman, 1996).

Fla. 9-20C is derived from northeastern USA germplasm which goes back to 'Chinese Cling' and its many derivatives ('J. H. Hale', 'Elberta', 'Fay Elberta', 'Belle of Georgia') (Scorza *et al.*, 1985). This is the same germplasm from which the early ripening nonmelting varieties 'Delta' (Louisiana) and 'Spring Baby' (USDA, Fresno, CA) were developed.

The Brazilian source for nonmelting flesh in the Florida program goes to a 1982 importation of 'Diamante' seed from Brazil. It is 50% derived from Brazilian local cultivars ('Abobora', 'Aldrighi'), 25% from the old variety, 'Amsden' (a white, melting flesh, mid-season variety selected in Missouri in 1868, see Hedrick, 1917), and 25% derived the 'Southland' x 'Hawaiian' germplasm from which the Florida program began (Sharpe, 1961). The Brazilian cultivar, 'Aldrighi' is a nonmelting type from which the brown rot resistant cultivar 'Bolinha' was selected (Bruckner, 1987; Feliciano *et al.*, 1987; Gradziel, 1994). 'Aldrighi' was initially selected from seed obtained from fruit imported from Argentina for processing in Pelotas. Once named, it was propagated by seed and finally reselected by Sergio Sachs in the 1950s. 'Abobora' is a small, firm orange clingstone with good flavor but an unknown origin. (Raseira, PC, 1996). Most likely the Mexican and Brazilian sources are derived from the same germplasm pool in Spain and Portugal. It was from this lot of seed that 'OroA' was selected which has been used extensively in recent crosses for its earlier ripening character (FDP = 85).

The Brazilian programs have worked for many years to develop nonmelting peaches for processing as well as dual purpose types that are useful for both fresh and processing markets. The program in Pelotas has a major objective to develop better processing cultivars and has released about 30 cultivars since beginning the program over 40 years ago (Bruckner, 1987). These cultivars are largely based (52%) on four Brazilian nonmelting processing cultivars (Table 17) of which 'Aldrighi', the parent of 'Bolinha' a brown rot resistant cultivar, and 'Ambrosio Perret' are the most prominent. 'Aldrighi' appears to be the parent of 'Intermediario'. 'Ambrosio Perret' was a selection of unknown origin found in the 1937 catalog of the Quinta Bom Retiro nursery. It was a firm, nonmelting yellow processing cultivar that transmitted good size to its progeny. Although it was used commercially in the 1970s, it lost favor because of inconsistent cropping and an apparent high susceptibility to brown rot. In addition, the cultivar, 'Lake City', which reportedly is of US origin has been used extensively. This cultivar is a very productive, orange fleshed, nonmelting peach in Brazil (Taquari, RS) which had a medium size, a conic shape with a tip, and an acid flavor (Raseira, PC, 1996).

The only reference to 'Lake City' in the USA was from Florida which described it as a nonmelting peach which ripens in July of Spanish origin via St. Augustine (Sherman, PC, 1996). It does not appear to be available any more in Florida.

Table 17. Percent contribution of the founding clones for the nonmelting low-chilling peach germplasm developed by Florida and Brazil

	Fla. 9-26c Aztec Gold	*Fla. 9-20C*	*OroA*	*Brazilian-Pelotas*
Founding clones	50% Mexican cling	25% J. H. Hale	25% Abobora	26% Aldrighi
	13% Hawaiian	16% St. John	25% Aldrighi	15% Ambrosio Perret
	8% J. H. Hale	12.5% Erly Red Fre	25% Amsden	6% Abobora
	6% Boston	12.5% Palomar	10% J. H. Hale	5% Intermediario
	6% Rio Oso Gem	12.5% Hawaiian	6% Hawaiian	15% Lake City
	6% St. John	3% Chinese Cling	6% St. John	6% Amsden
				5% Peento

The Mexican breeding programs at Chapingo and Queretaro are developing nonmelting flesh peaches for the fresh market (Perez, 1995; Rodriguez and Sherman, 1995). In Mexico, as in Spain, traditionally these peaches do not have a bright external red color but rather are yellow or orange (pavia types in Spain). In addition, the harvest is generally used completely because the fruit size is not as critical in the traditional Mexican market as compared to the US market. Thus, fruit not sold in the fresh market are processed into canned or other products.

Traditional peach orchards in Mexico are propagated by seed and thus since the Spanish explorers began to introduce peaches into Mexico in the 16th century, these have gone through many generations of sexual reproduction and selection for adaptation to a wide range of environments from subtropical to the cool highlands (1800 to 2450 meters above sea level) (Perez *et al.*, 1993). Although selection for adaptation and productivity is evident among the Mexican feral populations, there has been less selection for fruit size and other qualities. They are also late ripening since early ripening types were eliminated because of low seed germination. It is also probable there are few early ripening genes among the feral populations because no genotypes are produced that ripen in less than 120 days.

Early fruit researchers in Mexico such as A. Vega Leyva and F. Gutierrez (CONAFRUT and INIFAP, Aguascalientes), spent much effort selecting within and between these seedling orchards to find earlier ripening and the best fruit quality types. This work resulted in several releases such as 'CNF1', 'Lucero', 'S100', 'S66' and 'S165' (Okie, 1993). This work has continued with S. Perez in Queretaro with a collection from around Mexico and the release of seed propagated lines such as Rendidor, Prodagosto, and I8-16 (Perez, 1995).

Both Mexican breeding programs have used 'Diamante' in their breeding efforts (Table 18). This Brazilian cultivar is a major commercial cultivar in Mexico and has been used for its good adaptation, fruit quality, and moderate resistance to powdery mildew which is a major disease problem in the highland growing regions (1800 to 2450 meters above sea level) of Mexico.

The program based in Chapingo has worked closely with the Florida program and has used several of their selections (Fla. 9-20C and Fla. 9-26C) in their breeding and have made several joint releases ('OroA', 'Aztec Gold'). The major difference between these programs is that the selection pressure in Mexico is for adaptation to cooler growing season, resistance to powdery mildew (bacterial leaf spot resistance in Florida), little or no external blush (high external blush in Florida), and later ripening in Mexico (100-180 day FDP) versus Florida (60-100 day FDP) (Rodriguez and Sherman, 1995).

The program in Queretaro has used local peaches (criollos) from the traditional highland (1800 to 2450 meters above sea level) production regions of Mexico and intercrossed these among themselves and crossed them with 'Lucero' (a local selection named by A. Vega Leyva in Aguascalientes), two Brazilian cultivars ('Diamante' and 'Magno'), and other commercial types ('Allgold', 'Flordaprince', 'Oom Sarel'). About 38% of the present germplasm is from Mexican sources and 12% from Brazilian sources. About one quarter of the germplasm base is derived from the cultivar, 'Lucero' (Table 18). This selection is from an orchard in Aguascalientes but based on the phenotypic study by Perez *et al.* (1993), it was more similar to the South African material than the Mexican criollo collections. In fact, seedling populations of 'Lucero' and 'Neethling', a South African cultivar, are very similar. This would suggest that although 'Lucero' was selected in Mexico, it is actually derived from a South African processing cultivar (Perez, PC, 1996) which is derived from 'Kakamas' and 'Early Dawn' which trace back mainly to 'Chinese Cling'.

The work in Australia and South Africa is also largely based on local processing cultivars or selections ('Golden Queen' and 'Transvaal'/RP344 respectively) of unknown origins (Table 18). In Australia, the objective was to develop a series of 'Golden Queen' processing types that ripen over a period of several months. Crosses were made among mainly 'Golden Queen' (and its sports), 'Oom Sarel', 'Babygold 5' and 'Babygold 7'. All these, except perhaps 'Oom Sarel' which is a cultivar from South Africa, are as high-chilling as is 'Golden Queen' (800 CR) and consequently most of the selections were high-chilling as well. Although one, 'Tatura 204' ('Babygold 7' x 'Oom Sarel') is medium-chill (450 chill units), the other cultivars require between 650 and 950 chill units (Richards, PC, 1996). The other major founding clones are PI 35201, 'Early Rivers', an old white fleshed cultivar from England, and 'J. H. Hale'. PI 35201 is a large, freestone peach from Mengtsz, Yunnan, China (Hedrick, 1917) used by Drs. Hough and Bailey (Rutgers University, New Brunswick, NJ) in the development of the Babygold cultivar series.

Since one of the present goals of the Australian work is to lower the chilling requirement to ensure more consistent yields in years after mild winters they have been using 'Tatura 204' extensively in their crossing program (Richards *et al.*, 1995). A similar strategy is being explored in the breeding program at the University of California (Davis, CA). Tom Gradziel is currently crossing his California germplasm with lower chilling germplasm from Brazil, Mexico, and Florida and selecting for better bloom uniformity and set but against very early bloom. With this approach, he is trying to decrease year-to-year yield instability due to chilling quality problems (Gradziel, PC, 1996).

Table 18. Percent contribution of the founding clones and the cultivars used in breeding for four nonmelting medium- and transitional-chill peach breeding programs.

	Mexico - Chapingo[z]	Mexico - Queretaro[z]	Australia - Tatura[z]	South Africa[z]
Founding clones	20% Mexican NM Carlos Mexican cling	38% Mexican NM S100 871B Guanajuanto I7-4 Celaya Cr.	45% Golden Queen	19% Transvaal
				13% RP344
			13% PI35201	11% Early Rivers
	20% Brazilian Abobora Aldrighi		9% J. H. Hale	9% 5-6-1965
			8% Early Rivers	9% Connett
				9% Old Mixon Free
		27% Lucero		
	10% Amsden	12% Brazilian Abobora Aldrighi Ambrosio Perret		
	10% Hawaii			19% Chinese Cling Jonker Muir
	11% J. H. Hale			
	7% St. John			
		5% Amsden		
		7% Calanda		
Clones used in breeding	Diamante Sunred Fla. 9-20C Fla. 9-26C	Diamante Magno	Oom Sarel Babygold 5 Babygold 7	5-6-1965 Catherina Jonker Kakamas Oom Sarel Imperani

[z] Based on releases since 1976.

In South Africa, about 30% of the background of recently released processing peaches traces back to two local peaches: 'RP344', and 'Transvaal' (Table 18). They have also used several materials from the breeding program in New Jersey (5-6-1965 and 'Catherina'). At present, the commercial industry employs the older cultivars such as 'Kakamas', 'Black', 'Neethling', 'Walgant', and 'Oom Sarel' which trace back to 'Chinese Cling' and 'Early Rivers'. 'Chinese Cling' is in the parentage of many of the most important peaches involved in the breeding of fresh market peaches in the US such as 'Elberta', 'Belle of Georgia', and 'J. H. Hale' (Scorza *et al.*, 1985).

d. General conclusions on peach germplasm

The low and medium-chill germplasm appears to be quite diverse. All the programs in the USA have founding clones that are integral to the development of high-chill commercial peach cultivars ('J. H. Hale', 'Elberta', 'Fay Elberta', 'July Elberta',

'Boston', 'Mayflower', 'St. John') in their background because these were used as source of fruit quality (Scorza *et al.*, 1985; Scorza *et al.*, 1988). The low-chill types used were 'Peento' (early Californian programs), 'Okinawa', and 'Hawaiian'. All three of these are low quality, soft, small fruited, white fleshed peaches from south China. In Brazil, although these and a similar type, 'Suber' (a seedling of Peento), are among the founding clones, 6 local cultivars also figure prominently as founding clones. Local selections and cultivars also important among the founding clones for the breeding in Mexico (7 clones), South Africa (2 clones), and Australia (1 clone). The relationships among these 3 low-chill Chinese peaches and 16 local selections are not well understood. The clones from Latin America were brought over from Spain and Portugal beginning in the 1600s and propagated for many generations via seed. This resulted in a series of genetically uniform populations adapted to a wide range of environments ranging from tropical to medium to high-chilling highland climates.

Unfortunately there is little research on the variability within and among these populations. Perez *et al.* (1993) studied the morphological and phenological polymorphisms in peach germplasm from Mexico and compared it to peaches from USA, Brazil, Europe, and South Africa. This analysis clustered the Mexican and other central American criollos into one large group, indicating some cohesiveness but much variability. Other groups were formed by the South African peaches, the evergreen and related peaches, and the Brazilian with the low-chill US cultivars. The high-chilling cultivars from Europe and Pakistan clustered very loosely but apart from the low and medium-chill peaches studied.

Work with isozymes has found few polymorphisms among peaches (Arulsekar *et al.*, 1986; Durham *et al.*, 1987; Messeguer *et al.*, 1987; Byrne, 1990; Werner, 1992) although polymorphisms have been more common among those clones from China or other countries (Werner, 1992; Mowrey *et al.*, 1990) as compared to the commercially important melting flesh varieties of the USA. In a study in Spain, the nonmelting types had more polymorphisms than the US fresh market melting type varieties (Ibanez *et al.*, 1993; Messeguer *et al.*, 1987).

A recent study of peach germplasm using RAPDs generally confirms the conclusion arrived with isozyme data, that the peach germplasm from the USA is relatively homogeneous as compared to the germplasm collected from China and other parts of Asia (Warburton and Bliss, 1996). This data also indicates that the low and medium-chill germplasm derived from the Florida breeding program is more diverse than the germplasm developed in the higher chill areas, probably due to the more recent incorporation of exotic germplasm ('Okinawa' and 'Hawaiian').

7.2 Plum germplasm

All the plum breeding programs in the medium and low-chill zones work with the diploid Japanese plum. This crop is a multispecies complex that consists of *P. salicina* and its hybrids (Byrne, 1989a). Among the commercial cultivars of California its species composition is derived mainly from the two Chinese species: *P. salicina* (56%) and *P. simonii* (24%) and one North American species: *P. americana* (20% via 'Gaviota' and 'Santa Rosa'). The assumption that the *P. americana* was used to develop 'Gaviota' and

'Santa Rosa' is not strongly supported by either isozyme and RAPD data or visual plant morphology (Boonprakob, 1996; Byrne and Littleton, 1990; Byrne, 1993b).

Although all the clones of *P. simonii* (a northern Chinese species) in the USA are high-chilling (800 CU), *P. salicina* clones range from high-chilling to very low-chilling (100 CU). *Prunus salicina* is native to China and has a wide distribution ranging from the north to the south and Taiwan. In China, the commercial plum production and wild plum zones are north of latitude of 23° N. South of this latitude, there are only scattered plum trees which at times have all stages of growth simultaneously and do not appear well adapted (Zhang *et al.*, 1990). Plums imported as seed from Taiwan (latitude of 22° to 25° N) that were used by the Florida breeding program needed about 100 CU but had small (20-40 grams), non-astringent fruit (Sherman and Lyrene, 1985). Both isozyme and RAPD data indicates that these low-chill plums are quite distinct from the higher chill materials used in the breeding of the major commercial cultivars in the USA (Boonprakob, 1996; Byrne, 1993b).

In addition, other species such as *P. cerasifera* (via 'Methley'), and *P. angustifolia* (via 'Bruce', 'Six Weeks', and a clone from Georgia) have been used to incorporate better disease resistance into the plum cultivars developed in humid areas such as the southeastern USA (Byrne, 1989; Topp and Sherman, 1990) (Table 4). It is for this reason that the plums developed in the southeastern USA show much greater genetic diversity as compared to the California developed plums (Byrne, PC, 1996; Boonprakob, 1996).

Other species native to medium and low-chill areas that were used little in breeding include *P. texana* Dietr (Texas), *P. mexicana* S. Wats. (South Eastern USA to Mexico), southern form of *P. americana* (Marsh.), *P. umbellata* Ell. (North Carolina to Florida to Texas), *P. geniculata* R. M. Harper (Florida), *P. minutifolia* Engelm. (Texas), *P. fasiculata* (Torr.) Gray (California, NM, AZ), *P. andersonii* A. Gray (California), *P. fremontii* S. Wats. (California), and *P. harvardii* (W. Wight) S. C. Mason (Texas). All these appear to intercross readily (Layne and Sherman, 1986) but have small, poor quality fruit and consequently would not be used unless they offer other traits beyond low-chilling requirement such as disease resistance.

Hybrids between cultivated plum and *P. texana* (precocious, fruiting 1 year from seed) from south Texas have been produced in Florida and Texas. In Florida, all the F_2 seedlings were discarded because they had small, soft fruit which cracked at ripening and the trees were very susceptible to twig cankers incited by *Xanthomonas* (Sherman and Lyrene, 1985). The hybrids produced in Texas require about 400 CU, and are productive with small, soft fruit but are extremely precocious (fruit 2nd leaf, about 11 months, from seed). Backcrosses to high quality plum types are being done.

Another species that is being explored as a source of disease resistance is *Prunus mexicana*. Progenies between this species and cultivated types have been produced in both Texas and Georgia. In Texas, the hybrids are very healthy and many are very productive but all have small fruit. Open pollinated seed from the F_1 plants germinate poorly with substantial problems with unthrifty seedlings. These F_2 plants, although not fruiting, are segregating widely for morphological traits.

The Florida program began by crossing a low-chill plum from Taiwan (small, non astringent fruit) with better quality plums such as 'Ozark Premier', 'Burbank', 'Methley',

'Beauty', and 'Bruce'. Since these initial crosses, they have gone through five cycles of selection and polycrossing. From this they selected 25 plums that are low-chilling (200 to 400 CU) and have acceptable fruit size (35 to 60 gm), fruit quality and resistance to bacterial leaf spot and plum leaf scald (Sherman and Lyrene, 1985; Sherman, Topp, and Lyrene, 1992). These Florida selections appear to have a low-chilling requirement (*ca.* 200 CU) but a high heat requirement. Thus they flower with 'Bruce', 'Mariposa', and 'Excelsior' in north Florida, but 3 to 4 weeks earlier at Gainesville, in central Florida. The Florida selections also appear to fruit well in South Florida at Immokalee (Rouse, PC, 1997) and south Texas at Weslaco (Strong and Byrne, 1995). Work in Mexico (Chapingo) and Australia (NSW) with this low-chill plum germplasm has resulted in several releases (Rodriguez and Sherman, 1995b; Richards, PC, 1996).

The program in Georgia has employed several sources of moderate chilling and disease resistance in their breeding. These include a local selection of *Prunus angustifolia*, 'Mariposa', 'Ozark Premier', 'Morris', and 'Barstow'. These were intercrossed among themselves as well as crossed with standard plum cultivars used in California such as 'Santa Rosa', 'Gaviota', and 'Queen Ann' (Okie, 1995; Byrne, 1989a).

The plum breeding program in Campinas, Brazil selected a range of well adapted cultivars to extend the harvest season (October to beginning of February) and the area of production from the higher elevations to the lower elevations mainly from open pollinated seedlings of yellow fleshed plum 'Kelsey Paulista', a seedling of 'Kelsey'. Given the number of these that have red flesh, it is likely that many of these are out crosses with one of the local red fleshed cultivars some of which are derived from 'Satsuma' (Ojima *et al.*, 1992). Both 'Kelsey' and 'Satsuma' are *P. salicina* varieties from Japan which are reported to have good resistance to bacterial leaf spot (Topp *et al.*, 1989; Topp and Sherman, 1990a; Ritchie, 1995a). The parents of the plums named by the program in Pelotas were 'Satsuma', 'Santa Rosa' and 'The First' (Okie, 1995). Current work in Parana and Santa Catarina is employing some of the local cultivars such as 'Amarelinha'.

The South African program used 'Wickson' and 'Golden King' extensively in their program to develop better adapted, bacterial leaf spot resistant plums suitable for shipping. 'Wickson', a cross between 'Burbank' and *P. simonii*, is resistant to bacterial leaf spot (Topp *et al.*, 1989; Byrne, PC, 1996). 'Golden King' was imported into South Africa in 1936 but its origin is not clear (Hurter, 1971). These were intercrossed with each other but also with 'Methley', 'Gaviota', 'Eldorado' and 'Laroda'. Their cultivars range in chilling from about 600 to 750 CU.

In addition to these programs, higher chill programs in Alabama (USA) and Australia (Stanthorpe, Queensland), have also worked to develop plum germplasm with resistance to bacterial leaf spot (Topp, 1993; Topp and Russell, 1989) and plum leaf scorch (Okie, 1995). These programs have worked independently from the other programs described and have developed a unique source of resistant germplasm. This material, although it is high-chill (>750 CR), it is an important source of germplasm for resistance and fruit quality and is currently being used in the breeding efforts in Texas, Florida and Brazil. The Stanthorpe program recently began hybridizing between their high quality genotypes and the low-chill genotypes from Florida to produce low-chill, early ripening plums for warmer winter areas in Queensland and other similar climates.

Thus these programs have used different sources of germplasm for quality and sources of resistance. It appears that resistance to bacterial leaf spot is available from *P. angustifolia, P. cerasifera* and even *P. salicina* (Byrne, 1989; Topp and Sherman, 1990a). It is important to combine these different sources of germplasm to capitalize on the diversity of resistance available.

7.3 Apricot germplasm

The center of origin of the apricot is believed to be in north and northeast China with a secondary center in the Tien Shan Mountains in the Xinjiang autonomous region and the Dzhungar and Zailij Mountains in Central Asia (Wang, 1985; Zeven and de Wet, 1982). Over the centuries, this apricot germplasm spread by seed throughout Central Asia, Afghanistan, Kashmir, Iran, Turkey, and Trans-Caucasia and cultivated in seedling orchards (Mehlenbacher *et al.*, 1990). This resulted in the differentiation of the cultivated germplasm into four eco-geograpical groups: Dzhungar-Zailij, Central Asian, Irano-Caucasian, and European groups. There are four eco-geographic groups among the Chinese germplasm. Unfortunately little is known about this material, especially its southern range and adaptability to humid weather and fluctuating winter temperatures.

The medium and low-chill apricots reported from Israel, Egypt, Morocco, Tunisia, Canary Islands, Iraq, Syria, southern and eastern Turkey, Mexico, and Australia are generally late ripening with small, soft fruit (Mehlenbacher *et al.*, 1990; Gathercole, 1984; Perez, 1992; Perez, 1995; Sherman, PC, 1996). The north African apricots are in the Irano-Caucasian group of apricots which are generally less vigorous, longer lived, less hardy, and have thicker branches and twigs, larger and shinier leaves, and a lower chilling requirement than the Central Asian types. Most of these cultivars are self incompatible and probably spread via the trade routes throughout the region and maintained in seedling orchards. Thus selection among seedlings from open pollinated (and given the self incompatibility, out crossed seed) for adaptability has been practiced over many generations. This germplasm generally has less disease resistance, and paler flesh color (white and yellow versus yellow to orange) but higher sugars and lower acidity than the major cultivated group: the European group (Mehlenbacher *et al.*, 1990).

Although most of these lower chill types appear to require around 600 CU (as does 'Canino'), a few from north Africa (Tunisia and Egypt) bloom with the peaches that require 200 and 300 CU. Unfortunately even though these bloom and foliate well, they tend to fruit poorly in areas such as the southeastern USA, the southern production zones of California, southern Spain, and the lower altitudes of Mexico where the temperatures fluctuate widely during the winter months (Sherman, 1987; Okie, PC, 1997). Thus it seems that the apricot, although grown over an apparently wide geographic range, thrives only under specific environmental conditions: warm, dry summers and winter temperatures that do not fluctuate widely (Bailey and Hough, 1975). This presents problems for the development of apricots for many low-chill zones, since in most of these areas the temperatures fluctuate widely within a given winter as well as between different winters. The only exception to this might be the culture of apricots in tropical highlands where the temperature fluctuations are less. Apricots also have a low heat requirement for breaking dormancy, thus they are most susceptible to spring frost since they bloom earlier than most other stone fruit.

Apricots have been grown in the highlands (1800 to 2500 meters above sea level) of Mexico since apricot seeds were brought over by the Spanish explorers. These exhibit a wide range of adaptability and bear small (10 to 40 grams), pale yellow to orange fruit with no or little red overcolor (Perez, 1992; 1995). Limited trials in the medium-chill zone of Texas indicates that although these bloom with the peaches that require 400 to 700 CU, they do not fruit. This may be due to the greater temperature fluctuations in winter of the medium-chill zone of Texas as compared to the Mexican highlands.

The most active programs are in the transitional chilling zone and have primarily worked within the European germplasm with occasional introgression of germplasm from the Irano-Caucasian and Central Asian groups. The longest-lived program is in South Africa and their work has resulted in 6 releases: three each for the drying and fresh market industry (Table 5). About 70% of the genetic background of these releases is from two varieties from the European group: 'Royal' (='Blenhiem', a major commercial variety in California) and 'Bulida' (a major commercial cultivars grown in Valencia, Spain). The rest of the genetic background is from two local cultivars ('Verdun' and 'Alpha') and 'Canino'. The original apricots grown in South Africa are probably of the European group and derived from apricot seeds brought there by Jan Van Riebeeck in 1652 (Smith, PC, 1997).

The varieties that have been released from the transitional and medium-chill zones of California (Zaiger Genetics, Sunworld) frequently contain 'Perfection' in their background. The best selection thus far from the Georgia breeding effort is derived from this parentage as well although some other promising selections have a Japanese apricot cultivar ('Koshu-omi') in its parentage. It is possible that Japanese and Chinese apricot cultivars may provide a source of genetic adaptation to the medium and transitional chilling zones (Okie, PC, 1997). The Indian work in apricot breeding has resulted in three cultivars that were developed by intercrossing the locally selected cultivars (Kumar, PC, 1997) (Table 5). In addition to this, a joint French-Tunisian breeding program in the 1960s, developed three cultivars {'Priana' (='Ouardi'), 'Ferriana' (='Amal') and 'Beliana' (='Sayeb')} by intercrossing 'Canino' and 'Hamidi' (a Tunisian variety).

Probably the most interesting breeding is being done in South Australia where they are trying to incorporate the high soluble solids content (~26° brix) of the Turkish drying apricot 'Hasanbey' (oval shape, small, soft, pale, cracks) into apricots (12 to 16° brix) well adapted to their medium-chill climate. They have progressed to the second generation from the original cross and have been able to recover a few selections that have high soluble solids (~21° brix) with acceptable fruit size, color, shape, and firmness (Gathercole, PC, 1996). This is an important step in the development of higher quality apricots for both drying and fresh market use.

Still in the lower chill zones where the winter temperatures are variable and the summers are not always dry, apricots, even though they may bloom as if they need only 200 to 400 chilling units, tend to have major problems with flower bud drop, failure to set fruit on normal appearing flowers, and susceptibility to bacterial canker, bacterial leaf spot and other diseases. Two possible approaches to extending the adaptability of the apricot are to cross apricots with diploid plums (*Prunus salicina* and hybrids) or with a low-chill apricot relative, *Prunus mume* Sieb et Zucc.

Apricot - plum hybrids (*P. x dasycarpa* Ehrh.) occur naturally (Byrne, 1993a). In the beginning of this century Burbank released 11 plumcot cultivars (Howard, 1945) and since that time, many other plum - apricot hybrids have been made (Byrne and Littleton, 1989) in New Jersey, Georgia, Texas, Florida, and California and recently several plumcot cultivars have been released in California for commercial use (Cummins, 1991; 1994; 1995). These materials when created did not have the objective of developing apricots with wider adaptability but rather a unique type of fruit - the plumcot. To develop widely adapted apricots from these crosses, selection would have to be toward an apricot fruit type while maintaining the wider adaptability traits of the plum tree. Hybrids can be made between apricot and a wide range of diploid plums (Mehlenbacher *et al.*, 1990; Yoshida *et al.*, 1975) but most of the initial hybrids have a reduced fertility and it takes several generations of selection to regain full fertility and fruit set. Efforts at this approach in Florida have been unsuccessful (Sherman, PC, 1997).

The other approach is to use mume (*P. mume*) as a donor species for increased adaptability (Boonprakob and Byrne, 1990). This species is native to southern China and Taiwan in the more humid areas with milder winters that experience fluctuating winter temperatures (Yu, 1979; Chen, 1989). It has been cultivated in China for about 3,000 years and was introduced into Japan about 2,000 years ago (Chu and Ban, 1990; Yoshida, 1994). At present there are hundreds of ornamental (Chen, 1989) and fruiting cultivars (Chu and Ban, 1990; Yoshida, 1994; Shimada *et al.*, 1994). The fruiting types are generally classified according to fruit size (small, medium, and large), and interspecific hybridity (mume-apricot or mume-plum group) (Yoshida, 1994).Work with RAPDs indicates that these groups are genetically distinct from each other. Furthermore, a comparison of Taiwanese (small fruited) mume cultivars indicated that this group was developed independently from the small fruited and other Japanese mumes. The fruiting cultivars in Japan were most likely developed from the flowering types brought from China (Shimada *et al.*, 1994). How the Japanese, Thailand and Taiwanese mume varieties compare to the Chinese wild collections and local varieties still needs to be investigated.

Although probably better known as an ornamental outside Asia, mume is cultivated on about 80,000 ha in China, Taiwan, and Japan for its fruit (Chu, PC, 1993) which is processed into a variety of products (dried and salted, sugared and pickled, jelly, dried, wine, powdered for beverage) which are considered good for one's health (Chu and Ban, 1990). The fruit itself is generally smaller (15 to 50 gm) than an apricot fruit and has low soluble solids ($< 10°$ brix), high acidity (4 to 7% citric acid), and frequently are bitter although non bitter varieties exist (Ou, 1994; Yoshida, 1987; 1994; Chu, PC, 1993; Mehlenbacher *et al.*, 1990). The fruit ranges in flesh and skin color from green to yellow to red with or without a red blush. Mume appears to be resistant to numerous diseases although there are varieties that are susceptible to bacterial leaf spot (Mehlenbacher *et al.*, 1990). Its chilling requirement ranges from less than 100 to about 800 CU and it appears to have a low heat requirement for bloom (Byrne, PC, 1996; Ou, 1994; Sherman, PC, 1993). The tree is precocious (fruits in 2nd leaf) and long lived

(50 to 100 years is common and a tree with the age of 500 years is documented) in China (Chu, PC, 1993). Consequently, there is a wide range of mume germplasm to choose from for developing hybrids with apricot although it appears the lowest chilling genotypes are also the smallest fruited (Shimada *et al.*, 1994; Yoshida, 1987).

Mume is closely related to and crosses easily with apricot to yield fertile hybrids (Boonprakob and Byrne, 1990; Yoshida *et al.*, 1975). Mume-apricot hybrids are known to occur naturally and many are cultivated as mume having usually larger fruit size although their processing qualities are usually lower than pure mume (Yoshida, PC, 1996; Chu, PC, 1993). Recent apricot-mume hybrids made in Florida have shown excellent set and adaptability. They have backcrossed these hybrids twice back to apricot and have a genotype that produces small, green-fleshed, freestone fruit (Sherman, PC, 1997).

7.4 Cherry germplasm

Zaiger Genetics has worked with this germplasm from the northern Mediterranean region (specifically germplasm from Spain), to develop medium-chill sweet cherries (*Prunus avium*) adapted to the southern half of California. They have gone through 8 generations of breeding to develop their present sweet cherry selections that have good commercial characteristics and bloom about 14 days before 'Bing'. They estimate that their selections need 300 to 500 chill units to fruit well (Zaiger, PC, 1996) and are presently being evaluated under medium-chill conditions for productivity, fruit quality, and twinning (doubling) propensity.

Since there is no low-chill germplasm available for sweet cherry (Sherman, 1977), the only alternative is to go to another species of cherry for this trait. Several species have been used in crosses with sweet cherry with occasional success with *Prunus pleiocerasus* and *Prunus campanulata*. In 1957, W. E. Lammerts made a cross between *P. pleiocerasus* and *P. avium* 'Black Tartarian'. This hybrid is very low-chilling (<200 CU) but not self fruitful. The hybrid was repeatedly crossed with sweet cherry and *P. campanulata*. In the mid 1970s, the Florida program developed several seedlings by using mixed pollen (*P. campanulata* and 'Stella'). All the hybrids had pink blooms and thus were probably hybrids with *P. campanulata*. Several of these seedlings were fruitful. Although the size is still small, this germplasm is useful for the development of low-chill sweet cherries. Graeme Richards (University of West Sydney- Hawesburg, New South Wales, Australia) initiated a similar effort in 1987. His goal is to produce early ripening, low-chill (150 to 400 CU) sweet cherries. Beyond creating and evaluating hybrids derived from the Florida work, he has collected and is evaluating additional sweet cherry relatives from southeast Asia (Topp, 1993).

Work is also being done in California by Jim Doyle and Ted DeJong. They have been working with some *P. avium* x *P. pleiocerasus* hybrids that they originally obtained from Paul Hansche (Department of Pomology, University of California, Davis, CA). The clone of *P. pleiocerasus* was probably 'Tangsi' from China. They have done two backcrosses using *P. avium* (mainly 'Ranier' and some 'Brooks') and currently have about 30 seedlings. These bloom as if they need as little as 200 CU and are fruitful but the fruit tends to be soft, small, and forms much double fruit (Doyle, PC, 1996).

They plan to continue selecting for better yields, fruit quality, fruit size and against the tendency to form twin (double) fruits.

There has been no work done on the development of sour cherries for the medium and low-chill zones although it appears that this species can fruit with less doubling under mild winter, hot summer conditions than *P. avium* (Mowrey, PC, 1997). Nevertheless, the species is not well adapted to this environment. Another tetraploid cherry that is native to the medium-chill regions of the highlands of central and south America is *Prunus serotina* (*P. salicifolia or* capulin) (Popenoe, 1920). This species fruits precociously, is resistant to the common diseases of this zone, fruits consistently and very heavily, and produces a fruit that is bigger than most wild cherries but smaller than the cultivated species. Its fruit, collected from wild trees, is consumed fresh in the region and has a pleasant sweet flavor and sometimes overlaid with a distinct 'woody' flavor. Recently, Amy Iezzoni (Michigan State University, East Lansing, MI, USA) and Salvador Perez (Universidad Autonoma de Queretaro, Queretaro, Mexico) have begun to test its compatibility with *P. cerasus*. The interest is to combine the apparent disease (powdery mildew) resistance of *P. serotina* with the fruit quality of *P. cerasus*. If the two species are compatible, it should be possible to develop a sour cherry that is adapted to the medium-chill region of the world.

8 Conclusions

As with the development of many "new" crops, the breeding efforts of stone fruit breeders in medium and low-chill areas have been poorly supported as compared to those programs in areas with an established stone fruit production industry. Nevertheless, the early breeders who believed in the possibility of converting the existing low-chill peaches into a range of commercial cultivars have made tremendous advances over the last 50 years. These medium and low-chill peach and nectarine cultivars are the basis of a new industry that is developing in many countries throughout the world and with this ever increasing commercial success, more breeding efforts have been initiated. Although there is still much to do, the accomplishments in the breeding of low-chill peaches, and nectarines gives us great hope that similar accomplishments can be made in plums, apricots and cherries.

9 Contributors

Jean-Marc Audergon	INRA, Station de Recherches Fruitieres Mediterraneennes, Domaine Saint-Paul, Site Agroparc, 84914 Avignon, Cedex 9, France
Wilson Barbosa	Secao de Fruticultura de Clima Temperado, Caixa Postal 28, 13001-970 Campinas, SP, Brazil
Thomas Beckman	USDA-ARS, PO Box 87, Byron, GA 31008, USA
Unaroj Boonprakob	Dept. of Horticulture, Kasetsart University, Kampangsaen, Nakhonpathom 73140, Thailand

Mengyuan Chu	Department of Horticulture, Nanjing Agricultural University, Nanjing, Jiangsu, People's Republic of China
Jim Doyle	Kearney Agriculture Center, 9240 South Riverbend Ave., Parlier, CA 93648, USA
Frank Gathercole	Senior Research Officer - Stone Fruit, Primary Industries, South Australia, Loxton Centre, PO Box 411, Loxton, SA 5333, Australia
Joseph Goffreda	Cook College, Rutgers University, Dept. Plant Science, New Brunswick, NJ 08903-0231, USA
Thomas Gradziel	Department of Pomology, University of California, Davis, CA 95616, USA
Charles J. Graham	Louisiana Agricultural Experiment Station, Calhoun Research Center, 321 Hwy. 80 East, PO Box 539, Calhoun, LA 71225, USA
Roberto Hauagge	IAPAR, Polo de Curitiba, PO Box 2301, 80001-970 Curitiba, PR, Brazil
Xulan Hu	Yunnan Agricultural Extension Center, 3 Xinwen South Rd., Kunming, Yunnan 650032, People's Republic of China
Charles E. Johnson	Department of Horticulture, Louisiana State University, 137 J.C. Miller Hall, Baton Rouge, LA 70803, USA
Jocelyne Kervella	INRA, Station de Recherches Frutieres Mediterraneennes, Domaine Saint-Paul, Site Agroparc, 84914 Avignon, Cedex 9, France
Krisnan Kumar	Department of Fruit Breeding and Genetic Resources, Dr. Y.S. Parmar University of Horticulture and Forestry, NAUNI, Solan, H.P., India
Antonino Nicotra	Head of Breeding Department, Istituto Sperimentale per la Frutticoltura, Ciampino Aeroporto, 00040, Roma, Italy
Mario Ojima	Head, Secao de Fruticultura de Clima Temperado, Caixa Postal 28, 13001-970 Campinas, SP, Brazil
W. R. Okie	USDA-ARS, SE Fruit and Tree Nut Research Lab, 111 Dunbar Rd. Byron, GA 31008, USA
Shyi-Kuan Ou	Associate Horticulturist, Taiwan Agricultural Research Institute, Dept. of Horticulture, 189 Chung Cheng Rd., Wu- Feng, Taichung, 431, Taiwan
Salvador Perez	Prol. Zaragoza 408, Jardines de la Hacienda, Queretaro, Qro, 76180, Mexico
Gavin Porter	Plant Production Department, The University of Queensland - Gatton College, Lawes, Qld, 4343, Australia
David Ramming	USDA-ARS, Horticulture Crops Research Laboratory, 2021 South Peach Ave., Fresno, CA 93727, USA
Maria do C. Bassols Rasiera	CPACT-EMBRAPA, Caixa Postal 403, 96001-970 Pelotas, RS, Brazil

Graeme Richards — University of Western Sidney, Bourke Street, Richmond 2753, NSW, Australia

Susan Richards — Peach Breeding Officer, ISIA, Agriculture Victoria, State Public Offices, Welsford Street, PO Box 862, Shepparton, Victoria 3632, Australia

Jorge Rodriguez — Centro de Fruticultura, Colegio de Postgraduados, Chapingo, 56230, Mexico

Dougal Russell — District Experimentalist, Queensland Department of Primary Industries, Granite Belt Horticultural Research Station, PO Box 501, Stanthorpe, Qld 4380, Australia

Chris Smith — INFRUITEC, Bien Donne, Private Bag X01, Groot Drakenstein, 7680, South Africa

Bruce Topp — Queensland Department of Primary Industries, Granite Belt Horticultural Research Station, PO Box 501, Stanthorpe, Qld 4380, Australia

Burger W. Van Dyk — SAPO, Private Bag X5023, Stellenbosch 7599, South Africa

Lirong Wang — Zhengzhou Fruit Research Institute, Chinese Academy of Science, Zhengzhou, Henan, People's Republic of China

Jenny Witherspoon — Research Officer, Primary Industries, South Australia, Loxton Centre, PO Box 411, Loxton, SA 5333, Australia

Floyd Zaiger — Zaiger's Genetics, 1219 Grimes Ave., Modesto, CA 95358, USA

10 References

Ackerman, W.L. (1953) The evaluation of peach leaf curl in foreign and domestic peaches and nectarines grown at the U.S. Plant Introduction Garden, Chico, California. *Division Plant Exploration and Introduction, Bur. of Plant Ind., Soils and Agric. Eng., USDA (Mimeo)* p. 1-31.

Adaskaveg, J.E., Feliciano, A.J. and. Ogawa , J.M. (1991) Evaluation of the cuticle as a barrier to penetration by *Monilinia fructicola.* (Abstr.) *Phytopathology* 81, 1150.

Arulsekar, S., Parfitt, D.E. and Kester, D.E. (1986) Comparison of isozyme variability in peach and almond cultivars. *J. Hered.* 77, 272-274.

Bacon, T.A. and Byrne, D.H. (1990) The effects of mild winter conditions on peach fruit shape (Abstr.) *87th ASHS Meeting, Tucson, AZ.*

Bacon, T.A. and Byrne, D.H. (1995) Relationships of fruit development period, seed germination, seedling survival and percent dry weight of ovule in peach (Abstr.) *HortScience* 30, 833.

Bailey, C.H. and Hough, L.F. (1975) Apricots. In: *Advances in fruit breeding* (Janick, J. and Moore, J. N eds) Purdue Univ. Press, West Lafayette, Indiana. p. 367-383.

Bakarcic, M. and de Santis, M.S. (1969) Comportamiento a la escaldadura de la hoja de variedades locales de ciruelo. *Delta del Parana* 9, 35-42.

Barbosa, W., Campo-Dall'Orto, F.A., Ojima, M., Kalil, G.P.C., Lovate, A.A., Ribeiro, I.J.A., Martins, F.P.

and Nogueira, E.M.C. (1994) Incidência de ferrugem em folhas de pessegueiro e nectarineira do germoplasma IAC. *Sci. Agric.* (Piracicaba, Brazil) 51, 90-93.

Barbosa, W., Ojima, M., Campo-Dall'Orto, F.N., Rigitano, O., Martins, F.P., Ribeiro dos Santos, R. and Lopes de Castro, J. (1995) Melhoramento do pessegueiro para regiões de clima subtropical - temperado: *Realizações do Instituto Agronômico no período de 1950 a 1990. Doc. IAC* (Campinas, SP) No. 52.

Beckman, T.G. and Sherman, W.B. (1996) The non-melting semi-freestone peach. *Fruit Var. J.* 50, 189-193.

Bertrand, P.F. (1995) Rust, In: *Compendium of Stone Fruit Diseases* (Ogawa, J.M., Zehr, E. J., Bird, G.W., Ritchie, D.F., Uriu, K. and Uyemoto, J.K. eds) APS Press. p. 23-24.

Biggs, A.R. and Northover, J. (1989) Association of sweet cherry epidermal characters with resistance to *Monilinia fructicola. HortScience* 24, 126-127.

Bolkan, H.A., Ogawa, J.M., Michailides, T.J. and Kable, P.F. (1985) Physiological specialization in *Tranzschelia discolor. Plant Dis.* 69, 485-486.

Boonprakob, U. (1996) RAPD polymorphisms in diploid plums: Genetic relationships and genetic linkage maps. *Ph.D. Dissertation. Texas A&M University, College Station, TX.*

Boonprakob, U. and Byrne, D.H. (1990) Mume, a possible source of genes in apricot breeding. *Fruit Var. J.* 44, 108-113.

Boonprakob, U., Byrne, D.H. and Mueller, D.M.J. (1996) Anatomical differences of axillary bud development in blind nodes and normal nodes in peach. *HortScience* 31, 798-801.

Boonprakob, U., Byrne, D.H. and Rouse, R.E. (1994) A method for blind node evaluation. *Fruit Var. J.* 48, 213-215.

Bowen, H.H. (1980) 'EarliGrande' peach. *HortScience* 15, 207.

Boyhan, G.E., Abrahams, B.R., Norton, J.D. and Huang, H. (1996) Budding method affects transmission of *Xylella fastidiosa* in plum. *HortScience* 31, 89-90.

Boyhan, G.E., Tangsukkasemsan, B., Norton, J.D. and Himelrick, D.G. (1997) Incidence of *Xylella fastidiosa* Wells et al. on plum and peach in Alabama. *Fruit Var. J.* 51, 31-35.

Brooks, R.M. and Olmo, H.P. (1952) Register of new fruit and nut varieties. List 7. *Proc. Amer. Soc. Hort. Sci.* 60, 497-504.

Brooks, R.M. and Olmo, H.P. (1962) Register of new fruit and nut varieties. List 17. *Proc. Amer. Soc. Hort. Sci.* 81, 568-596.

Brooks, R.M. and Olmo, H.P. (1972) Register of new fruit and nut varieties. *Univ. California Press, Berkeley.*

Brovelli, E.A., Bracht, J.K., Sherman, W.B. and Sims, C.A. (1995) Quality profile of fresh market melting and nonmelting peach fruit. *Proc. Fla. State Hort. Soc.* 108, 309-311.

Bruckner, C.H. (1987) Melhoramento genético do pessegueiro no Brasil. *Rev. Bras. Frutic.* (Cruz das Almas) 9, 7-19.

Byrne, D.H. and Littleton, T.G. (1988) Electrophoretic characterization of diploid plums of the southeastern United States. *J. Amer. Soc. Hort. Sci.* 113, 918-924.

Byrne, D.H. (1989) Inbreeding, coancestry and founding clones of Japanese-type plums of California and the Southeastern United States. *J. Amer. Soc. Hort. Sci.* 114, 699-705.

Byrne, D.H. (1990) Isozyme variability in four diploid stone fruits compared with other woody perennial plants. *J. Hered.* 81, 68-71.

Byrne, D.H. (1993a) Isozyme phenotypes support the interspecific hybrid origin of *Prunus xdasycarpa* Ehrh. *Fruit Var. J.* 47, 143-145.

Byrne, D.H. (1993b) Isozyme variability in the Japanese plum (Abstr.) *HortScience* 28, 455.

Byrne, D.H. and Bacon, T.A. (1989) Peach cultivar and advanced selection evaluation in the medium-chill region of Texas. *Fruit Var. J.* 43, 59-66.

Byrne, D.H. and Littleton, T.G. (1989) Interspecific hybrid verification of plum x apricot hybrids via isozyme analysis. *HortScience* 24, 132-134.

Byrne, D.H. and Bacon, T.A. (1994) Plum cultivar evaluation based on adaptation to medium chill zones. *The Texas Horticulturist* 21, 6-7.

Byrne, D.H., Nikolic, A.N. and Burns, E.E. (1991) Variability in sugars, acids, firmness and color characteristics of 12 peach genotypes. *J. Amer. Soc. Hort. Sci.* 116, 1004-1006.

Campo Dall'Orto, F.A., Ojima, M., Ribeiro, I.J.A., Rigitano, O. and Alves, S. (1980) Resistência varietal da ameixeira à ferrugem (*Tranzschelia* spp.) nas condições de Monte Alegre do Sul, S.P. *Bragantia* (Campinas) 39, 171-174.

Campo Dall'Orto, F.A., Ojima, M., Ribeiro, I.J.A., Rigitano, O. and Veiga, A.A. (1979) Resistência varietal de ameixeira (*Prunus salicina* Lindl.) à ferrugem (*Tranzschelia* Spp.) nas condições de Tietê-SP. *Congresso Brasileiro de Fruticultura, 5.* Pelotas. 2, 789-796.

Carraut, A. and Crossa-Raynaud, P. (1974) L'améioration des variétés d'abricotier en Tunisie. (In French) *Ann. Inst. Nat. Rech. Agr. Tunisie* 47, 3-33.

Chen, J. (1989) Chinese mei flower cultivars (Chinese) *China Forestry Publishing House. Beijing, People's Republic of China.*

Chu, M. and Ban, J. (1990) Mume production in China, now and in the future. *Chinese Hort.* 6, 33-35 (In Chinese).

Civerolo, E.L., Sasser, M., Helkie, C. and Burbage, D. (1982) Selective medium for *Xanthomonas campestris* pv. *pruni*. *Plant Dis.* 66, 39-43.

Connors, C.H. (1922) Inheritance of foliar glands of the peach. *Proc. Amer. Soc. Hort. Sci.* 18, 20-26.

Crossa-Raynaud, P.H. (1969) Evaluating resistance to *Monilinia laxa* (Aderh. & Ruhl.) Honey of varieties and hybrids of apricots and almonds using mean growth rate of cankers on young branches as a criterion of susceptibility. *J. Amer. Soc. Hort. Sci.* 94, 282-284.

Cummins, J.N. (1991) Register of new fruits and nut varieties. Brooks and Olmo. List 35. *HortScience* 26, 951-986.

Cummins, J.N. (1994) Register of new fruit and nut varieties. Brooks and Olmo. List 36. *HortScience* 29, 942-969.

Cummins, J.N. (1995) Register of new fruit and nut varieties. Brooks and Olmo. List 37. *HortScience* 30, 1135-1150.

Daniell, J.W. and Chandler, W.A. (1982) Field resistance of peach cultivars to gummosis disease. *HortScience* 17, 375-376.

Dhatt, A.S., Chanana, Y.R., Nijjar, G.S., Minhas, P.P.S., Bindra, A.S., Dhillon, D.S. and Uppal, D.K. (1992) Studies on the performance of a new table plum suitable for northern plains. *Indian J. Hort.* 49, 172-174.

D'Bov, S. (1983) Inheritance of peach resistance to powdery mildew, *Sphaerotheca pannosa* var. *persicae*, III: leaf resistance in F_1 of cultivars 'J.H. Hale' x 'Nectarine-Ferganensis-2' (Russian) *Genet. Sel.* 16, 146-150 (Biol. Abstr. B1AbA477038418, (1984))

Decker, P. and Sherman, W.B. (1976) Reactions of plums to rust, *Transchelia pruni-spinosae* (Pers.) Diet., in Florida. *Proc. Fla. State Hort. Soc.* 89, 254-255.

Diaz, D.H., Martinez, J.J. and Sherman, W.B. (1986) Apple and peach production in warm climates of northwest Mexico. *Fruit Var. J.* 40, 121-125.

Dirlewanger, E., Pascal, T., Zuger, C. and Kervella, J. (1996) Analysis of molecular markers associated with powdery mildew resistance genes in peach (*Prunus persica* (L.) Batsch) x *Prunus davidiana* hybrids. *Theor. Appl. Genet.* 93, 909-919.

Durham, R.E., Moore, G.A. and Sherman, W.B. (1987) Isozyme banding patterns and their usefulness as genetic markers in peach. *J. Amer. Soc. Hort. Sci.* 112, 1013-1018.

Edwards, G.R. (1987) Temperature in relation to peach culture in the tropics. *Acta Hort.* 199, 61-62.

El-Agamy, S.Z. and Sherman, W.B. (1987) Sequence of pollination in relation to fruit set and progeny in peach (*Prunus persica* (L.) Batsch) *J. Hort. Sci.* 62, 469-473.

English, H. (1947) Powdery mildew on cherry fruit in Washington. *Phytopathology* 37, 421-424.

Evreinoff, V.A. (1939) *Amygdalus petunnikowii* Litw., a remarkable peach rootstock. *Fruit Belge* 7, 129-30. (from Abstr. no. 895, Knight, 1969).

Feliciano, A., Feliciano, A.J. and Ogawa, J.M. (1987) *Monilinia fructicola* resistance in the peach cultivar Bolinha. *Phytopathology* 77, 776-780.

Fernández, M.V. and Bakarcic, M. (1963) Escaldadura del borde de las hojas del ciruelo. *Delta del Paraná* 3, 39-47.

Fernández, M.V. and Bakarcic, M. (1954) Nuevas enfermedades del ciruelo en el Delta del Paraná. *Informe Invest. Agric.* (*IDIA*) 94, 2-6. (Inst. Nac. Tecn. Agropec. (INTA), Buenos Aires, Argentina).

Fideghelli, C., Della Strada, G. and Quarta, R. (1983) A source of immunity for the peach leaf curl [*Taphrina deformans* (Berk.) Tul.]. *Acta Hort.* 140, 129-132.

Flory, W.S. (1941) Varietal rating of plums with reference to canker resistance. *Texas Agr. Exp. Sta. Progr. Rep. No. 753.*

Flory, W.A. (1937) The case for plums in Texas. *Texas Agr. Exp. Sta. Progr. Report No. 446.*

Fogle, H.W. and Faust, M. (1975) Ultrastructure of nectarine fruit surfaces. *J. Amer. Soc. Hort. Sci.* 100, 74-77.

French, W.J. and Feliciano, A. (1982) Distribution and severity of plum leaf scald disease in Brazil. *Plant Disease Reporter* 66, 515-517.

French, W.J. and Kitajima, E.W. (1978) Occurrence of plum leaf scald in Brazil and Paraguay. *Plant Disease Reporter* 62, 1035-1038.

Gathercole, F.J. (1984) Report on apricot collecting expedition to Syria and Turkey. *Dept. Agric., South Australia Technical Rept. No. 54.*

George, A.P., Nissen, R.J. and Sherman, W.B. (1988) Overlapping double and early single cropping of low-chill peaches in Australia. *Fruit Var. J.* 42, 91-95.

Goffreda, J.C., Scopel A. and Voordeckers, A.M. (1992) Ripening behavior of the stony hard ripening mutant. *Mimeograph-prelim. report passed out at Prunus Breeders Meeting. 1992.*

Gradziel, T.M. (1994) Changes in susceptibility to brown rot with ripening in three clingstone peach genotypes. *J. Amer. Soc. Hort. Sci.* 119, 101-105.

Gradziel, T.M., Beres, W. and Pelletreau, K. (1993) Inbreeding in California canning clingstone peach cultivars. *Fruit Var. J.* 47, 160-168.

Gradziel, T.M. and Wang, D. (1993) Evaluation of brown rot resistance and its relation to enzymatic browning in clingstone peach germplasm. *J. Amer. Soc. Hort. Sci.* 118, 675-679.

Grove, G.G. (1995) Powdery mildew. In: *Compendium of Stone Fruit Diseases* (Ogawa, J.M., Zehr, E.J., Bird, G.W., Ritchie, D.F., Uriu, K. and Uyemoto, J.K. eds). *APS Press*, St. Paul, MN. p. 12-14.

Guerriero, R., Massai, R. and Scalabrelli, G. (1992) Classificazione delle cultivar di albicocco in base al fabbisogno in freddo e in caldo delle gemme a legno, . In: *Germplasma frutticolo: Salvaguardia e valorizzazione delle risorse genetiche.* (C. Delfino ed) C.N.R., Inst. per lo Studio dei Problemi Bio-Agronomici delle Colture Arboree Mediterranee, Sassari, Italy. p. 335-341

Gupta, A.K. and Kashyap, A.S. (1991) Evaluation of apricot and almond fruit germplasm against bacterial gummosis under cold desert conditions of Himachal Pradesh. *Agr. Sci. Digest Karnal* 11, 230-232.

Hammerschlag, F.A. (1984) Optical evidence for an effect of cultivar filtrates of *Xanthomonas campestris* pv. *pruni* on peach mesophyll cell membranes. *Plant Sci. Letters* 34, 295-304.

Hammerschlag, F.A. (1988a) Screening peaches in vitro for resistance to *Xanthomonas compestris* pv. *pruni. J. Amer. Soc. Hort. Sci.* 113, 164-166.

Hammerschlag, F.A. (1988b) Selection of peach cells for insensitivity to culture filtrates of *Xanthomonas campestris* pv. *pruni* and regeneration of resistant plants. *Theor. Appl. Genet.* 76, 865-869.

Hansche, P.E., Hesse, C.O. and Beres, V. (1972) Estimates of genetic and environmental effects on several traits in peach. *J. Amer. Soc. Hort. Sci.* 97, 76-79.

Hattingh, M.J. and Ross, I.M.M. (1995) Bacterial canker In: *Compendium of Stone Fruit Diseases* (Ogawa, J.M., Zehr, E. J., Bird, G.W., Ritchie, D.F., Uriu, K. and Uyemoto, J.K. eds) APS Press, St. Paul, MN. p. 48-50.

Hauagge, R., Tsuneta, M. and Hissano, Z. (1991) IAPAR 49-Irati, uma nova cultivar precoce de ameixeira. *Rev. bras. Frutic.* (Cruz das Almas) 13, 285-291.

Hedrick U.P. (1917) The peaches of New York. *Rpt. N.Y. Agric. Expt. Sta. 1916.*

Herter, F.G. and Raseira, M.C.B. (1990) Cerejeira e sua perspectiva de cultivo no sul do Brasil. *Hort. Sul.* 1, 40-43.

Hesse, C.O. and Griggs, W.H. (1950) The effect of gland type on the wettability and water retention of peach leaves. *Proc. Amer. Soc. Hort. Sci.* 56, 173-180.

Hopkins, D.L. (1989) *Xylella fastidiosa*: Xylem-limited bacterial pathogen of plants. *Annu. Rev. Phytopathol.* 27, 271-290.

Horne, W.T., Weldon, G.P. and Babcock, E.B. (1926) Resistance of peach hybrids to an obscure disease in Southern California. *J. Hered.* 17, 98-104.

Hough, L.F. (1985) Perspective for peach breeding for the cultivars for 2000 A.D. *Acta Hort.* 173, 11-20.

Hough, L.F. (1989) Apricot breeding in the United States. *Unpublished manuscript prepared for Dr. J. Couranjou, INRA, France by Fruit Germ Plasm International.* Personal communication.

Howard, W.L. (1945) Luther Burbank's plant contributions. *Calif. Agr. Expt. Sta. Bul. 691.*

Hume, H.H. (1902) Peen-to peach group. *Fla. Exp. Sta. Bul. 62.*

Hurter, N. (1971) Origin and characteristics of plum and prune cultivars. *FFTRI Inform. Bull. No. 31.* Stellenbosch, South Africa.

Ibanez, M.A., di Renzo, M.A. and Poverene, M.M. (1993) Isozyme diversity among and within peach groups: freestone, clingstone and nectarines. *Scientia Hort.* 53, 281-288.

Iezzoni, A., Schmidt, H. and Albertini, A. (1990) Cherries (*Prunus*). *Acta Hort.* 290, 111-173.

Issell, L. and Jerie, P.H. (1992a) 'Golden Blaze' peach. *HortScience* 27, 932-933.

Issell, L. and Jerie, P.H. (1992b) 'Golden Charm' peach. *HortScience* 27, 933-934.

Kable, P.F., Ellison, P.J. and Bambach, R.M. (1986) Physiologic specialization of *Tranzschelia discolor* in Australia. *Plant Dis.* 70, 202-204.

Keil, H.L. and Fogle, H.W. (1974) Orchard susceptibility of some apricots, peach and plum cultivars and selections to *Xanthomonas pruni. Fruit Var. J.* 28, 16-19.

Kfoury, L. and Massonié, G. (1995) Caractéristiques de la résistance du cultivar de pêcher Rubira à *Myzus persicae* Sulzer. *Agronomie* 15, 277-284.

King, L., Norton, R.A. and Moulton, G.A. (1988) Relative susceptibility of certain peach cultivars to summer infection of leaf curl. *Fruit Var. J.* 42, 28-29.

Knight, R.L. (1969) Abstract bibliography of fruit breeding and genetics to 1965. *Prunus Commonwealth Agr. Bur. Tech. Comm. 31.*

Ko, Y. and Sun, S.-K. (1992) Peach gummosis disease caused by *Botryosphaeria dothidea* in Taiwan. *Plant Pathol. Bull.* 1, 70-78.

Latham, A.J. and Norton, J.D. (1980) Incidence of plum leaf scald in Alabama. *Alabama Agr. Exp. Sta. Bull. 525.*

Layne, R.E.C. (1966) Susceptibility of apricots to bacterial fruit infection of foliage and fruit. *Plant Disease Reporter* 50, 112-115.

Layne, R.E.C. and Sherman, W.B. (1986) Interspecific hybridization of Prunus. *HortScience* 21, 48-51.

Lesley, J.W. (1939a) A genetic study of saucer fruit shape and other characters in the peach. *Proc. Amer. Soc. Hort. Sci.* 37, 218-222.

Lesley, J.W. (1939b) Five new peach varieties especially adapted to mild winters. *Univ. Calif. Agric. Exp. Stn. Bul. 632.*

Lesley, J.W. and Winslow, M.M. (1953) Peach varieties for a warm winter climate. *Calif. Agric. Exp. Sta. Circ. 406.*

Li, H-Y., Cao, R-B. and Mu, Y-T. (1995) In vitro inhibition of *Botryosphaeria dothidea* and *Lasiodiplodia theobromae*, and chemical control of gummosis disease of Japanese apricot and peach trees in Zhejiang Province, China. *Crop Protection* 14, 187-191.

Lownsbery, B.F., English, H., Noel, G.R. and Schick, F.J. (1977) Influence of Nemaguard and Lovell rootstocks and *Macroposthonia xenoplax* on bacterial canker of peach. *J. Nematol.* 9, 221-224.

Malone, M.T. and Snelling, C.M. (1997) Fruit characteristics of Asian peaches grown under New Zealand conditions. *Fruit Var. J.* 51, 83-87.

Massonié, G., Maison, P., Monet, R., Grasselly, C. (1982) Résistance au puceron vert du pêcher, *Myzus persicae* Sulzer (Homoptera Aphididae) chez *Prunus persicae* (L.) Batsch et d´autres espèces de *Prunus*. *Agronomie* 2, 63-70.

Martins, O.M. (1996) Evaluation of virulence of strains of Xanthomonas campestris pv. pruni on peach and plum cultivars. *Fruit Var. J.* 50, 221-225.

Martins, O.M. and Raseira, M.C.B. (1996) Sources of bacterial spot resistance in plum cultivars. *Fruit Var. J.* 50, 156-159.

Meagher, R.L. and Werner, D. (1992) Variability in resistance of peach plant introductions to *Tetranychus urticae* (Acari: Tetranychidoe) infestation. *Fruit Var. J.* 46, 108-111.

Medeiros, A.R.M. (1979) Efeito de temperatura controlada na germinacâo dos grãos de pólen e crescimento do tubo polinico em pessegueiro (*Prunus persica* (L.) Batsch) *Congr. Bras. Fruticultura* (Pelotas) 5, 1-8.

Mehlenbacher, S.A., Cociu, V. and Hough, L.F. (1990) Apricots. *Acta Hortculturae* 290, 65-107.

Messeguer, R., Arus, P., and Carrera, M. (1987) Identification of peach cultivars with pollen isozymes. *Scientia Hortic.* 31, 107-117.

Micke, W.C., Doyle, J.F. and Yeager, J.T. (1983) Doubling potential of sweet cherry cultivars. *Calif. Agr.* 1983, 24-25.

Millikan, D.F. and Hibbard, A.D. (1964) Bacterial spot on apricots. *Plant Disease Reporter* 48, 900-901.

Monet, R. (1979) Transmission genetique due caractere "fruit doux" chez le pecher incidence sur la selection pour la qualite. *Eucarpia Symp. Fruit Sec. Tree Fruit Breeding*. p. 273-276.

Monet, R. (1985) Heredity of the resistance to leaf curl (*Taphrina deformans*) and green aphid (*Myzus persicae*) in the peach. *Acta Hort.* 173, 21-23.

Monet, R. and Massonie, G. (1994) Genetic determinism of resistance to the green aphid (*Myzus persicae*) in the peach. Further results. *Agronomie* 14, 177-182.

Mowrey, B.D., Werner, D.J. and Byrne, D.H. (1990) Inheritance of isocitrate dehydrogenase, malate dehydrogenase, and shikimate dehydrogenase in peach and peach x almond hybrids. *J. Amer. Soc. Hort. Sci.* 115, 312-319.

Nakasu, B.H., Bassols, M. C. and Feliciano, A.J. (1981) Temperate fruit breeding in Brazil. *Fruit Var. J.* 35, 114-122.

Nakasu, B.H. and de Castro, L.A. (1990) Indicaçao de cultivares de ameixeiras para o sul do Brasil.

Horti Sul 1, 24-28.

Norton, J. (1972) Crimson and Purple, two disease resistant plums. *Highlights of Agric. Res.* 19, 2.

Norton, J. (1986) Characteristics of japanese type plums evaluated in Alabama. *Unpublished manuscript presented at the Southern Region ASHS meetings.* Personal communication.

Norton, J. (1992) Crimson and Purple, two disease resistant plums. *Highlights of Agric. Res. (Alabama)* 19, 1

Ojima, M., Campo-Dall'Orto, F.A., Barbosa, W. and Rigitano, O. (1992) Cultivares de ameixeira para o estado de São Paulo. Inst. Agr. Campinas (SP, Brazil) *Bol. Techn. No. 144.*

Okie, W.R. (1993) "Peach and nectarine varieties" - A hypertext index. *HortScience* 28, 1186-1187.

Okie, W.R. (1994) Notice to fruit growers and nurserymen: the naming and release of Black Ruby, a japanese-type plum. *USDA Notice. May 2, 1994.*

Okie, W.R. (1995) Il miglioramento genetico del susino (Plum breeding and genetics) In: *International Symposium on State of the Art and Perspectives of World Genetic Improvement of Fruit Tree Species.* (E. Bellini ed) Faenza, Italy. p. 29-66.

Okie, W.R., Beckman, T.G., Nyczepir, A.P., Reighard, G.L., Newall, Jr. W.C. and Zehr E.I. (1994b) BY520-9, a peach rootstock for the Southeastern United States that increases scion longevity. *HortScience* 29, 705-706.

Okie, W.R., Reighard, G.L., Beckman, T.G., Nyczepir, A.P., Reilly, C.C., Zehr, E.I., Newall, Jr., W.C. and Cain, D.W. (1994a) Field-screening *Prunus* for longevity in the Southeastern United States. *HortScience* 29, 673-677.

Okie, W.R. and Reilly, C.C. (1983) Reaction of peach and nectarine cultivars and selections to infection by *Botryosphaeria dothidea. J. Amer. Soc. Hort. Sci.* 108, 176-179.

Okie, W.R., Thompson, J.M., Reilly, C.C., Meredith, F.I., Robertson J. A. and Lyon, B.G. (1992) 'Segundo', 'Byrongold' and 'Rubysweet' plums and BY69-1637P plumcot - fruits for the southeastern United States. *Fruit Var. J.* 46, 102-107.

Ou, S-K. (1994) Physiological studies on flowering and fruiting of mume trees. *J. Chinese Soc. Hort. Sci.* 40, 60-70 (in Chinese).

Palmiter, D.H. and Hickey, K.D. (1970) Relative resistance of 26 peach cultivars to bacterial spot and *Valsa* canker. *Plant Disease Reporter* 54, 395-399.

Pascal, T., Levigneron, A., Kervella, J. and Nguyen-The, C. (1994) Evaluation of two screening methods for resistance of apricot, plum and peach to *Monilinia laxa. Euphytica* 77, 19-23.

Patten, K., Nimr, G. and Neuendorff, E. (1989) Fruit doubling of peaches as affected by water stress. *Acta Hort.* 254, 319-321.

Paunovic, S.A. and Ogasanovic, D. (1975) Leaf serration as an indicator of mildew susceptibility in peach hybrids. *Hort. Abstracts* 46, 6507.

Paunovic, S.A., Ogasanovic, D., and Plazinid, R. (1976) The effect of breeding parents on the susceptibility of F1 progeny towards powdery mildew [*Sphaerotheca pannosa* (Wallr:Fr.) Lev.]. *J. Yugoslav Pomology* 37/38, 331-340.

Perez, S. (1989) Characterization of Mexican peach populations from tropical and subtropical regions. *Acta Hort.* 254, 139-144.

Perez, S. (1992) Associations among morphological and phenological characters representing apricot germplasm in Central Mexico. *J. Amer. Soc. Hort. Sci.* 117, 486-490.

Perez, S. (1995) Dinamica en la adopción de variedades de duraznero, ciruelo y albaricoque en México, In: *Reunion Nacional e´ Internacional Sobre Produccion de Durazno, Chabacano y Ciruelo.* May 1995. *Universidad Autónoma de Queretaro. Mexico.* p. 45-59.

Perez, S. and Gonzalez, S. (1992) Herencia de la resistencia a cenicilla en duraznero, In: *XIV Congreso Nacional de Fitogenética,* Tuxtla Gutiérrez, Chis. Mexico. Soc. Mex. de Fitogenética, Escuela de Ciencias

Agronómicas, Univ. Autónoma de Chiapas. p. 156.

Perez, S., Montes, S., and Mejía, C. (1993) Analysis of peach germplasm in Mexico. *J. Amer. Soc. Hort. Sci.* 118, 519-524.

Pisani, P.L. and Roselli, G. (1983) Interspecific hybridization of *Prunus persica* x *P. davidiana* to obtain new peach rootstocks. *Genet. Agr.* 37, 197-198.

du Plessis, H.J. (1988) Differential virulence of *Xanthomonas campestris* pv. *pruni* to peach, plum and apricot cultivars. *Phytopathology* 78, 1312-1315.

Popenoe, W. (1920) The capulin, In: *Manual of Tropical and Subtropical Fruits.* Hafner Press, London. p. 268-269.

Pscheidt, J.W. (1995) Leaf curl. In: *Compendium of Stone Fruit Diseases.* (Ogawa, J.M., Zehr, E.I., Bird, G.W., Ritchie, D.F., Uriu, K. and Uyemoto J.K. eds) APS Press, St. Paul, Minn. p. 22.

Pukanova, Z.G., Gatina, E. Sh., and Sokolab, S.A. (1980) Inheritance of resistance to powdery mildew in the F_2 of peach (Russian) *Referativnyi Zhurnal* 2.65.137, 108-115.

Pusey, P.L. (1989) Availability and dispersal of ascospores and conidia of *Botryosphaeria* in peach orchards. *Phytopathology* 79, 635-639.

Pusey, P.L., Kitajima, H., and Wu, Y. (1995) Fungal gummosis. In: *Compendium of Stone Fruit Diseases.* (Ogawa, J.M., Zehr, E.I., Bird, G.W., Ritchie, D.F., Uriu, K. and Uyemoto J.K. eds) APS Press, St. Paul, Minn. p. 33-34.

Puterka, G.J., Scorza, R. and Brown, M.W. (1993) Reduced incidence of lesser peachtree borer and *Leucostoma* canker in peach-almond hybrids. *J. Amer. Soc. Hort. Soc.* 118, 864-867.

Raju, B.C., Wells, J.M., Nyland, G., Brlansky R.H. and Lowe S.K. (1982) Plum leaf scald: Isolation, culture and pathogenicity of the causal agent. *Phytopathology* 72, 1460-1466.

Ramming, D.W. and Cociu, V. (1990) Plums. *Acta Horticulturae* 290, 235-287.

Randhawa, P.S. and Civerolo, E.L. (1985) A detached-leaf bioassay for *Xanthomonas campstris* pv. *pruni. Phytopathology* 75, 1060-1063.

Rasiera, M.C.B. (1988) Peach cultivars, breeding and culture in Brazil, In: *The Peach*: (Childers, N.F. and Sherman W.B. eds) Hort. Publications. Gainesville, FL. p. 146-154.

Rasiera, M.C.B., Nakasu, B.H., Santos, A.M., Fortes, J.F., Martins, O.M., Rasiera, A. and Bernardi, J. (1992) The CNPFT/EMBRAPA fruit breeding program in Brazil. *HortScience* 27, 1154-1157.

Reilly, C.C. and Okie, W.R. (1982) Distribution in the southeastern United States of peach tree fungal gummosis caused by *Botryosphaeria dothidea. Plant Dis.* 66, 158-161.

Richards, G.D., Porter, G.W., Rodriguez, J. and Sherman, W.B. (1994) Incidence of blind nodes in low-chill peach and nectarine germplasm. *Fruit Var. J.* 48, 199-202.

Richards, S., Issell, L., Jerie, P., Murphy, K. and Edge, N. (1995) Variety improvement program for yellow fleshed canning peaches. *ISIA Report 1993-95.* Tatura, Australia.

Ritchie, D.F. (1995a) Bacterial spot, In: *Compendium of Stone Fruit Diseases.* (Ogawa, J.M., Zehr, E.I., Bird, G.W., Ritchie, D.F., Uriu, K. and Uyemoto J.K. eds) APS Press, p. 50-52.

Ritchie, D.F. (1995b) Plum leaf scald, In: *Compendium of Stone Fruit Diseases.* (Ogawa, J.M., Zehr, E.I., Bird, G.W., Ritchie, D.F., Uriu, K. and Uyemoto J.K. eds) APS Press, p. 54-55.

Ritchie, D.F. and Werner, D.J. (1981) Susceptibility and inheritance of susceptibility to peach leaf curl in peach and nectarine cultivars. *Plant Dis.* 65, 731-734.

Robertson, J.A., Horvat, R.J., Lyon, B.G., Meredith, F.I., Senter, S.D. and Okie, W.R. (1990) Comparison of quality characteristics of selected yellow- and white-flesh peach cultivars. *J. Food Sci.* 55, 1308-1311.

Robertson, J.A., Meredith, F.I., and Scorza, R. (1988) Characteristics of fruit from high- and low-quality peach cultivars. *HortScience* 23, 1032-1034.

Robertson, J.A., Meredith, F.I., Lyon, B.G., Chapman, G.W. and Sherman, W.B. (1992) Ripening and cold storage changes in the quality characteristics of nonmelting clingstone peaches (FLA9-20C) *J. Food Sci.* 57, 462-465.

Rodriguez, J. (1988) The peach industry and breeding program, Central Mexico, In: *The Peach* . (Childers, N.F. and Sherman, W.B. eds) Hortic. Publ., Gainesville, FL. p. 140-141.

Rodriguez, J., Jasso, J. and Sherman, W.B. (1992) Evaluation of peach and nectarine germplasm for powdery mildew resistance [*Sphaerotheca pannosa* (Wallr.) Lev.]. *Acta Hort.* 315, 163-169.

Rodriguez, J. and Sherman, W.B. (1995a) Nuevas variedades de duraznos de pulpa firme para zonas subtropicales y de invierno moderado, In: *Reunión Nacional e Internacional Sobre Producción de Durazno, Chabacano y Ciruelo.* Univ. Autónoma de Queretaro, Mexico. p. 72-75.

Rodriguez, J. and Sherman, W.B. (1995b) 'Corazon Rojo' y 'Cirena I', dos nuevas variedades de ciruela japones (*Prunus* spp) de bajo requerimiento de frio (resumen) *XL 'Reunión Anual. Interamerican Soc. for Trop. Hort.(1994*). Campeche, Mexico.

Rodriguez, J., Sherman, W.B. and Lyrene, P.M. (1986) High-density nursery system for breeding peach and nectarine: A 10-year analysis. *J. Amer. Soc. Hort. Sci.* 111, 311-315.

Roselli, G. and Bellini, E. (1976) Indagine sulla suscettibilità all'oidio *Sphaerotheca pannosa* (Wallr.) Lev. in una collezione di cultivar di pesco. *Riv. Ortoflorofrutticoltura Italiana* 60, 40-55.

Rouse, R.E. (1989) Peaches and nectarines for Texas' subtropical lower Rio Grande Valley. *Fruit Var. J.* 43, 52-57.

Rouse, R.E. and Byrne, D.H. (1990) 'Valle Grande': An early maturing, low-chilling peach for subtropical climates. *HortScience* 25, 714-715.

Rouse, R.E. and Sherman, W.B. (1987) 'Tropic Sweet': A freestone peach for subtropical climates. *HortScience* 22, 169-170.

Rouse, R.E., Sherman, W.B. and Sharpe, R.H. (1985) 'Florda Grande': A peach for subtropical climates. *HortScience* 20, 304-305.

Russell, D. (1993) Plums. *Queensland Department of Primary Industries.*

Scorza, R. (1992) Evaluation of foreign peach and nectarine introductions in the U.S. for resistance to leaf curl [*Taphrina deformans* (Berks) Tul.]. *Fruit Var. J.* 46, 141-145.

Scorza, R., Mehlenbacher, S.A. and Lightner, G.W. (1985) Inbreeding and coancestry of freestone peach cultivars of the eastern United States and implications for peach germplasm improvement. *J. Amer. Soc. Hort. Sci.* 110, 547-552.

Scorza, R. and Okie, W.R. (1990) Peaches. *Acta Horticulturae* 290, 177-231.

Scorza, R. and Sherman, W.B. (1996) Peaches, In: *Fruit Breeding, Tree and Tropical Fruits*, (Janick, J. and Moore, J.N. eds) Wiley & Son, Inc. Vol. 1 p. 325-440.

Scorza, R., Sherman, W.B. and Lightner, G.W. (1988) Inbreeding and coancestry of low chill short fruit development period freestone peaches and nectarines produced by the University of Florida breeding program. *Fruit Var. J.* 42, 79-85.

Sharma, I.M. and Badiyala, S.D. (1994) Susceptibility of peaches to *Taphrina deformans* in relation to blooming, environmental factors and genetic inheritance. *Indian Phytopathology* 47, 65-71.

Sharma, J.N. and Josan, J.S. (1993) Varietal performance of plum under arid-irrigated region of Punjab. *Indian J. Hort.* 50, 301-305.

Sharma, R.L. and Kumar, K. (1994) Temperate fruit crop improvement in India, In: *Progress in Temperate Fruit Breeding.* (Schmidt, H. and Kellerhals, M. eds) Kluwer Academic Publishers, Netherlands. p. 149-156.

Sharpe, R.H. (1961) Developing new peach varieties for Florida. *Proc. Fla. State Hort. Soc.* 74, 348-352.

Sharpe, R.H. (1969) Sub-tropical peaches and nectarines. *Proc. Fla. State Hort. Soc.* 82, 302-306.

Sharpe, R.H. and J.B. Aitken. 1971. Progress of the nectarine. *Proc. Fla. State Hort. Soc.* 84, 338-345.

Sherman, W.B. 1977. Attempts at cherry breeding in Florida. *Fruit Var. J.* 31, 60-61.

Sherman, W.B. (1987) Where can we go from here in low-chill stonefruit breeding and production, In: *Proc. 1st National Low-Chill Stonefruit Conference.* (Skinner, I. ed) Exotic Fruit Growers' Assn. NSW Australia. p. 169-175.

Sherman, W.B. and Lyrene, P.M. 1981. Bacterial spot susceptibility in low chilling peaches. *Fruit Var. J.* 35, 74-77.

Sherman, W.B. and Lyrene, P.M. (1984) Biannual peaches in the tropics. *Fruit Var. J.* 38, 37-39.

Sherman, W.B. and Lyrene, P.M. (1985) Progress in low-chill plum breeding. *Proc. Fla. State Hort. Soc.* 98, 164-165.

Sherman, W.B. and Lyrene, P.M. (1997) 'UFGold' Peach. *Fruit Var. J.* 51, 76-77.

Sherman, W.B. and Rodriguez, J. (1987) Breeding of low-chill peach and nectarine for mild winters. *HortScience* 22, 1233-1236.

Sherman, W.B., Rodriguez, J. and Miller, E.P. (1984) Progress in low-chill peaches and nectarines from Florida. *Proc. Fla. State Hort. Soc.* 97, 320-322.

Sherman, W.B., Rodriguez, J. and Topp, B.L. (1992) Peaches and nectarines developed but not released by the University of Florida. *Fruit Var. J.* 46, 124-127.

Sherman, W.B. and Rouse, R.E. (1989) Five new low-chill peach cultivars. *Fruit Var. J.* 43, 120.

Sherman, W.B., Sharpe, R.H. and Janick, J. (1973) The fruiting nursery: ultra high density for evaluation of blueberry and peach seedlings. *HortScience* 8, 170-172.

Sherman, W.B., Topp, B.L. and Lyrene, P.M. (1990) Non-melting flesh for fresh market peaches. *Proc. Fla. State Hort. Soc.* 103, 293-294.

Sherman, W.B., Topp, B.L.and Lyrene, P.M. (1992) Japanese-type plums for subtropical climates. *Acta Hort.* 317, 149-153.

Sherman, W.B., Yonce, C.E., Okie, W.R. and Beckman, T.G. (1989) Paradoxes surrounding our understanding of plum leaf scald. *Fruit Var. J.* 43, 147-151.

Shimada, T., Haji, T., Yamaguchi, M., Takeda, T., Nomura, K. and Yoshida, M. (1994) Classification of mume (*Prunus mume* Sieb. et Zucc.) by RAPD assay. *J. Jap. Soc. Hort. Sci.* 63, 543-551.

Simeone, A.M. (1985) Study on peach and nectarine cultivars susceptibility to the main fungus and bacteria. *Acta Hort.* 173, 541-551.

Southwick, S.M., Shackel, K.A., Yeager, J.T., Asai, W.K. and Katacich, M. (1991) Over-tree sprinkling reduces abnormal shapes in 'Bing' sweet cherries. *Calif. Agr.* 45, 24-26.

Souza, V.A. de, Byrne, D.H. and Taylor, J.F. (1998) Heritability, genetic and phenotypic correlations, and predicted selection response of quantitative traits in peach: I. An analysis of several fruit traits. *J. Amer. Soc. Hort. Sci.* 123, 604-611.

Strong, K.J. and Byrne, D.H. (1995) Performance of low-chill japanese plums in subtropical Texas. *Subtropical Plant Sci.* 47, 30-33.

Thompson, J.M. 1981. The plum industry in the southeastern United States. *Fruit Var. J.* 35, 53-55.

Thompson, J.M. and Prince, V.E. (1981a) 'Explorer' plum. *HortScience* 16, 797-8.

Thompson, J.M. and Prince, V.E. (1982a) 'Robusto' plum. *HortScience* 17, 91.

Thompson, J.M. and Prince, V.E. (1982b) Robusto - A new vigorous and precocious plum for the southeastern United States. *Fruit Var. J.* 36, 112-114.

Thompson, M.M. (1993) Exploration and exploitation of new fruit and nut germplasm. In: *New Crops.* (Janick, J. and Simon, J. F. eds) John Wiley. NY. p. 155-160.

Topp, B.L. (1993) The nature of the nineties. *Pome and Stonefruit Seminar. Stonefruit breeding in Australia. (Mimeograph).*

Topp, B.L., Heaton, J.B. and Russell, D.M. (1989) Introduction and evaluation of stonefruit varieties for the Granite Belt of Queensland. *Acta Hort.* 240, 39-42.

Topp, B.L. and Russell, D.M. (1989) Breeding early ripening Japanese plums. *Acta Hort.* 240, 27-30.

Topp, B.L. and Sherman, W.B. (1989a) The relationship between temperature and bloom-to-ripening period in low chill peach. *Fruit Var. J.* 43, 155-158.

Topp, B.L. and Sherman, W.B. (1989b) Location influences on fruit traits of low-chill peaches in Australia. *Proc. Fla. State Hort. Sci.* 102, 195-199.

Topp, B.L. and Sherman, W.B. (1990a) Sources of bacterial spot resistance in Japanese-type plum cultivars. *Fruit Var. J.* 44, 32-35.

Topp, B.L. and Sherman, W.B. (1990b) Potential for low-chill Japanese plums in Florida. *Proc. Fla. State Hort. Soc.* 103, 294-298.

Topp, B.L., Sherman, W.B., Huber, D.A. and Linda, S.B. (1993a) Combining abilities of five Japanese plum cultivars for resistance to *Xanthomonas* stem canker. *HortScience* 28, 727-729.

Topp, B.L., Sherman, W.B. and Stall, R.E. (1991) Comparison of rating methods for bacterial spot resistance in Japanese-type plum. *Fruit Var. J.* 45, 70-74.

Topp, B.L., Sherman, W.B., Stall, R.E., Minsavage, G.V. and Wilcox, C.J. (1993b) Comparison of greenhouse methods for assessing resistance to bacterial leaf spot in plum. *J. Amer. Soc. Hort. Sci.* 118, 667-671.

Tucker, L.R. (1934) Notes on sweet cherry doubling. *Proc. Amer. Soc. Hort. Sci.* 32, 300-302.

Tuz, A.S. (1958) *Prunus ferganensis* (Russian) *Sad I Ogorad* 10, 57-58. (Plant Breeding Abstracts 29, 316).

Uppal, D.K., Chopora, S.K. and Chanana, Y.R. (1993) Improvement of temperate fruits for subtropical climate, In: *Advances in Horticulture Fruit Crops* (Chadha, K.L. and Pareek, O.P. eds) Malhotra Publishing House, New Delhi, India Vol. I. Part 1. p. 445-461.

Van Der Heyden, C.R., Holford, P. and Richards, G.D. (1997) A new source of peach germplasm containing semi-freestone nonmelting flesh types. *HortScience* 32, 288-289.

Van Rooyen, P.C. (1988) Peach situation in South Africa breeding objectives, In: *The Peach.* (Childers, N. F. and Sherman, W.B. eds) Horticultural Press. Homestead, FL. p. 220-226.

Ved, P., Joon, M.S., Dahiya, S.S. and Parkash, V. (1994) Physico-chemical characteristics of some promising Japanese plum cultivars grown in the plains of Haryana. *Intern. J. Trop. Agr.* 12, 123-125.

Wang, Y-L. (1985) Peach growing and germplasm in China. *Acta Hort.* 173, 51-55.

Wang, B.P., Qian, Y.C., Zhu, Y.F. and Chen H. (1989) The effect of meteorological factors on the pollination, fertilization, and fruit setting of the peach (In Chinese, English summary) *Acta Horticulturae Sinica* 16, 11-16.

Warburton, M.L. and Bliss, F.A. (1996) Genetic diversity in peach (*Prunus persica* L. Batsch) revealed by RAPD markers and compared to inbreeding coefficients. *J. Amer. Soc. Hort. Sci.* 12, 1012-1019.

Weaver, D.J. (1974) A gummosis disease of peach trees caused by *Botryosphaeria dothidea. Phytopathology* 64, 1429-1432.

Weaver, D.J., Doud, S.L. and Wehunt, E.J. 1979. Evaluation of peach seedling rootstocks for susceptibility to bacterial canker caused by *Pseudomonas syringae. Plant Disease Reporter* 63, 364-367.

Weinbaum, S.A., Parfitt, D.E. and Polito, V.S. (1984) Differential cold sensitivity of pollen-grain germination in two *Prunus* species. *Euphytica* 33, 419-426.

Weinberger, J.H. 1967. Studies in flower bud drop in peaches. *Proc. Amer. Soc. Hort. Sci.* 91, 78-83.

Weinberger, J.H. (1975) Plums, In: *Advances in fruit breeding* (Janick, J. and Moore, J.N. eds) Purdue Univ. Press. West Lafayette, Indiana. p. 336-347.

Weinhold, A.R. 1961. Temperature and moisture requirements for germination of conidia of *Sphaerotheca pannosa* from peach. *Phytopathology* 51, 699-703.

Weldon, G.P. (1934) Fifteen year's study of delayed foliation of deciduous fruit trees in Southern California. *Cal. State Dept. Agr. Mo. Bul.* 23, 160-181.

Wells, J.M. (1995) Phony peach, In: *Compendium of Stone Fruit Diseases.* (Ogawa, J.M., Zehr, E.I., Bird, G.W., Ritchie, D.F., Uriu, K. and Uyemoto J.K. eds) APS Press, p. 53-54.

Wells, J.M., Raju, B.C., Thompson, J.M. and Lowe, S.K. (1981) Etiology of phony peach and plum leaf scald diseases. *Phytopathology* 71, 1156-1161.

Wen, I-C., Koch, K.E. and Sherman, W.B. (1995a) Comparing fruit and tree characteristics of two peaches and their nectarine mutants. *J. Amer. Soc. Hort. Sci.* 120, 101-106.

Wen, I-C., Sherman, W.B. and Koch, K.E. (1995b) Heritable pleiotropic effects of the nectarine mutant from peach. *J. Amer. Soc. Hort. Sci.* 120, 721-725.

Werner, D.J. (1992) Catalase polymorphism and inheritance in peach. *HortScience* 27, 41-43.

Werner, D.J., Mowrey, B.D. and Chaparro, J.X. (1988) Variability in flower bud number among peach and nectarine cultivars. *HortScience* 23, 578-580.

Werner, D.J., Mowrey, B.D. and Young, E. (1988) Chilling requirement and post-rest heat accumulation as related to difference in time of bloom between peach and western sand cherry. *J. Amer. Soc. Hort. Sci.* 113, 775-778.

Werner, D.J., Ritchie, D.F., Cain, D.W. and Zehr, E.I. (1986) Susceptibility of peaches and nectarines, plant introductions and other *Prunus* species to bacterial spot. *HortScience* 21, 127-130.

Wills, R.B.H., Scriven, F.M. and Greenfield, H. (1983) Nutrient compositions of stone fruit (*Prunus* spp.) cultivars: apricot, cherry, nectarine, peach and plum. *J. Sci. Food Agric.* 34, 1383-1389.

Yoshida, M. (1976) Genetic studies on the fruit quality of peach varieties. *Bull. Fruit Trees Res. Stn.* 3, 1-16.

Yoshida, M. (1987) Roots of fruit trees (4) Mume. *Kajitsa Nippon* (*Fruit Japan*) 42, 54-59. (In Japanese).

Yoshida, M. (1994) Mume, plum, and cherry. In: *Horticulture in Japan.* (Konishi K. ed) 24th Intern. Hort. Cong., Kyoto, Japan. 37-40.

Yoshida, M., Kyotani, H. and Yasuno, M. (1975) Studies on interspecific crossing in *Prunus* spp. I. Cross compatibility (In Japanese, English Summary) *Japan. J. Breed.* 25, 17-23.

Yu, D. (1979) Stone Fruit, In: *Taxonomy of Chinese fruit trees* (Chinese) Agric. Publ. House. Beijing. p. 25-86.

Zegbe, J.A. and Rumayor, A.F. (1995) Selecciones de durazno criollo con floración tardía. In: *Reuníon Nat. e Internac. Sobre Prod. de Durazno, Chabacano y Ciruelo.* Univ. Autónoma de Queretaro, Mexico., p. 104-112.

Zeven, A.C. and de Wet, J.M.J. (1982) Dictionary of cultivated plants and their regions of diversity. *Centre Agr. Publ. Docum., Wageningen, The Netherlands*, pp. 259.

Zhang, J., He, Y., Li, T., Pen, X., Guo, Z. and Li, F. (1990) Natural resources and geographic distribution of *Prunus* germplasms in China tropic and subtropic zones, In: *Proc. Int. Symp. Hort. Germplasm, Cultivated and Wild.* China Soc. Hort. Sci., Beijing. p. 504-508.

9 STONE FRUIT SPECIES UNDER WARM SUBTROPICAL AND TROPICAL CLIMATES

ALAN P. GEORGE [1] *and* AMNON EREZ [2]

[1] *Maroochy Horticultural Research Station, P O Box 5083, SCMC, Nambour, Qld., Australia*
[2] *Institute of Horticulture, The Volcani Center, Bet Dagan 50250, Israel*

1 Introduction

In warm subtropical regions of the world, stonefruit production has rapidly increased during the past 10 years due to the ready availability of high-quality low-chill stonefruit germplasm (Sharpe *et al.*, 1990; Sherman, *et al.*, 1978, 1984; Sherman, 1987, [See Chapter 8] (Table 1). Within the tropics and subtropics, suitable agroclimatological conditions exist within many countries, particularly in elevated regions, to grow temperate fruits. Based on an increasing awareness of the opportunities to grow these fruits, new industries are being developed in regions where none previously existed. In countries which already grow temperate fruits in higher-chill regions, early-season production from low-chill stonefruit in warmer subtropical regions constitutes a valuable extension of the marketing period.

Fruit quality and productivity of stonefruit grown under warm subtropical conditions will be maximised by selecting high-quality varieties, matching these correctly to the environment under which they are to be grown, and by optimising cultural and management inputs. While some management requirements are similar to those used under high-chill conditions others are not. However the main management objectives for both groups of cultivars remain the same; these are:

1. maximising yield per hectare;
2. obtaining high fruit quality;
3. reducing production costs associated with fertiliser and pesticide use;
4. increasing labour use efficiency.

In addition, under warm growing climates a 5th requirement is the manipulation of rest, flowering and consequently the fruit maturity period.

Considerable progress has already been made in developing new techniques for growing temperate fruits under tropical and subtropical conditions. A description of the most appropriate managements techniques to use with successful culture of stonefruit under subtropical and tropical conditions will be presented in this chapter.

Amnon Erez (ed.), Temperate Fruit Crops in Warm Climates, 231-265.
© 2000 *Kluwer Academic Publishers. Printed in the Netherlands.*

Table 1. Low- and medium-chill stonefruit production in selected warm, subtropical regions and countries of the world (1996)

Region	*Country*	*Area (000 ha)*	*Production (000 tonnes)*
Asia	Southern China	47	400
	Pakistan	13.2	155.1
	Nepal	6.5	41.1
	Thailand	0.24	0.05
North America	Florida	0.6	na
	California	3.4	50.4
	Texas	0.4	3.2
	Mexico	45	na
South America	Brazil	2.5	20
	Chile	1.8	na
Middle East	Egypt	12.3	300
	Libya	3.5	6.2
	Israel	2.6	43
	Yemen	1.2	8.1
Oceania	Australia	2.1	20.2
Africa	South Africa	1.8	29.2
Europe	Spain	3.2	48.6

na = no figures available

2 Cropping systems and growth cycles

2.1 Description of cropping systems

Two production systems can be used for growing stonefruit in sub-tropical and tropical regions. These are the dormancy-induced, annual cropping system and the dormancy-avoided, continuous cropping system (Notodimedjo *et al.*, 1981a, b; Sherman and Lyrene, 1984; Edwards, 1987a). With the dormancy-induced, annual cropping system, trees enter a definite dormancy period where there is no active growth. Chilling must be sufficient to release buds from dormancy. Trees produce only one flowering wave and crop annually. With the dormancy-avoided system, continuously cropping trees do not enter endo-dormancy. Specialised practices of defoliation, water stress, dormancy breaking chemicals and branch bending are needed to induce flowering and growth before buds enter endo-dormancy (Notodimedjo *et al.*, 1981a, b; Erez and Lavi, 1984; Edwards, 1987b; Edwards and Notodimedjo, 1987; Edwards, 1990). This system has been used successfully with grapes and apples and to a lesser extent with stonefruit in a number of countries including Indonesia (Kusumo, 1972; Saure, 1973; Janick, 1974; Verheij, 1985; Verheij, 1990), Kenya (Erez and Lavi, 1984), Philippines (Edward, 1987a, 1987b; Llanes *et al.*, 1987), Peru (Duarte and Franciosi, 1974), India (Javaraya, 1943) and Mexico (Diaz and Alvarez, 1987).

2.2.1 *Phenological cycles (annual cropping system)*

Management techniques which have been developed for high-chill cultivars are not always applicable to low-chill cultivars as the rate of fruit development and phenological cycle are different for both groups of cultivars (George *et al.*, 1992a; Allan *et al.*, 1993a) (Fig. 1). Although the shorter fruit development period (FDP) of low-chill cultivars is a major disadvantage in terms of maximising fruit size, the comparatively longer period of vegetative growth after harvest of low-chill cultivars can be exploited using management techniques such as post-harvest topping. Some understanding of the comparative phenological cycles of both low and high-chill cultivars is necessary for the selection and correct timing of specific management practises, for modifying practises used with high-chill cultivars and for manipulating flowering and fruit maturity.

2.2.1.1 *Vegetative-reproductive competition*

The major difference between low-chill, mostly early maturing and high-chill, late maturing cultivars is in their growth cycle. Most low-chill cultivars make two vegetative growth flushes per year compared with only one for late maturing cultivars (George and Nissen, 1987; George *et al.*, 1992a; Allan *et al.*, 1993a). The first growth flush, which is usually the strongest, occurs during the spring and the second after the crop is removed. Furthermore, maximum vegetative growth of low-chill cultivars grown under warm climatic conditions continues through to after fruit harvesting with strong competition occurring between shoots, roots and fruit for nutrients during the final stage of fruit development (Allan *et al.*, 1993a) (Fig. 1). In contrast, with higher-chill, later maturing cultivars, shoot extension is completed well before the final stage of fruit growth. Vegetative growth of low-chill cultivars can also be up to 10 times greater than that of high-chill cultivars depending on prevailing temperatures.

2.2.1.2 *Fruit development stages*

High-chill cultivars exhibit 3 distinct stages of fruit growth, in contrast, with low-chill, early maturing cultivars, the second stage is relatively short, and may show a reduced rate of volume expansion rather than a complete stop (Allan, *et al.*, 1993a; Grossman and DeJong, 1995a, b). Due to rapid seed and fruit development the sealing of the stone (endocarp) of low-chill cultivars is often incomplete and this may lead to split pits at fruit maturity.

2.2.1.3 *Root growth*

Like vegetative growth, there are two distinct root flushes. Each precedes the vegetative growth flushes (Cockroft and Olsson, 1972; Chalmers and Van den Ende, 1975a; Allan *et al.*, 1993a). Root flushing commences in late winter when soil temperatures are about 12°C or higher. With early-maturing cultivars there is often strong competition between fruit, shoot and root (Allan *et al.*, 1993a). Time of root flushing may also influence the most effective time to apply soil applied growth manipulators such as paclobutrazol (Cultar) and N.

In regions receiving less than 400 chill units, low-chill rootstocks such as Okinawa or Flordaguard must be used otherwise trees will show signs of sporadic vegetative budbreak. The rootstock may have a chilling requirement, just as the scion, but this is yet to be conclusively proven.

2.2.1.4 *Floral initiation*

In late summer, at about the same time as the second growth flush stops, floral initiation occurs and continues into the autumn (Notodimedjo *et al.*, 1981b). With young or vigorous trees floral initiation and bud development will be delayed. Floral bud development can be advanced by both mild water stress and/or through the application of growth retardants such as Cultar (George and Nissen, 1987; George *et al.*, 1992a, b).

2.2.1.5 *Dormancy*

During late summer, trees enter para-dormancy (Lang *et al.*,1987). Growth of buds is inhibited by growth regulators released from the apically dominant terminal buds. During mid-autumn, buds enter endo-dormancy (George and Nissen, 1988a, 1993; Lloyd and Firth, 1990). Under low latitudes, small changes in daylength and lack of autumn chilling delay onset of endo-dormancy. During autumn, nutrients are retranslocated from the leaves to the shoots, trunk and roots for storage over-winter. The length of the dormant period for low-chill cultivars is shorter than it is for high-chill ones. The application of Cultar in the late summer, water stress in the autumn and early winter pruning appear to shorten the length and depth of dormancy (George *et al.*, 1992b) but strong, vigorous vegetative growth will lengthen dormancy (Erez, 1987).

2.2.2 *Phenological cycles (continuous cropping system)*

Management techniques for growing stone fruit species under the continuous cropping system are unique and may vary between regions depending on their specific climatic patterns. The basic element of this system is to prevent the buds from entering into endo-dormancy. This can be achieved by inducing a new growth cycle prior to the onset of endo-dormancy. At the time of induction of a new cycle, differentiation and initial development of flower buds must have taken place. A period of recovery must be allowed between cycles so to increase the trees carbohydrate reserves. The length of the cropping cycle can vary between 6-8 months under tropical conditions, having typically very little seasonal change. No endo-dormancy will develop in the buds enabling strong response to stimuli that will break para-dormancy *e.g.* defoliation (Edwards, 1987a), water stress (Duarte and

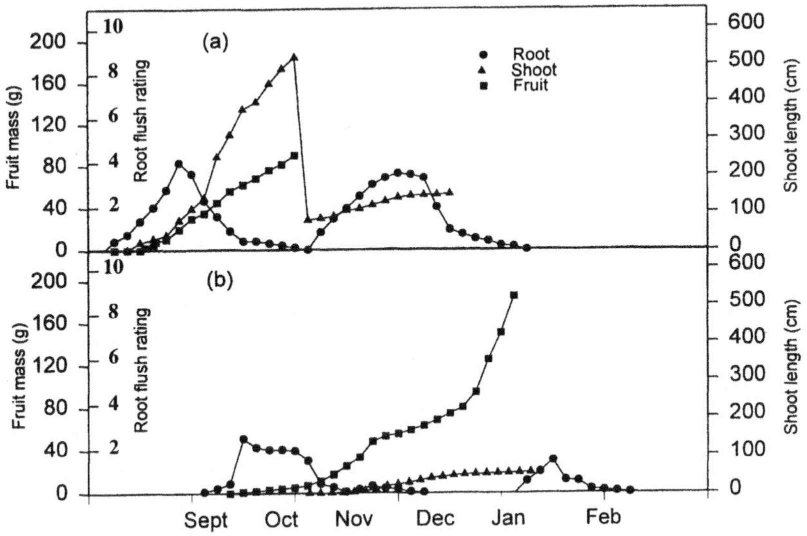

Figure 1. Comparison of phenological cycles for low (a) and high (b) chill cultivars in Australia.

Franciosi, 1974) and dormancy breaking agents (Erez and Lavi, 1984; Erez and Lerner, 1990) [see also Chapter 2].

Little work has been done on the requirements for growing stonefruit species under a continuous cropping system, but a few generalisation may be drawn: vegetative growth is relatively restricted as the temperatures are rather mild year round. This is very beneficial for good reproductive development due to limited vegetative competition. In many regions between 1.5 to 2.0 crops a year are possible. Early maturing, low-chill cultivars are especially adapted allowing completion of a cycle from bud break to fruit ripening within <4 months leaving another 2 months for recovery prior to defoliation and initiation of a new cycle. Although the drain in carbohydrates and minerals reserves by carrying 2 crops a year is high, this is compensated for by young to mature leaves photosynthesising for 11 months in a year as against 8-9 in the subtropics and 6-7 in the temperate zone (Erez, 1986). Premature loss of leaves or a reduction in leaf health will reduce the buildup of reserves in the tree. Adequate tree nutrition, particularly N, will be essential to maintain leaf health; unfortunately leaf nutrient levels are not available for stone fruits grown under the continuous cropping system.

Successful stonefruit production using the continuous cropping system has been demonstrated in Venezuela, Kenya, Ecuador (Erez and Lavi, 1984; Erez, personal observations; Edwards, 1987a) and Thailand (Subhadrabandhu, pers. comm.). However, the system appears to have been more successful with apples and grapes (Edwards, 1987a).

3 Ecology

3.1 Climatic requirements for the dormancy-induced, annual cropping system

3.1.1 *Temperature effects on flowering, pollination and fruit growth*

Temperature is the most important environmental variable affecting distribution of stonefruit in the subtropics. It can influence survival, growth, flower bud induction, flowering, fruit set and fruit quality of low-chill stonefruit. Stone fruit exhibit a differential response of the vegetative and reproductive organs to temperature. Providing bud break is satisfactory, vegetative growth exerts a stronger sink strength than the reproductive sink with increasing temperature (Erez *et al.* 1998); consequently, fruit set and fruit growth are reduced whilst canopy shading is increased.

High temperatures during the growing period may also interfere with normal flower bud differentiation. Warm temperatures in autumn and early winter can induce abnormal ovule development leading to multiple ovules, double or triple pistils at bloom and consequently to double or triple fruits fused together at their bases. These problems can also be accentuated by severe post-harvest water stress (Johnson *et al.,* 1992). Warm winter conditions will prevent normal flower bud break and are also often associated with necrosis and death of floral buds. Compared with apples, where floral buds may remain viable after staying dormant for a whole year, stone fruit floral buds will always die if they do not open in spring.

The temperature at which stonefruit fruit buds are injured by cold temperatures depends primarily on their stage of development (Westwood, 1978) with flowers and fruitlets being most susceptible to damage at temperatures between -2 and -3°C (Sharpe *et al.*, 1990). Because of their early flowering behaviour, low-chill cultivars can not be grown in areas subjected to regular late winter or spring frosts without some method of frost protection.

High day (>25°C) and/or night temperatures (>18°C) may reduce or completely prevent fruit set of stonefruit (Edwards, 1987c; Erez *et al.*, 1993). This response appears to be the major reason why the continuous cropping system for stonefruit has not been as successful as it has been with apples. Apples appear to be able to set fruit satisfactorily at higher temperatures (up to 28°C). The ideal temperature for fruit growth of both stonefruit and apples appears to be between 22 and 28°C (Tan and Buttery, 1986). At high temperatures (>32°C) increased respiration may reduce carbon accumulation (Flore, 1994). Field observations indicate that peaches may develop well at 30°C without reduction of fruit quality but under extremely high temperature (>35°C) such as experienced in Southern Spain and North Western Mexico, fruit development may be retarded.

Time of ripening in peach depends on time of bloom and length of the FDP. Time of bloom is regulated by the chilling requirement of the variety (see below) and the heat accumulation after chilling is satisfied, whereas the FDP, though varying with cultivar, is regulated by heat accumulation from blossom to ripening (Munoz *et al.*, 1986). Heat accumulation is measured by calculating the total number of hours between 4 and 25°C (Richardson *et al.*, 1974). Temperatures >25°C receive the same value as for 25°C. For low-chill stonefruit grown in subtropical Australia, a 1°C reduction in mean monthly temperature during FDP may delay fruit maturity by 5 days (Topp and Sherman, 1989a).

3.1.2 *Dormancy and chilling*

In regions where temperate fruit cultivars enter endo-dormancy, chilling is necessary to release resting floral and vegetative buds from dormancy. If chilling is insufficient, flowering and vegetative budbreak following dormancy may be sporadic and fruit set and yields severely reduced (Black, 1952; Saure, 1985; Edwards, 1987d). The chilling requirements is usually expressed in 'chill units' which, in the past, has been defined as one hour below 7.2°C (Weinberger, 1950). More recent studies have shown that this definition of a 'chill unit' was not correct and that actual chilling accumulation was underestimated by at least 50% (Erez, *et al.*, 1979a, b). The reason for this was that most early studies on dormancy were conducted in high-chill regions where temperatures up to 7°C only were regarded as effective in breaking dormancy but studies in subtropical regions showed that higher temperatures, up to 13°C, were also partially effective in breaking dormancy (Erez *et al.*, 1979a, b). To further complicate matters, high temperatures (>20°C) were shown to negate chilling effects; this being one of the reasons why in warm-subtropical regions less chilling is accumulated than is expected (Couvillon and Erez, 1985). [See Chapter 2 on dormancy in this book].

Many climatic models have been developed to determine chilling accumulation. Models at best can give only an approximate idea of which cultivars can be grown. Many other factors such as irradiance, water stress, tree vigour and tree age can act alone or interactively to reduce or increase the chilling requirement by up to half (Batten, 1988; George, 1995, unpublished data).

Breeders normally describe the chilling requirements of varieties by the number of 'chill units' they require to break dormancy. The main advantage of using such values is that it allows easy comparison to known reference cultivars. The chilling requirements for stonefruit varies between 75-1 500 chill units with sweet cherries having the highest. With all stone fruit species, such as peaches, the environmental range where they can be grown is wide due to the availability of both low- and high-chill cultivars. In most subtropical regions only low-medium chill cultivars (between 75-600 chill units) can be grown (low-chill cultivars can be defined as those which have a chill unit requirements <400, medium chill, 400-600, high-chill >600).

3.1.3 *Daylength, solar radiation and rainfall*

Low solar radiation or irradiance levels may affect fruit set, yield, fruit quality, and tree physiological processes such as photosynthesis and tree growth. Many studies with temperate fruits have shown a minimum threshold level of 20% total irradiance to achieve good fruit colour and sugar levels (George *et al.*, 1988a; Corelli-Grappadelli and Coston, 1991) [see also Chapter 1 in this book].

In the southern hemisphere, slopes with a north-easterly aspect (reverse for northern hemisphere) will receive greater sunlight exposure. Management techniques such as summer pruning and trellising can also be used to improve light penetration into the tree canopy. In tropical regions, the cropping cycle of the dormancy-avoided system can be manipulated so that critical stages such as flowering occur during the dry season. In subtropical regions, using an annual cropping system, a much smaller degree of growth cycle manipulation is possible, usually by advancing bloom and subsequent fruit maturity with dormancy breaking agents [see Chapter 2 for more details].

In regions with uniform yearly rainfall distribution patterns, tree growth throughout the year will be more vigorous necessitating the use of regulated deficit irrigation, growth retardants or other means of controlling tree growth..

3.2 Climatic requirements for dormancy-avoidance, continuous cropping system

Multiple cropping systems are used in regions that receive no winter chilling. For successful multiple cropping of apple and peach, Edwards *et al.* (1988) and Edwards (1990) have described maximum and minimum temperature limits throughout the year which help select areas as being suitable for this system. A computer program has been developed so that temperature data from selected sites can be compared with the monthly temperature limits for proven peach and apple production. For peach, the values are maximum, 20-25°C, minimum 10-17°C. Erez and Lavi (1984) have successfully grown peaches and plums at slightly higher maximum temperatures than reported by Edwards *et al.* (1988) and Edwards (1990). Such conditions are found in the tropics at high elevations of 1500-2500 m above sea level. At these elevations humidity is normally low and diurnal temperature variation large, even on the equator.

Stonefruit species are more affected by fungal and bacterial diseases under high humidity climates than pomefruits; a major problem being premature shedding of leaves in the summer and autumn due to these causes. This early leaf defoliation reduces the build up of starch reserves in the tree for the following season, consequently trees are rapidly weakened after 2-3 years of cropping (see section on leaf health). Tree health and fruit quality will be best in regions receiving <1500 mm of summer rainfall per year. Therefore best production of these species in the tropics is in high elevation, dry climates. Low precipitation levels year round is especially important particularly if using a continuous cropping system. Heavy rains may therefore limit production potential. Thus a prerequisite for a successful operation of continuous cropping is regulated irrigation, with water stress used as one means to induce the initiation of a new growing cycle.

4 Factors affecting productivity

4.1 Yield

Tree yield is dependent on two variables, the number of fruits and their final fruit size. Generally, percentage fruit set of most low-chill peach cultivars is high (50-80%) under conditions favourable for fruit set. However, late winter frosts at or just after fruit set may reduce fruit set. In warm frost-free areas, excessive vegetative growth and consequently heavy shading in the lower canopy may have a detrimental effect on reducing floral initiation in the next season (George and Nissen, 1987). Under controlled conditions Erez *et al.* (1997) found a negative correlation between high day temperature and fruit adherence. This response was due to increased vegetative competition particularly in the temperature range from 23-29°C. Studies on greenhouse production have also verified the negative effects of high temperatures before and during bloom on reducing fruit set and consequently yield (Erez *et al.*, 1998).

Due to their short FDP, yields of very early maturing, low-chill cultivars when planted at standard densities are low and often only one half of the maximum yields recorded for later maturing, higher-chill cultivars planted at similar densities (early maturing cultivars 15-25 t/ha; late maturing cultivars 40-60 t/ha) (Chalmers and Van den Ende, 1975a, b, Van den Ende and Chalmers, 1982; McDermott *et al.*, 1987, George *et al.*, 1992a). Based on field observational and experimental data (George *et al.*, 1992a), prospects of increasing yields per hectare, to the levels established for higher-chill cultivars, does not appear feasible. Yield benchmark studies in Australia and elsewhere have established preliminary target potentials that can be achieved under good management conditions (Table 2).

Table 2. Maximum yield potential recorded from low chill stonefruit grown under different production systems

Training system	*Cultivar*	*Number of trees per hectare*	*Maximum yield recorded* (t/hectare)*	*Reference*
Vase	Flordaprince	660	24	George and Nissen (1992)
Palmette	Flordaprince	800	28	George, Huett and Slack (1995, unpublished data)
Hedgerow post-harvest topped	Flordagold	6 000	32	Young and Crocker (1982)
Vase	Flordagold	600	51	George, Huett and Slack (1995, unpublished data)
Meadow	Earligrande	13 000	54	Erez (1982)

* Note higher yields for cultivars with longer fruit development periods

4.2 Fruit quality

Fruit size, shape and colour are the most important fruit quality characteristics readily identified by the consumer; most of these characteristics are culturally conditioned.

In some countries (China) there appears to be a preference for large size, white-fleshed, sub-acid fruit; whilst in others (Australia, USA) there is a preference for yellow fleshed, medium size fruit with acid flavour.

Small fruit size is the major quality defect of low-chill stonefruit especially nectarines, plums and apricots. Fruit size is adversely affected by the short FDP (70-110 days), heavy crop loads (George *et al.*, 1988b), excessive vegetative growth (Allan *et al.*, 1993a), low temperatures during fruit development (Topp and Sherman, 1989a, b) and 'carbohydrate stress' during the early post-budbreak period at the time of cell division (George *et al.*, 1996; Hieke *et al.*, 1997). Crop load can be adjusted for by thinning (discussed later) and excessive vegetative growth controlled by the use of growth retardants, pruning and girdling (discussed later). Environmental modification techniques such as the use of plastic covers can increase fruit size of peach (Hieke *et al.*, 1997), plum and sweet cherry (Erez, unpublished data) by 20-30%.

A major problem with very low-chill, early-maturing cultivars especially when grown in higher rainfall, subtropical regions is their poor sugar and flavour. Generally, later cultivars do not exhibit this problem. Fruit with better flavour and sugar levels are produced in drier regions and on well aerated soil types. George *et al.* (1988a) have shown that heavy rainfall prior to harvest can cause fruit sugar levels to drop as much as 1.5° Brix. This fall in Brix appears to be due to a dilution of sugar levels in the cells as a consequence of rapid cell expansion. Sugar levels increase exponentially during the final 3 weeks of stage III of fruit growth (Chalmers and Van den Ende, 1975a). Climatic conditions during this period are therefore critical. A minimal acceptable sugar level at harvest for most low-chill stonefruit cultivars would be 9° Brix.

Besides environmental factors, many cultural practises have been shown to affect fruit sugar levels. Fruit sugar levels have been shown to decrease as crop load per tree increase (George *et al.*, 1988b). Fruit sugar levels may also be improved by girdling (Allan *et al.*, 1993b), spring and summer pruning to remove water shoots and excessive vegetative growth 2-3 weeks prior to harvest, heavier applications of potassium (90 kg per hectare) prior to harvest (George *et al.*, 1988b) and enclosing trees under plastic cover during the FDP (Hieke *et al.*, 1997). Non-melting flesh cultivars may also reach higher sugar levels by being left on the tree longer, without a corresponding softening and ripening of the flesh.

Attractive skin colour is one of the key quality characteristic in selling any fruit. With few exceptions most early maturing cultivars from the University of Florida breeding program are highly coloured. Many studies with high-chill peaches have indicated a minimum threshold of 20% total irradiance or sunlight to achieve good fruit colour (Erez and Flore, 1986; Corelli-Grappadelli and Coston, 1991). This level of irradiance can only be achieved by obtaining good light penetration into the tree canopy. Pre-harvest spring pruning (about 3 weeks prior to harvest) improves skin colour (Day *et al.*, 1989; Corelli-Grappadelli and Coston, 1991). Training and trellising systems such as palmette and Tatura should also improve skin colour provided vegetative growth is controlled. In contrast, post-harvest topping may reduce fruit colour in the next harvest season by about 10 -15%, unless strong vegetative regrowth is controlled with Cultar (Mc Dermott *et al.*, 1987, George *et al.*, 1993). In Japan and Israel, reflective mulch has been used to improve fruit colour of apples and stonefruit (Moreshet, *et al.*, 1975).

Inadequate chilling may result in some cultivars producing fruit with pronounced tips and sutures. Stylar tip has been shown to be mildly correlated ($r = -0.65$, $P < 0.05$) with mean temperature during the FDP (Topp and Sherman, 1989a, b). Use of gibberellin biosynthesis inhibitors may improve fruit form, reduce the tip size and produce a more roundish fruit (Erez, 1984a, b).

5 Orchard management to increase productivity

5.1 Thinning

5.1.1 *Fruit set and yield*

Many early maturing cultivars such as Flordaprince were specifically bred to produce abundant flowers so as to ensure fruit set after winter freezes (Sherman, 1987; George *et al.*, 1992a). These cultivars are capable of setting in excess of 3 000 fruit on standard size trees (George *et al.*, 1992a). Early fruit thinning is essential if good fruit size is to be achieved. Delaying thinning reduces starch reserves that could be utilised by the remaining developing fruit (Allan *et al.*, 1992). With early maturing stonefruit, because of their short FDP, early fruit thinning is more important than it is with late maturing cultivars with longer FDPs as fruit suffering 'carbohydrate stress' during the early FDP are unable to recover potential growth lost during resource-limited periods (Grossman and DeJong, 1995a, b; George *et al.*, 1996).

5.1.2 *Thinning level*

The level of fruit thinning will depend on many factors including: level of fruit set, market prices received for different fruit size grades, tree size, and the need to avoid overcropping and consequently reduced tree health (George *et al.*, 1992a). Lower levels of thinning and smaller fruit size grades may be more acceptable and commercially more profitable for early-season fruit when demand and prices are higher. Whilst leaving heavier crop loads on the tree may initially prove commercially profitable, continued heavy cropping may reduce trunk and root starch reserves that are normally available for the next season's fruit growth (George *et al.*, 1992a). This is especially true with spur bearing species like sweet cherries or plums where flower bud differentiation occurs during the FDP. With peaches, especially early ripening cultivars, new flower buds may be produced after harvest preventing the biennial bearing phenomenon. Consequently, accurate adjustment of crop load is important for maintaining tree health and sustaining long-term productivity.

Various methods have been used to estimate the severity of thinning. These include: fruit number per tree, fruit number per cm of trunk circumference, number of fruiting laterals per trunk circumference, number of fruiting laterals per m^3 of canopy, fruit number per cm of lateral. None have proven to be entirely satisfactory. Still, the basic unit related to inter-fruit competition is the branch and very little compensating effect will be obtained from an overthinned branch situated adjacent to an underthinned one.

Overriding is final number of fruits to be left on the tree after thinning. This will depend on the level of management, the desired date of harvest and size range required. Under better management larger size grades can be obtained with heavier crop loads.

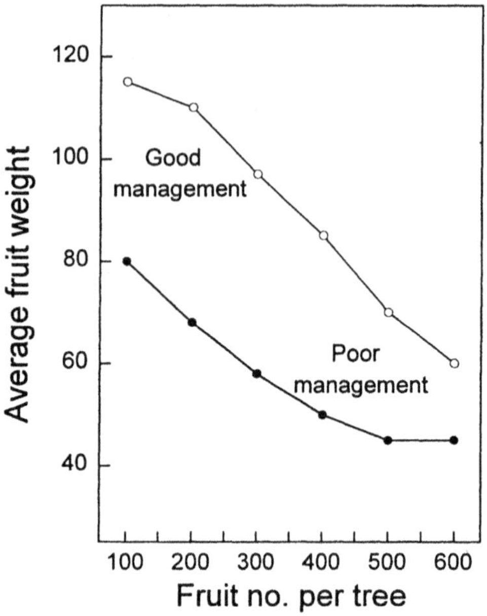

Figure 2. Effect of crop load on mean fruit weight of the peach cv Flordaprince under poor and high management regimes.

5.1.3 *Thinning methods and timing*

Fruit thinning is the traditional method used to thin both early and late maturing stonefruit (Westwood, 1978). Blossom thinning and stub thinning have also been used successfully and these methods where applicable offer clear advantages over fruit thinning in getting better compensation in size of the remaining fruits (Stadler, 1975; Byers and Lyons, 1984; Allan *et al.*, 1992). Depending on cultivar, blossom thinning involves removal of between 50 to 90% of blossoms on the tree, leaving between 4 -8 blossoms per 30 cm lateral. About half of these flowers will set fruit. Generally a final fruit thin will be necessary to achieve the required fruit number. In frost-prone areas, the risk of frost discourages growers from bloom thinning. Similarly, under semi-tropical conditions, high temperatures at bloom may be detrimental to *fruit set* on early blooming low-chill peaches and nectarines. Reduced fruit set may occur at day temperatures of >25°C.

An alternative to both blossom and fruit thinning is stub thinning which involves pruning back laterals by about 50-70% to leave only sufficient flowers to set a crop.

Allan *et al.* (1992) has shown that stub thinning promotes compensatory regrowth which delays fruit maturity. However, if stub thinning is combined with a soil application of Cultar, then compensatory regrowth can be controlled. Both blossom and stub thinning techniques can give substantial increases in fruit size, *e.g.* with *cv.* Flordaprince, both blossom thinning and stub thinning when combined with Cultar application produced the greatest percentage of large fruit (70 and 78% of fruit >90g, respectively) compared with fruit thinning (61%) (Allan *et al.*, 1992). The advantages of stub thinning are that labour costs associated with hand-thinning can be reduced and the operation could be mechanised, particularly if trees were trained onto a palmette, Tatura or Y trellis system. Currently, there are no reliable methods of chemically thinning low-chill stonefruit.

5.2 Nutrition

5.2.1 *Nutritional effects on yield and fruit quality*

Few definitive studies have been made on the effects of nutrition on yield or fruit quality of early maturing low-chill temperate fruits. Studies on high-chill cultivars are often not directly applicable for early maturing cultivars because the growth cycle and growth rates for the two groups are very different. Nitrogen is the key element affecting fruit set, with inadequate N shown to reduce fruit set by as much as 48% (George and Nissen, 1992). Nitrogen also appears to be the key element affecting fruit size. Excessive N application increases vegetative growth and shading and results in a reduction in fruit size and quality. However, studies by George and Nissen. (1992; 1996, unpublished data) have shown that when excessive growth is controlled with Cultar, increasing N application may actually increase fruit size by as much as 39%.

Potassium is the other major element affecting fruit quality. Leaf K levels are often negatively correlated with tree crop load (George *et al.,* 1997) indicating that fruit may exert a major sink force for this nutrient. On low K soils, soil applications of K were shown to increase fruit size of low-chill stonefruit by as much as 8% (George *et al.*, 1988) but on soils where K is adequate K foliarly applied during the FDP did not improve size or quality (Campbell, unpublished data, 1992). Excessive application of K should be avoided as this may induce Mg deficiencies (Huett *et al.*, 1997).

5.2.2 *Nutrient replacement*

Stonefruit remove significant quantities of the major elements N and K and the minor elements B and Zn (George *et al.*, 1997, Slack *et al.*, 1996). Whilst it is desirable to base fertiliser rates on the amounts of nutrients removed by the crop (Huett *et al.*, 1997) many other factors such as soil type, climate (leaching and fixation) and method of application must be taken into account. Consequently the rates of nutrients to apply for different soil types, locations/regions may vary by a factor of 2-3; thus both leaf and soil nutrient analyses are an essential tool for adjusting fertiliser rates. [See also Chapter 4].

In high rainfall regions deficiencies of minor nutrients B, Cu and Zn are common. These can be best corrected through late summer applications rather than by foliar applications during the flowering and fruit development period.

The aim with newly planted trees is to maximise their canopy surface area as quickly as possible. Trees up to 2 years of age can be fertilised every 2-3 months with a complete fertiliser such as 15:4:12 N:P:K formulation. In modern orchards fertilisers are applied through the irrigation system, with many different formulations used [for details see Chapter 4 of this book].

5.2.3 *Soil analyses*

Both soil and leaf analyses are required to monitor the nutritional status of a low-chill temperate fruit orchard. Soil analysis is generally less useful than a leaf analysis in the preparation of a fertiliser program but soil analyses are particularly useful in determining the cause of nutrient imbalances as indicated by leaf symptoms or leaf analyses. Soil analyses is an integral part of fertiliser management. The soil sample should be taken at 0-30 cm depth. About 20 samples are required to adequately represent an orchard block. On very low fertility soils (K < 0.5, Mg < 1.0 meq/100 g), leaf deficiencies of potassium and magnesium may be common.

5.2.4 *Leaf nutrient analyses-sampling times*

Few studies have been conducted to define the most suitable leaf nutrient sampling times and establish tentative critical levels for early ripening stonefruit. George and Nissen (1987) and Allan, *et al.* (1993a) have shown that the phenological pattern of growth of early and late ripening cultivars, grown in their respective climatic environments, are different. (see also above 2.2.1). Because of these differences in growth patterns, the seasonal patterns of leaf nutrients may also be different to those established for high-chill cultivars (Leece and Gilmour, 1974). Generally, leaf N, P and K exhibit continuous decline throughout the growing season while leaf Ca and Mg increase (Leece and Gilmour, 1974). With early ripening low-chill cultivars, the seasonal patterns are more variable due to vegetative flushing and management factors such as summer pruning (George *et al.*, 1997). Leaf sampling is normally carried out when most nutrients have stabilised. Early maturing low-chill cultivars, appear to be most stable 2-3 weeks after harvest (prior to summer pruning); this appears to be the most desirable time to leaf sample (George *et al.*, 1997; Huett *et al.*, 1997). The leaf sample is collected by taking mature leaves from the mid-portion, current season's shoot. Approximately 10 trees in every 100 should be sampled.

5.2.5 *Leaf nutrient analyses-standard levels*

In Australia, several surveys to establish leaf nutrient standards for low-chill early maturing stonefruit have indicated that leaf N and K levels may need to be maintained in the higher optimum range established for late maturing cultivars in Australia or for early-maturing cultivars in Florida (Table 3) (Arnold and Crocker, 1982; Crocker and Sherman, 1986; George *et al.*, 1997, Huett *et al.*, 1997). Based on surveys of low-chill stonefruit orchards in subtropical Australia (Huett *et al.*, 1997), N levels need to be maintained between 3.0 to 3.5%, at post-harvest, the currently recommended sampling time. However, current research on light-textured soils in West Australia suggests that it may be better to assess N status at fruit set because higher leaf N levels (between 4.0-4.5%) during this period appear to be associated with higher yields of larger size fruit (Goldspink, 1997, pers comm.).

Table 3. Leaf nutrient standard range for low- and high-chill peach in Florida and Australia

Element		*Australian high chill optimum range*	*Florida low-chill optimum range*	**Australian low-chill optimum range*
Nitrogen	N %	3.00 - 3.50	3.00 - 3.50	3.5 - 3.7
Phosphorus	P %	0.14 - 0.25	0.17 - 0.29	0.23 - 0.32
Potassium	K %	2.00 - 3.00	1.10 - 2.00	2.2 - 2.6
Calcium	Ca %	1.80 - 2.70	0.90 - 1.50	1.2 - 1.6
Magnesium	Mg %	0.30 - 0.80	0.30 - 1.00	0.33 - 0.37
Zinc	Zn mg/kg	20 - 50	17 - 60	28 - 32
Manganese	Mn mg/kg	40 - 160	40 - 100	46 - 162
Copper	Cu mg/kg	5 - 16	7 - 18	8 - 11
Iron	Fe mg/kg	100 - 250	50 - 100	45 - 160
Boron	B mg/kg	20 - 60	25 - 80	29 - 47

* Australian low-chill range based on survey data (Huett *et al.*, 1997; Slack *et al.*, 1996)

5.3 Irrigation

Detailed discussion of this topic is found in Chapter 3. In this section we discuss only the controlling effect of irrigation on growth. Regulated deficit irrigation employs a predetermined water deficit to control vigour and has been successfully used with very late-season cultivars to control growth (up to 75%) without loss of yield or fruit size (Chalmers *et al.*, 1981, 1984, 1985). Moderate stress is applied to trees in the period from fruit set to stone-hardening (predawn leaf water potential measured with the pressure chamber, -0.5MPa-0.3MPa). After stone-hardening, trees are fully irrigated (pre-dawn leaf water potentials -0.2MPa). The effects of pre-stone-hardening water stress on early maturing cultivars has not been evaluated. Obviously, the technique may have limited value in wet subtropical regions because of the high summer rainfall. With early maturing cultivars, moderate water stress applied after harvest may also prove beneficial by restricting vegetative growth in the period up to senescence. Observations have shown that moderate stress may advance floral bud development and reduce the chilling requirement substantially but severe stress may damage the floral buds.

5.4 Spring and summer pruning

Summer pruning of high-chill cultivars 3 weeks prior to harvest has been shown to increase fruit size by about 20% (Day *et al.,* 1989). With low-chill cultivars, due to their greater growth rates, 2 to 3 pre-harvest prunings, commencing prior to stone-hardening can increase fruit size. With high-chill stonefruit cultivars, shading and associated reduction in tree photosynthesis during the final stage (stage III) of fruit growth may reduce fruit growth more than during other stages of fruit development (Marini, 1985). In contrast, for the low-chill *cv.* Flordaprince, shading during stage I of fruit growth reduced fruit growth, indicating that cell division may be adversely affected, but shading during stage III reduced fruit quality (George *et al.,* 1996). Peaches developing in the shaded part of the tree canopy are dependent on translocated photosynthate compared with fruit developing in well illuminated parts of the tree canopy, where photosynthates are supplied from adjacent leaves. A photomorphogenetic effect of leaf shade on reproductive development can not be ignored too (Erez, 1971).

5.5 Girdling

Many studies on both low and high-chill stonefruit have shown that girdling can increase fruit size, (particularly the larger fruit grades) by between 5 and 15% and advance fruit maturity by 3-5 days (Dann *et al.*, 1984; North *et al.*, 1988; Allan *et al.*, 1993b). This response is a result of a temporary increase in shoot carbohydrates during the later stages of fruit development. The most suitable time to girdle appears to be just prior to stone-hardening (DeVilliers *et al.*, 1990; Allan *et al.*, 1993b). It is important to keep the girdling wound open till fruit ripening to get maximum benefit. Both whole trunk and branch girdling have been shown to be effective (Dann *et al.*, 1984). Girdling may increase the incidence of split pit and woolliness (a cool storage disorder) (North *et al.*, 1988) and reduce long-term tree health. Studies in Spain suggest that some cultivars are less tolerant of girdling and that the girdling width be restricted to less than 10 mm and preferably of 5 mm for rapid healing (Fernandez *et al.*, 1987).

5.6 Maintain foliage

Early autumn defoliation has been shown to reduce fruit growth of low-chill stonefruit *cvs.* Flordaprince and Flordagold by 21 and 6%, respectively (Lloyd and Firth, 1990). Premature defoliation may result from several factors: water stress; low leaf N (<1.5%); wind damage and leaf diseases (shot hole, rust, etc). Some of these factors are interactive *e.g.* low leaf N levels predispose leaves to the build up of leaf fungal diseases such as shot hole.

Early autumn defoliation reduces the reserves of both starch and N in tree structures such as trunks and shoots (Castagnoli *et al.*, 1990; George and Nissen., 1993). A typical response of the prematurely defoliated tree is that some floral buds flower prematurely by breaking dormancy via the dormancy avoidance system [see Chapter 2 in this book]. As the chances for fruit set and/or good fruit development of this early flowering is poor during mid-winter, there is a further loss of reserves.

The longer-term consequence of premature defoliation is a rapid decline in tree health, usually about 3-4 years after planting out. The reason for this is that once trees commence heavy cropping, the starch reserves available for root growth are depleted and the tree becomes more susceptible to stress. With low crop loads, starch reserves are initially depleted from branches but with increasing crop loads they are depleted from trunks and roots as well (George *et al.*, 1993).

Late summer and autumn leaf health and photosynthesis can be maintained by increasing leaf N levels through fertilising when vegetative growth stops. Photosynthesis rates are highly correlated with leaf N levels (DeJong and Doyle, 1985). Soil applications of Cultar in late summer have been shown to increase leaf N levels in the autumn (George *et al.*, 1993). During the late summer period, both leaf N per unit leaf area and leaf dry weight per unit leaf area have been shown to be better indices of photosynthetic capacity than leaf N (Weinbaum *et al.*, 1989); these may prove to be useful monitors of the effects of management practises. Compared with lightly summer-pruned trees, post-harvest topped trees may exhibit better leaf health in the autumn due to regeneration of leaf area free from disease. Consequently this technique may be particularly useful in regions using minimal or no fungicidal spray programs to control leaf diseases post-harvest (Young and Crocker, 1982).

6 Management techniques to modulate flowering and fruit maturity

6.1 Management techniques to reduce the chilling requirement

High prices are paid for early maturing stonefruit. Consequently many techniques have been developed to achieve this aim. Management techniques which advance fruit maturity can be divided into 4 categories:

1. those which advance flowering by avoiding the chilling requirement (discussed in section 2.2.2);
2. those which advance flowering by reducing the chilling requirement of the cultivar;
3. those that advance flowering by enhancing chilling;
4. those that increase heat summation during the FDP.

Besides the use of dormancy release chemicals there are many other techniques such as early pruning, defoliation and use of growth retardants which will reduce the chilling requirement and which will advance fruit maturity by between 1 to 3 months in very low-chill cultivars (<200 chill units) grown under warm subtropical conditions (either with or without environmental modification) (Sherman and Lyrene, 1984; Edwards, 1985; George *et al.*, 1988c; George *et al.*, 1992b). The location selected should meet the chilling unit requirement of the cultivar, although this can be markedly reduced as a consequence of management practices. The subsequent fruit development rate will depend on heat accumulation after flowering (Munoz *et al.*, 1986; Topp and Sherman, 1989a). Many of these can be incorporated into a single management package.

However there are environmental constraints for successful modulation of flowering. Although it may be feasible to induce flowering of low-chill cultivars in early autumn, fruit set appears to be reduced at night temperatures >12-14°C (Edwards, 1987a, b, c, d) and/or at day temperatures >26°C (Erez *et al.*, 1998). Similarly, if trees are induced to flower during the coldest period of the year, young fruit are killed at -2°C (Sharpe *et al.*, 1990).

6.1.1 *Dormancy-release chemicals*

Many studies have shown that Dormex (hydrogen cyanamide) is highly effective in breaking the dormancy of grape buds (George *et al.*, 1988a; George and Nissen, 1990; George *et al.*, 1992b). In contrast, with stonefruit, in particular, the very low-chill cultivars appear to be particularly sensitive to this chemical. Severe floral abscission (between 10-90%) can occur if the chemical is applied at the wrong dose rate and within 4 weeks of flowering (George and Nissen, 1988a; George *et al.*, 1992b; George and Nissen, 1993). Studies with higher chill stonefruit cultivars have not shown similar high rates of floral bud abscission (Erez, 1987b) which suggests that these cultivars may be less sensitive to the phytotoxic effects of cyanamide than low-chill cultivars such as Flordaprince or that low-chill cultivars may be in shallower state of dormancy at the time of spraying. The risk with low-chill cultivars (<200 chill units) of being out of the endo-dormant state is too great to warrant using Dormex.

Exogenous applications of nitrogenous compounds containing either the NH_4 or NO_3 ion have been shown to increase arginine levels which may promote flowering. In apple it was shown that exogenous application of the NH_4 ion rather than the NO_3 ion promoted flowering (Edwards, 1986). However with stonefruit, the NO_3 ion appears to have been more effective in advancing floral budbreak of very low-chill cultivars (Erez and Lavi, 1974; Ross and Pharis, 1985; George and Nissen, 1988; George and Nissen, 1993). Application of potassium nitrate in the late autumn, as trees are shedding their leaves, (early dormancy) can advance flowering, fruit set and fruit maturity by 10-21 days. Later applications when trees are fully dormant are also effective but the advance in maturity time will be less. For autumn application, multiple, sequential applications of 5% potassium nitrate are more effective than a single application but fruit set and yield may be reduced when more than 3 sprays are used (George and Nissen, 1993).

More recently, the efficacy of potassium nitrate has been improved considerably by applying in combination with alkoxylated amines (Armobreak, AKZO, The Netherlands) (North, 1995, Erez, 1995). This lipophilic agent appears to enhance uptake of the potassium nitrate at the bud site. This combination will prove more ecologically friendly than the more toxic dormancy-breaking chemicals such as DNOC and Dormex which are currently being used. [For a more detailed description see Chapter 2].

The correct rate of application of dormancy-release chemicals appears to be dependent on the chilling requirement of the cultivar with higher rates required for higher chill cultivars. Still without sufficient chilling, poor response will be obtained. The chilling requirements of the specific cultivar and the level of modulation required determines the appropriate mix of agrotechnical dormancy-modifiers. In the future, the time of dormancy termination and flowering will be able to be more accurately predicted using bioassay techniques and phenoclimatological models. This will also allow more accurate timing of dormancy-breaking chemicals.

6.1.2 *Rootstock effects on chilling requirement*

Rootstocks may influence time of bud break by three means:

1. by having a vigour control effect which indirectly affects bud dormancy;
2. by having a direct effect through earlier root development and sap flow containing hormones that may affect bud break;
3. having a chilling requirement that needs to satisfied just as the scion cultivar has a chilling requirement. In Australia, low-chill rootstocks such as Nemasun and Okinawa must be used in regions receiving <400 chill units to avoid delayed and sporadic budbreak problems.

6.1.3 *Defoliation*

The understanding of the role of leaves on depth of dormancy has been confused by contradicting results after mechanical and/or chemical defoliation (Saure, 1985; Lloyd and Firth, 1990). Studies on the effects of leaf retention on depth of dormancy are contradicting. In some experiments bud development and budbreak has been advanced (Fuchigami, 1977; Walser *et al.*, 1981; Mannini and Ryugo, 1983), in others, delayed (Broome and Zimmerman, 1976; Fuchigami, 1977), whilst in others no effects were observed (Broome and Zimmerman, 1976). Lloyd and Firth (1990) showed that depth of dormancy increases subsequent to defoliation, but the extent of this effect was dependent upon both cultivar and date of defoliation. With the very low-chill peach 'Flordaprince', buds were consistently less dormant than a higher-chill *cv.* Flordagold and overcame dormancy at an earlier date. Lloyd and Firth (1990) also found that whilst a lower depth of dormancy was translated into earlier budbreak for 'Flordaprince' the opposite was true for 'Flordagold'. Their results indicated that although early defoliation reduces depth of dormancy, the growth of buds after dormancy requirements are met may be variable and dependent on cultivar.

Conflicting results in relation to time of autumn defoliation may also be due to differences in autumn climates between regions. In low latitude regions, early defoliation in autumn, is a strong stimulus for early bud break due to the late onset of short days and chilling which will delay the buds from entering into the endo-dormancy; consequently a response similar to dormancy avoidance will be obtained. (Erez,1987a, Erez and Lerner, 1990). If trees are already in endo-dormancy, weak or no effects will be seen.

Lloyd and Firth (1990) hypothesise that the reduction in depth of dormancy with early defoliation may be due to a reduction in growth inhibiting substances, most probably abscissic acid, which is translocated from the abscissing foliage to the floral buds. Another possibility is a drop in the inhibitor naringenin, accumulating in dormant peach buds through translocation of its glycoside prunin from the leaves (Erez and Lavi, 1969).

Defoliation of low-chill cultivars in the late autumn may be beneficial in reducing the depth of dormancy and advancing flowering and fruit maturity. Complete defoliation can be achieved with a foliar application of 5% zinc sulphate. Two other chemicals which have been reported to effectively defoliate deciduous fruit trees are magnesium chlorate (0.36%) and cyanamides (2%) (Erez, 1985a). The latter chemical will also break dormancy effectively.

6.1.4 *Pruning*

Late autumn and early winter pruning of very low-chill cultivars such as Flordaprince has been shown to advance bloom and hence fruit maturity by about one week (George *et al.*, 1992b). Weaker laterals (swingers) with base diameters (2-4 mm) produce the early winter set fruit while the stronger laterals (base diameters 4-8 mm) flower at the normal time. In contrast, post-harvest topping, a technique to control tree size with low chill stonefruit unless performed before mid-November in Australia (early summer), will result in strong compensatory regrowth and delayed floral bud development. However, if Cultar is applied shortly after post-harvest topping is carried out then rate of floral bud development will be normal.

6.1.5 *Early cessation of growth*

Gibberellin biosynthesis inhibitors (GBI) like Cultar will cause an early cessation of growth in late summer. This is important for two reasons; floral bud differentiation is advanced and enhanced and there is increased accumulation of photosynthate reserves in the tree which are available for early fruit and shoot development. Late summer and autumn growth cessation can be induced by applying Cultar and water stress, either singly, or in combination. The application of Cultar has been shown to advance bud break (Erez, 1984b) and fruit maturity by as much as 2 weeks (George *et al.*, 1993) and when combined with autumn defoliation and/or dormancy release chemicals by as much as 8 weeks. Alternatively, under warm, dry summers, vegetative growth may be controlled post-harvest with regulated deficit irrigation without affecting trees health. [See Chapter 4]. When used in conjunction with drip irrigation limited water supplies can be conserved. However, extreme water stress, post-harvest, should be avoided as this will lead to premature defoliation and rest avoidance [see Chapter 2].

6.2 Management techniques to enhance chilling - overhead sprinkling

High day temperatures (>18°C) have been shown to negate chill accumulation. In both Florida and Israel, overhead intermittent sprinkling, to provide evaporative cooling on days when temperatures >18°C, has been shown to reduce bud temperatures 3-5°C and to advance flowering by as much as 20 days (Gilreath and Buchanan, 1981; Erez and Couvillon, 1983, Erez *et al.*, 1993).

6.3 Increasing heat summation during the FDP

Fruit maturity can be advanced if temperatures during the FDP are increased. For example, the *cv.* Flordaprince has a heat summation requirement of 1190°C days above 3°C, requiring an average maximum of 18°C for 80 days (Munoz *et al.*, 1986). High average maximum temperatures can be achieved by growing stonefruit under plastic. Many trials have been conducted in Italy, France, Spain, Australia, Israel, New Zealand and Japan (Bellini *et al.*, 1980; Caruso *et al.*, 1989; Rageau and Ridray, 1989; Fuss *et al.*, 1990; Erez *et al.*, 1993; Hieke *et al.*, 1997). The critical factor in greenhouse-grown orchards is the need to satisfy the chilling requirement of the cultivar prior to elevating the ambient temperature. This means that for part of the season the orchard has to be uncovered and the chilling requirement of the cultivar fully satisfied.

Low-chill early maturing peaches and nectarines were the first to be grown in such system (Bellini *et al.*, 1980). The extra costs of inputs in the orchard were balanced by higher returns due mainly to obtaining extremely early fruit maturity. Early peaches and nectarines can be ripened in Israel by the end of March (Table 4) and in Sicily by mid-April (Caruso *et al.*, 1980). Fruit maturity of apricots and sweet cherries can similarly be advanced (Erez, unpublished data).

High daytime temperatures >25°C, either prior to or during bloom and fruit set, may reduce fruit set and produce excessive vegetative growth. In order to obtain good fruit set, competition between the reproductive and vegetative sinks should be minimised using vegetative growth inhibitors like Cultar (Hieke *et al.*, 1997) or by avoiding excessive high temperature in the greenhouse. Apart from advancing fruit maturity, fruit size of sweet cherries, plums and peach may also be improved (Erez, unpublished data; Hieke *et al.*, 1997).

Table 4. The effect of evaporative cooling followed by greenhouse conditions on bud break and time of fruit harvest in peach and nectarine cultivars in comparison to non-treated control (Bet Dagan, Israel, 1995)

Cultivar	evaporative cooling + greenhouse	time of full bloom	time of harvest
Mayglo	-	Feb 22	May 13 -20
	+	Jan 7	Mar 28 - Apr 8
Fiesta red	-	Feb 27	May 28
	+	Jan 15	Apr 14 -19
9-11N	-	Feb 24	May 28 -Jun 6
	+	Jan 13	Apr 14 -23
Sunblaze	-	Feb. 27	Jun 6
	+	Jan 13	Apr 19 - May 2
Flordastar	-	Feb 27	May 2 -Jun 6
	+	Jun 15	Apr 14
Maravilha	-	Feb 25	May 9 -13
	+	Dec 31	Mar 23 - Apr 7
Earligrande	-	Feb 17	May 19 -29
	+	Jan 15	Apr 14 -19

7 Tree training

7.1 Current systems

Due to increasing labour costs for thinning, pruning and harvesting, alternative methods of management and training of excessively vigorous low-chill stonefruit are urgently required. Currently orchardists have a choice of using a wide range of management systems. However, only 4 systems are currently used commercially under warm climates; central leader, standard vase, palmette, and Tatura. Most orchards, including those trained on trellis systems, are planted at low densities (<600 trees per hectare).

No comprehensive field experiments have been conducted comparing different trellising systems for low-chill stonefruit. Observations have indicated that yield, but particularly fruit quality, can be increased on palmette or Tatura trellising systems provided excessive vegetative growth is controlled (George and Nissen, 1987). Reduced labour input on palmette system has made this the preferred system to use in Australia. Many variations of the palmette system are being trialled. Various training and trellising systems currently used or which have potential for use with low-chill stonefruit are presented in Table 5.

Table 5. Training and trellising systems for low-medium chill stonefruit

Training and/or trellising system	*[1]Density (trees/ha)*	*Management requirements and suitability for different management combinations*			
		[2]Growth control (Cultar/RDI)	*[3]Moderate summer pruning or post harvest topping or*	*Cloche*	*[4]Mechanisation*
Vase	400	-	-	-	-
Tatura	400	-	-	-	+
Palmette	400	-	-	-	+
Vase	400-600	+	-	-	-
Tatura	500-700	+	-	-	+
Palmette	500-700	+	-	-	+
Mini-Vase	700-1000	+	-	+	-
Mini-Tatura	700-1200	+	+	+	+
Mini-Palmette	700-1200	+	+	+	+
Mini-Palmette - (2 leader slender spindle)	1000-1500	+	+	+	+
Hedgerow	1200-1800	+	+	+	+
Meadow	2000-10000	+	+	+	+

[1] Maximum tree density that can be used will depend on many factors. Under dry, cool subtropical conditions and on low fertility sandy soils, the highest densities can be used

[2] Higher density systems require growth control under warm subtropical climates RDI = regulated deficit irrigation applied post-harvest

[3] These systems require more intensive spring and summer pruning to control growth

[4] These systems may be able to be mechanically pruned, stub thinned and harvested particularly if trees are trellised

7.2 Medium density systems

In modelling approaches to optimise orchard design, it has been shown that high quality fruit are produced in well illuminated parts of the tree canopy where photosynthetic rates are at a maximum (Jackson, 1980; Marini and Sowers, 1990) [see also Chapter 1]. Smaller trees have greater surface to volume ratios and consequently greater % light interception. With these trees, canopy depth should be no greater than 1 m so that fruit produced at the centre of the canopy receive at least 20% full sunlight. Tree height should also be restricted to less than 2.5 m if labour costs associated with pruning, thinning and harvesting are to be reduced.

The ideal training system should be relatively inexpensive to establish, maximise labour use efficiency and be suitable for mechanisation. Given these considerations, a mini-palmette (free standing) (Fig. 3) or Tatura trellis system may offer the best approach to orchard design. Such systems are now being evaluated in subtropical Australia. Trees can be produced inexpensively from softwood cuttings and planted at densities of 800-1500 trees per hectare (Fig. 3). If trees are planted in early summer, they can produce their first crop within 12 months after planting. Excessive tree growth can be controlled with several very low dose rate applications of Cultar within 6 months after planting, or with regulated deficit irrigation and summer pruning. The aim in the early stages of tree development is to prevent excessive shading and loss of fruiting wood.

Figure 3. Left: Low-chill nectarine cultivar Sunwright trained to the palmette system and planted at moderate densities of 800 trees/hectare. Right: A softwood cutting of Sunwright nectarine ready for field planting 6 weeks after removal from the tree. Cuttings are used in medium density (1200 trees/hectare) orchards in subtropical Australia. They produce their first commercial crop 6 months after planting out.

7.3 Meadow orchards

Meadow orchard systems (*ca.* 10 000 trees per hectare) have been tried in a few locations in subtropical climates. Their main advantages are the reduction in perennial scaffold branches which are 'parasitic' on the rest of the tree, their high productivity (60 t/hectare) and the suitability of the system for mechanisation (Erez, 1982) (Fig. 4). Their major disadvantage is that growth must be strictly controlled; still this system is being tried whenever protected orchards are grown due to its low stature. However, the sustainability of these systems is still in question and few growers have established commercial plantings (Erez, 1982, 1985a, b, c, Menzies and Richens, 1989).

Figure 4. Earligrande meadow orchard propagated from hard wood cutting, 2nd leaf. Southern coasted plain, Israel.

7.4 Container grown trees

Progressively more complex systems of training and management can be developed. In Israel, container grown peach trees receive their winter chilling in cool rooms or in the high mountains before transfer to the field or glasshouse in the spring (Erez, 1990a, b). Most low-chill cultivars require only 2-3 weeks in the cool room to satisfy their chilling. Fruit maturity could be advanced by 1-2 months depending on the cultivar. With tree spacings of 0.5 by 2.0 m, yields of 20-25 t/ha could be obtained. Similar responses have also been obtained with apricots and sweet cherries. This system allows the growing of stonefruits in sites devoid of any adequate winter chilling (Erez, 1990a, b). The extra costs associated with this system must be paid for by producing a high value, high quality product (Fig. 5).

Figure 5. Earligrande peach trees grown on their own roots in vulcanic tuff filling 15 L plastic bags. Picture taken March 15 in Bet-Dagan, Israel.

8 Control of tree size and vegetative growth

8.1 Methods of controlling tree size

With low-chill stonefruit, yield of high quality fruit per management and labour input will determine economic viability. The level of labour input will be highly dependent on maintaining small tree size (trees less than 2.5 m in height) and the ability to mechanise cultural operations. Many methods have been used to control trees size in stonefruit including:

1. rootstock selection (Rom, 1988; Murase *et al.*, 1990);
2. regulated deficit irrigation (Chalmers *et al.*, 1981; Chalmers *et al.*, 1984; Chalmers *et al.*, 1985);
3. pruning (Leuty and Pree, 1980; Young and Crocker, 1982; Erez, 1984a; Marini, 1985);
4. use of growth retardants (Williams, 1984; Erez, 1984a, b; Marini, 1986,1987; Martin *et al.*, 1987).
5. girdling (Noel, 1970; Dann *et al.*, 1984; Fernandez-Escobar, 1987; De Villiers *et al.*, 1990); These methods have been previously reviewed by George and Nissen (1987).

8.2 Dwarfing rootstocks

Ideally tree size would be best controlled using dwarfing rootstocks as is the case with apples. Until recently most dwarfing rootstocks tested on stonefruit were weak and produced unstable graft unions. In Japan, recent studies (Murase, 1990) have shown that specific clones of *Prunus tomentosa* and *P. japonica* can successfully dwarf peach trees and produce commercially acceptable yields.

8.3 Regulated deficit irrigation

Regulated deficit irrigation has been used with higher-chill peach cultivars to control tree growth. Studies on container grown peaches has shown that when regulated deficit irrigation is combined with Cultar, tree size can be reduced by as much as 46% (George and Nissen, 1992). However, regulated deficit irrigation may be more difficult to implement in the field under summer rainfall. In dry growing conditions, with low-chill, early maturing cultivars, water stress is best applied post-harvest rather than during stage II of fruit growth, which is the case with high-chill cultivars. The reason for this is the short FDP for early maturing cultivars. Post-harvest regulated deficit irrigation will control growth, advance floral and vegetative bud development and save water (see section 5.4). The level of post-harvest stress applied is critical and should not reach a threshold level, above which physiological damage may result. Excessively high stress levels will also cause pre-mature leaf drop.

8.4 Pruning and post-harvest topping

In subtropical Australia, tree size can be controlled though a combination of both dormant and summer pruning alone, without using growth retardants. However, the severity of pruning appears to waste tree resources put into vegetative growth, when compared to systems where growth retardant are used. Using pruning alone, trees are normally dormant pruned, and spring pruned (2-3 times) prior to harvest and summer pruned after harvest. The severity of the post-harvest pruning varies with the management system required, and in its most severe form is called topping, where the tree is cut back to the main structural framework (Crocker *et al.*, 1982). In regions with high summer rainfall and where no fungicides sprays are used to control leaf diseases, it may be a way of regenerating a healthy, photosyntheticly active leaf surface. Trials with early maturing peach meadow orchards in Israel have also shown that trees that are severely post-harvest pruned can regenerate the tree frame in the same year without a decline in tree health (Erez, 1982)

Although high yields can be obtained with post-harvest topping systems (Young and Crocker, 1982), the major disadvantage is the strong compensatory regrowth which follows topping which subsequently delays floral bud differentiation, flowering and fruit maturity (Fig. 6). Limited observational studies in Australia indicate that if post-harvest topping is carried out before mid-November (early summer), then trees will have sufficient time to regenerate new fruiting wood without delay in flowering or fruiting in the following season. If trees are pruned after November flowering may be delayed by 5-12 days; this delay can avoided by applying Cultar after pruning (McDermott *et al.*, 1987; George *et al.*, 1992,1993). Similar results were obtained in Israel with meadow orchard peaches (Erez, 1982; Erez, 1985).

Figure 6. Strong regrowth on trees severely post-harvest topped. Trees need to be treated with Cultar at this stage if floral bud differentiation is not to be delayed.

Post-harvest topping techniques would be suitable to use with very early maturing varieties and for trees that are planted at higher densities than normal. The major production advantages of this system are greater fruit production in the first 2 years after planting and the system can be mechanised. Summer pruning and stub thinning operations could be mechanised particularly if trees were trellised (George *et al.*,1992).

8.5 Growth inhibitors

Both soil and foliar applications of Cultar have proven to be highly effective in controlling tree growth in low-chill stonefruit, reducing pruning weights and time involved in pruning (Williams, 1984; Erez, 1984a, b; Mika, 1986; Marini, 1986, 87; George and Nissen, 1987; George *et al.*, 1993, 1994). Apart from its effect on reducing vegetative growth Cultar also reduces the top/root ratio (Avidan and Erez, 1995) and changes new branch growth to a more plagiotropic orientation (Erez, 1984). This enhances flower bud differentiation and further reduces vegetative growth.

8.5.1 *Rates of soil applied Cultar*

Rates and timing of application will depend on many factors such as tree age, soil type, climate. These factors have been previously reviewed by George and Nissen (1987) and George *et al.* (1997). Depending on tree growth rates, trees planted in high-density systems under warm subtropical condition, may need to be treated with Cultar within 6-18 months after planting out. If trees are not treated earlier enough, before inter-and intra-tree competition commences, severe shading of the lower canopy can result in reduced floral bud initiation and weakening of fruiting wood. Initial rates of Cultar to apply in the first 2 or 3 season after planting are normally higher than in subsequent seasons due to changes in growth habit of treated trees and also because of a buildup of soil residues. Rates to apply on sandy soils will be much lower that they are on highly fertile loams. Similarly, lower rates will be required under dry growing conditions and where trees are drip irrigated due to slower growth rates. Because many variables affect the tree's response to Cultar, growers will need to establish the response dose rates for their own orchard and also evaluate the long term effects of Cultar on tree growth and health.

8.5.2 *Seasonal timing of soil applied Cultar*

In Australia, split applications may prove to be more effective than a single application per season. Limited experimental evidence indicates that 2 times of application may be appropriate:

1. in spring at the time of thinning (particularly if stub thinning is used);
2. post-harvest about 2-4 weeks after post-harvest summer pruning or topping.

The rationale for using these times of application is as follows. With low-chill stonefruit fruit development often occurs at the time of strongest vegetative growth so that control of spring growth with Cultar can increase fruit size substantially (up to 20%) (Allan *et al.*, 1993a, George *et al.*, 1994; Hieke *et al.*, 1997). Late summer or early autumn application will control growth after the post-harvest pruning, advance floral bud development and secure uptake and some growth control in the next season. Cultar application has also been shown to increase fruit set and advance floral bud development, flowering and fruit maturity (George and Nissen, 1987; Martin *et al.*, 1987).

8.5.3 *Cultar and other management practises*

Cultar alters the tree's response to other management practises such fertilising, pruning and timing and rates of dormancy-release chemicals (George and Nissen, 1987). For example, on heavy textured soils, trees treated with high rates of N during the FDP exhibit excessive vegetative growth. However when N and Cultar are combined, leaf N levels, photosynthesis rates and starch reserves are increased; these responses are associated with increases in yield, fruit size and quality (George *et al.*, 1992, 1996, unpublished data).

9 References

Allan, P., George, A.P. and Nissen, R.J. (1992). Effects of different methods of thinning on 'Flordaprince' peach. *Journal South African Society of Horticultural Science*, 2, 24-27.

Allan, P., George, A.P., Nissen, R.J., Rasmussen, T.S. and Morley-Bunker, M.J. (1993a). Effects of paclobutrazol on phenological cycling of low-chill 'Flordaprince' peach in subtropical Australia. *Scientia Horticulturae*, 53, 73-84

Allan, P., George, A.P. and Nissen, R.J. (1993b). Effects of girdling time on growth, yield, and fruit maturity of low-chill Flordaprince peach. *Australian Journal of Experimental Agriculture*, 33, 781-785.

Arnold, C.E. and Crocker, T.E. (1982). Peaches and nectarines in Florida. Cooperative Extension Service, University of Florida, Circular 299-B.

Avidan, B. and Erez, A. (1995). Studies of the response of peach and nectarine plants to gibberellin biosynthesis inhibitors in a hydroponic system. *Plant Growth Regulation*, 17, 73-80.

Balandier, P. and Rageau, R. (1989). Greenhouse production of nectarines for early harvest in France: a cultivation system with shallow dormancy or no dormancy. *Forest Tree Physiology*, 46, 242-245.

Batten, D. (1988). The effect of tree age on chilling requirements of low-chill stonefruit. In '*Proceedings of the 4th Australasian Conference on Tree and Nut Crops*', (ed. D Batten.) (NSW Agriculture) pp. 274-276.

Bederski, K. (1987). Selection and dormancy management of temperate zone deciduous fruit tree cultivars in coastal valleys of Peru. *Acta Horticulturae*, 199, 33-38.

Bellini, M., Cimato, A. and Mariotti P. (1980). Un trienno di observazioni sul pescheto-prato in cultura protetta. *Culture Protette*, 12, 27-32.

Black, M.W. (1952). The problem of prolonged rest in deciduous fruit trees. *Proceedings of the 13th International Horticultural Congress, London'* (ed. P. Synge) Volume 2, pp. 1122-1131.

Bowen, H.H. and Derickson, G.W. (1978). Relationship of endogeneous flower bud abscissic acid to peach chilling requirements, bloom dates and applied gibberellic acid. *HortScience*, 13, 694-696.

Broome, O.C. and Zimmerman, R.H. (1976). Breaking bud dormancy in tea crabapple (*Malus hupehensis* (Pamp.) Rehd.) with cytokinins. *Journal of the American Society for Horticultural Science*, 101, 28-30.

Byers, R.E. and Lyons, C.G. (1984). Flower thinning of peach with desiccating chemicals. *HortScience*, 19, 545-546.

Caruso, T., Di Marco, L., Giovannini, G. and Motisi, A. (1989). Trials on training systems for greenhouse cultivated peach trees. *Acta Horticulturae*,. 243, 345-352.

Castagnoli, S.P., DeJong, T.M., Weinbaum, S.A. and Johnson, R.S. (1990). Autumn foliage applications of $ZnSO_4$ reduced leaf nitrogen remobilisation in peach and nectarine. *Journal of the American Society for Horticultural Science*, 115, 79-83.

Chalmers, D., Mitchell, P.D. and van Heek, L. (1981). Control of peach tree growth and productivity by regulated water supply, tree density and summer pruning. *Journal of the American Society of Horticultural Science*, 106, 307-312.

Chalmers, D.J., Mitchell, P.D. and Jerie, P.H. (1984). The physiology of growth of peach and pear trees using reduced irrigation. *Acta Horticulturae*, 146, 143-149.

Chalmers, D.J., Mitchell, P.D. and Jerie, P.H. (1985). The relation between irrigation, growth and productivity of peach trees. *Acta Horticulturae*, 173, 283-288.

Chalmers, D.J. and Van den Ende, B. (1975a). Productivity of peach trees: factors affecting dry weight distribution during tree growth. *Annals of Botany*, 39, 423-433.

Chalmers, D. and Van den Ende, B. (1975b). The 'Tatura trellis' - a new design for high yielding orchards. *Victorian Journal of Agriculture*, 73, 473-476.

Cockroft, B. and Olsson, K.A. (1972). Pattern of new root production in peach trees under irrigation. *Australian Journal of Agricultural Research*, 23, 1021-1025.

Corelli-Grappadelli, L., and Coston, D.C. (1991). Thinning pattern and light environment in peach tree canopies influence fruit quality. *HortScience*, 26, 464-466.

Couvillon, G.A. and Erez, A. (1985). Effect of level and duration of high temperatures on rest in the peach. *Journal of the American Society for Horticultural Science*, 110, 579-581.

Crocker, T.E. and Sherman, W.B. (1986). Peaches and nectarines in Florida. Circular 299-C, Institute of Food and Agricultural Sciences, University of Florida, Gainesville.

Crocker, T.E., Young, M.J. and Norden, D.E. (1982). Effects of time of severe topping on yield, shoot growth and bud set of high - density peaches. *Proceedings of the Florida State Horticultural Society*, 95, 97-98.

Dann, I.R., Wildes, R.A. and Chalmers, D.J. (1984). Effects of limb girdling on growth and development of competing fruit and vegetative tissues of peach trees *Australian Journal of Plant Physiology*, 11, 49-58

Day, K.R., DeJong, T.M. and Hewitt, A.A. (1989). Postharvest and preharvest summer pruning of 'Firebrite' nectarine trees. *HortScience*, 24, 238-240.

De Villiers, H., Cutting, J.G.M., Jacobs, G. and Strydom, D.K. (1990). The effect of girdling on fruit growth and internal quality of Culemborg peach. *Journal of Horticultural Science*, 65, 151-155.

DeJong, T.M. (1982). Leaf nitrogen content and CO_2 assimilation capacity in peach. *Journal of the American Society for Horticultural Science*, 107, 955-959.

DeJong, T.M. and Doyle, J.F. (1985). Seasonal relationships between leaf nitrogen content (photosynthetic capacity) and leaf canopy light exposure in peach (*Prunus persica*). *Plant, cell and environment*, 8, 701-706.

Diaz, D.H. and Alvarez, A. (1987). Cultural and chemical practices to induce uniform budbreak of peach and apple under warm climate of Mexico. *Acta Horticulturae*, 199, 61-62.

Duarte, O. And Franciosa, R. (1974). Temperate fruit production in Peru, a special situation. In *Proceedings of the 19th Horticultural Congress, Warsaw,* 1B, pp. 519.

Edwards, G.R. (1985). Changes in endogenous hormones in apple during bud burst induced by defoliation, *Acta Horticulturae*, 158, 203-210.

Edwards, G.R. (1986). Ammonia, arginine, polyamines and flower initiation in apple. *Acta Horticulturae*., 179, 363-368.

Edwards, G.R. (1987a). Producing temperate-zone fruit at low latitudes: avoiding rest and the chilling requirement. *HortScience*, 26, 236-239.

Edwards, G.R. (1987b). Conditions of growth, dormancy and rest to produce temperate zone fruits under tropical conditions. *Acta Horticulturae*, 199, 57-61.

Edwards, G.R. (1987c). Temperature in relation to peach culture in the tropics. *Acta Horticulturae*, 199, 61-62.

Edwards, G.R. (1987d). The chilling requirement: How does it arise? How is it overcome? In *Proceedings of the First National Low Chill Stonefruit Conference, N.S.W. Australia.* (ed. I Skinner) NSW Agriculture, pp. 19-29.

Edwards, G.R. (1990). Ten years experience with temperate fruits in the tropics. *Acta Horticulturae*, 279, 47-53.

Edwards, G.R., Chapman, K.R. and O'Rourke, P.K., (1988). Computer assisted selection of locations in the tropics for the continuous cropping of apples. *Acta Horticulturae*, 232, 95-100.

Edwards, G.R, and Notodimedjo, S. (1987). Defoliation, bending and tip pruning of apple under tropical conditions. *Acta Horticulturae*, 199, 125-127.

Erez A. (1982). Peach meadow orchard: Two feasible systems. *HortScience*, 17, 138-142.

Erez, A. (1984a). Dwarfing peaches by pruning and paclobutrazol *Acta Horticulturae*, 146, 235-238.

Erez, A. (1984b). Effect of soil-applied paclobutrazol in drip irrigated peach orchards. *Acta Horticulturae*, 179, 513-520.

Erez A. (1985a). Defoliation of deciduous fruit trees with magnesium chlorate and cyanamide. *HortScience* 20, 452-453.

Erez A. (1985b). Growth control of peaches grown in a meadow orchard by paclobutrazol. *Acta Horticulturae.* 160, 217-224.

Erez A. (1985c). Peach meadow orchards. In '*First Symposium on "The Peach"*'. Acta Horticulturae, 173, 405-411.

Erez A. (1986). The significance of the leafless stage in deciduous fruit trees on fruit production. In *Proceedings of the Symposium on "Regulation of photosynthetic efficiency in fruit trees."* New York State Agricultural Experiment Station, Geneva, New York, pp. 23..

Erez A. (1987a). Use of the rest avoidance technique in peaches in Israel. *Acta Horticulturae*, 199, 137-144.

Erez, A. (1987b). Chemical control of budbreak. *HortScience*, 22, 1240-1243.

Erez A. (1990a). Container grown trees: a new option for an intensive culture of temperate-zone fruit trees in the tropics. *Acta Horticulturae*, 279, 159-166.

Erez A. (1990b). Off-season production of deciduous fruits by manipulation of the rest period. In *"Off Season Production of Horticultural Crops"* (ed. J Bay-Peterson) FFTC Book Series 41 Taiwan ISBN 957 9539 022. pp. 1-9.

Erez, A. (1995). Means to compensate for insufficient chilling to improve bloom and leafing. *Acta Horticulturae*, 395:81-95.

Erez, A. and Couvillon, G.A. (1983). Evaporative cooling to improve rest breaking of nectarine buds by counteracting high daytime temperatures. *HortScience*, 18, 480-481.

Erez, A., Couvillon, G.A. and Hendershott, C.H. (1979a). Quantitative chilling enhancement and negation in peach buds by high temperature in a daily cycle. *Journal of the American Society for Horticultural Science*, 104, 536-540.

Erez, A., Couvillon, G.A. and Hendershott, C.H. (1979b). The effect of cycle length on chilling negation by high temperatures in dormant peach leaf buds. *Journal of the American Society for Horticultural Science*, 104, 573-576.

Erez, A., Fishman, S., Linsley-Noakes, G.C. and Allan, P., (1989). The dynamic model for rest completion in peach buds. *Acta Horticulturae*, 276, 165-175.

Erez, A. and Flore, J.A. (1986). The quantitative effect of solar radiation on anthocyanin production in Redhaven peach fruit skin colour. *HortScience*, 21, 1424

Erez, A., and Lavee, S. (1969). Prunin identification, biological activity and quantitative change in comparison to naringenin in dormant peach buds. *Plant Physiology*, 44, 342-346.

Erez, A., Lavee, S. and Samish, R.M. (1971). Improved methods for breaking rest in the peach and other deciduous fruit species. *Journal of the American Society for Horticultural Science*, 96, 519-522.

Erez, A. and Lavee, S. (1974). Breaking the dormancy of deciduous fruit trees in subtropical climates. In *'Proceedings of the 19th International Horticultural Congress, Warszawa.'* (eds. R. Antoszewski, L. Harrison and J. Nowosielski) (Wroclawaska Drukamia Naukowa Warszawa).Volume III pp. 69-78

Erez, A. and Lavi B. (1984). Breaking bud rest of several deciduous fruit tree species in the Kenyan highlands. *Acta Horticulturae* 158, 239-248.

Erez, A.and Lerner H. (1990). Means to improve leafing using rest-avoidance technique in peaches in Israel. *Acta Horticulturae*, 279, 239-246.

Erez, A. Yablowitz, Z. and Korcinski, R. (1993). High density plantings for protected cultivation of fruit crops; Plastic cover for peach and nectarine. *Acta Horticulturae*, 349, 95-98.

Erez, A., Yablowitz Z. and Korcinski R. (1998). Greenhouse peach growing *Acta Horticulturae* 465, 593-600.

Erez, A. and Zur, A. (1981). Breaking the rest of apple buds by narrow-distillation range oil and dinitro-o-cresol. *Scientia Horticulturae*, 14, 47-54.

Evans, E.P. (1977). Pruning and crop control of peach trees. Department of Agricultural Services, Pretoria, South Africa. Technical Publication B.I. pp. 1-7

Fernandez-Escobar, R., Martin, R., Lopez-Rivares, P. and Paz Suarez, M. (1987). Girdling as a means of increasing fruit size and earliness in peach and nectarine cultivars. *Journal of Horticultural Science*, 62, 463-467.

Flore, J.A. (1994). Stonefruit. In *Handbook of Environmental Physiology of Fruit Crops. Volume 1. Temperate Fruits.* (eds. B. Schaffer and P. C. Andersen) CRC Press, USA. pp. 233 -271

Fuchigami, L.H. (1977). Ethephon-induced defoliation and delay of spring growth in *Cornus stolonifera* Michx. *Journal of the American Society for Horticultural Science*, 102, 444-452.

Fuss, A. M., Burne, P.M., Coombe, B.G. and Sedgley M. (1990). Cultural manipulation of out-of-season peach production under glass. *Scientia Horticulturae*, 43, 15-27.

George, A.P., Campbell, J.A., and Nissen, R.J. (1992a). Orchard management - An overview. In *Proceedings of the 2nd National Low-chill Stonefruit Conference,* Ballina Beach Resort, Ballina, NSW (ed. J.M. Slack) pp. 83-99.

George, A.P., Hieke, S., Rasmussen, T., and Ludders, P. (1996). Early shading reduces yield and late shading reduces quality in low chill peach in subtropical Australia. *Journal of Horticultural Science*, 71, 561-571.

George, A.P. and Nissen, R.J. (1987). Growth control of low chill stonefruit using growth retardants and other management techniques. In *Proceedings of the First National Low Chill Stonefruit Conference, N.S.W. Australia.* (ed. I. Skinner). pp.132-133.

George, A.P. and Nissen, R.J. (1988a). Chemical methods of breaking dormancy of low-chill nectarines: preliminary evaluations in subtropical Queensland. *Australian Journal of Experimental Agriculture*, 28, 425-9.

George, A.P. and Nissen, R.J. (1988b). Effects of paclobutrazol on reproductive and floral development of Fla 6-3 and Sunred nectarines. *Acta Horticulturae*, 189, 41-45.

George, A.P. and Nissen, R.J. (1990). Effects of hydrogen cyanamide on yield, growth and dormancy release of table grapes in subtropical Australia. *Acta Horticulturae*, 279, 427-436.

George, A.P. and Nissen, R.J. (1992). Effects of water stress, nitrogen and paclobutrazol on flowering, yield and fruit quality of the low-chill peach cultivar 'Flordaprince' *Scientia Horticulturae*, 49, 197-209

George, A.P. and Nissen, R.J., (1993). Effects of growth regulants on defoliation, flowering, and fruit maturity of the low chill peach cultivar Flordaprince in subtropical Australia. *Australian Journal of Experimental Agriculture*, 33, 787-795.

George, A.P., Nissen, R.J. and Baker, J.A. (1986). Low-chill peach and nectarine cultivars. *Queensland Agricultural Journal*, 112, (1), 27-32.

George, A.P., Nissen, R.J. and Baker, J. (1988a). Effects of hydrogen cyanamide in manipulating budburst and fruit maturity of table grapes in south-eastern Queensland. *Australian Journal of Experimental Agriculture*, 28, 533-538.

George, A.P., Nissen, R.J. and Campbell, J.A. (1992b). Control of flowering and fruit maturity in low-chill stonefruit using different management techniques. In *Proceedings of the 2nd National Low-chill Stonefruit Conference, Ballina Beach Resort, Ballina NSW* (ed: J. Slack) pp. 83-90.

George, A.P., Nissen, R.J., Campbell, J.A., Rassmussen, T. and Allan, P. (1994). Effects of paclobutrazol on growth and yield of low chill peaches in subtropical Australia. *Acta Horticulturae*,409, 109-117.

George, A.P., Nissen, R.J. and Campbell, J.A. 1995. Nutritional studies in low-chill stonefruit. *Acta Horticulturae*, 409, 99-107.

George, A.P., Nissen, R.J., Lloyd, J. and Richens, K. (1988b). Factors affecting fruit quality of low chill stonefruit in subtropical Australia. *Acta Horticulturae*, 279, 559-571

George, A.P., Nissen, R.J. and Rasmussen, T. (1993). Effects of post-harvest topping, autumn cincturing and paclobutrazol on growth, yield and fruit quality of the low-chill nectarine *cv.* Sundowner in subtropical Australia. *Australian Journal of Experimental Agriculture*, 33, 353-362.

George, A.P., Nissen, R.J. and Sherman, W.B. (1988c). Overlapping double and early single cropping of low-chill peaches in Australia. *Fruit Varieties Journal*, 42, 91-95.

Gilreath, P.R. and Buchanan, D.W. (1981). Floral and vegetative bud development of 'Sungold' and 'Sunlite' nectarine as influenced by evaporative cooling by overhead sprinkling during rest. *Journal of the American Society for Horticultural Science*, 106, 321-324.

Grossman, Y and DeJong, T.M. (1995a). Maximum fruit growth potential and seasonal patterns of resource dynamics during peach growth. *Annals of Botany*, 75, 553-560.

Grossman, Y and DeJong, T.M. (1995b). Maximum fruit growth potential following resource limitation during peach growth. *Annals of Botany*, 75, 561-567.

Hieke, S., George, A.P., Rasmussen, T. and Ludders. P. (1997). Effects of plastic soil mulch and cloches on fruit growth and development of low-chill peach *cv.* Flordaprince (*Prunus persica* (L.) Batsch) in Australia. *Journal of Horticultural Science*, 72, 187-193.

Huett, D.O., George, A.P., Slack, J.M. and Morris, S.C. (1997). Diagnostic leaf nutrient standards for low-chill peaches in subtropical Australia. *Australian Journal of Experimental Agriculture*, 37, 119-126.

Jackson, J.E. (1980). Light interception and utilisation by orchard systems In '*Horticultural Reviews*', (ed. J. Janick), 2, 208-266.

Janick, J. (1974). The apple in Java. *HortScience*, 9, 13-15.

Javaraya, H.C. (1943). Biannual cropping of apple in Bangalore. *Indian Journal of Horticulture*, 1, 31-34.

Johnson R.S., Handley D.F. and De Jong T.M. (1992). Long term response of early maturing peach trees to post-harvest water deficit. *Journal of the American Society of Horticultural Science*, 117, 881-886.

Kusumo, S. (1972). Cultivation of apple in Indonesia. Department of Agriculture, Malang (in Indonesian).

Lang, G.A., Early, J.D., Martin, G.C. and Darnell, R.L. (1987). Endo-, para-, and eco-dormancy: physiological terminology and classification for dormancy research. *HortScience*, 22, 371-377.

Leece, D.R. and Gilmour, A.R. (1974). Seasonal changes in leaf composition of peach. *Australian Journal of Experimental Agriculture and Animal Husbandry*, 14, 822-827.

Leuty, S.J. and Pree, D.J. (1980). The influence of tree population and summer pruning on productivity, growth and quality of peaches. *Journal of the American Society for Horticultural Science*, 105, 702-705.

Llanes, R., de las Alas, R.A., Otom, E. and Edwards, G.R. (1987). Apples and peaches in tropical lowlands of the Philippines. *Acta Horticulturae*, 199, 64-68.

Lloyd, J. and Firth, D. (1990). Effect of defoliation time on depth of dormancy and bloom time for low chill peaches. *HortScience*, 25, 1575-1578.

Mannini, F., and Ryugo, K. (1983). Effect of 2-chloro-ethylphosphonic acid (ethephon) on the endogeneous levels of gibberellin-like substances and abscissic acid in buds of developing shoots of three grape varieties. *Journal of Enology and Viticulture*, 33, 164-167.

Marini, R.P. (1985). Vegetative growth, field and fruit quality of peach as influenced by dormant pruning, summer pruning and summer topping. *Journal of the American Society for Horticultural Science*, 110, 733-739.

Marini, R.P. (1986). Growth and cropping of 'Redhaven' peach trees following foliar applications of flurprimidol and paclobutrazol. *Journal of the American Society for Horticultural Science*, 111, 849-853.

Marini, R.P. (1987). Growth and cropping of 'Redhaven' peach trees following soil application of paclobutrazol. *Journal of the American Society for Horticultural Science*, 112, 18-21.

Marini, R.P. and Sowers, D.L. (1990). Net photosynthesis, specific leaf weight, and flowering of peach as influenced by shade. *HortScience*, 25, 331-334.

Martin, G.C., Yoshikawa, F. and La Rue, J.H. (1987). Effect of soil applications of paclobutrazol on vegetative growth, pruning time, flowering, yield, and quality of 'Flavorcrest' peach. *Journal of the American Society for Horticultural Science*, 112, 915-921.

McDermott, L.G., Sherman, W.B. and Crocker, T.E. (1987). Yield and fruit quality evaluations of seven peach clones in two orchard systems. *Proceedings of the Florida State Horticultural Society*, 100, 315-319.

Menzies, A.R. and Richens, K. (1989). Manipulation of the peach meadow orchard. *Acta Horticulturae*, 240, 115-118.

Mika, A.L. (1986). Physiological response of fruit trees to pruning. *Horticultural Review*, 8, 337-378.

Mika, A.L. (1968). Effects of shoot-bending and pruning on growth and fruit bud formation in young apple trees. *Horticultural Research*, 9, 93-102.

Moreshet, M., Stanhill, G. and Fuchs, M. (1975). Aluminum mulch increases quality and yield of Orleans apples. *HortScience*, 10, 390-391.

Munoz, C., Sepulveda, G., Garcia-Huidobro, J. and Sherman, W.B. (1986). Determining thermal time and base temperature required for fruit development in low-chilling peaches. *HortScience*, 21, 520-522

Murase, S., Yamazaki, T., Inomata, Y. and Suzuki, K. (1990). Dwarfing rootstock for peach. *Japan Agricultural Research Quarterly*, 23, 294-300.

Noel, A.R.A. (1970). The girdled tree. *Botanical Review*, 36, 162-195.

North, M.S. (1995). New rest-breaking agents for the control of delayed foliation of apples. *Acta Horticulturae*, 409, 151-154.

North, M., Stadler, J.D. and De Villiers, H. (1988). Effect of cyanamide and girdling on yield, fruit size and quality of De Wet peach. *South African Journal of Botany* 5, 224-226.

Notodimedjo, S., Danoesastro, H., Sastrosumarto, S. and Edwards, G.R. (1981a). Shoot growth, flower initiation and dormancy of apple in the tropics. *Acta Horticulturae*, 120, 179-186.

Notodimedjo, S., Danoesastro, H., Sastrosumarto, S. and Edwards, G.R. (1981b). Growth periodicity in apples under tropical conditions. *Acta Horticulturae*, 120, 256.

Rageau, R. and Ridray, G. (1989). Nectarine cultivation in greenhouse for early harvest in France: A management system with rest avoidance. *Acta Horticulturae*, 254, 145-150.

Richardson, E.A., Seeley, S.D. and Walker, D.R. (1974). A model for estimating the completion of rest for 'Redhaven' and 'Elberta' peach trees. *HortScience*, 9, 331-332.

Rom, R.C. (1988). Progress in peach rootstocks. In *The Peach* (eds. N.F. Childers and W.B. Sherman) Horticultural Publications, Gainsville, Florida. pp. 302-308.

Ross, S.D. and Pharis, R.P. (1985). Promotion of flowering in tree crops: different mechanisms and techniques, with special reference to conifers. In *'Attributes of Trees as Crop Plants'* (eds. M.G.R. Cannell and J.E. Jackson.) Institute of Terrestial Ecology, UK. pp. 383-397.

Saure, M. (1973). Successful apple growing in Indonesia. *Fruit Varieties Journal*, 27, 44-45.

Saure, M.C. (1985). Dormancy release in deciduous fruit trees. *Horticultural Reviews*, 7, 256-259.

Sharpe, R.H., Sherman, W.B. and Martsolf, J.D. (1990). Peach cultivars in Florida and their chilling requirements. *Acta Horticulturae*, 279, 191-197.

Sherman, W.B. (1987). The low-chill peach improvement program: why, how, where. In *'Proceedings of the First National Low-chill Stonefruit Conference'*. Exotic Fruit Growers Association, Lismore, NSW. pp.16-18.

Sherman, W.B., Knight, R.J. and Crocker, T.E. (1978). Peach and nectarine breeding and testing in warm parts of the world In *Proceedings of the Tropical Region of the American Society for Horticultural Science*, 22, 103-107.

Sherman, W.B. and Lyrene, P.M. (1984). Biannual peaches in the tropics. *Fruit Varieties Journal*, 38, 37-39.

Sherman, W.B., Lyrene, P.M. and Miller, E.P. (1984). Progress in low-chill peaches and nectarines from Florida. *Proceedings of the Florida State Horticultural Society*, 97, 320-322.

Slack, J.M., Huett, D.O. and George, A. (1996). Fertilising low-chill stonefruit. A guide to nutrient management of low-chill peaches and nectariens in Australia. NSW Agriculture.

Stadler, J.D. (1975). Thinning of stonefruit. Farming in South Africa. Stone Fruit Series. Leaflet No C3. Fruit and Nut Technological Research Institute, Stellenbosh.

Stiles, W.C. (1984). Effects of pruning on growth and size of trees. *Acta Horticulturae*, 146, 97-104.

Tan, C.S. and Buttery, B.R. (1986). Photosynthesis, stomatal conductance and leaf water potential in response to temperature and light in peach. *HortScience*, 21, 1180-1188.

Topp, B.L. and Sherman, W.B. (1989a). The relationship between temperature and bloom-to- ripening period in low-chill peach. *Fruit Varieties Journal*, 43, 155-158.

Topp, B.L. and Sherman, W.B. (1989b). Location influences on fruit traits of low-chill peaches in Australia. *Proceedings of the Florida State Horticultural Society*, 102, 195-199.

Tukey, L.D. (1964). *Dwarfed Fruit Trees*. Macmillan: New York.

Van den Ende, B. and Chalmers, D.J. (1982). An evaluation of commercial experience with the Tatura trellis for growing peaches, *HortScience*, 17, 218-220.

Verheij, E.W.M. (1985). Apple growing in east Java: the scope for improvement. *Acta Horticulturae*, 158, 47-51.

Verheij, E.W.M. (1990). Adaptations of apple to tropical conditions. *Acta Horticulturae*, 279, 83-90.

Walser, R.H., Walker, D.R. and Seeley, S.D. (1981). Effect of temperature, fall defoliation and gibberellic acid on the rest period of peach buds. *Journal of the American Society Horticultural Science*, 106, 91-94.

Westwood, M.H. (1978). Temperate-zone Pomology. Freeman and Company, San Francisco.

Weinbaum, S.A., Southwick, S.M., Shackel, K.A., Muraoka, T.T., Krueger, W. and Yeager, J.T. (1989). Photosynthetic photon flux influences macroelement weight and leaf dry weight per unit of leaf area in prune tree canopies. *Journal of the American Society for Horticultural Science*, 114, 720-723.

Weinberger, J.H. (1950). Chilling requirements of peach varieties. *Proceedings of the American Society of Horticultural Science*, 56, 122-128.

Williams, M.W. (1984). Use of bioregulators to control vegetative growth of fruit trees and improve cropping efficiency. *Acta Horticulturae*, 146, 97-104.

Young, M.J. and Crocker, T.E. (1982). Severe post-harvest topping of high density peaches and nectarines. *HortScience*, 17, 220-222.

10 POME FRUIT GENETIC POOL FOR PRODUCTION IN WARM CLIMATES

ROBERTO HAUAGGE [1] *and* JAMES N. CUMMINS [2]

[1] *IAPAR - Instituto Agronômico do Paraná, P.O. Box 2301 80.001-970 - Curitiba, Pr., Brazil*
[2] *Department of Horticultural Sciences, New York State Agricultural Experiment Station, Cornell University, Geneva, NY 14456 USA*

1 Introduction

Two basic factors determine the adaptation of pome fruit to warm climates:
1. the ability of a given species/cultivar to break bud, bloom, set fruit, and grow satisfactorily, naturally or under proper cultural practices; and
2. the ability to produce quality fruit in temperatures during growth and ripening that are higher than the optimum for most cultivars.

The first factor is mostly determined by the genotype's chilling requirement (CR) to break dormancy. In both cases there is genetic variability that can be used as the basis for selection for successful production in warm climates.

2 Dormancy

The cultivar's CR is determined by its genetic makeup, and the chilling units (CU) to satisfy that requirement are accumulated by the environment. Given these two fixed determinants, there are three basic pome fruit production in warm areas:
(a) use of cultivars that have sufficiently low CR to break dormancy naturally;
(b) application of rest-breaking chemicals;
(c) dormancy avoidance observed in certain cultivars when grown in environment that does not induce endodormancy.

2.1 Cultivar chill requirement

Wide variation for the length of the bud dormancy period among cultivars, wild species and interspecific hybrids under field conditions has been reported for apples (Hill and Campbell, 1949; Black, 1952; Darrow, 1953; Crossa-Reynaud, 1955; Chandler and Brown, 1957; Boynton, 1959; Overcash and Murphy, 1959; Chandler, 1960; Samish and Lavee, 1962; Mazzocchi and Pucci, 1963; Nesterov, 1957; Fregoni, 1968; Popenoe, 1970; Popenoe and Ben'tez, 1965; Lloréns, 1967; Popenoe, 1970; Kemp, 1971; Lecrenier, 1972; Ruck, 1975; D'az and Sadowska, 1979; Erez, 1980; Hunter, 1981; Walheim and Stebbins, 1981; Miller and Baker, 1982; Seth and Ghildiyal, 1984; Bederski, 1988; Nichols, 1994) and pears

Amnon Erez (ed.), Temperate Fruit Crops in Warm Climates, 267-303.
© 2000 *Kluwer Academic Publishers. Printed in the Netherlands.*

(Crossa-Raynaud, 1955; Chandler and Brown, 1957; Mazzocchi and Pucci, 1963; Tabuenca, 1964; Phillipe, 1965; Sherman and Crocker, 1982; Subhadrabandhu and Punsri, 1987; Ghariani and Stebbins, 1994). In only a few cases have variability studies been conducted under controlled conditions (Eggert, 1951; Pasternak, 1983; Petropoulou, 1985; Young and Werner, 1985; Hauagge and Cummins, 1991a; Ghariani and Stebbins, 1994).

Genotypic sequence of budbreak/bloom is related to the genotype CR growing in a chill-deficient environment. This relationship has been used to establish the CR for peach (Weinberger, 1967; Sharpe and Sherman, 1990), apple (Miller and Baker, 1982; Hauagge and Cummins, 1991a) and pear cultivars growing in warm climates. Regarding different species, it seems that peaches bloom earlier than apples and pears for the same CR level (Fig. 1) (Hauagge, unpublished).

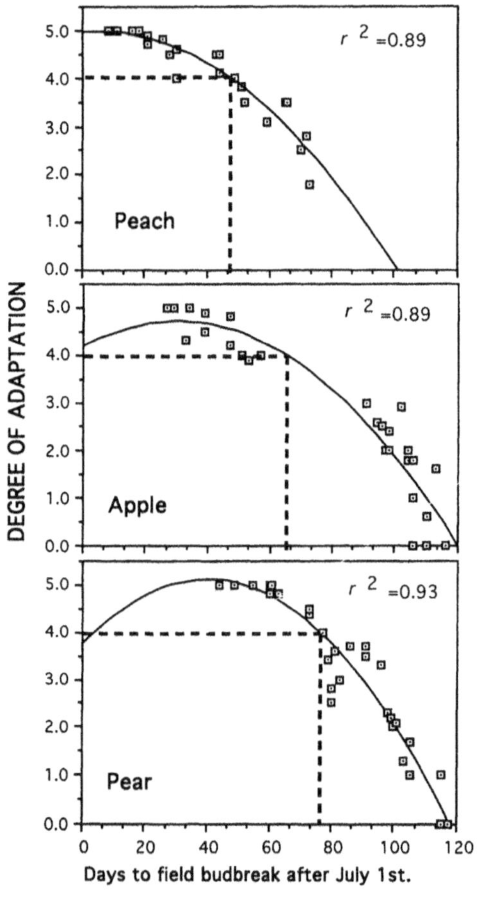

Figure 1. Field budbreak sequence and degree of adaptation of peach, apple and pear cultivar collections at IAPAR in Piraquara, Pr., Brazil; average of two seasons. Annual effectively accumulated CU ranged from 350 to 550. Degree of adaptation ranged from 0 (absolutely non-adapted) to 5 (completely adapted), where Sungold nectarine was rated 4 (full crop, full leafing, but not quite uniform leafing).

Besides bud CR, there are other attributes that may play roles in genotypic adaptation to warm climates. These include: the general pattern of intensity of bud dormancy; variability in the maximum bud dormancy intensity; and the level of changes in bud dormancy intensity caused by the same amount of chilling (Hauagge and Cummins, 1991b). Characteristically, very low *CR* apple cultivars (less than 400 CU) have a shallow depth of dormancy with very little alteration throughout the year. Budbreak may be stimulated even by a short water deficit, without any CU accumulation. As long as flower buds have formed, this type of cultivar responds well to the application of chemicals that break dormancy at any given time (Fig. 2). Because of this, they are more reliable producers.

Figure 2. 'Golden Delicious' (right) and 'Anabela' (IAPAR 76) (left) midsummer response to the application of rest-breaking chemicals at Piraquara, Pr., Brazil; plants had not experienced prior water stress; dormancy breaking was made 180 days after a previous similar treatment during winter.

2.2 Dormancy avoidance

Dormancy avoidance is observed in some high chill cultivars when grown in warm environments that will not induce endodormancy (Notodimedjo *et al.*, 1981; Erez and Lavi, 1984; Edwards, 1987). Terminal bud dormancy is avoided by defoliation after harvest. No CR is apparent and dormancy of most lateral buds is not broken by defoliation. Lateral bud burst is stimulated by shoot bending, pruning, renewed irrigation, and/or application of rest-breaking chemicals. 'Rome Beauty' apple has been widely and successfully grown in this production system.

2.3 Dormancy and productivity

Although it is possible to break bud dormancy with chemicals, final results of growth, production and quality under this production system are generally inferior to that obtained with environmentally adapted cultivars. Under those circumstances, productivity is much more dramatically affected in pears (Fig. 3b) than in apples (Fig. 3a). This is because in pear there is a strong correlation between lack of chilling and flower bud abortion (Fig. 4) before and during budbreak (Fig. 3c). It is therefore possible to produce pears efficiently only with environmentally adapted cultivars.

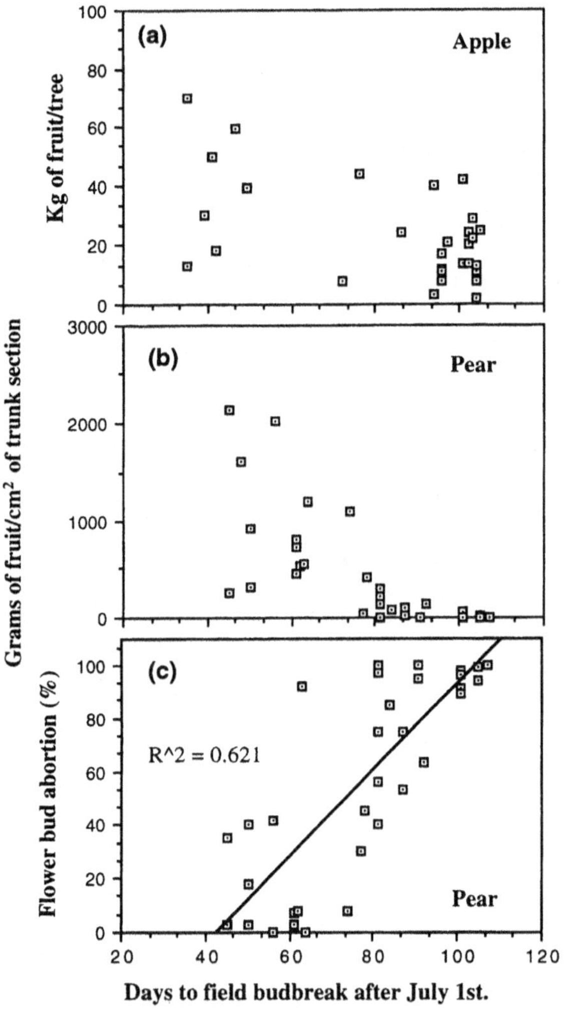

Figure 3. Relationship between time of budbreak (as related to CR) and productivity of apple and pear cultivar collections at IAPAR in Piraquara, Pr. Brazil. (a) Average kg of fruit per tree from 5th to 7th year after planting; trees grafted on MM.106 rootstocks. (b) Accumulated production through the 7th year of planting; trees grafted on 'Garber' rooted cuttings. (c) Average of 3 seasons; annual cumulative CU ranged from 350 to 550; dormancy was artificially broken on non-adapted cultivars at the start of green tip bud stage.

Figure 4. Early spring normal bloom of a low chill European type selection (upper photograph) compared to a severe flower bud abortion (lower photograph) in 'Bartlet' pear later in the spring (IAPAR in Piraquara, Pr., Brazil, 400 CU accumulated).

Besides delayed foliation, other problems may also result from insufficient chilling. These problems include: stunted growth; shortening of internodes; shoot stiffening; shoot suckering; upright growth; zinc deficiency-like leaf chlorosis; shorter tree life; higher susceptibility to latent viruses; poor foliage development; flower bud abortion; lesser amount of pollen produced; lower pollen viability; low fruit set; flattening and other shape abnormalities in fruits; short fruit peduncle; poorer fruit flavor; excessive fruit acidity; higher fruit russeting susceptibility; and lower fruit storage capacity.

3 Fruit quality

For apple, the amount of chilling received during winter and the temperature regime during fruit growth are major determinants of fruit quality characteristics such as shape, size, flavor, firmness, sugar/acid ratio, total soluble solids, acidity, color, russeting susceptibility; and storage potential. It is difficult to dissociate these temperature effects because they are related to each other in higher altitudes.

It is possible to identify apple genotypes in which fruit quality is differently affected by these environmental factors (Table 1). For example, in the case of 'Fuji' apple the ratio length: diameter may vary in a given year between 0.90±0.09 (about 600 CU location), and 0.77±0.05 (250 CU location). However, 'Fuji Frey', a lower chill mutant of 'Fuji' apple found in Brazil, the ratio in the same 0.97±0.08.

Night temperatures around 12°C have been found to be the optimum for color development for 'McIntosh' apples. Cultivars such as 'Beacon' and a few of its derivatives will color well even at much higher temperatures. In Brazil, the best color development for 'Gala' and 'Fuji' apples is found in locations where minimum temperatures a month prior to ripening are below 15.5°C and 13°C, respectively. In Florida, with temperature minimal are rarely under 22°C in the month preceding harvest, the ground color of apples is often green or washed-out yellow and striping is fainter (Miller and Baker, 1982).

Pear fruit quality is not as highly affected as apple. It is possible to observe, however, a consistent increase in both yellow color and smoothness of skin in colder areas.

4 Apples

Only a few apple cultivars with CR of 600 CU or lower can be found within the wide range of cultivars developed in temperate areas. Within wild species, the low chill character is found in *Malus arnoldiana, M. baccata, M. brevipes, M. floribunda, M. halliana, M. magdeburgensis, M. pallasiana, M. platycarpa, M. prunifolia*; *M. x purpurea, M. rockii, M. robusta, M. scheideckeri, M. sieboldii, M. spectabilis, M. sylvestris*, and *M. zumi calocarpa* (Nesterov, 1957; Petropoulou, 1985; Hauagge and Cummins, 1991a), and their hybrids. The generally poor fruit quality of these species makes them of little value for immediate cultivar development.

Table 1. Intensity of changes of the main fruit quality parameters and productivity observed in high chill apple cultivars grown in different location which accumulates between 250 and 800 hours below 7.2°C in Paraná State, Brazil (0 = no changes, 1 = little changes, 2 = significant changes, 3 = intensive changes)

Cultivar	*Size*	*Shape*	*Color*	*Flavor*	*Productivity*
Blackjohn	2	2	1	3	2
Braeburn	2	1	3	3	2
Cox Orange Pippin	3	3	3	3	3
Delicious (colored strains)	3	3	1-2	2	3
Elstar	3	3	2	1	2
Empress	1	1	1	1	2
Fred Hough	2	0-1	2	1	1
Fuji	3	3	3	1-2	2
Gala	1-2	1	1	0-1	1-2
Golden Delicious	0-1	0-1	1-2	0-1	1-2
Granny Smith	0-1	0-1	0	1	1
Jonagold	1	2	2	3	2-3
Jonamac	1	1	1	1-2	2
Melrose	0-1	1	0-1	2	1
Mollie's Delicious	0-1	0-1	0-1	0-1	1
NJ 55	0-1	0-1	1	1	1-2
Rainha	0-1	0-1	1	1	0-1
Rome Beauty	1-2	1	2	1-2	2
Sansa	2	3	1	1	3
Vista Bella	1	0-1	0-1	1	2
William's Pride	0-1	1	1	1	2
Winter Banana	0-1	1	2	2	1-2

There are several low-chilling cultivars of poor quality in Mediterranean countries. These include Douce de Sfax, Aigre de Sfax, Mestiri, Halouca (Crossa-Raynaud, 1955, 1960); Khashabi (Samish and Lavee, 1962); Hammuri, Kadduri, Meschi (Mazzocchi and Pucci, 1963); Bou Tabgaia (Lecrenier, 1972); Abdu Khora, Adjemi, Ahmar Seyfi, Bayyudi, Beiri, Drashawi, Firiki, Hammuri, Sakarjii, Sherabi, Sukkari, Khashabi (Ruck, 1975). A few of them were used in the apple breeding programs in Israel.

There are small fruited highly colored low chill apple cultivars locally grown in South of China (G. Richards, 1997, University of Western Sydney, Australia, personal communication). They have not been introduced yet in other countries, and they should be useful in breeding.

4.1 Apple breeding for low chilling requirement (LCR)

Breeding for LCR apples was started as early as 1939 in Israel by Oppenheimer and Slor. Very LCR local low fruit quality Arabic cultivars 'Succari', 'Biari' and an unnamed selection were initially crossed with 'Calville St. Sauveur' and 'Red Astrachan'. 'Vered', 'Yael', 'Rachel', 'Naomi' and 'Tamar', which come from a second generation of these crosses, were released in the mid-1950s. 'Michal', 'Maayan', and 'Shlomit', involving modified backcrosses with 'Delicious' and 'Jonathan', were released in 1967 (Oppenheimer and Slor, 1962; 1968). 'Slor', an open-pollinated seedling of 'Shlomit', was released in 1982 (Slor, 1982). 'Slor' and 'Elah' are the two most interesting cultivars originated from the work of Oppenheimer and Slor.

'Anna' was the first commercial LCR apple widely grown (Fig. 5). It was bred by Abba Stein from Ein Shemer, Israel (Brooks and Olmo, 1972), being a cross of 'Red Hadassiya' and 'Golden Delicious'. 'Ein Shemer', a cross between 'Zabidani' and 'Golden Delicious' was released about the same time.

Possibly derived from LCR material from the middle East, 'Dorsett Golden' was first selected in 1959 by Mrs. Irene Dorsett of Nassau, Bahamas (Miller and Sherman, 1980). 'Anna' and 'Dorsett Golden' are far superior to any previously developed LCR apple cultivar. They have been widely grown commercially as well as used as parents in many breeding programs.

Instituto Agronômico de Campinas (IAC), São Paulo, was the first to start breeding apples in Brazil. Among the released cultivars, 'Rainha', introduced in 1975 by O. Rigitano, M. Ojima, and A. C. Dall'Orto, has achieved economic importance in eastern and southern Brazil. 'Galícia' (Gala x Anna) and 'Soberana' (Anna open-pollinated seedling), both with CR in the range of 200-300 CU, have been recently released and are undergoing testing (Dall'Orto *et al.*, 1987).

At EPAGRI, Santa Catarina, Brazil, the apple breeding program carried out by F. Denardi, A. Camilo and the late L. F. Hough is focusing on the development of apple cultivars for areas that accumulate about 500 hours below 7.2°C and that are resistant to scab and powdery mildew. The cultivars 'Primicia', 'Princesa' (Fig. 6), 'Fred Hough', 'Imperatriz' (EPAGRI 404) (Fig. 7) and 'Baronesa' (EPAGRI 405) have been released recently. 'Princesa' needs about 450 CU, and 'Baronesa' 500-600 CU. The other releases are in the range of 600-800 CU, have excellent response to rest-breaking chemicals, and have good productivity even in much warmer locations. They are recommended by EPAGRI as pollinizers to other commercial cultivars. 'Fred Hough' and 'Primicia' are immune to scab (*Vf* gene).

The apple breeding program underway at Instituto Agronômico do Paraná (IAPAR), Paraná, Brazil aims to develop cultivars that need less than 400 hours of chilling and are early ripening and disease resistant. The cultivars 'Eva' (IAPAR 75) (Fig. 8), 'Anabela' (IAPAR 76) and 'Carícia' (IAPAR 77) (Fig. 9) were released in 1995. All of them are resistant to apple leaf scorch caused by *Glomerella* sp. All need between 300-400 CU. 'Carícia' (IAPAR 77) also carries *Vf* gene for immunity to scab.

The LCR early ripening apple 'Tropic Sweet' was recently released by W. B. Sherman and P. M. Lyrene at University of Florida, Gainesville. Fruits are firmer, sweeter and tastier than Anna (Sherman and Lyrene, 1996). The selection M-60-39, also from

Figure 5. Heavy setting in last year wood of 'Anna' apple.

Figure 6. 'Princesa' LCR apple.

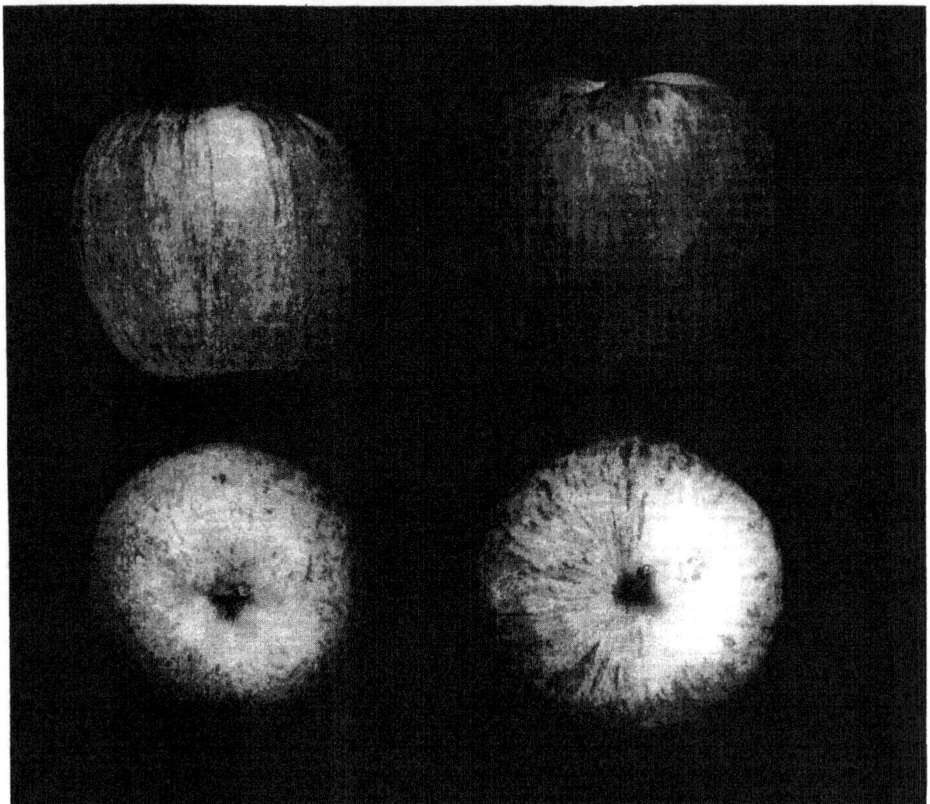

Figure 7. 'Imperatriz' (EPAGRI 404) medium chilling requirement apple.

Figure 8. 'Eva' (IAPAR 75) LCR apple.

Figure 9. 'Carícia' (IAPAR 77) low chill disease resistant apple.

Florida's apple breeding program, is a very firm 'Dorsett Golden' type that has excellent dessert qualities in cool night climates.

Western Australia has unique climatic conditions with longer growing season than other traditional apple producing countries in Europe and North America, and high temperatures till late in the season (Tancred *et al.*, 1994). An apple breeding program led by John Cripps aimed at development of varieties that would mature very late, as the cool nights of autumn set in. From the 'Lady Williams' x 'Golden Delicious' family, Cripps introduced 'Pink Lady' and 'Sundowner', which mature four and six weeks after 'Granny Smith', respectively. These cultivars have a CR of about 700 CU, but still they have the ability to perform well under warm growing conditions. Other Australian introductions with similar behavior are 'Adina' and 'Goldina' (both developed by Henry Franklin) and GB 63-43 (from Granite Belt Horticultural Research Station, Applethorpe).

4.2 Apple cultivars for warm climates

Some apple cultivars that may have value for fruit production in locations that are warmer than the typical production areas are listed below alphabetically:

Adina (Frankad, Satark® Adina™). Originated at Stanthorpe, Queensland, Australia, by H. Franklin. Fruit: large, color red to purplish red, occasional stripe; shape round and conic; flesh firm, very juicy; sweet, with distinctive hint of cinnamon; superior eating quality; harvest season very early. Tree: vigorous; precocious; low CR (350 h); grows in Hardiness Zones 6 to 9 in the United States. Dorsett Golden is a suitable pollenizer (Cummins, 1991).

Anabela (IAPAR 76). Originated at IAPAR, Paraná, Brazil by R. Hauagge and M. Tsuneta. Released in 1995. Cross 'Gala' x 'Anna'. Fruit: medium size; cylindrical-conical; 30-40% light red color over cream/yellow background; fruit needs to be well exposed for coloring; resembles 'Anna'; juicy and sweet; very good dessert quality; firm; can be cold stored for 2-4 months; harvest 1-2 weeks after 'Anna'. Tree: vigorous growth and productive; needs 300 CU; blooms 1 weeks after 'Anna'; does better on semi-dwarfing and dwarfing rootstocks.

Anna. Originated at Ein Shemer, Israel, by Abba Stein. Introduced in 1963. Cross 'Red Hadassiya' and 'Golden Delicious' (Brooks and Olmo, 1972). Fruit: large, red striped on light green background; susceptible to calyx end breakdown. Trees: annually productive (80-120 tons/ha have been recorded); excellent fruit set in one year old shoots; susceptible to powdery mildew and scab; self unfruitful.

Baronesa (EPAGRI 405). Originated at EPAGRI, Caçador, SC, Brazil, by F. Denardi and A. P. Camilo. Cross 'Princesa' x 'Fuji'. Introduced in 1997; tested as M-57/91. Fruit: harvested with 'Fuji'; 130 g average; round, uniform; red blush over green-yellow background, similar to 'Fuji'; sweet and crisp, quality similar to 'Fuji'; storage potential up to 10 months in CA. Tree: vigorous; highly productive; good field resistance to scab, powdery mildew and bitter rot; needs 500-600 CU; recommended as pollenizer for 'Fuji' at Caçador.

Beverly Hills. Originated at Los Angeles, at University of California by W. H. Chandler. Introduced in 1945. Cross 'Melba' x 'Early McIntosh' (Brooks and Olmo, 1972). Fruit: resembles McIntosh. Tree: Needs 600-800 CU; once widely grown in subtropical areas; lower fruit quality than the newer LCR cultivars.

Blenheim Orange. English variety; synonymous with 'Emilia' in Ecuador and 'Pennsylvania' in Colombia (Darrow, 1953). It is a triploid variety that requires cross pollination; requires less chilling than most other European and American cultivars; once grown in many Latin American Countries.

Brasil. Originated in Piedade, SP., Brazil, by A. Bruckner. Synonymous with 'Bruckner' and 'Bruckner do Brasil'. Introduced in the 1940s. Fruit: large; resembles standard 'Delicious'; subacid; stores poorly. Tree: vigorous; initially upright, then spreading when mature; high CR; self-fruitful, but pollination improves fruit shape, texture and shelf life. It was the most widely planted apple cultivar in São Paulo, Brazil till late 1970s.

Carícia (IAPAR 77). Originated at IAPAR, Paraná, Brazil, by R. Hauagge. Released in 1995. Cross 'Prima' x 'Anna'. Fruit: harvested 3-4 weeks after 'Anna'; uneven ripening; resembles 'Prima', but larger and redder; flavor subacid, good dessert quality; firm; can be cold-stored for 2 months. Tree: vigorous; spur growth habit; resistant to scab (*Vf* gene), powdery mildew and leaf scorch caused by *Glomerella* sp.; productive; needs 350 CU; blooms 2-4 weeks after 'Anna'.

Delicia de Viscas. Cultivar grown in Peru, where both chemical defoliation and chemical budbreak induction are used to give multiple cropping. Graft-compatible on quince. (Nichols, 1994)

Delicious, (Red Delicious). High chill cultivar; successfully grown only in highlands where chilling accumulation is close to the CR, otherwise productivity, fruit size and uniformity of fruit are highly affected; lower chill mutants have being found in Mexico (Aguilar, 1979).

Dorsett Golden. Originated at Nassau, Bahamas, by Mrs. Irene Dorsett. First selected in 1959. Possibility derived from original LCR material from some Middle East country. Fruit: yellow skin, 10-40% light red nonstriped blush; sweet, subacid, good quality; medium size. Tree: vigorous, spreading; blooms with 'Anna' to 7 days ahead; subject to alternate bearing if not thinned adequately. (Miller and Sherman, 1980).

Drakenstein. Originated at FFTRI, South Africa by Hurter and van Tonder. Introduced in 1981. Cross ('Wemmershoek' x 'Granny Smith') x 'Richared Delicious'. Fruit: round; bright red; harvested 1 month before 'Starking Delicious'; good early dessert apple. Tree: vigorous, spreading growth habit; relatively low CR; tendency to alternate bearing; self-compatible (Hunter, 1981). For local markets in South Africa.

Ein Shemer. Originated at Ein Shemer, Israel, by Abba Stein, in 1963. Cross 'Zabidani' x 'Golden Delicious' (Brooks and Olmo, 1972). Fruit: medium; conical to round; uniform; skin greenish-yellow; crisp; subacid. Tree: semi-compact; productive; needs 350 CU; surpassed by more recently developed LCR cultivars.

Empress. Originated at Geneva, N. Y., by R. D. Way and coworkers. Introduced in 1988. Cross 'Jonamac' x 'Vista Bella'. Fruit: dark red over most of surface; resembles 'Empire' in appearance; fair dessert quality; very short shelf life. Ripens just after 'Vista Bella'. Tree: high CR but does well at 500-700 CU in southern Brazil with artificial dormancy breaking treatment.

Elah. Originated at Rehovot, Israel, by C. Oppenheimer. Fruit: medium, round/conical; skin light green/cream; sweet, subacid flavor; firm; gets mealy when fully ripe. Tree: medium vigor; relatively open tree; surpassed by more recently developed LCR cultivars.

Eva (IAPAR 75). Originated at IAPAR, Paraná, Brazil by R. Hauagge and M. Tsuneta. Cross 'Gala' x 'Anna'. Fruit: medium, firm, conical, attractive, up to 70% bright red striped over cream/yellow background; susceptible to stem end cracking; flavor sweet/subacid; good eating quality for the season; can be cold stored for 2-4 months; harvested 2 weeks after 'Anna'. Tree: compact growth, semispur habit; excellent fruit set even on one year old wood; very productive; needs heavy thinning for sizing; needs 350 CU; blooms 1-3 weeks after Anna.

Fred Hough. Originated at EPAGRI, Caçador, SC, Brazil, by F. Denardi and A. P. Camilo. Released in 1994. Cross NJ 76 x Coop 14. Fruit: medium to small; red striped over green-yellow background; color tends to develop closer to maturation; very low acid content, somewhat flat flavor; good storage potential in CA. Tree: compact growth; immune to scab; needs 600-700 hr. of chilling; has better adaptation to mild climatic conditions than Fuji and Gala (Denardi and Camilo, 1994).

Fuji Frey. Originated at Pomifrai Orchards, Fraiburgo, Santa Catarina, Brazil. Introduced in 1995. Bud mutation of Fuji. Fruit: at location that accumulates 400 CU (Fig. 10), considerable improvement in size and shape over original Fuji. Fruit color will be related to night temperature before ripening as with original 'Fuji'. Tree: needs about 450 CU. Otherwise identical to 'Fuji'

Fuji. Originated in Japan, by Tohoku Hort. Res. Sta., Morioka City, Japan. Introduced in 1962. Cross 'Ralls Janet' x 'Delicious' (Brooks and Olmo, 1972). Fruit: sweet and crispy in a wide range of environments; color, size and fruit shape are highly dependent on chilling accumulation and cool temperatures before ripening; tendency to be flat in warm climates; red stripes over greenish-cream background; firm; can be stored in CA up to 10 months; Tree: tall; spreading; slow to get into bearing; susceptible to *Alternaria* spot, scab and fire blight; responsive to chemical rest-breaking agents; grown commercially in southern Brazil in areas that accumulate at least 600 CU.

Figure 10. The low chill 'Fuji' mutant 'Fuji Frey' (shoots and fruit on the left) compared to the original 'Fuji' (two shoots on the right) grown in a 450 CU location without the application of rest-breaking chemicals.

Gala (Kidd's D.8). Originated in New Zealand, by J. H. Kidd in 1960. Cross 'Kidd's Orange' x 'Golden Delicious'. Fruit: medium, conical red striped over cream background very juicy, very sweet, quality excellent; Tree: precocious; yield heavy and regular (Brooks and Olmo, 1972). Gala performs well in hot climatic conditions during ripening, while many other early fall varieties are too soft and do not hold in storage; responds well to chemicals to break dormancy (Fig. 11). May have problems with fruit set in locations that accumulate less than 500 CU; the mutant 'Imperial Gala' shows better budbreak than 'Royal Gala' under marginal chilling conditions.

Figure 11. 'Gala' apple grown at Porto Amazonas, Pr., Brazil, with the application of rest-breaking chemicals (accumulation of about 450 CU).

Galícia (IAC 276/2). Originated at 'Instituto Agronômico de Campinas, Campinas, SP, Brazil, by Dall'Orto *et al.*, in 1987. Cross 'Gala' x 'Anna'. Fruit: small, cylindrical, yellow/striped red; crispy and juicy, flavor resembles Gala. Tree: very productive; requires 200-300 CU; mostly for breeding or local market.

Golden Delicious. High chill apple cultivar that performs very well upon the application rest-breaking agents; marked tip bearer under such circumstances; very good eating quality in a wide range of climates; fruits are greener and more susceptible to russeting as chilling deficit increases; LCR mutants are under evaluation by Aguilar (1979); 'Agua Neueva II' is a mutant found in Mexico in 1975 that requires 500 chilling hours (Ramirez and Saavedra, 1990). 'Davilla Golden Delicious', a bud mutation also found in Mexico that needs about 600 CU (Miller and Baker, 1982) has shown to be more susceptible to russeting than the original 'Golden Delicious' at Paraná, Brazil. Most of the newer cultivars that perform well in subtropical areas have 'Golden Delicious' in their parentage.

Hatsuaki. Originated at Morioka, Iwate, Japan, by Y. Yoshida and coworkers. Introduced in 1976. Cross 'Jonathan' x 'Golden Delicious'. Fruit: medium (Cummins, 1991); skin light rose-red on yellow-green background. Rather thick, dull finish; juicy and sweet, good eating quality; high CR; in southern Brazil performs well with the use of rest-breaking agents in areas that accumulate at least 500 CU.

Imperatriz (EPAGRI 404). Originated at EPAGRI, Caçador, SC, Brazil, by F. Denardi and A. P. Camilo. Introduced in 1997. Cross 'Gala' x 'Mollie's Delicious'. Tested as M-1/89. Fruit: harvested with 'Gala'; 150 g average, larger than 'Gala'; attractive red striped skin over yellow background; sweet, subacid, higher sugar content than 'Gala'; storage potential 2 months longer than 'Gala'. Tree: Average vigor; resistant to powdery mildew; as productive as 'Gala'; needs 600-650 CU; recommended also as pollenizer for 'Gala'.

Jonagold. Originated at Geneva, N. Y., by NYSAES in 1968. Cross 'Golden Delicious' x 'Jonathan'. Outstanding dessert quality. Tendency to produce a reduced number of very large greenish fruit when dormancy is not broken naturally; High CR; performs better with the help of rest-breaking agents in areas that accumulate at least 600 CU. Flavor is not as rich as observed in Europe, Japan and USA. 'Jonagored' is an attractive striped red strain that has shown excellent color in Southern Brazil.

Jonamac. Originated at Geneva, N. Y. by R. D. Way in 1972. Cross 'McIntosh' x 'Jonathan'. High CR; in southern Brazil performs well with the use of rest-breaking agents in areas that accumulate more than 500 CU. Good McIntosh flavor.

Maayan. Originated at Rehovot, Israel, by C. Oppenheimer. Introduced in 1967. F2 selection from the cross (Calville St. Sauveur x Damascus) x Delicious (Brooks and Olmo, 1972). Fruit: medium, round, greenish skin, covered with 20-50% dark red; medium juicy; average firmness; sweet, acid; average quality; 4-5 weeks after 'Anna';

Tree: large, spread; shallow leaf coverage; bears on spurs; needs 450 CU; surpassed by more recently developed LCR cultivars.

Melrose. Originated at Wooster, Ohio, by F. S. Howlett in 1944. Cross 'Jonathan' x 'Delicious'. High CR; in southern Brazil very productive with the help of rest-breaking agents in areas that accumulate more than 400 CU. Susceptible to alternate bearing. Relatively long storage capacity.

Michal. Originated at Rehovot, Israel, by C. Oppenheimer. Introduced in 1967. F2 selection (Calville St. Sauveur x Damascus) x Delicious (Brooks and Olmo, 1972). Fruit: medium, round, greenish skin, light red blush; juicy; average firmness; sweet, subacid; average quality; 4-5 weeks after 'Anna'; Tree: large, spread; productive; needs 450 CU; surpassed by more recently developed LCR cultivars.

Mollie's Delicious. Originated at New Brunswick, N. J. by L. F. Hough and C. H. Bailey in 1966. Cross ('Golden Delicious' x 'Edgewood') x ('Red Gravestein x Close). Fruit: large, conic; crisp; sweet, resembling 'Delicious', good quality; short shelf life. Medium - high CR; in southern Brazil very productive with the help of rest-breaking agents in areas that accumulate 400 CU. Susceptible to alternate bearing.

Primicia. Originated at Caçador, Santa Catarina, Brazil, by F. Denardi, L. F. Hough, and A. P. Camilo. Cross: D1R101T117 X D1R103T245; selected in 1982. Fruit medium/large, attractive bright red over yellow ground; moderately juicy; flavor mildly acid, only fair dessert quality; ripens 2 weeks before 'Gala' and 6 weeks before 'Golden Delicious'; tree vigorous, open spreading; resistant to *apple* scab (*Vf* gene) and powdery mildew; needs 600-700 CU (Denardi *et al.*, 1988a). Softens quickly in warm areas.

Princesa. Originated at Caçador, Santa Catarina, Brazil, by F. Denardi, L. F. Hough, and A. P. Camilo. Introduced in 1988. Cross NJ 56 (NJ440249 x NJ 39) x 'Anna'. Fruit: average weight 160 g; attractive, mostly dark red over a cream-yellow background; flavor sweet, subacid; harvest 3-4 weeks after 'Anna'. Tree: moderate vigor; *CR* 450 CU; flowering 3-4 weeks after 'Anna'; precocious cropping; (Denardi *et al.*, 1988b). Very susceptible to powdery mildew and apple moldy core; low fruit set may be a problem in some location.

Rainha (IAC 8/3). Originated at IAC, Campinas, SP, Brazil, by O. Rigitano, M. Ojima, and A. C. Dall'Orto. Released in 1975. Cross 'Golden Delicious' x 'Valinhense' (Rigitano *et al.*, 1975). Fruit: large, uniform, 'Golden Delicious' shape; sweet, mildly subacid in flavor; dark red striped over light green background; suffers from severe preharvest drop; has a tendency to get 'oily' after 30 days of cold storage. Tree: moderate vigor, moderately spreading; highly productive and consistent bearer; needs 600-700 CU; slow to get into dormancy in subtropical climates; good response to dormancy breaking agents; highly susceptible to leaf blotch caused by *Glomerella sp.* under the humid and warm conditions of southern Brazil.

Rome Beauty. Originated in Ohio, USA, by H. N. Gillett in 1848. Chance seedling. Fruit: round to round-conical; ground color green/pale yellow; covered with bright red in mature fruit; tough skin; rather coarse texture; sweet, sub-acid. Tree: moderately vigorous, spreading; marked tip bearing in subtropical conditions. High chill cultivar; slow to get into dormancy in tropical highlands; widely grown in subtropical/tropical areas; many times produces without cross pollination; generally heavy cropper; more difficult to grow than LCR cultivars.

Selection Malus 1/92. Named recently 'Condessa' originated at EPAGRI, Caçador, SC, Brazil, by F. Denardi and A. P. Camilo. Cross 'Gala' x Selection M-41 (NJ 46 x 'Anna'). Fruit: medium/small, conic; red striped over cream background, resembles 'Gala'; firm; sweet, subacid, quality good; ripens 25 days before 'Gala' at Caçador. Tree: spur vigorous type, resemble 'Anna'; productive; blooms a few days after 'Princesa'; better resistance to scab and mildew than 'Princesa'; susceptible to apple leaf scorch caused by *Glomerela* sp.; needs 400 CU.

Selection Malus 4. Originated at EPAGRI, Caçador, SC, Brazil, by F. Denardi and A. P. Camilo. Cross 'Prima' x 'Anna'. Fruit: medium, attractive bright red over cream ground; moderately juicy; sweet, flat flavor; gets mealy quickly after ripening; ripens 2 weeks after 'Anna'; tree vigorous, open spreading; immune to apple scab ('*Vf*' gene), powdery mildew and apple leaf scorch caused by *Glomerella sp;* good horizontal resistance to other summer diseases; needs 300 CU; useful for breeding LCR disease resistant apples.

Selection Malus 41. Originated at EPAGRI, Caçador, SC, Brazil, by F. Denardi and A. P. Camilo. Cross NJ 56 x 'Anna'. Fruit: medium size round, solid bright red, very attractive; firm; sweet, tart. Tree: better resistance to scab and powdery mildew than 'Princesa'; productive; needs 400 CU; transmits red fruit color to its descendants.

Selection Malus 44. Named recently 'Duquesa' originated at EPAGRI, Caçador, SC, Brazil, by F. Denardi and A. P. Camilo. Cross D1R100T147 x Anna. Fruit: round, uniform; dark red stripes over green background; sweet, tart; Tree: medium vigor; very productive; carries '*Vf*' gene for scab immunity; needs 500 CU.

Selection M-60-39. Originated at Gainesville, Fla. at Univ. of Florida, by W. B. Sherman. Fruit: resembles Dorsett Golden, but is redder and larger; juicy, sweet and slightly subacid. Excellent flavor when grown in areas where minimal temperatures are below 15°C during the month preceding ripening, as in Paraná, Brazil; excellent fruit firmness. Tree: vigorous, spreading and productive. Fruit is highly susceptible to rot in humid climatic conditions; needs 250 CU.

Selection IAPAR 26-80-144. Originated at IAPAR, Curitiba, Pr., Brazil, by R. Hauagge. Selected in 1988. Cross 'Golden Delicious' and 'Anna'; Fruit: large, round and uniform; 20-60% dull red color over cream background; juicy and crisp; very good sweet flavor; will not color when night temperatures in the month preceding harvest are above 15°C;

can be stored for 1-2 months; harvest 6-7 weeks after 'Anna'. Tree: medium vigor, spur type growth; highly productive; needs 300 CU; blooms 1-2 weeks after 'Anna'; resistant to *Glomerella sp.* leaf scorch.

Selection IAPAR 6-80-28. Originated at IAPAR, Curitiba, Pr., Brazil, by R. Hauagge. First selected in 1988. Cross 'Golden Delicious' and 'Anna'; Fruit: medium/large, uniform, resembles 'Golden Delicious'; attractive, 40-70% bright red color over cream background; conspicuous lenticels; juicy; sweet flavor; good tolerance to handling; can be stored for 1-2 months; harvest 5-6 weeks after 'Anna'. Tree: medium vigor spur type growth; highly productive; needs 300 CU; blooms 1-3 weeks after 'Anna'; highly susceptible to *Glomerella* sp. leaf scorch; may be useful in cool dry areas where *Glomerella* is not a problem.

Slor. Originated at Bet Dagan, Israel, by E. Slor. Introduced in 1982. Cross Selection 6/13 x Shlomith (Slor, 1982). Fruit: large, conic; dark red stripes covering most of light green background; uniform; soft texture; sweet, subacid; medium firmness; harvest about 5 weeks after 'Anna'. Tree: average vigor; productive. 'Slor' and 'Elah' are possibly the two best cultivars released by Oppenheimer and Slor's apple breeding program.

Tropical Beauty. Originated at Maidstone, South Africa, by M. B. Strapp. Seedling of unknown parentage planted about 1930. Rated as a very good eating quality apple, resembling 'Rome Beauty' in Australia. In Missouri, New York, and Paraná, fruits are large; irregular in form; skin dull red; soft flesh, and poor quality; needs 450 CU.

Tropic Sweet. Originated at Gainesville, Fla. at Univ. of Florida, by W. B. Sherman and P. M. Lyrene. Polycross from selections from Jerseymac x Anna. Introduced in 1996; tested as Fla. 90-3. Fruit: medium, 7 cm in diameter; slightly less red overcolor than 'Anna'; firmer and better flavor than 'Anna'; low acidity; 105-110 days cycle, ripens 5-7 days before 'Anna'. Tree: moderately vigorous, semi-spreading; semi-spur type; precocious flowering; resistant to powdery mildew and fire blight; self-unfruitful, pollinated by 'Anna' and 'Dorsett Golden' (Sherman and Lyrene, 1996).

Valinhense. Originated at Valinhos, SP, Brazil. Unknown parentage. Fruit: round, medium size; uniform bright dark red covering most of light green background; very firm tart apple. Tree: medium vigor; upright, then spreading when mature; pale green leaves and shoots; highly productive; needs about 600 CU; slow to become dormant under subtropical conditions; responds well to rest-breaking chemicals; good pollenizer; a processing apple basically; still widely grown in São Paulo State, Brazil.

Williams' Pride. Originated at Indiana, by J. Janick and coworkers. Introduced in 1987. Tested as Co-op 23. Fruit: medium-large, oblate to round, uniform shape; green skin ground color under 90% ruby red, smooth, slightly waxy, no russeting; flesh firm, crispy and juicy; flavor moderately subacid; tree immune to apple scab (*Vf* gene), apparently resistant to cedar apple rust and moderately resistant to fire blight and powdery mildew

(Cummins, 1991); needs high chill; responds well to chemical agents to break dormancy; does well in southern Brazil in locations that accumulate at least 600 CU.

Willie Sharpe. Originated in New Zealand by H. E. Sharpe in early 1900s. Resultant from crosses among English varieties. Fruit: smooth light green skin; round, uniform, medium size; sweet, subacid; good flavor; melting texture; becomes mealy quickly after harvest; highly susceptible to bitter rot caused by *Glomerela* sp; Tree: medium/low vigor; light green leaves; productive; resistant to woolly aphid and to apple leaf scorch caused by *Glomerela* sp.. blooms heavily on one year wood; good pollen producer; used as pollenizer for 'Gala' and 'Fuji' in southern Brazil; needs about 700 CU; responds well to rest-breaking agents.

Winter Banana. Originated at Cass County, Indiana, USA by D. Flory. Introduced in 1876. Fruit: medium; conical/flat; yellow smooth skin, with bright red blush; crisp and juicy; sweet, subacid; average dessert quality; gets mealy quickly; gets oily upon storage; susceptible to bitter rot. Tree: vigorous; spreading; highly productive in subtropical-/tropical areas; once widely grown in Tropical America and many other warm countries; need 600 CU; lower fruit quality than the newer LCR cultivars.

4.3 Apple rootstocks

The English Malling (M) and Malling Merton (MM) apple rootstock series have been the most widely used in temperate areas. Most of MM and M rootstock series are high to very high CR. Besides, high humidity and acid soils are many times associated with high temperatures in subtropical areas. As a consequence, there have been either physiological or pathological problems many times associated with them. Dwarf rootstocks so far have had little commercial success in hot climates.

Chill requirement of apple rootstock is an important factor affecting 'Gala' apple; there are major differences in rootstock effects on both percentage of budbreak and scaffold angle of primary branches in southern Brazil (accumulation between 250-600 CU). Budbreak tends to be delayed on MM111, M9 and M26, compared to MM106 and M7, and accelerated on 'Anna' seedlings.

We have observed another rootstock-related phenomenon near Manizales, Colombia, where nurseries produce trees in containers, packed very densely under shade. Trees are grown until 150-200 cm tall; water is withheld to induce growth cessation. When trees are planted in the orchard and the tops are cut back to 40-50 cm. Buds of trees on 'Anna' seedling rootstock begin normal, vigorous growth while buds of trees on MM.106 remain dormant.

Scaffold angles of 'Gala' are considerably wider when grafted on 'Maruba-kaido' and many LCR Cornell-Geneva rootstocks, than on Malling-Merton rootstock series. Many of the Cornell-Geneva rootstocks, including G.30, CG.179, CG.202 and CG.210, have the LCR rootstock Robusta 5 as a parent and appear to have relatively low chilling requirements.

In addition, a physiological phenomenon observed in the first 3-6 weeks after budbreak is generally associated with excessively vigorous growth in the previous season, with low chill accumulation during winter, and with high chill rootstocks such as MM111, M26 and M9. A variable percentage of growing tips show thinner leaves with marginal reddening, which may later become dead tissue. At the same time, the tip end of young fruits may show variable degrees of necrosis, followed by calyx end cracking.

Burrknots have been a major problem in many orchards in Brazil and Chile, especially on the widely used MM106 rootstock. This genetic disorder is a potential in all the English rootstocks of the Malling and Malling-Merton series because the first selection criterion was usually for ease of rooting. The climatic conditions of the humid subtropics exacerbates the burrknot problem.

'Maruba-kaido' has been shown to be quite successful both in replant situation and in acidic and low fertility soils in South of Brazil. It shows certain ability to exclude manganese and higher calcium uptake under such situation. The use of M.9 interstock is showing to be one possible alternative for dwarfing rootstocks.

Rootstocks specifically adapted to hot climates have been recently developed in Israel (Assaf, 1995a; Assaf, 1995b). The Hashabi clones MH 13.4, MH 14.5, and MH 15.6 were shown to have superior performance than MM106 and MM104. They were selected from a population of Hashabi seedling rootstock in orchards at Israel. LCR Hashabi and Italian Doucin seedlings have been widely used in Israel as apple rootstocks. In Israel, apples on MH series rootstocks are planted in density ranging from 1250 to 2500 trees per hectare for standard and spur cultivars respectively. Apple trees on MH series show good equilibrium between fruiting and vegetative growth, high productivity and fruit quality. MH 10.1 is the least vigorous clone. MH 15.6 may be the most interesting one under hot climates.

5 Pears

5.1 Variation in chilling requirement in pear

Within the genus *Pyrus*, *P. pashia*, *P. kawakamii* and *P. calleryana* are best adapted to warm climates (Bell *et al.*, 1982). *Pyrus kawakamii* behaves as an 'evergreen' species in southern California but is partially deciduous in colder zones. *Pyrus pyrifolia* and *P. calleryana* grown in the wet low land of the Yangtze River Valley in South China are reported to have LCR (Zai-Long, 1990). At IAPAR, Brazil, introductions of *P. calleryana*, *P. betulifolia*, and *P. pyrifolia* have shown low CR, in the range of 200-600 (Hauagge, unpublished). Although they have low CR and fruits may measure up to 70 mm in diameter, *P. pyrifolia* seedlings from Taiwan (called "Taiwan-nashi") have bitter fruit and therefore have little potential as parents. All the commercial cultivars of *P. pyrifolia* and *P. communis* introduced at IAPAR, Brazil need chilling greater than 700 CU, and most of them above 1000 (Hauagge, unpublished).On the other hand, introduced cultivars of *P. bretschneideri* have CR between 300 and 500 CU. Seeds of *P. amygdaliformis*, *P. calleryana*, *P. fauriei*, and *P. pashia* need very low chilling, less than 35 days for seed stratification to break their dormancy (Westwood and Bjornstad, 1968).

5.2 Pear breeding for low chilling

5.2.1 *European type pear*

Oriental pears were introduced into the United States from Europe between 1820 and 1840 (Hedrick *et al.*, 1921). LCR was "accidentally" introduced into buttery flesh pear germplasm when resistance to fire blight has been sought by crossing European and oriental pears. Hybrids between oriental and European pears were selected, such as 'Le Conte' (1846), 'Kieffer' (1873) and 'Garber' (1880), and in spite of their low quality, these and other pear hybrids were widely grown in the eastern and midwestern USA due to their resistance to fire blight. In addition, in many warm parts in the World, they were the only type of pear to bear fruits. These hybrids have served as the initial bases for the development of low chill European pears.

Compared to apple, little effort has been made to develop pear cultivars with low CR through breeding. However, considerable advances have been achieved in the improvement of European-type pears in the 200-400 CU range at the University of Florida at Gainesville; Instituto Agronômico de Campinas, Brazil; and EMBRAPA/CNFCT, Pelotas, Brazil. From them, the cultivars 'Flordahome' (Fig. 12), 'Seleta' (IAC 18/28) (Fig. 13), 'Triunfo' (IAC 16/34), 'Primorosa' (IAC 9/3), and Selection IAC 16/30 (Fig. 14), have proved at IAPAR, Brazil to have very good texture and flavor. On the other hand, the following problems need to be overcome in this group of cultivars: Inconsistent productivity, pollen sterility, fruit browning after ripening, intolerance of fruit handling, short storage life, and susceptibility to *Fabrea* leaf spot.

'Carrick', which originated in Tennessee (Deyton and Cummins, 1991), is a backcross to *P. communis* with a good desert quality. It needs about 550 CU.

Figure 12. 'Flordahome' LCR European type pear.

Figure 13. 'Seleta' LCR European type pear.

Figure 14. Selection IAC-16/30 pear.

5.2.2 *Oriental type pear*

Asian pears do not have the buttery smooth texture of European pear, tend to have crispy texture, juicy, sweet, and may have abundant stone cells in the flesh, with little or no aroma.

Oriental high quality pears developed in Japan in general have CR of above 700 CU. As in *P. communis*, *P. pyrifolia* also shows accentuated flower abortion on warm winter condition. 'Hosui' which needs about 700 CU, is less prone to flower bud abortion than other Asian pears, and do well in locations that accumulate 400 CU in southern Brazil, when budbreak is stimulated artificially.

The cultivars 'Okusankichi', 'Atago' and 'Kosui' show some rest avoidance at Presidente Prudente, São Paulo, Brazil, where they are grown commercially. In this area, fruit appears to be superior to locations where deep dormancy is observed.

Breeding Asian pears for adaptation to subtropical areas of South China has been underway since 1955 at Zhejiang Agricultural University (Zai-Long, 1990). Crosses involve mainly local LCR *P. pyrifolia*, besides *P. bretscheneideri, P. ussuriensis*, and the cultivars 'Kieffer', 'Le Conte' and 'Shaw Li'. Two cultivars from the Zhejiang program, 'Huanghua' and 'Hangzhou Green', have been released and are being grown commercially in China. They are reported to have sweet, fine texture and crisp flesh. 'Hangzhou Green' is reported to be of higher quality, but somewhat less productive.

5.3 Sterility problems

Either total or partial pollen sterility was widely found in hybrid pear cultivars and selections at IAPAR, Brazil (Hauagge, unpublished). It is possible that male cytoplasm sterility and genetic sterility (Thompson *et al.*, 1976) play an important role. A very high proportion of extremely weak seedlings resulting from crosses among LCR hybrids (Hauagge, unpublished) may indicate the presence of unbalanced chromosomes.

5.4 Pears cultivars for warm climates

A number of pear cultivars that may have value for fruit production in locations that are warmer than the typical pear production areas are listed below alphabetically:

Africana. Packham's Triumph clone of unknown origin introduced in Minas Gerais, Brazil. Identical to Packham's Triumph, but show better productivity than the later at EPAGRI, Caçador, SC (I. D. Faoro, 1995, personal communication) and lower flower bud abortion intensity in IAPAR, at Palmas, PR, Brazil. Other lower chilling requirement mutants of Packham's Triumph were also found in Australia (G. Richards, 1997, University of Western Sydney, Australia, personal communication).

Ayres. Originated in Tennessee, by B. D. Drain. Introduced in 1954. Cross Garber x Anjou. Fruit: skin golden russet with a rose tint; flesh juicy, sweet; good for eating fresh and average for canning. Tree: resistant to fireblight; needs 600 CU.

Baldwin. Unknown origin, grown in southern United States. Fruit: medium to large and oblong in shape; light green skin color overlaid with light russet; texture not as smooth as 'Hood'; canning and fresh consumption (Andrews and Sherman, 1979); similar to 'Carnes' and 'Orient'; highly susceptible to *Fabrea* leaf spot.

Carnes. Unknown origin. Fruit round shape, medium, yellowish green with some russet in Florida; coarse texture; flavorful; more for canning.

Carrick. Originated in Tennessee, by B. D. Drain and L. M. Safley. Introduced in 1957; cross between 'Seckel' and 'Garber'. Fruit: medium to large; skin, yellow, very russeted, slightly blushed, usually reddish brown in Paraná; many observed lenticels; flesh firm, tender, juicy, crisp; flavor sweet-subacid, sprightly, good. Tree: large; vigorous; spreading; some resistance to fire blight (Brooks and Olmo, 1972). Needs about 500 CU in southern Brazil; good compatibility with Portugal quince.

Cascatense. Originated at CNFCT/EMBRAPA, Pelotas, RS, Brazil, by B. Nakasu and D. L. Leite, in 1996. Cross Packham's Triumph x LeConte; Fruit: 140-220 g; pyriform shape; skin greenish yellow to yellow when quite ripe; russeting near peduncle only; firm; good buttery texture; some stone cells present in the flesh; reasonable pear flavor, quality similar to 'Carrick'. Tree: vigorous, spreading; productive at Pelotas.

Centenaria (IAC 9/47). Originated at 'Instituto Agronômico de Campinas, Campinas, SP, Brazil, by M. Ojima and coworkers. Released in 1987; cross between Packham's Triumph x Hood; Fruit: large; smooth yellow skin color, small lenticels, resembles 'D'Anjou' in shape; firm; buttery texture, a bit dry with some stone cells near to the core; average quality; Tree: vigorous; regularly productive; compatible on quince; needs 250 CU (Ojima *et al.*, 1987a);

Ceres. Ripens 10 days before 'Packham's Triumph' and has the same keeping quality and productivity as 'Packham's Triumph' in South Africa (van Zyl, 1982).

Coscia. originated in Italy prior to 1800. Chance seedling (Postman, 1992). Considered to be a LCR cultivar in North of Africa.

D'Agua. Generic name given to hybrid pear selections in Brazil, mostly chance seedlings. Included are 'D'Agua', 'Branca', 'Bela Aliança', 'Joaquina', 'D'Agua de Valinhos', 'Branca de São Roque', 'Bronger', 'Madame Sieboldt ', 'Francesa', 'Branca Francesa', 'Agua Bronger', 'D'Agua de Outono', and 'Dadico'. Fruit: generally green/light green/cream-yellow; round to pyriform shape; mostly crisp, juicy and sweet; some eventually will melt; some are good oriental type; Tree: mostly vigorous; chill between 150-400 CU; many extremely productive; some clones have the potential to produce more than 60 ton/ha in 400 CU environments (Fig. 15).

Eldorado. Originated in California by Robert Patterson. Introduced in 1945; Bartlett open-pollinated seedling. Fruit: has the same dessert, canning and drying qualities as

Figure 15. Fruit set and fruit characteristics of a clone of D'Agua hybrid pear at IAPAR in Piraquara, Pr., Brazil.

'Williams' but ripens a few days later (Brooks and Olmo, 1972). Due to its lower CR, it is far more productive than William's and seems to have a slightly better keeping quality in South Africa (van Zyl, 1982).

Fanstil. Hybrid cultivar originated in Texas. Fruit: medium size; yellow, with red blush; crisp, coarse flesh is juicy but gritty. Tree: resistant to fire blight; sensitive to extreme temperatures; low CR (Walheim and Stebbins, 1981).

Flemish Beauty. Originated in Belgium by Van Mons about 1800. Chance seedling. Fruit: large, round; yellow with red blush; good dessert quality. Tree: branches are spreading and drooping; very susceptible to pear scab (Postman, 1992); more productive than Packham's Triumph at EPAGRI, Caçador, SC, Brazil, where it accumulates around 600 CU (I. D. Faoro, 1995, EPAGRI, Caçador, SC, Brazil, personal communication).

Flordahome. Originated at Gainesville, Fla. by Univ. of Florida, Dept. of Fruit Crops by W. B. Sherman, C. P. Andrews, P. M. Lyrene and R. H. Sharpe. Cross of Tenn. 37-20 ('Garber' x 'Anjou') x 'Hood'; tested as Fla. 41-116; released in 1981 Fruit: precocious, ripens 15 days before Hood; large, round-ovate; skin green, tender; buttery texture; need to be harvested at hard green stage for maximum aroma and flavor. In Paraná, fruit is sometimes astringent. Tree: semi-compact; highly productive; resistant to fire blight and moderately resistant to leaf spot; can be pollinated by 'Hood' and 'Pineapple'; requires 250 CU. (Sherman *et al.*, 1982).

Hakucho. Asian pear cultivar grown in South Brazil of unknown origin. Fruit: harvested at least 4 weeks before 'Ya-Li', which it closely resembles; crisp; sweeter and coarser texture than 'Ya-Li'; medium size stone cells, mainly close to the core. Tree: vigorous; productive; gets into production considerably earlier than 'Ya-Li'; needs 400-450 CU; useful as pollenizer for 'Ya-Li'.

Hood. Floridian variety of unknown Origin. Fruit: early ripening; large, yellow-green; buttery texture, some stone cells; sweet mild pear flavor. Trees: highly resistant to fire blight and moderately resistant to *Fabrea* leaf spot; mature trees do not require cross-pollination (Sherman and Crocker, 1982). Parent of many European type LCR cultivars.

Hosui. Asian pear originated in Japan by the National Fruit Tree Research Station. Cross of (Kikisui x Yagumo) x Yagumo; released in 1972; Fruit: large, round, sweet, oriental type, very juicy, with a refreshing taste; skin is golden brown, fully russeted. Tree: vigorous, producing large shoots that tend to bend and droop, and consequent lateral vertical shoots may develop during summer (Brooks and Olmo, 1972). Needs 700 CU; if chilling is not enough, delayed foliation occurs. Fruit size is then irregular and may crack. Reasonably productive with application of rest-breaking agents where at least 450 CU accumulate, as in southern Brazil.

Huang-Sang-Li. Low *CR* Asian *pear* from Taiwan, that produces low quality fruit (Byrne *et al.*, 1995).

Kieffer. Originated in Pennsylvania by Peter Kieffer in 1873. Cross *P. pyrifolia* x Bartlett. Fruit: late ripening; large; flesh is coarse, watery, sweet, but little flavor, large stone cells; mostly for processing (Postman, 1992); Tree: moderate vigor; productive; moderately resistant to fire blight; moderately resistant to leaf spot in Florida (Sherman and Crocker, 1982); higher CR than 'Baldwin', 'Hood' and 'Orient' in Florida. 'Kieffer' was once the basis for the pear industry of the eastern United States.

LeConte. Originated in United States in 1846; hybrid between oriental and European pears but origin unknown. Fruit: mid-season; large; yellow when ripe; flesh is firm, watery, sweet, oriental type; may develop some flavor in cool production areas; moderate quantity of stone cells. Tree: vigorous; productive; slow to get into bearing; moderately resistant to leaf spot in southern Brazil; needs about 500 CU.

Niau-Li. Low CR Asian pear from Taiwan; produces low quality fruit (Byrne *et al.*, 1995).

Nijisseiki. Asian pear, originated in China; chance seedling, introduced in 1898. Fruit: round, medium size; light green skin; bruises easily; not as sweet as 'Hosui'; stores well; Tree: medium size; upright; more susceptible to fire blight than other Japanese pears. Needs 900 CU; reasonably productive with application of rest-breaking agents where at least 700 CU accumulates, as in southern Brazil; less productive than 'Hosui' under the same circumstances.

Okusankichi. Asian pear, old Japanese cultivar originated at Nigata Province, Japan, probably as a open-pollinated seedling of 'Wasesankichi'. Fruit: extremely large; round, dark green/brown; abundant stone cells, coarse texture; very juicy, crispy and sweet; only fair quality; Trees: moderately vigorous; upright; high CR; exhibits dormancy avoidance in some warm areas of Southeastern Brazil.

Orient. Originated at Chico, California, by Warter Van Fleet. Introduced in 1945. *P. communis* x *Pyrus* sp. from China; Fruit: nearly round; roughened skin; flesh firm, juicy, slightly sweet, lacking in flavor; good for canning; resistant to fire blight. (Brooks and Olmo, 1972). Highly resistance to leaf spot in Florida (Sherman and Crocker, 1982). Recommended for planting in north Florida.

Packham's Triumph. Originated at Molong, New South Wales, Australia, by C. H. Packham. Introduced at the turn of the 20th century. Uvedale St. Germain (Bell) x Williams Bon Chretien (Bartlett). Fruit: large; usually obtuse pyriform, surface slightly uneven; skin thin, lemon-yellow when ripe; flesh texture very fine, very juicy, melting, flavor particularly rich and pleasing; long storage potential. Tree: upright; moderately vigorous; good, consistent bearer; highly susceptible to fire blight; sensitive to stony-pit virus (Brooks and Olmo, 1972). It is commonly referred as a cultivar capable to do well in LCR climates but really does better than most European cultivars only in areas just slightly warmer than the traditional pear production areas.

Pineapple. Unknown origin. Fruit: large, coarse skin, abundant large stone cells, non melting, and no pear flavor. Tree: resistant to leaf spot and fire blight (Sherman and Crocker, 1982); blooms 2-3 days before Kieffer in Florida; requires cross-pollination.

Primorosa (IAC 9/3). Originated at the Instituto Agronômico de Campinas, SP, Brazil by M. Ojima and coworkers. Cross of Packham's Triumph x Hood. Introduced in 1987. Fruit: large; pyriform; smooth, light green to yellow skin; small whitish lenticels; russeting near peduncle; buttery texture; excellent European type flavor; early ripening (Ojima *et al.*, 1987b). Tree: average/low vigor; willowy; compatible on Portugal quince seedlings; susceptible to *Fabrea* leaf spot; irregular bearing; pollen sterile; needs 250-300 CU; blooms at the same time as 'Tenra' and 'Centenaria' in São Paulo, which are used as pollenizers.

Selection IAC 16/30. Originated at Instituto Agronômico de Campinas, Campinas, SP, Brazil, by O. Rigitano and M. Ojima. Cross of Packham's Triumph x Hood. Fruit; large; firm; smooth cream skin; buttery texture, good European flavor; good quality. Tree: vigorous; slow to commence bearing; pollen sterile, needs cross pollination; moderately resistant to *Fabrea* leaf spot; requires about 300 CU.

Seleta (IAC 16/28). Originated at the 'Instituto Agronômico de Campinas, SP, Brazil, by Rigitano and Ojima. Cross of Packham's Triumph x Hood. Introduced in 1972. Identical to 'Bento Johbire'. Fruit: medium to small; shape resembles 'D'Anjou'; skin light green/cream; buttery texture, excellent European flavor; bruises easily; Tree: low vigor, open tree; sensitive to latent viruses; pollen sterile; inconsistent bearer; pollenized by ‘Pera D’Agua’ and ‘Tenra’; needs 250 CU. (Dall’Orto *et al.*, 1995).

Smith. Hybrid of unknown origin. Fruit: early maturing; fruit light green/cream skin; flesh generally non-buttery texture; “watery” type. Tree: vigorous; upright; shiny leaves; slow to begin production; heavy bearer; needs 250 CU.

Spadona. Originated in Italy. Grown in Israel, where is said to be of very low CR (Sherman and Crocker, 1982). Vigorous tree, Fruit green, small and juicy. Stores relatively well. Performing well also in Hermosillo, Mexico.

Tang-Li. Low CR Asian pear from Taiwan; produces low quality fruit (Byrne *et al.*, 1995).

Tenn. Originated in Knoxville, Tennessee by Brooks D. Drain (Andrews and Sherman, 1979). Probably a cross of Garber x Anjou. Fruit: small to medium; variable red blush over green ground color; sweet, subacid; crisp, turns into buttery texture when quite ripe; some stone cells; fair pear flavor. Fruits are somewhat astringent in some areas of southern Brazil. Tree: highly resistant to fire blight and moderately resistant to leaf spot in Florida (Sherman and Crocker, 1982). Highly productive.

Tenra (IAC 15/20). Originated at the Instituto Agronômico de Campinas, SP, Brazil, by O. Rigitano and M. Ojima. Cross of Packham’s Triumph x Madame Sieboldt. Introduced in 1972. Fruit: average size; thick green skin; sweet; closer to oriental type. Tree: vigorous, upright initially but spreads as it gets into production; productive; mostly used as pollenizer; compatible on Portugal *quince*; needs 300 CU.

Triunfo (IAC 16/24). Originated at the Instituto Agronômico de Campinas, SP, Brazil, by Rigitano and Ojima. Cross ‘Packham’s Triumph’ x ‘Hood’. Introduced in 1972. Fruit: medium to large; dark green thick skin that is relatively rough; not very attractive; firm; buttery texture; some small stone cells near to the core; good pear flavor. Tree: vigorous, productive, slow to get into bearing; high proportion of sterile pollen; good tolerance to *Fabrea* leaf spot; needs 300 CU.

Tsu-Li. Asian pear, old Chinese cultivar. Fruit: very large; pyriform, light green skin, large lenticels; pronounced russeting near peduncle; sweet, crisp and soft, with few stone cells. Tree: very large, upright; moderate fire blight and leaf spot resistance in Florida (Sherman and Crocker, 1982); needs 500-600 CU; slow to begin bearing in Brazil.

Waite. Introduced in 1938 by M. B. Waite, Arlington, Virginia, USA, from the USDA breeding program. Adapted to the middle and upper south of USA. Resembles ‘Moonglow’, but smaller; quality is fair to good; almost free of grit cells; ripe with ‘Kieffer’. Tree: resistant to fire blight; not very productive.

Wan-Li. Asian pear from China of unknown origin, grown in southern Brazil. Fruit: large; long pear shape, somewhat irregular; creamy skin with some large green/whitish lenticels; firm; crisp; sweet; large numbers of large stone cells in the flesh; coarse texture; Tree: medium vigor, long weeping branches; very early bearing; very productive at IAPAR/Curitiba, Pr., Brazil; needs 300 CU, lowest CR among the Chinese cultivars grown at IAPAR/Curitiba, Pr., Brazil; good pollenizer.

Wilder - grown in North of Africa (Tunisia, Libya). Good early pear, relatively resistant to transport. Possibly lower CR than other common European cultivars.

Ya-Li. Asian pear, old Chinese cultivar. Fruit: large; pyriform, cream-yellow; large lenticels; pronounced russeting near peduncle; sweet; quality is improved by delaying harvest, otherwise it may have low sugar content; crisp and soft; excellent oriental texture; good storage potential; Tree: very large, spreading; slow to get into production in warm areas; needs limb bending; needs 450-550 CU. Widely grown in Paraná, Brazil

Other European type pear cultivars such as 'Spadona d' Estate' (Mazzocchi and Pucci, 1963); Gentil Blanca, Gamusina, Donguindo, Agua de Aranjuez, Ternal de Invierno, San Antonio, Epargne, Ercolini, Bergamota de Verano, Roma y Castell (Tabuenca, 1964); Muskavi, Mustafakey (Phillipe, 1965); Meski, Saint Jean (Crossa-Raynaud, 1955); have also been referred as having lower chilling requirement.

Yokoyama Wase, Pien Pu and Song Mao from Taiwan; and Pathanak from India, all Oriental, needs lower chill than the Japanese high quality cultivars, but all have inferior fruit quality (Subhadrabandhu and Punsri, 1987).

5.5 Pear rootstocks

That rootstock affects budbreak and growth has been reported in pear as well as in apple (Westwood and Chestnut, 1964). Almost no observations have been made on the effects of recently developed rootstocks on budbreak and fruiting in subtropical conditions.

Latent virus infection is widespread in pears, and virus-free scionwood of most cultivars is not widely available. This may affect the performance of a given scion/rootstock combination, especially when quince or *P. betulifolia* are used as rootstocks.

Besides the usual horticultural attributes needed when selecting rootstocks, transplanting ability is a major concern in areas where warm/hot weather conditions may occur during the transplanting of bare-rooted trees.

In Europe and in some areas of North America, quince rootstocks have given outstanding results with European pears. This has not been repeated in mild climatic conditions. In most cases quince is reported to be incompatible with subtropical pear cultivars. In southern Brazil, some clones of local 'watery pears', 'Carrick' and 'Tenra' have been compatible on both BA-29 and Portugal quince. (See also chapter 6).

BP-1, BP-2 and BP-3 clones are introductions developed in South Africa. They probably have 'Kieffer' in their backgrounds. All are propagated by hardwood cuttings. All are susceptible to fire blight. BP-3 is at present most favored in South Africa.

Many Old Home x Farmingdale (OhxF) clones were introduced in southern Brazil. With high CR, they have been difficult to grow and propagate in areas that accumulate less than 600 CU.

Pyrus communis seedlings are commonly used as pear rootstocks, especially in the United States. *Pyrus communis* seeds are not readily available in subtropical areas. Nonetheless it can be used as rootstocks where *P. communis* cultivars can be successfully grown. It is not a good choice for both Asian and hybrid pear cultivars because they often produce fruit of smaller size.

In many areas around the world, seedlings of hybrid pear cultivars such as 'Kieffer', 'Garber', 'Orient', 'Pineapple', and many others are presently used. Seeds of these cultivars are readily available. In humid climatic conditions seedlings of these cultivars (as in *P. communis)* are highly susceptible to *Fabrea* leaf spot in the nursery. Also they vary widely in vigor in the nursery. Having limited numbers of lateral roots may be a serious problem for transplanting. Grafted trees may be slow growing in the first 2-3 seasons. Mature trees are very large in size, and slow to get into bearing. Kieffer's seedlings are considered to be a promising rootstock in New Zealand.

Pyrus calleryana and P. betulifolia seedlings are commonly used as rootstock in many parts in the world. In southern Brazil they have proved to be a better choice than seedlings of hybrid cultivars. Seedlings of *P. betulifolia* may still be too vigorous. And sometimes show more problems with latent viruses, but fruit size, especially of Asian cultivars, tends to be larger.

Seedlings of *P. pyrifolia* from Taiwan, commonly called Taiwan Nashi, have been recently used in Paraná State, southern Brazil. Seeds can be produced satisfactory in areas that accumulate 150-500 CU. Dormant cuttings of young seedlings also root easily. Trees have excellent early growth after transplanting. Vigor and precocity have been intermediate between *P. calleryana* and *P. betulifolia. Pyrus pyrifolia cv.* Hengshan, a low CR, is commonly used as rootstock for already non-dormant high quality Japanese cultivars in Taiwan (Sansavini, 1991).

In Guatemala, hawthorne, probably *Crataegus pubescens,* is used as rootstock for pears (Darrow, 1953). *Crataegus oxycantha* is used to a limited extent in the southern United States.

6 Quince

Compared with apples and pears, quince CR has a narrower range and shows considerably less variability. In Caçador, SC, Brazil, variability on bloom dates within apple cultivars varies to more than 120 days. In the same circumstances, bloom dates among quince cultivars varies about 10 days (F. Denardi, at EPAGRI, Caçador, SC, Brazil, personal communication).

Economic importance of quince is quite limited nowadays. Few cultivars are grown in warm climate countries, and there is very little demand for quince fruit. (Brooks and Olmo, 1972; Ruck, 1975) (Table 2).

Table 2. Countries and quince cultivars

Location	*Cultivars (Comments)*
Zimbabwe	Cape Selected, Meeche's Prolific; crops well areas above 1200 m elevation.
Tropical America	Pineapple, Orange (Honduras, Guatemala at 1,200 to 2,400 m; Equator, 2,300 to 2,700 m) (Darrow, 1953).
Santa Catarina, Brazil	De Patras, Apple, Costantinopla, Pineapple, Van Deman, Orange, Provence, Smyrna, D'Angers; decreasing order of productivity (F. Denardi, at EPAGRI, Caçador, SC, Brazil, personal communication).
São Paulo, Brazil	Portugal, Smyrna, Mendoza-INTA 37 and Provence (100-200 CU accumulation) (Dall'Orto *et al.*, 1986).
South Africa	Portugal, Meeche's Prolific EM, Meeche's Prolific E2, Apple Shape, Orange, Pineapple, Van Deman, Raes Mammoth, all have reasonable characteristics of production and fruit quality for canning; Borrie, Pear Shape, Quince A, Quince B, and Quince C have no perspectives either because of fruit quality or poor bearing (Allison, 1974).

7 Disease Problems

Besides the traditionally important pome fruit diseases observed in temperate areas, some other diseases can be quite destructive in subtropical environments. This happens because of the longer growing season, higher temperatures and higher humidity during the growing season, and lack of subfreezing temperatures during the winter.

In apples, bitter rot of fruit (*Glomerella* sp.) and leaf scorch (incited by *Glomerella* sp.) can become destructive in these conditions. 'Gala' and 'Golden Delicious' are highly susceptible.

Fire blight, incited by the bacterium *Erwinia amylovora,* is a highly destructive pear disease under warm and humid climates when present, such as in the southeastern United States. Fortunately this disease has not yet been introduced in most districts of South America and Central America; fire blight is rampant in Egypt, where it is very destructive of pears.

Fabraea leaf spot, caused by *Fabraea maculata* Atk., causes severe defoliation and even surface lesions on the fruit. Early defoliation will induce summer/fall blooming and foliation on low CR pears. Frequent infection will lead pear to decline in absence of chemical control on susceptible cultivars.

8 Prospects for improvement by breeding

The successes already achieved by the apple breeding programs in Israel, Brazil and United States and by the pear breeding work in Brazil and United States indicate that major improvements can be made in the repertoire of cultivars suitable for culture in the subtropics. Exploration of the genetic potential in *Malus* has already made substantial progress and at least some potential parents have been identified for almost all the important attributes.

8.1 Fruiting Cultivars

The way for immediate improvement of fruiting varieties is now open, and rapid progress can be expected in the current breeding programs. In the next generation, it will be possible to bring together the high quality of the best temperate region cultivars, good handling tolerance, productivity, scab-immunity and fire blight-resistance of *Malus floribunda*, and low-chilling requirement. Similar advances, though not so rapid, can be anticipated in pears.

8.2 Rootstocks

Horticultural Attributes

For apple growing in temperate climates, a wide spectrum of tree-size controlling rootstocks is commercially available, ranging from the diminutive Malling 27 to the super-vigorous 'Maruba-kaido'. In temperate regions, almost all apple rootstocks are more or less readily propagated in layerbeds or stoolbeds. There are major differences in capacity of clonal rootstocks to induce precocity and productivity of the scion cultivars they support, and inefficient clones are rapidly weeded out. Almost all the apple rootstocks in commercial use are well-anchored, and most are relatively free of suckering. To our knowledge, though, none of the traditional apple rootstocks used commercially in temperate regions has a CR low enough to be considered well adapted to subtropical apple production. The newer MH series developed by Assaf in Israel are the first ones with this characteristic.

Similarly, asexually propagated pear rootstocks have been selected for tree size control, for inducing precocity and productivity, for good anchorage, for freedom from suckering, and for propagability. Like apple rootstocks, the standard clonal pear rootstocks have relatively high CRs and have but limited potential for the subtropics.

Disease and Pest Susceptibilities

The apple and pear rootstocks of the temperate regions are usually acceptably resistant to *Phytophthora*-incited root rots and some are resistant to fire blight. Some are resistant to apple woolly aphids, especially the Malling-Merton apple clones. Preliminary work, mostly by Sutton in North Carolina, has shown that within *Malus* there is resistance/tolerance to several root pathogens endemic in warmer climates, including *Corticum galactinum, Xylaria mali,* and *Armillaria* spp. J. Boneti in Brazil has been systematically searching for resistance to *Dematophthora*, thus far with no success. Similarly, no significant resistance has been reported for cotton root rot or *Sclerotium rolfsii*.

Likewise, no sources of resistance to the subtropical root diseases has been reported for *Pyrus*. Most of the root diseases found in apple are also damaging for pear.

9 Acknowledgements

The authors acknowledges support of F. Denardi and I. D. Faoro, from EPAGRI, Estação Experimental de Caçador, Santa Catarina, Brazil. Thanks are also, due to the Brazilian Institution IAPAR - Instituto Agronômico do Paraná in which facilities most of the apple and pear germoplasm evaluation reported in this chapter were carried out.

10 References

Aguilar, A.L. (1979). Fenologia de mutantes de manzano de bajo requerimiento de frio. *Proc. Tropical Reg. Amer. Soc. Hort. Sci.* 23, 30-33.

Allison, E. (1974). Quince cultivation. Fruit and Fruit Technology Research Inst., Stellenbosch. *Information Bulletin 247.* Republic of South Africa. 6 pp.

Andrews, C.P. and Sherman, W.B (1979). Hybrid pear cultivars for Florida. Proc. Fla. Hort. Soc. 92, 266-267.

Assaf, R. (1995a). Les Hashabi, nouveaux porte-graffes pour la culture du pommier en Israel. I. Comportement sous différents climats. *Fruits* 50(1), 65-75.

Assaf, R. (1995b). Les Hashabi, nouveaux porte-graffes pour la culture du pommier en Israel. II. Comportement dans différentes conditions de culture. *Fruits* 50(2), 133-143.

Bederski, K.A. (1988). Apple growing in the Coastal areas of Peru. *Acta Hort.* 232, 51-55.

Bell, R.L., Moore, J.N., and Ballington, J.R. Jr. (1982). Pears *(Pyrus).* Genetic Resources of Temperate Fruit and Nut Crops II. *Acta Hort.* 290, 657-97.

Black, M.W. (1952). The problem of prolonged rest in deciduous fruit trees. *Proc. 13th Intern. Hort. Congr.* London 2, 1122-1131.

Boynton, D.O (1959). Observations on the temperature limitations of the apple in Tropical America. *Proc. Amer. Soc. Hort. Sci. Caribbean Region* 7, 69-85.

Brooks, R.M., and Olmo, H.P. (1972). *Register of new fruit and nut varieties.* Berkley. Univ. of California Press. 708 pp.

Byrne, D.H., Ou, S., and Storey, J.B. (1995). Annual grafting of nashi pear in low chill areas of Taiwan. *Fruit Var. J.* 49(2), 103-106.

Chandler, W.H. (1960). Some studies of rest in apple trees. *Proc. Amer. Soc. Hort. Sci.* 76, 1-9.

Chandler, W.H., and Brown, D.S. (1957). Deciduous orchards in California winters. *Calif. Agr. Ext. Serv. Circ.* 179.

Crossa-Raynaud, P. (1955). Effects des hivers doux sur le comportement des arbres fruitiers a feuilles caduques. *Ann. du Service Bot. Agron. Tunisie.* 28, 1-22.

Crossa-Raynaud, P. (1960). Problemes d'arboriculture fruitiere en Tunisie. *Ann. Inst. National Recherche Agron. Tunisie* 33, 1-263.

Cummins, J.N. (1991). Register of new fruit and nut varieties. List 35. *HortScience* 26(8), 951-986.

Dall'Orto F.A.C., Ojima, M., Barbosa, W, Martins, F.P., Sabino, J.C., Castro, J.L., Santos, J.L., and Rigitano, O. 1987. Maçã Galícia (IAC 276-2). *O Agronômico.* Campinas 39 (2), 109-110

Dall'Orto, F.A.C., Ojima, M., Barbosa.,W., and Rigitano, O. (1986). Marmelo. *B. Inst. Agron. Campinas,* 200, 145-146.

Dall'Orto, F.A.C., Ojima, M., Barbosa, W., Rigitano, O., Martins, F.P., Castro, J.L., Santos, R.R. and Sabino, J.C. (1995). Variedades de pera para o Estado de São Paulo. *B. Técn. Inst. Agron. Campinas* 164, 1-34.

Darrow, G.M. (1953). Deciduous fruits in Central America, Colombia and Ecuador. *Ceiba* 4, 69-81.

Denardi, F. and Camilo, A. (1994). Fred Hough - a new apple cultivar resistant to scab. *Rev. Bras. Frut.* 16(1), 1-6.

Denardi, F., and Camilo, A.P. (1997). - Novas cultivares de macieira. *Agropecuaria Catarinense,* 10(2), 25-30.

Denardi, F., and Camilo, A.P. (1998a). 'Epagri 408 - condessa: Nova cultivar de macieira de baixa exigeancia em frio hibernal. *Agropecuaric Catarinense* 11(2), 12-15.

Denardi, F., and Camilo, A.P. (1998b). Duguesa: Nova cultivar de macieira de baixa exigeancia em frio hibernal et alta resistencia a sarna. *Agropecuaria Catarinense* 11(4), 19-21.

Denardi, F., Hough, L.F., and Camilo, A.P. (1988a). 'Primicia' Apple. *HortScience* 23(3), 632.

Denardi, F., Hough, L.F., and Camilo, A.P. (1988b). 'Princesa' apple. *HortScience* 23(4), 787.

Deyton, D.E. and Cummins, J.C. (1991). History of pear breeding in Tennessee. *Fruit Var. J.* 45(3), 143-146.

Diaz, D.H. and Sadowska, A. (1979). Comportamiento de cultivares de manzano con bajo requerimiento de frio al metodo de cultivo superintensivo. *Proc. Trop. Reg. Amer. Soc. Hort. Sci.* 23, 43-45.

Edwards, G.D. (1987). Producing temperate-zone fruit at low latitudes: avoiding rest and chilling requirement. *HortScience* 22(6), 1236-1240.

Eggert, F.P. (1951). A study of rest in several varieties of apple and in other fruit species grown in New York State. *Proc. Amer. Soc. Hort. Sci.* 57, 169-178.

Erez, A. (1980). Deciduous fruit growing in Kenya. Hort. Research and Development Project. Consultant Report n#2. FAO/Natl. Hort. Research Station Thika-Kenya.

Erez, A., and Lavi, B. (1984). Breaking bud rest of several deciduous fruit tree species in the Kenyan highlands. *Acta Hortic.* 158, 239-248.

Fregoni, M. (1968). Aspetti della frutti-viticoltura di um paese dell'America Latina. L'Ecuador. *Ann. Fac. Agrar. S. Cuore,* Milan 8, 299-379.

Ghariani, K. and Stebbins, R.L. (1994). Chilling requirements of apple and pear cultivars. *Fruit Var. J.* 48(4), 215-222.

Hauagge, R., and Cummins, J.N. (1991a). Phenotypic variation of length of bud dormancy in apple cultivars and related *Malus* species. *J. Amer. Soc. Hort. Sci.* 116(1), 100-106.

Hauagge, R., and Cummins, J.N. (1991b). Seasonal variation in intensity of bud dormancy in apple cultivars and related *Malus* species. *J. Amer. Soc. Hort. Sci.* 116(1), 107-115

Hedrick, U.P., Howe, G.H., Taylor, O.M., Francis, E.H., and H.B. Tukey. (1921). *The pears of New York.* Department of Agriculture. 29th Annu. Rpt. Vol.2. Part II.

Hill, A.G.G. and Campbell, G.K.G. (1949). Prolonged dormancy of deciduous fruit-trees in warm climates. *Empire J. Exper. Agric.* 68, 259-264.

Hunter, N. (1981). A new early red apple cultivar, Drakenstein. *Agroplantae* 13, 17.

Kemp, P. (1971). *Deciduous fruit in Rhodesia.* Tech. Bull.*13.* Min. Agric. Rhodesia.

Lecrenier, A. (1972). *Situation des arbres fruitiers 'a pépins en Algérie.* 1ére partie. Projet Algérie *30.* Formation professionelle. FAO, Rome, Italy.

Lloréns, J.A.S.C. (1967). Contribución al conocimiento de nedesidades en frio invernal de variedades fruticolas. *Levante Agricola* 6, 26-29.

Mazzochi, G. and Pucci, E. (1963). Indagine preliminare sul fabbisogno di freddo di diverse specie arboree da frutto in Tripolitania. *Rev. Agric. Subtrop. Trop.* 57, 370-394.

Miller, E.P. and Baker, L.H. (1982). An evaluation of apple cultivars for Central and North Florida. *Proc. Fla. State Hort. Soc.*, 95, 88-90.

Miller, E.P., and Sherman, W.B. (1980). Origin and description of 'Dorsett Golden' apple. *Proc. Fla. State Hort. Soc.* 93, 108-109.

Nesterov, Ja. S. (1957). Dormancy and hardiness in fruit trees. Dokl. *Akad. Nauk SSSR*. 117, 507-510. (in Russian)

Nichols, M. (1994). Cultivos. Manzanas tropicales. *Agricultura de las Americas*, (Sep./Oct.): 21-33.

Notodimedjo, S., Danoesatro, H., Sastrosumarto, S., and Edwards, G.R. (1991). Shoot growth, flower initiation and dormancy of apple in the tropics. *Acta Hort.* 120, 179-186.

Ojima, M., Dall'Orto, F.A.C., Martins, F.P., Castro, J.L., Sabino, J.C., Rigitano, O. (1987a). Pêra 'Centenária' (IAC 9-47). *O Agronômico*, Campinas 39(2), 115-116.

Ojima, M., Dall'Orto, F.A.C., Martins, F. P., Castro, J.L., Sabino, J.C., Rigitano, O. (1987b). Pêra 'Primorosa' (IAC 9-3). *O Agronômico* Campinas 39(2), 116.

Oppenheimer, C. (1962). Breeding of apples for a subtropical climate. *Proc. 16th Intern. Hort. Congr.* Brussels 3, 18-24.

Oppenheimer, C.H. and Slor, E. (1968). Breeding apples for subtropical climate II. Analysis of two F_2 and nine backcross populations. *Theor. Appl. Genet.* 38, 97-102.

Overcash, J.P. and Murphy, B.C. (1959). Apple varieties react differently to mild winters. *Mississipi Farm Res.* 22(1), 5.

Pasternak, G.K. (1983). *Chilling requirement of buds and seeds of apple cultivars and associated hormonal changes in seeds during breaking of dormancy.* Ph.D. Thesis. Cornell University.

Petropoulou, S.P. (1985). *Temperature related factors as selection criteria in apple breeding.* Ph.D. Thesis. University of London.

Philippe, J.M. (1965). *Proposals for a research programme on fruit trees.* II. Rep. Agric. Res. Stat. Proj., UNSF Damascus, Syrian Arab Rep.

Popenoe, W. (1970). Aclimatación de frutas de zonas templadas en los trópicos. *Hacienda* N.Y. 65, 56-58.

Popenoe, W., and Ben'tez, J. M. (1965). Varieties of temperate zone fruits for tropical America. *Proc. Carib. Reg. Amer. Soc. Hort. Sci.* 8, 118-129.

Postman, J. (1992). Pears from the past preserved for posterity. *Fruit Var. J.* 46(4), 255-262.

Ramirez, H. and Saavedra, L.L. (1990). A low chill requirement Golden Delicious apple mutant from Northeast Mexico. *Acta Hortic.* 279, 67-73.

Rigitano, O., Ojima, M., Dall'Orto, A. C. (1975). Novos cultivares de maçã para o clima Paulista. *Inst. Agron. Campinas, Brazil, Boletim Technico* 31, 11 p.

Ruck, H. C. (1975). *Deciduous fruit tree cultivars for tropical and sub-tropical regions.* Hort. Rev. N/ 3. Commonwealth Bureau of Hort. Plantation Crops. East Malling, Maidstone, Kent, 91 pp.

Samish, R.M. and Lavee, S. (1962). The chilling requirement of fruit trees. *Proc. 16th Intern. Hort. Congr.*, Brussels, 5, 372-388.

Sansavini, S. (1991). Tecniche manipolative del ciclo di fruttificazione in aree subtropicali. *Revista di Frutticoltura*, LIII(2), 5-17.

Seth, J.N. and Ghildiyal, P.C. (1984). Some promising apple hybrids developed at Chaubattia India. *Prog. Hort.*, *15*, 239-243.

Sharpe, R.H., and Sherman, W.B. (1990). Peach cultivars in Florida and their chilling requirements. *Acta Hort.* 279, 191-197.

Sherman, W.B. and Lyrene, P.M. (1996). 'TropicSweet' apple. *Fruit Var. J.* 50, 4, 217.

Sherman, W.B., and Crocker, T.E. (1982). Low chill pears, in The Pear. Culture, Varieties, Breeding, Propagation, Nutrition, Prunning and Training. Diseases and Insects. Harvesting, Storage and Marketing (eds T. van der Zwet and N.F. Childers). pp 130-135.

Sherman, W.B., Andrews, C.P., Lyrene, P.M. and Sharpe, R.H. (1982). 'Flordahome' pear. *HortScience* 17(2), 270.

Slor, E. (1982). 'Slor', a new subtropical apple cultivar. *Hassadeh* 62, 1580-1582. (in Hebrew)

Subhadrabandhu, S. and Punsri, P. (1987). Deciduous fruit trees as an alternative to opium popy in Northern Thailand.. *Acta Hort.* 199, 39-44.

Tabuenca, C. (1964). Necessidades de frio invernal de variedades de albaricoquero, melocotonera y peral. *Estación Experimental de Aula Dei, Zaragoza, Anales* 7, 113-132.

Tancred, S.J., Zeppa, A.G., and Cummins, J. N. (1994). Thirty-seven apple varieties of Australian origin. *Fruit Var. J.* 48(2), 118-125.

Thompson, J.M., Van Der Zwet, T., Draper, A.D., and Blake, R.C. (1976). Evidence of cytoplasmic and genetic male sterility in pears. *J. Hered.* 67, 339-346.

van Zyl, H.J. (1982). Pear culture in the Republic of South Africa. 1982, in The Pear. Culture, Varieties, Breeding, Propagation, Nutrition, Prunning and Training. Diseases and Insects. Harvesting, Storage and Marketing (eds T. van der Zwet and N. F. Childers). pp 38-51.

Walheim, L. and Stebbins, R.S. (1981). *Western Fruit Berries and Nuts. How to select, grow and enjoy.* H.P. Books 118-122.

Weinberger, J.H. (1967). Some temperature relations in natural breaking of rest of peach flower buds in the San Joaquin Valley, California. *Proc. Amer. Soc. Hort. Sci.* 91, 84-89.

Westwood, M, N. and Chestnut., N. E. (1964). Rest period chilling requirement of Bartlett pear as related to Pyrus calleryana and P. communis rootstocks. *Proc. Amer. Soc. Hort. Sci.* 84, 82-87.

Westwood, M.N., and Bjornstad, H.O. (1968). Chilling requirements of dormant seeds of 14 pear species as related to their climatic adaptation. *Proc. Amer. Soc. Hort. Sci.* 92, 141-149.

Young, E. and Werner, D.J. (1985). Chill unit and growing degree hour requirements for vegetative budbreak in six apple rootstocks. *J. Amer. Soc. Hort. Sci.* 110, 411-413.

Zai-Long, Li. (1990). Advances of pear breeding program at Zhejiang Agricultural University. *Acta Hortic.*, 279, 167-184.

11 APPLE PRODUCTION AT LOW LATITUDES

JOHN E. JACKSON

Horticultural Research Centre, P.O. Box 810
Marondera, Zimbabwe

1 Introduction

The Apple, *Malus* domestica Borkh, is the most important of the deciduous tree fruits in terms of production (Westwood 1993). In the past it was generally regarded as a crop of the temperate zones but is increasingly cultivated under sub-tropical and even tropical conditions (Luckwill 1984). This has been made possible by selection of adapted cultivars, described in chapter 10, and by both technological innovations and modifications to standard temperate-zone practices.

Apples are now grown at almost every latitude from the equator to more than 60° N with a significant shift to warm temperate and even subtropical climates over the last forty years.

Production statistics are most readily available on a national rather than a latitudinal basis. However, it is clear from these (Table 1) that whereas in 1951 87% per cent of world apple production was in Europe, the USSR, the USA and Canada with by far the greater part of their production from north of 35° N; by 1994 only 47% of world production was in these areas. This is because, although there has been a rise in production in Europe, the USSR and the USA, it has been much less dramatic than that elsewhere.

Much of the increase in world production has been in countries at the same latitudes as apple producing areas in the USA and southern Europe, *e.g.* in Turkey, Japan, the two Koreas and temperate zone areas in Argentina and Chile and in intermediate-climate countries such as Iran. However, there has also been a spectacular increase in production in countries which are completely or almost completely within 35 degrees of the equator, the major producers in this zone in 1994 being India (1,238,000 t), Brazil (743,000 t), South Africa (551,000 t), Mexico (522,000 t), Pakistan (370,000 t) and Egypt (300,000 t), the combined production in these six countries being 75% of that in the USA. The relative increase in the output of these tropical and subtropical countries has been progressive over the years and has been greatest in the last two decades.

Cultural practice has had to be modified to meet the different environmental conditions but the general factors controlling the production technologies for apple are virtually universal wherever apples are grown for sale.

In general, apples are marketed by cultivar name and according to strict standards of perceived quality. Apples failing to meet these standards are virtually unsaleable or saleable only for juice at a low price. The capital cost of establishing an apple orchard is very high,

Amnon Erez (ed.), Temperate Fruit Crops in Warm Climates, 305-342.
© 2000 *Kluwer Academic Publishers. Printed in the Netherlands.*

being 43,600 guilders per hectare in The Netherlands in 1994/95 (Joose and Besseling 1994/95) with the kind of high density planting now being adopted worldwide. This cost is mainly for trees but is also for tree support structures, soil preparation, irrigation etc.

These two facts lead to a global focus of the technologies of production on:

1. Provision of high quality fruit of named cultivars, which is generally dependent on true-to-name, clonally propagated cultivars grafted on rootstocks.
2. Tree pruning and management so as to give good light penetration to fruits and spur leaves so as to enable optimal fruit colour and size development, fruit thinning to improve fruit size, treatments to control blemishes due to pest and disease, russetting etc and emphasis on producing fruits with the potential for long-term storage.
3. Use of high-density planting systems. These can give high yields soon after planting, so enabling rapid repayment of establishment costs, but only if the trees fill their allotted space, *i.e.* attain full canopy development, within a few years of planting and are precocious in cropping. These objectives are usually achieved by use of precocity-inducing rootstocks and appropriate tree pruning and training methods.

Table 1. Changing world apple production from 1954 to 1994 (thousand tons)

	1954	*1964*	*1974*	*1984*	*1994*
World	16,658	22,220	27,423	39,798	48,890
Europe	10,383	12,504	12,179	15,760	13,200
USSR	1,480	2,639	5,077	7,100	4,327*
USA	2,436	2,866	2,984	3,779	4,948
Canada	296	409	406	434	500
Oceania	296	477	487	513	862
Argentina	242	371	786	872	1,000
Chile	84	106	120	410	810
Brazil	8	11	14	128	743
Mexico	56	124	232	459	522
China	174	302	1,160	2,955	12,007
India	112	110	592	986	1,238
Iran	44	68	98	1,200	1,690
Japan	450	1,090	850	812	1,048
Korea DPR	77	85	290	560	665
Korea Rep	50	125	297	528	631
Pakistan	3	13	56	143	370
Turkey	161	337	950	1,900	2,080
Egypt	4	7	8	28	300
Morocco	2	17	20	30	270
South Africa	92	149	284	484	551
Other Countries	208	410	533	717	1,128

Data from F.A.O. World Crop and Livestock Statistics 1948-1985 and F.A.O. Yearbook Production Vol. 48, 1994.
* Contries of the former USSR

2 Major relevant effects of latitude on environment

The adaptations in apple production technology in relation to latitude are dependent on the effects of this on a number of key environmental parameters. These are shown in detail in most atlases and textbooks of geography and climatology.

Effects of latitude are modified by effects of altitude, proximity to large bodies of water, and prevailing winds, which affect temperatures, cloudiness, rainfall etc, but if these local effects are ignored it can be summarised as follows:-

2.1 Effects of latitude on temperature

In general, the length of the season during which temperatures are adequate for plant growth increases from the poles to the equator, where they are adequate throughout the year. This progressive shortening of the season of winter dormancy affects a wide range of practices from propagation technology and season of planting to the number of crops which can be obtained each year or the length of the growing season for a single crop, which determines demand for water.

The progressive shortening, and reduction in severity, of winter conditions with decreasing latitude results in reduction in winter chilling, whether defined as hours below 7.2°C (Weinberger 1950) or chill units giving different weighting to different (low) temperatures and a negative (de-chilling) effect of high temperatures (Richardson, Seeley and Walker 1974, Fishman, Erez and Couvillon 1987). This in turn has a major effect on bud break of both vegetative and flower buds. The reduced winter chilling at low latitudes is perhaps their most distinctive feature with respect to apple production.

The incidence of spring frosts and autumn/winter freezes also decreases with decreasing latitude with a consequent reduction in the inputs used to avoid or overcome these catastrophic, as opposed to incremental events. Hail, on the other hand, can be a frequent and serious problem at low latitudes.

The highest daily maximum temperatures are reached in summer at intermediate latitudes rather than at the equator. In general, night temperatures during the fruit production season are higher at low latitudes than in the temperate zones but this is by no means always the case, *e.g.* mean monthly minima at Pillaro, Tungurahua, Equador range from 7.8 to 10°C (Niegel 1988) which are cooler than those of the main season of fruit development at East Malling, Kent which is near the northern limit of commercial fruit production (Wickenden 1982).

The fact that much of the apple production at low latitudes is at high altitude has a major effect on temperatures in the apple growing areas since, on average, temperature declines by 5°C per 1000 m and high altitude is also frequently associated with cloudiness. Some relevant data on temperatures is given in Table 2.

Soil temperatures are primarily controlled by receipt of solar radiation (Richards, Hagan and McCalla 1952) so show large effects of latitude. The surface layers respond most rapidly to changes in solar energy input. Temperatures in the upper 5 cm, for example, can be much higher than air temperatures when the latter are high and lower than them when the latter are low. Temperatures in the main rooting zones are reasonably similar to those of air temperature. Typical results from a temperate zone site given in

Table 2. Temperatures at some typical subtropical sites compared with cool and warm temperate sites

Site	*N. Hem* *S. Hem*	*July* *Jan*	*Aug* *Feb*	*Sept* *Mar*	*Oct* *Apr*	*Nov* *May*	*Dec* *June*	*Jan* *July*	*Feb* *Aug*	*Mar* *Sept*	*April* *Oct*	*May* *Nov*	*June* *Dec*
1	Max ° C	20.5	22.1	19.6	12.6	11.1	4.3	8.0	4.6	11.5	11.8	15.6	17.6
	Min ° C	11.9	11.6	9.9	5.6	4.6	-0.9	2.1	-0.1	6.2	4.7	7.8	9.6
2	Max ° C	23.1	23.2	21.5	18.7	15.7	13.4	12.6	13.5	15.6	17.8	19.7	21.7
	Min ° C	11.6	11.6	10.5	7.6	4.5	1.9	1.4	2.5	4.8	6.8	8.6	10.6
3	Max ° C	25.9	24.9	23.8	21.2	16.9	14.2	13.4	15.1	18.0	21.0	23.5	25.8
	Min ° C	15.1	14.9	13.3	9.2	6.1	2.7	1.4	2.8	5.5	9.0	11.4	13.6
4	Max ° C	26.7	26.7	23.5	22.4	20.2	18.0	17.7	19.9	21.0	22.9	24.8	25.6
	Min ° C	15.8	15.7	13.9	10.9	7.6	6.6	6.3	7.6	9.6	11.2	12.6	14.5
5	Max ° C	36.9	35.8	36.1	35.8	32.2	25.5	22.7	28.3	32.8	38.9	42.2	39.4
	Min ° C	26.7	25.8	24.4	19.2	13.3	8.9	8.3	10.6	16.7	21.1	25.5	27.2

Site 1. A cool temperate site: East Malling England 51° 16'N (Source Wickenden 1982)
Site 2. A warm temperate site: Riwaka, New Zealand 41° 65'S (Source Dr Palmer)
Site 3. Subtropical Australia: Applethorpe Queensland 28° 37'S (Source Dr Middleton)
Site 4. Subtropical Brazil: Cacador Santa Catarina 26° 42' S (Bernardi 1988)
Site 5. Subtropical Pakistan: Tando Jam, Sindh approx 25° 24' N (Panhwar 1995)

Table 3, data for which were extracted from Wickenden (1982), show the relationship between soil and air temperature and solar radiation.

In Egypt, Taylor and Burns (1924) found average July soil maxima to be 41.7°C at 10 cm and 35.1°C at 20 cm while in an apple orchard in Mexico when air temperature was 34°C soil temperatures at 20 to 120 cm depth were from about 30 to about 28°C and when air temperatures were 45°C were from about 36 to about 28°C (Diaz and Romo 1988).

Table 3. The relationship between air and soil temperature 51° 16' N in 1981

Month	*Solar Radiation (mW h/cm² Total)*	*Mean screen temperature (°C)*	*Soil temperature (°C)*		
			10 cm	*20 cm*	*50 cm*
January	2083	5.1	3.3	4.1	5.0
February	3545	2.3	2.0	2.9	4.2
March	4938	8.9	7.2	7.2	7.2
April	9946	8.3	8.3	8.5	9.5
May	11470	11.7	12.1	11.5	11.9
June	12951	13.6	15.5	14.9	15.2
July	12640	16.2	17.2	16.7	17.0
August	12775	16.9	18.0	18.0	18.6
September	8657	14.8	14.9	15.1	16.4
October	5446	9.1	8.5	9.5	11.7
November	2374	7.9	6.8	7.6	9.1
December	1537	1.7	1.6	2.6	4.7

Data from Wickenden 1982.
N.B. Blossoming of apples was approx 25 April to 25 May.

2.2 Effects of latitude on light (photosynthetically active radiation) and daylength

This is reviewed in Chapter 1. In general annual radiant energy inputs are much higher at low latitudes than in the temperate zone with the highest peak incidence in the summer season in the subtropics and around the tropics but with equatorial regions having the longest (all year) season with high light intensities. Despite high light intensities the physical laws governing light penetration into leaf canopies ensure that within-canopy shade remains a constraint on the production of high quality fruits and management systems of pruning and training must optimise light penetration.

Although daylength is markedly affected by latitude this factor does not appear to be important as such for apple, which is not known to respond to photoperiod.

2.3 Effects of latitude on rainfall and hail

In broad terms there are zones of low rainfall centred on the tropic of Cancer in the northern hemisphere (c.f. the Saharan and Arabian deserts) and the tropic of Capricorn in the southern hemisphere (c.f. the Australian and Kalahari deserts), although there are some areas of higher rainfall at these latitudes. Equatorial areas are, in general, areas of higher precipitation than the subtropical and temperate zones. Unlike the temperate zones, where seasons of production are controlled by temperature, natural seasonality in the tropics is controlled by both temperature and rainfall. In practice, however, apples are often produced under irrigated conditions so the importance of rainfall lies in its effect in influencing irrigation demand and in facilitating the spread of some diseases. It should be noted that a much higher proportion of the rainfall in the tropics tends to be concentrated in short duration storms of high intensity than is the case in temperate zones (Jackson 1989). This reduces the effectiveness of any given rainfall level because in agricultural terms the effective rainfall is that entering the soil and remaining within the rooting depth. Heavy, high-intensity, storms may result in much more loss by surface run-off and drainage beyond root range than the typically more moderate temperate area rainfall. This problem is compounded by the fact that such storms may be widely separated in time, allowing the soil to dry out and the plants to suffer moisture stress. For these reasons, as well as because of higher evaporative demand in sub-tropical and tropical climates as a result of the higher incoming radiation, a level of annual or seasonal rainfall in the tropics or subtropics which would be adequate for fruit production in temperate areas will need to be accompanied by supplementary irrigation.

Although hail can be a serious problem in temperate-zone apple production areas, especially in northern Italy and other areas where production is in mountain valleys, it can be especially severe in summer rainfall areas at low latitudes *e.g.* in Zimbabwe and Queensland, Australia.

2.4 Effects of latitude on evaporation

Evaporation, on the assumption that water is available, depends primarily on solar radiation, (discussed in chapter 3) with higher levels of solar radiation supplying more energy for evaporation, wind, which removes moist air from the layer above or around the evaporating surfaces, relative humidity because the higher the relative humidity of the air the lower its ability to absorb water, and temperature. The higher the temperature the more rapid evaporation both because of its energy input and because the capacity of air to absorb water vapour increases as its temperature rises. As a result of these factors, evaporative demand is generally much higher at low latitudes than higher latitudes except in cloudy and humid areas. Evaporative demand under specific conditions can be calculated by methods outlined by Doorenbos and Pruitt (1984) but is also frequently estimated by use of evaporation pans and crop demand calculated from this using a specific crop factor.

2.5 Effects of latitude on soils

Soil characteristics are determined by the nature of the parent rocks and by climatic factors influencing weathering, mineralisation of organic substances and leaching so that broad generalisations are of limited value. However, the following do seem widely true (Weischet and Caviedes 1993).

In the humid tropics soils tend to be of low weatherable mineral content and cation exchange capacity and to be subjected to high mineralisation rates of organic substances. At higher altitudes within these areas there is reduced hydrolytic mineral weathering and a lessened mineralisation rate of organic matter due to lower temperature and moisture levels, which improves the soil nutrient content.

In the outer tropical zones with climates with specific wet and dry seasons the soils tend to increased mineral content and cation exchange capacity therefore much higher potential fertility. The soil structure is, however, often such as to make them very liable to erosion.

In terms of chemical properties phosphorus deficiency is probably the most important characteristic of soils at low latitudes. This is a consequence of the fact that total phosphorus content decreases with increasing weathering intensity. For example whereas the total content of phosphorus in representative soils in the US Midwest is around 3000 ppm, tropical oxisols, ultisols and alfisols usually have less than 200 ppm (Sanchez 1976, Weischet and Caviedes 1993). Moreover soil phosphorus is held partly in inorganic form as adsorbed anions on the soil colloids and partly as organic phosphorus contained in the humus and is slowly released from both these sources to maintain an equilibrium level in the soil solution. In the inorganic form the phosphorus is held as compounds of calcium, iron and aluminium. The compounds with iron and aluminium are, however, covered by a coat of inert material so are effectively inert. This fixation or immobilisation of phosphorus is general in acid tropical ferralitic soils with abundant iron and aluminium.

Soils in tropical and sub-tropical areas are commonly acidic (*e.g.* Nyamapfene, 1991; Kotzé and du Preez, 1988; Basso and Wilms, 1988). This has a major effect on calcium availability. Kotzé and Deist (1972) showed that exchangeable aluminium increased rapidly as pH decreased below 5.0 ($CaCl_2$ scale) and Kotzé (1976) showed that calcium uptake was strongly suppressed by the presence of aluminium.

Tropical and subtropical soils are also frequently deficient in micronutrients (Tagwira, 1995), levels of zinc and boron being particularly poor on granite-derived sandy ferralitic soils and on some basaltic soils and Kalahari sands. There is however also substantial apple production on calcareous soils with high pH in Mexico (Diaz and Romo 1988). High pH soils in warm, dry, conditions are discussed in Chapter 4.

3 Effects of latitude-dependent environmental factors on apple-tree growth and cropping

3.1 Temperature effects on apple production

3.1.1 *Dormancy*

Dormancy can be generally defined as the temporary suspension of visible growth of any plant structure containing a meristem (Lang 1987). Apples, in common with a wide range of winter-deciduous trees, when grown in temperate regions undergo a period of winter dormancy or rest, which characteristically involves three stages. First comes the initiation of dormancy in late summer or autumn, then winter 'endodormancy' which is internally controlled so that growth will not take place even with higher temperatures, then late winter 'ecodormancy' when growth is prevented by low temperatures.

The length of the period of winter endodormancy varies greatly between cultivars, presumably as an adaptation to their environments of genetic origin since it is obvious that the period of dormancy should last until there is no longer serious risk of lethal frosts. A cultivar which could be stimulated into growth by a few days of warm weather in mid winter would be unlikely to survive natural and human selection. The duration of this period of endodormancy was originally found to be most accurately defined not in days but in chilling-hours or hours below 7.2°C (Chandler *et al.* 1937, Lamb, 1948). Subsequently it was noted that warm days could reverse the effects of chilling and Richardson, Seeley and Walker (1974) and others developed the concept of chilling units with each hour below 1.5°C or between 12.5 and 15.9°C being ineffective, hours between 1.5 and 2.4°C and 9.2 and 12.4°C contributing 0.5 of a chill unit, and hours between 2.5 and 9.1°C contributing 1 chill unit while hours between 16 and 18°C had a -0.5 effect, between 18.1 and 21°C a -1 effect and between 21.1 and 23.0°C a -2 effect. Further refinements to these models have been made on the basis of the concept that the positive effect of chilling can be maximized by particular moderate temperatures immediately following chilling and that negative, de-chilling, effects only apply for a limited period after a chilling event, when the dormancy-breaking factor produced by chilling is still capable of being destroyed by high temperature (Fishman, Erez and Couvillon 1987, Linsley-Noakes, Allan and Matthee 1994). The de-chilling effects of high daytime temperatures are in general likely to be a more important factor in relation to the meeting of chilling requirements in locations where these high daytime temperatures commonly occur in winter, which is the case at low latitudes especially those without much winter cloud. This opens up technological approaches to meeting winter-chilling requirements under these conditions. Chapter 2 deals with dormancy problems in more detail.

3.1.2 *Growth*

In general, the growth of apple shoots is positively related to temperatures up to about 25-30°C but at higher temperatures growth may be limited by other factors (Lakso 1994). Gur, Bravdo and Mizrahi (1972) found that root temperatures above 30°C caused reduction in shoot as well as root growth with severe effects on leaves at root temperatures of 35°C. Rooting of apple rootstock cuttings was low (less than 20%) at 11.7°C but rose to around 70-88% at 21.7°C (Howard 1968). Rootstocks vary in their response to high soil temperatures. This is discussed later.

A major effect of temperature on growth of all organs, *i.e.* on total tree carbon, is a result of the differential effects of temperature on photosynthesis and respiration. Photosynthesis of apple leaves does not appear to have a strong response to temperature over a fairly wide range of temperatures from 15 to 35°C with temperature optima near 30°C whereas respiration responds to temperature logarithmically over the range of 0-42°C according to the equation.

$$R = ae^{kT}$$

where R is respiration rate, a = R at T = 0°C (the intercept of ln R versus T), k is the temperature co-efficient of R (the slope of ln R versus T) and T = temperature in °C (Lakso 1994). Lakso and Johnson (1990) and Wagenmakers (1994) both concluded that major site differences in potential production, including latitudinal differences, arise from the effects of night temperatures on respiratory losses.

3.1.3 *Flower bud development*

Flowering in apple can be divided into two major developmental processes, which in the temperate zone occur in two successive growing seasons: the initiation and development of flower buds occurring during the summer and autumn of one season and flowering itself in the spring and summer of the second season. However fruit trees must have attained the right developmental stage before they can flower, the pre-flowering stage being referred to as the juvenile phase. The transition from the juvenile to the flowering phase is dependent on the attainment of a critical size or number of nodes (Zimmerman 1971) and is therefore dependent on the rate of growth of juvenile plants (Visser, Verhaegh and deVries 1976). A long growing season with optimal temperatures for growth is thus likely to result in a shorter juvenile period. Very rapid growth of apple seedlings has been followed by precocious cropping under the excellent growing conditions of high-veld Zimbabwe (author observation).

Faust (1989) noted that grafted trees in their first 2-3 years after grafting return to their non-flowering status, apparently passing into a vegetative although non-juvenile phase. Under Zimbabwean and presumably other low-latitude conditions, this is not evident, with many cultivars flowering profusely in the nursery and the first orchard year, possibly as a consequence of their rapid growth.

Once apple trees are no longer in the juvenile or vegetative phases, in order for a vegetative bud to develop into a flower bud it must be at a certain stage. The apple flower bud is essentially a shortened axis bearing typically 21 leaf formations; nine bud scales followed by three transition leaves, six true leaves and three bracts (Bijhouwer 1924) and only after a critical node number, 16 for Golden Delicious or 20 for Cox, has been reached does flower induction begin. The first flower primordium to develop is in the shoot apex itself (the King flower primordium) and then come flower primordia in the axils of the bracts and true leaves. The rate of node production and the length of the growing season determine whether a bud will produce flowers or remain vegetative. Luckwill (1974) calculated that a Cox apple apex with only 6 nodes at the beginning of the season needs about 100 days to attain the 20 node stage for flower induction to commence: the time interval between induction of successive primordia being about 7 days (Fulford 1966). At high latitudes with short growing seasons flower buds may develop only or mainly on pre-existent spurs.

At low latitudes with much longer growing seasons flower buds commonly develop in terminal or axillary buds of the shoot which has developed in the early part of the season.

Warm conditions during the mid-winter to bud-break period have been associated with low yields in both England (Beattie and Folley 1978, Jackson and Hamer 1980, Jackson, Hamer and Wickenden (1983) and the USA (Lakso 1994). Jackson, Hamer and Wickenden (1983) showed that potted trees transferred to cold stores at 5°C in February, March and April had a much higher fruit set when hand pollinated than did outdoor controls or trees maintained at 10°C and also that mist irrigation from 13 February to 15 April or later for evaporative cooling increased fruit set of hand-pollinated Cox blossom from 5.2% in the controls to an average of 21.3% for evaporatively-cooled trees. This has particularly important implications for apple growing in warm-winter areas.

3.1.4 *Pollination, fertilisation and fruit set*

Pollen tube growth in apple is curvi-linearly related to mean daily temperatures with growth per day of only 8% of the depth of the style in a day at 5°C, 14% at 10°C, 20% at 12°C but 25%, 35 and 50% at 13, 14 and 15°C respectively (Williams 1970). Since the ovules have a limited fertile life (approximately from 5 to 14 days dependent on cv and other factors) then there is an effective pollination period, EPP, which equals the life of the ovule before degeneration minus the time required for the pollen tubes to reach the embryo sac, which is strongly temperature dependent as noted above. Since self-pollen grows more slowly down the style than foreign pollen (Nucci, L.S. and Gurnevski, V. 1989) then high temperature conditions increase the probability of self-fertilisation. They also generally increase the production of parthenocarpic fruits (Faust 1989) and it is generally observed in Zimbabwe that elongated seedless fruits of Anna apple are obtained from flowers which open before those of any pollinators.

High (19°C) temperatures post-pollination can however reduce ovule longevity and fruit set in comparison with 13°C (Tromp and Borsoom 1994). Also high night temperatures (20°C) for only four days around four weeks after full bloom can cause severe fruit shed (Kondo and Takahashi 1987, 1989).

3.1.5 *Temperature effects on fruit growth*

Fruit growth is by cell division and cell enlargement, following a sigmoid growth curve with cell divisions dominating growth for the first 50 days, mainly taking place in the first 4-5 weeks from anthesis, and cell enlargement thereafter (Faust 1989). Tukey (1956, 1960) found greater fruit growth and development at 24°C versus 9 or 31°C with the latter (night) temperature being particularly deleterious. Tamura *et al.* (1981) found that raising night temperatures to 23°C was stimulatory for the first 3 weeks after bloom but by 30-35 days all the fruits were the same size. Tromp (1975) compared the effects of 19/14°C day/night temperatures from full bloom onwards with 24/19°C temperatures and found fruit growth rates up to harvest (c 115 days from full bloom) were greater at the higher temperatures and the fruits much larger at harvest (23.9 to 13.6 g *dry* wt per fruit).

3.1.6 *Temperature effects on fruit shape and colour*

Fruit shape, in particular the length to diameter ratio, is very sensitive to temperature with more elongated fruits being produced when the temperatures as measured by degree days above 5°C between full bloom and harvest are relatively low and rounder fruits are produced in warmer climates (Westwood 1993). This has marketing implications for cultivars such as Red Delicious for which an elongated shape is considered desirable.

Apples are generally found to develop 'better' red coloration in their skins when days are bright and sunny and nights are cool. The effect of the cool nights is thought to be as a result of reducing respiration loss of the carbohydrates which are essential precursors to the red anthocyanin pigment idaein (cyanidin-3-galactoside).

3.1.7 *Temperature effects on mineral nutrition*

Temperature has very marked effects on calcium and potassium uptake of a type which is likely to influence the incidence of fruit storage disorders.

Tromp (1980) found that at a root temperature of 8°C much less potassium but as much or more calcium was taken up as at 18°C. The ratio of Ca/K was twice as high at 8°C root temperature as at 18°C, being 2.01 compared with 0.99 for whole plants and 1.05 compared with 0.59 for shoots. In another trial over a wider range of temperatures he found the ratio of Ca/K to decline from 6 to 24°C, from 1.89 to 1.08 for M.9 and 2.55 to 0.90 to MM.106 with temperatures at 12 and 18°C giving intermediate effects and 30°C having inconsistent effects between rootstocks. The effects of the differences in soil temperatures at different latitudes on calcium uptake are likely to be particularly important early in the season, *e.g.* April and May in the northern hemisphere, because at least in many areas the period of rapid uptake of Ca into the fruits is during only 4-6 weeks after blossoming. (Faust 1989). Gur, Hepner and Shulman (1979) found that at very high soil temperatures (36°C) relative calcium uptake (shown as % dry weight) was greater than at 22 or 29°C but the increased concentration was only found in the roots while there was no effect on leaf Ca. The uptake of calcium into Cox's apple was also greater at the higher of these temperatures but that into Golden Delicious was greater at the lower temperature and when fruits of the same weight were compared those grown at 24°C contained less Ca than those grown at 19°C. Moreover, in the majority of studies (Tromp 1975, Faust 1989) it has been found that calcium movement into fruits is virtually confined to the early part of the season and that subsequent fruit growth results in progressive dilution of fruit calcium concentration to the extent that Terblanche *et al.* (1980) have found that under South African conditions apples could not exceed 61 mm in diameter without risk of bitter-pit, which is a calcium deficiency disorder.

High soil temperatures reduced the zinc content of apple rootstock clones grown in sand culture by Gur, Hepner and Shulman (1979), in all cases the leaf zinc content with a root temperature of 35°C was lower than at a temperature of 25°C. In a separate trial on M.2 rootstock grown in clay the zinc content of the plants declined markedly as soil temperatures were raised from 22 to 29 to 36°C.

Phosphorus, magnesium, potassium and nitrogen uptake into fruits (expressed as amount per fruit) was greater at 24 than at 19°C (Tromp 1975). This was almost completely a result of differences in fruit growth although at the higher temperature the amount of each of these elements per unit dry weight was consistently slightly higher. The increase in the amounts of these elements in fruits continued throughout fruit growth.

3.2 Light effects on apple production

These have been reviewed in an earlier chapter.

3.3 Drought and water deficit effects on apple production

Although apple trees show a number of responses to water deficits which have adaptive value (Lakso 1994) extension shoot growth is more or less a function of rainfall for rain-fed apple trees. When stress develops new leaf production slows or ceases while existing leaves show osmotic adjustment. In general in temperate regions irrigation has been found to have the greatest effect on shoot growth early in the season but this information should be interpreted with caution at low latitudes where extension growth can be active over a longer season.

Excessive rainfall or irrigation which maintains extension growth late into the summer may reduce flower bud formation but this is inhibited under moderate drought and nearly completely prevented under severe stress (Salter and Goode 1967). Drought stress can also cause as much as a 3-fold reduction in fruitlet retention after pollination (Powell 1974). It does not generally seem to influence the cell division phase of fruit growth but later water stress, of duration sufficient to prevent fruit enlargement for a week, results in irreversible loss of fruit size (Wample 1982).

Rootstocks show appreciable variability in their response to soil moisture. Trees on MM.111, which has many fine roots, are reported to have survived summer drought better than those of the same cultivar on M.2 (Ferree and Carlson 1987) and MM.111 the most tolerant of the clonal rootstocks to drought while M.9, M.26, M.1 and M.13 are intolerant.

3.4 Effects of soil physical conditions and waterlogging on apple production

The adverse effects of waterlogging are due to lack of oxygen and possibly the accumulation of soil CO_2. Apple is moderately tolerant (Lakso 1994) but this tolerance varies with season. Flooding for periods of less than a month when the trees are dormant has little noticeable effect but sensitivity to flooding increases dramatically once growth begin, when even a week of flooding can have very adverse effects. Early leafing-out, characteristic of lower latitudes, is thus a potential hazard if it occurs while soils are still waterlogged from winter rains. Different rootstocks have differing sensitivities to waterlogging. Rom and Brown (1979) found M.7 and M.26 most tolerant of flooding during the growing seasons and MM 111 most subject to death as a result, and other studies (Ferree and Carlson 1987) have shown MM.106 to be more tolerant of flooding than many other rootstocks. Gur, Moyal and Gabai (1993) and Lee (1985) found that apple rootstocks resistant to flooding produce less ethanol under conditions of flooding than susceptible rootstock clones.

3.5 Effects of mineral nutrient supply on apple production

3.5.1 *Apple nutrient requirements*

Greenham (1980) reviewed studies of total nutrient uptake by apple orchards and found this to be in general very low compared with field crops and vegetables, usually less than 20% of that by the latter crops. Even a heavy cropping orchard, yielding 44.8 t ha^{-1}, had a net uptake (into the tree framework and fruits) of only 39.2 kg N, 10.5 kg P, 70.9 kg K, 50.2 kg Ca and 4.5 kg Mg per hectare. Apple tree fertiliser and soil management practice is controlled by the ability of the soil in an area of production to meet these relatively modest needs and also by the need to optimise the mineral content and growth rates of the fruits so as to maximise fruit quality and especially storage life. Fertiliser practice is generally based on soil, leaf and fruit analysis. Desirable leaf levels are shown in Table 4.

Table 4. Standard values for interpretation of leaf analysis of apples

Nutrient	*Insufficient*	*Below normal*	*Normal*	*Above normal*	*Excess*
N (%)	<1.70	1.70 - 1.99	2.00 - 2.50	2.51 - 3.00	>3.00
P (%)	<0.10	0.10 - 0.14	0.15 - 0.30	>0.30	-
K (%)	<0.80	0.80 - 1.19	1.20 - 1.50	1.51 - 2.00	>2.00
Ca (%)	<0.80	0.80 - 1.09	1.10 - 1.70	>1.70	-
Mg (%)	<0.20	0.20 - 0.24	0.25 - 0.45	>0.45	-
Fe (ppm)	-	50	50 - 250	>250	-
Mn (ppm)	<20	20 - 29	30 - 130	131 -200	>200
Zn (ppm)	<15	15 - 19	20 - 100	>100	-
Cu (ppm)	<3	3 - 4	5 - 30	31 -50	>50
B (ppm)	<20	20 - 30	25 - 50	51 -140	>140

From Basso and Wilms (1988).

3.5.2 *Effects of nitrogen supply*

Inadequate nitrogen supply is evidenced by chlorotic leaves and limited growth (less than 20-30 cm of terminal shoot growth on non-spur cultivars) and also an increased tendency to biennial bearing. Excessive nitrogen levels reduce fruit colour and fruit firmness though increasing fruit size. Also, at N/Ca ratios of 10 or less in the flesh of apples the fruits can be stored for a long time in good condition. If N/Ca ratios are increased to 30 the fruits suffer breakdown with bitter pit and cork spot (Shear 1974) and cannot be stored. The main target of nitrogen supply management is therefore to maximise the growth of newly planted trees, to obtain a full cropping canopy as quickly as possible and to obtain adequate shoot growth early in the season but then to allow nitrogen levels to fall later to optimise fruit colour and firmness and storage potential. The ways in which this can be achieved are very dependent on soil characteristics, particularly organic matter content and its mineralisation. In England the supply of nitrogen by mineralisation of organic matter in the soil may be more than enough to meet the N needs of an apple orchard (Greenham 1980), indeed 'grassing-down' was practised in the past to reduce nitrogen availability as a result of grass competition and improve fruit colour and firmness. At low latitudes also the requirement for added fertiliser N is dependent on the organic matter content of the soil. In the Cape region of South Africa where the organic matter content of the soil is very low (0.2% to 1% organic carbon) mineralisation is therefore also low (Kotzé and du Preez 1988) and nitrogen fertiliser must be applied. In southern Brazil where soils have good to high organic matter levels a survey by Bassa and Wilms (1988) found more than 70% of the orchards to have above normal or excessive leaf nitrogen levels, an indication of fertiliser treatments well above the necessary levels.

The timing of nitrogen application is generally considered to be important because early spring growth and fruit set draw on the nitrogen reserves within the plant. Application of N after harvest is therefore recommended (Faust 1989). The implementation of this poses real questions for farmers at low latitudes whose trees do not become dormant soon after harvest (see later).

3.5.3 *Effects of phosphorus supply*

The frequently low levels of P in tropical and sub-tropical soils do not usually pose a problem for apple production because of the very low requirement for this element by apple trees. Enough can usually be given at planting to last for the life of the orchard. An exception is in Western Australia where the soils fix large amounts of this element and heavy applications are required on virgin soil (Cripps 1978) and subsequently (Middleton 1986).

3.5.4 *Effects of potassium supply*

Deficiency results in leaf scorch with loss of green colour and necrosis, reduced growth and poor fruit colour. Excess K increases the incidence of bitterpit, which is in partially predictable from the K/Ca or K+Mg/Ca ratio in the leaves (Delver 1980).

3.5.5 *Effects of calcium supply*

Calcium is the most important mineral element in determining the storage life of apples. Calcium-deficient fruits suffer from cork spot, water core, bitter pit, Jonathan spot, deep cracking, lenticel spot, gleosporium rot and raised lenticels. Calcium is relatively immobile in the plant and the large quantities present in other tissues may not contribute to fruit calcium supply, which naturally comes through the xylem. Calcium uptake is primarily in the first period of fruit growth and the total amount of Ca in the fruits generally remains fairly constant thereafter. The concentration in the tissues therefore depends on the rate of subsequent fruit growth which may be much more rapid and have a greater tendency to give large, low calcium fruits at low than at high latitudes. Emphasis on fruit calcium management is therefore particularly important at low latitudes.

3.5.6 *Effects of boron supply*

The boron deficiency in many tropical and sub-tropical soils also has major implication for apple production. The flowers of apple are naturally high in B (Crassweller, Ferree and Stang 1981) and this seems to reflect a high need for this element for fruit set. If the B content is insufficient, flowers wilt and die but persist on the tree. Boron deficiency interacts with calcium deficiency in the promotion of cork spot and bitter pit (Bramlage, Drake and Lord 1980) and leads to deep fruit cracking. It can also lead to death of growing points with growth from lateral buds forming 'witches brooms'.

There is some evidence (Sadowski, Alcalde-Blanco and Enriques-Reyes 1980) that low levels of boron increase the incidence and severity of internal bark necrosis caused by Manganese toxicity so could contribute to the severity of this in the tropics and subtropics.

3.5.7 *Effects of zinc deficiency*

The zinc deficiency in many subtropical and tropical soils is potentially very serious. Zinc deficiency in the tree leads to rosetting of leaves at the shoot tips, the leaves produced being greyish or mottled, very small and narrow, hence the name of "little-leaf" disease. The terminals may then die and spurs lose their leaves and become unfruitful or produce only very small fruits. One of the cultivars most adapted to production at low latitudes, Anna, shows severe symptoms of zinc deficiency under soil conditions where other cultivars do not show this (author observation).

3.5.8 *Effects of Manganese supply*

Manganese deficiency leads to chlorotic leaves which drop early and to reduction in growth and fruiting as a result. Manganese toxicity as a result of an excess of available Manganese in acid soils is a common problem which shows up particularly on Delicious apple types such as Starking. The characteristic symptom is internal bark necrosis (IBN) commonly called measles. This first appears as raised pimples on the bark, then these crack and progress to give cracked islands of bark and rough scaly bark with dead, underlying tissue surrounded by waterlogged areas. Other effects include interveinal chlorosis, yellowing and early abscission of leaves and reduced growth such that trees affected early in their lives never become productive. (Ferree and Thompson 1970).

3.5.9 *Effects of magnesium supply*

Magnesium deficiencies result in a characteristic chlorosis of the leaves, lack of bud break of basal buds on leaders and laterals and small fruits. They are very likely to occur on the acid soils typical of many tropical and sub-tropical regions.

4 Production technology in relation to low-latitude environmental factors

4.1 The choice of scion cultivar re chilling requirements and market demand

The main factors governing cultivar choice at low latitudes are winter-chilling requirements, which range from less than 50 to 1700 chilling hours below 7.2°C (Childers 1983), and market demand.

Judged by earliness of blossoming under conditions of sub-optimal winter-chilling the Israeli cultivar Anna had the lowest chilling requirement out of a range of cultivars tested by Bepete and Jackson (1995). This cultivar, together with Dorsett Golden, (not tested by Bepete and Jackson) is perhaps the most widely planted under conditions of little winter chilling *e.g.* in Egypt (Stino 1995), Mexico (Diaz and Romo 1988), Thailand (Subhadrabandhu 1995) and Zimbabwe (authors observation). Other cultivars selected specifically for low chilling-requirements include the Israeli cultivars Maayan, Michal and Elah (Bepete and Jackson 1995) and the Brazilian *cvs.* Princesa and Primicia (Denardi, Hough and Bonetti, 1988). Under Zimbabwean conditions Primicia appears to have appreciably higher chilling-requirement than the Israeli cultivars. In Zimbabwe, Mutsu (Bepete and Jackson 1995) and, surprisingly, Braeburn also flower and crop well in an environment with less than 400 chilling hours. Rome Beauty was traditionally classed as the most suitable cultivar for warm winter areas but is being replaced by the above. Interestingly although Bernardi (1988) records Mutsu as having a medium chilling requirement in Brazil it is the second earliest in blooming after Anna, although, as in Zimbabwe, it is two to three months later so is clearly in a different category.

Cultivars such as Granny Smith, Golden Delicious, (Red) Delicious and increasingly, Fuji and Gala/Royal Gala which are not classed as "low-chill" cultivars are also grown at low latitudes because of market demand, especially that for export markets, and adequate cropping is being induced by the use of dormancy breaking sprays. Not only yield but also quality may suffer if they are grown too far outside of their natural range, however. Bernardi (1988) reports that cultivars with chilling requirements similar to Golden Delicious show a decrease in fruit size when grown in regions with less than 700 hours of chilling below 7.2°C.

The extent to which management practices for these cultivars need to be adapted to any particular local environment, especially their needs for chemical dormancy breaking depends on three factors:

a. The chilling requirement of the cultivar. Jackson and Bepete (1995) found a good relationship between the date of full bloom of a cultivar in the sub-optimal chilling environment of Zimbabwe and the extent to which its "unaided" cropping level fell short of that obtained after use of a dormancy breaking spray. Mutsu was the best adapted of the 'mainstream' cultivars tested, being early flowering and giving high yields both with and without a dormancy breaking spray whereas untreated Granny Smith and Ohinemuri trees flowered much later than those given dormancy breaking sprays and gave only a third of the yields of treated trees. Other commercial cultivars fell between these extremes (Table 5).
b. The degree of winter chilling. Obviously a cultivar which may have its chilling requirement met in one environment will receive inadequate chilling in another with fewer chill units: even Anna has been found to show typical symptoms of inadequate winter chilling and a strong response to Dormex sprays in Thailand (Subhadrabandhu 1995).
c. The possibility of continuous cropping by branch bending and defoliation of suitable cultivars. Cultivars with a very low chilling requirements growing in mild-winter climates with no imposed ecodormancy can break bud and flower at inappropriate times.

Firstly bud break can take place in summer especially following spells of cool cloudy weather in high-altitude summer rainfall climates when soil moisture supplies are adequate. Jackson (1990) reported that under Zimbabwean conditions flowering of Anna can take place at the same time as the main crop is harvested in December giving a second crop before the onset of the cold conditions. This second flowering season occurs without any defoliation either natural or artificial. In Zimbabwe this second crop did not appear to stress the trees and was accompanied by a delay in leaf senescence but led to the carry-over of apple scab so the process was discontinued by removal of second-crop flowers.

Secondly it can occur very early in winter, once the chilling requirement is met. This is very undesirable if it results in flowering during winter frosts. Bernardi (1988) noted that in Santa Catarina, Brazil, cultivars such as Anna, Ein Shemer and Vered give budburst and bloom from June to August, during which period frosts commonly occur and damage most flowers. Williams and Menegazzo (1988) noted similar problems in the Guatemalan highlands where the trees often break dormancy early in the season and a great percentage of the blossoms are killed by frost and hail. They state that pruning can be used to delay flowering and produce stronger flowers. In some climates it may be possible to delay bud-break, after the chilling requirement has been met, by controlling water availability but there is real need for generally applicable technologies to prolong dormancy of low-chilling requirement cultivars grown in areas of variable and unpredictable winter chilling.

It is therefore obvious that there cannot be any overall blanket recommendations for cultural practices for apples at low latitudes. Instead there are general tendencies and guidelines which will need to be followed to varying extents depending on cultivar and on the local environment with respect to winter chilling. There are, in particular, major differences in production technology depending on whether or not the cultivar is being grown, for marketing reasons, in an environment which does not meet its winter-chilling requirements.

4.2 Plant propagation

4.2.1 *Choice of rootstock*

Seedling rootstocks are still widely used at low latitudes, *e.g. Malus sylvestris* Mill in Peru (Bederski 1988), *Malus* seedlings and local crab apples in Egypt (Stino 1995). High soil temperatures have adverse effects on M.9, the main clonal rootstock used in Europe (Gur, Mizrahi and Samish, 1976) and also on M.1, M.2 and MM.104. The rootstocks M.7, M.16, M.25, MM.109 and seedling are relatively resistant to high soil temperatures. The semi-dwarfing MM.106 is being widely used *e.g.* in Brazil (Bernardi 1988), Egypt (Stino 1995) and Zimbabwe while Merton 793 (vigorous) is popular in Brazil (Middleton 1986) and South Africa. The prevalence of woolly apple aphid (*Eriosoma lanigerum* Hausum) in warm climates makes the resistance of the MM. rootstock series, derived from their Northern Spy parentage (Ferree and Carlson 1987), to this pest particularly valuable in tropical and subtropical regions. The characteristic development of many lateral fruit buds on young trees on M.9 and MM.106 rootstock appears to be accentuated under tropical and sub-tropical conditions and the consequent very heavy fruiting on young trees can result in trees on MM.106 being severely dwarfed (Cummins, Aldwinkle and Hauagge 1988). Fouad, Khalil, Mohamed and Hussein (1995) in fact found that in Egypt the number of flowers per one year old shoot of Anna apple was higher on MM.106 and MM.111 than on M.9. The number of flowers on spurs per branch was highest on MM.106 and so was the yield efficiency. Experience in Zimbabwe has shown that trees of, *e.g.* Golden Delicious, on MM.106 are no more vigorous than trees of the same cultivar on M.9 in northern Europe and that trees of precocious, heavy fruiting cultivars such as Anna and Elah on MM.106 can be so checked by cropping that they remain less than 2 m high.

4.2.2 *Rooting of rootstock cuttings*

High soil temperatures in tropical and subtropical areas enable the rooting of leafless (dormant) rootstock cuttings directly in soil without the need for heated rooting bins. Cuttings of MM.106 rootstock are routinely rooted directly in nursery soil in Zimbabwe, the percentage of successful establishment ranging from 57% to 88% in the years 1974/5 to 1977/8 (DR and SS 1979). This technique can be expected to give the highest success where temperatures in the rooting zone reach at least 17°C, at which temperature Howard (1968) showed around 70% rooting of Crab C rootstocks in heated bin experiments.

4.2.3 *Budding and grafting*

Some propagation methods depend on the bark 'slipping' as a result of actively dividing cambial cells. The initiation of cambial activity in the spring results from the onset of bud-development which tends to start earlier the lower the latitude. Thus Kuden and Kaska (1995) found that budding could be successful in 'spring' defined as between the end of February and the end of April in sub-tropical Turkey. The length of growing season available after budding has a major impact on the quality of the fruit tree that can be produced in a single season.

4.3 Induction of branches on young trees

One consequence of inadequate winter chilling is that the lateral buds on nursery trees may fail to break, those on unbranched 'whips' planted out in the orchard may remain dormant and the main axis growing in the first year may remain unbranched.

There are two approaches to this problem; artificial chilling and chemical treatment.

In Brazil, experiments on the effect of pre-planting cold storage on young trees of cultivars Fuji and Gala showed subsequent lateral bud break to be greater when the trees had been stored at 2°C rather than 6°C and when stored for 6 weeks rather than for shorter periods. The response was adequate for practical purposes after 45 days but not after 30 days (Petri and Stuker 1988).

Petri and Stuker (1988) also tested chemical treatment with 4% mineral oil plus 0.16% DNBP. This has some influence on bud break but the effect was relatively small and many of the buds formed spurs instead of developing into shoots.

Other chemicals appear more effective. Promalin (GA_{4+7}+BA), when used on the leading shoot as this grows after initial heading back, can stimulate lateral bud break from it and should presumably be effective also in the nursery, while Dormex (hydrogen cyanamide) can induce release of already formed lateral buds from dormancy (Jackson 1997). This latter treatment is now routinely used, with 3% Dormex (1.5% hydrogen cyanamide) applied by paintbrush to induce lateral branch formation or 4% mineral oil plus 0.5% Dormex used for the same purpose (Ebert and Raasch 1988).

4.4 Tree training and pruning

4.4.1 *Centre leader trees in high-density systems*

In common with most of the apple-growing world, fruit growers at low latitudes have the architectural model of a tree which is pyramidal in shape or a hedgerow which is triangular in section. This is to optimise light distribution and fruit quality and to prevent the upper portion of the tree "shading out' the lower part, with the fruitful zone becoming progressively higher above the ground. For such trees to develop and to be adequately furnished with fruiting wood there has to be a large number of lateral branches emerging from the main vertical axis, the central leader, and these laterals in turn must bear fruiting branches, whether spurs or short sub-laterals, as well as having, for some cultivars, fruits borne directly from axillary blossoms on the central leader and the laterals.

At low latitudes this architectural model is made more difficult to achieve because, in addition to the poor vegetative and floral bud break as a result of inadequate winter chilling, the trees tend to very strong apical dominance, with suppression of lateral bud break, and very upright growth of such lateral branches that do emerge. These laterals tend to have very narrow crotch angles, which are mechanically weak so that the branches break off from the central leader when bearing heavy crops. To overcome these problems the following practices are adopted:-

In South Africa. (Jacobs and Strydom 1993) the newly planted tree is initially headed back to 90 cm above ground level. At the first sign of bud break in spring trees are headed back again to 70 cm above ground level. The shoots emerging at nodes 2 and 3 from the point of heading, which would otherwise compete strongly with the central leader, are weakened by regular pinching during the growing season. The reason for their retention in weakened form rather than complete removal is that their presence results in wider crotch angles of the shoots emerging from lower down the central leader. Despite this the crotch angles may still be too narrow. This is overcome by physically bending laterals to achieve a crotch angle of about 70°C to the vertical in late summer or early autumn, keeping them at the desired angle by use of weights, by tying them to pegs in the ground or, in Zimbabwe, to wires just above ground level running at about 50 cm to each side of the centre line of the row. If there are insufficient laterals these may be induced by notching above buds.

Side shoots which are more than one-third of the diameter of the central leader at the point of insertion, and other lateral shoots in excess of six, are removed in the first winter after planting.

At the start of the second growing season chemical rest-breaking agents are applied (in Zimbabwe 24 August is the standard date) and the central leader headed back twice leaving 70 to 120 cm of one year old wood. The first set of side scaffolds (lateral branches) is also headed back twice. The uppermost side shoots are pinched to prevent excessive competition with the central leader and lateral branches tied down as before. By the end of the second growing season the trees should have filled their allotted space and minimal pruning, consisting mainly of removal of strong upright-growing shoots and shortening back laterals, is practised. Tying down shoots to a near-horizontal angle to check their growth and induce bud break and fruitfulness continues to be a key management practice.

In Brazil. (Ebert and Raasch 1988) the maiden apple trees are headed back to about 120 cm when removed from the nursery and kept for 45 days in a cold store at 2°C to improve bud break. Shortly after planting they are treated with a dormancy-breaking agent (see above). After bud-break, when the new shoots near the heading-back cut reach 2-3 cm in length the trees are headed back to a height of around 80 cm. It has been found in Brazil that heading back to 50 cm results in the emergence of only a few, too-strong, laterals which compete with the central leader while maiden apple trees headed back to 100 cm developed many side shoots but had a weak central leader. A few weeks after bud-break all shoots between ground level and 50 cm above this are removed.

During the first summer after planting the angle of lateral shoots to the central leader is widened by use of toothpicks or clothes pegs when the side shoots are 8 cm-10 cm in length. These can be removed after a week. At the end of the first summer, the lateral branches are tied down (to the base of the trunk using plastic strips) to 90°C to the vertical in high-density plantings to reduce branch vigour and increase flower bud formation and to 60°C to the vertical in lower density plantings (less than 1000 trees per hectare) where more vigorous individual tree growth is needed.

In the first winter after planting any side branches of diameter similar to that of the central leader are removed, the objective of the branch spreading by toothpicks having been to ensure that the diameter of lateral branches never exceeds 50% of the diameter of the vertical axis. In lower density plantings the number of lateral branches is reduced to 4 or 5 but in high-density plantings all lateral branches are retained.

In the second spring after planting, *i.e.* at the start of the second growing season, winter-oil plus Dormex is applied and seven to fifteen days after application of this dormancy-breaking spray the central leader and the lateral branches are shortened (headed-back) by one-third of their length. During this second growing season the first bud beneath the central leader shoot is removed to prevent it competing. When the side shoots reach 8 cm to 10 cm in length toothpicks or clothes pegs are inserted between them and the central leader to widen the branch angles. If buds on the central leader have failed to break, the bark is ringed above them. Shoots growing vertically from lateral branches are bent over when they reach 30 cm in length and are tied or weighted down: if there are too many of these some are summer-pruned to stub cuts.

In the third growing season, seven to fifteen days after applying the dormancy-breaking spray, the central leader and the lateral branches of the second tier are shortened by one third of their length and the new laterals arising from the fresh growth of the central leader and vertically growing laterals on the main branches bent down. The lower branches are allowed to bear a crop of 2 to 3 kg of fruits per tree: all fruits are removed from terminal buds and from the main axis so as to encourage new branch formation. In the fourth and subsequent seasons a similar pruning and training pattern is followed although once the trees have reached their desired height (3.5 to 4 m) the central leader is cut back to a suitable weaker lateral. If summer pruning is necessary to control excess vigour, especially in low crop years, this is done by cutting back strong current-seasons shoots to 3 fully developed leaves and weaker shoots to 4-5 fully developed leaves combined with branch thinning-out if need be.

4.4.2 *Training and pruning under strictly tropical conditions*

Irrespective of the production system the key component is invariably the bending of branches towards the horizontal to induce bud-break. Alum and Magherini (1995), working in Nigeria and Cameroon noted that without bending the apple trees developed long whippy branches with growth concentrated at the tips whereas inclination of branches to the horizontal terminated the endodormancy of 40% to the lateral buds within 30 days in one trial. They used trees trained in vase shapes, palm shapes and spindlebushes.

Janick (1974), describes a system used in Java in which trees grown are basically as 'vase' systems following heading back to 80 cm. Only selected uppermost branches are allowed to develop and these are tied down to stakes using plastic strips so as to get the branches horizontal and stimulate upright spurs evenly along the branch. One month after harvest all leaves are stripped to stimulate bud break. This enables two crops to be produced in a year. In India and the Philippines two crops a year are also achieved (Javaraya 1943, Edwards 1987). In Equador this is done with Anna and Dorsett Golden, using copper at 2% to defoliate and then applying 6% mineral oil + 0.25% DNOC 3-4 weeks after leaf fall to break dormancy (Niegel, 1988).

4.5 Control of fruit bud development, flowering, and fruit set

The first step towards adequate fruit bud development, flowering and set is to carry out the procedures defined above to produce the necessary lateral branches, side shoots and spurs by appropriate heading back of the young tree, tying down branches, notching and use of bud-breaking agents.

However, even if the cropping framework is in place and the effects of apical dominance on budbreak minimised by having near-horizontal branches the inadequacy of winter-chilling in relation to cultivar requirements may still seriously inhibit budbreak. Under these circumstances this may be induced by two separate technologies.

4.5.1 *Use of dormancy-breaking sprays*

Erez and Lavee (1974) reviewed research and practice in the use of dormancy breaking sprays up to that time. Mineral oils were first used for this purpose and then Samish (1945) introduced the use of combined oil plus DNOC and this became the most widely used treatment. Erez and Zur (1981) showed that there was no increased response to DNOC with concentrations above 0.12% whereas the effect of the oil was proportional to its concentration. Although the effectiveness of winter oil + DNOC mixtures is very sensitive to temperature, being ineffective at 12°C or less and having activity markedly enhanced by short spells of above 24°C (Erez 1979) and varying with the physiological stage of bud development, late winter applications of up to 6% oil plus 0.12% DNOC became standard in warm-winter areas. In South America dinitrophenols have been widely used instead of DNOC, *e.g.* in Brazil (Petri and Dittrich 1984) and in Equador a single application of 1000 ppm DNBP + 3% oil about 3 weeks before flowering was recommended by Soria (1988).

There has, however, been legislation against the use of the dinitro compounds and the threat of widespread bans. Thus DNSBP which was widely used in Mexico was officially prohibited for agricultural purposes (Ramirez, Galvan and Salazar 1990) and attention there, as elsewhere, focused on alternative chemicals.

Of these the most effective had earlier been found to be hydrogen cyanamide (Dormex), which generally gave enhanced bud break compared with the dinitro compounds (Erez 1987). This is a scorching chemical which can act as a defoliant and can cause damage to partly opened buds. It should therefore not be used too close to flowering. It does, however, appear to be effective as a dormancy breaking agent throughout the rest period from the onset of rest (dormancy) onwards (Fuchigami and Nee 1987). This presumably explains why when applied to a range of cultivars at a single date, and therefore are very different times from their natural budbreak, it can induce budbreak of all virtually at the same time (Jackson and Bepete 1995). Some yield responses are shown in Table 5.

Hydrogen cyanamide is very effective at 1.5% concentration (3% Dormex) but at this concentration, applied at high enough volume to wet all the buds, it is very expensive. The wetting of all buds is necessary because the effect does not appear to be translocated. Attention has therefore been focused on reducing the concentration needed and admixture with mineral oil, has proved beneficial in this respect.

Table 5. Effects of application of 1.5% hydrogen cyanamide in 1988, 1989, 1990 and 1991 on the 4-year accumulated yield of apple cultivars at HRC Marondera Zimbabwe

Cultivar	*Average date of Unsprayed tree full bloom*	*Accumulated yield of controls kg tree^{-1}*	*Accumulated yield of Cyanamide treated trees*
Granny Smith	19 Oct	33.4	99.0
Starking	15 Oct	90.7	182.5
Mutsu	25 Sept	179.0	176.1
SE cv x cyanamide			11.1
LSD cv x cyanamide			18.9

From Jackson and Bepete 1995

In Brazil Petri, Pola and Stuker (1990) found that a range of concentrations from 0.25% to 1% hydrogen cyanamide were equally effective when applied with 4% mineral oil. They raised the percentage bud burst of Golden Delicious and Fuji from less than 20% (Fuji) or less than 40% (Golden Delicious) to around 80% for Fuji and 80%-90% for Golden Delicious. The 0.75% and 1% concentrations resulted in some phytotoxicity on Fuji shoots but no loss of yield.

North (1995), in South Africa, found 0.5% and 0.75% hydrogen cyanamide combined with 2% or 3% mineral oil to induce budbreak, advance blossoming by 20-25 days and increase fruit set of Golden Delicious. His data also show, however, a decrease in fruit size with cyanamide application, although this may have been solely or mainly a consequence of the heavier crop load. In another trial he found 0.5% hydrogen cyanamide combined with 5% mineral oil to increase budbreak and fruit set although 1.25% hydrogen cyanamide had a greater effect.

An alternative approach to reducing the cost of treatment is to paint individual buds with hydrogen cyanamide or a cyanamide-oil mixture. This is highly effective and is particularly appropriate for young trees in the first few years after planting and grown in high-density systems such that the individual trees are small.

Although hydrogen cyanamide breaks down to non-toxic residual products it is highly toxic itself and there is now a search for 'softer' chemical approaches. Current interest centres on the use of adjuvants to increase the penetration and effectiveness of some chemicals long known to have a mild dormancy-breaking action, especially potassium nitrate (North 1995). Initial results at some sites are promising but it is too early to draw definite conclusions.

Artificial induction of budbreak by use of Dormex to attain uniformity of flowering and catch early-season markets can also lead to problems of yield reduction due to advancement of blossoming into unfavourably cool conditions even in Egypt (Hasseeb and Elezaby 1995)

It should be noted that because dormancy-breaking sprays are used firstly to create the tree framework in terms of branch structure, secondly, to created the sites for fruit and leaf bud development in terms of spurs and other shoots, and thirdly to induce budbreak both of fruit buds and canopy-creating leaf buds, it is not really appropriate to evaluate the effects of dormancy-breaking sprays as single-season responses. In practical terms the main benefit is increase in yield, which, because the treatment is overcoming the effects of a deficiency, can be expected to vary with the severity of the problem and also synchronisation of flowering and advancement of cropping. For more details see also chapter 2.

4.5.2 *Physical approaches to reducing de-chilling*

The potential of these methods lies in exploitation of the observation that hours between 1.5°C and 12.5°C have a chilling effect, in terms of release of buds from dormancy, and that temperatures above 16°C have a progressively more severe de-chilling effect if they occur within a limited period after the chilling event (see earlier). As long as the night temperatures are low enough to have a chilling effect then reduction in day time temperatures should help to conserve this. There are two obvious approaches.

The first is the use of shading. This has the obvious limitation that it will not reduce bud temperatures to below the level of shade air temperature but in many circumstances this, or an approach to it, should be enough to have an effect. The temperatures of apple buds under conditions of high insolation can be about 10°C warmer than the ambient air temperatures for several hours in the afternoon (Anderson, *et al.* 1975) although Erez and Couvillon (1983) found maximum differences of only about 3°C. A reduction of only 2°C might be expected to have a major effect at, for example, Cacador, Santa Catarina, Brazil where the mean (shade) maximum for the two coldest months is around 18°C (Bernardi, 1988) or Applethorpe, Queensland, where the mean maxima for three consecutive months are below 16°C (Table 2) so that only extremes need to be prevented. This approach is likely to be effective only in such marginal conditions and where the costs of support structures are partly or entirely met by other benefits, *e.g.* the support of hail netting.

The second, and much more powerful, approach is by evaporative cooling. This has the potential to cool buds down to the wet-bulb temperature. Obviously the degree of cooling which can be achieved with this will vary with environmental conditions of temperature, humidity and windspeed but Anderson, *et al.* (1975) demonstrated a fall of sprinkled bud temperature down to about 10°C below ambient (shade) air temperature and Erez and Couvillon (1983) a cooling of 1°C to 2°C below air temperature and up to 5°C below unsprinkled bud temperatures. Erez, *et al.* (1993) found a cooling effect (on peach buds) of more than 10°C in comparison with air and dry bud temperatures and a consequent induction of budbreak. The wetting of the buds for cooling can be achieved by use of microjets inserted at the top of each tree or could equally well be done by microjets suspended from an irrigation line above a hedgerow which could then be returned to ground level if desired for conventional microjet irrigation later in the season. Evaporative cooling by sprinkling has been found effective in inducing budbreak of kiwi-fruit in a commercial orchard in Zimbabwe (Lamb, personal communication).

4.6 Fruit thinning

4.6.1 *Fruit thinning to improve vegetative growth*

Both with low-chilling requirement cultivars and with conventional cultivars treated with a dormancy breaking spray the very high proportion of buds which become flower buds under low latitude conditions leads to a thinning requirement in the year of planting and the subsequent year which is much greater than in temperate areas. It is very easy for a tree to become stunted as a result of initial overcropping if the flowers or fruitlets on young trees on rootstocks such as MM.106 are not removed. Clearly this is a question for judgement depending on the vigour of the trees at planting and the spacing, but overcropping is more likely to be a problem than undercropping with cultivars such as Anna. This was shown in an extreme form in Zimbabwe when an attempt was made to produce 'Solen' trained trees with branches trained horizontally on wires. Almost all buds on these lateral framework branches were fruit buds, and leaving too many to fruit resulted in a serious check to growth and reduced the production of cropping laterals.

4.6.2 *Fruit thinning to improve fruit size*

Problems of fruit size arise in two ways at low latitudes. When winter chilling is inadequate and there is poor bud and leaf development then the fruits can be very small as a result of inadequate carbohydrate supply (Ebert and Kreuz 1988). With adapted cultivars, or with heavy flowering induced or synchronised by a dormancy-breaking spray, then fruit set may be excessive in relation to the size of the trees with negative effects on fruit size. Thus Jackson and Bepete (1995) found the heavier cropping on cyanamide-sprayed trees to be accompanied by a reduction in fruit size compared with unsprayed trees and concluded that all the treated trees would have benefited from fruit thinning. However, although a positive effect of fruit thinning, at the earliest possible stage, on the growth of the remaining fruits and their final size at harvest is almost invariably shown this is usually accompanied by some loss of total yield. The economics of thinning for fruit size improvement thus depend on the relative prices obtained for different sized fruits as well as the yield of these. Under Brazilian conditions Ebert (1988) demonstrated improvements in fruit size but not profitability. For Gala he found hand thinning preferable to use of NAA-m at 5 days after full bloom and for Fuji a spray of NAA plus mineral oil 10 days after full bloom was the best treatment.

In South Africa where large fruits are desired for an export-oriented marketing system the general approach is to calculate back from the production capacity of the orchard, based on experience, in metric tons, work out the number of fruits which can be left on the trees to give the desired average fruit weight and thin accordingly. The actual thinning methodology is to some extent cultivar-specific and involves different combinations of pruning, chemical thinning and hand thinning. Because the effect of thinning on the growth of the remaining fruits is greater the earlier it is carried out it is done as early as is possible. In general because the smallest fruitlets within the population at any given time will give the smallest fruits at harvest these are selectively removed. Also thinning fruits to one fruit per cluster permits more even colour development, which is important in red-coloured cultivars. Thinning also has benefits in terms of controlling biennial bearing. It should be noted that a prolonged flowering season results in the need to remove late flowers and fruitlets.

4.7 Fruit nutrition

This is covered in more detail in Chapter 4. The following emphasises work specifically on apple with particular attention to southern Africa.

4.7.1 *Pre-planting treatments*

As in temperate zones the first task is soil analysis and to adjust pH and nutrient status accordingly.

The pH adjustment is mainly to minimise the risk of nutrient imbalances and in South Africa the target pH is 5.5 to 6 on the 0.01 M $CaCl_2$ scale. The liming requirement is calculated from exchange acidity (Kotzé and Joubert 1979) and the choice of calcitic or dolomitic limestone determined by the Mg saturation of the exchange complex.

Where the soils are saline, gypsum is used to rectify this.

Phosphorus levels are increased to Bray II extractable P of 30 mg/kg. In acid soils known to be zinc or boron deficient it is useful to apply these at planting to prevent the problems developing. The high sensitivity of Anna, one of the most successful cultivars under conditions of low winter-chilling, to zinc deficiency renders pre-planting incorporation of this element into Zn deficient soils very important where this cultivar is to be grown. Because of the high cost of zinc and boron compounds these can be incorporated in the soil of the tree row only, whereas liming and phosphate application should be over the entire orchard.

As noted earlier calcium deficiency is one of the most serious problems of apple growing at low latitudes. The first step to ensuring high fruit calcium status is to prepare the soil properly prior to planting by adjustment of the pH if it is acidic and modification of the relative calcium and magnesium levels in the soil.

In Brazil Basso and Wilms (1988) found that over 3 years in Santa Catarina State 79, 70 and 77% of all orchards sampled had below normal (under 1.10%) leaf calcium as did 77 and 79% of samples received in two years in Paraná State and 77% of those received in the single year of testing in Rio Grande do Sul. They concluded that the problem was not one of shortage of Ca in the soil but of excess K and Mg reducing its uptake. Wilms and Basso (1988) concluded that the most important measure to improve the Ca supply of apple orchards in Brazil would be to increase the Ca/Mg ratio in the soils and recommended liming acid soil to water pH 6 (approx Calcium chloride pH 5) and increasing the Ca^{2+} Mg^{2+} in the soil to 10-12 meq/100 g and the Ca/Mg ratio to at least 5, using dolomitic or calcitic lime or gypsum depending on initial soil conditions. They found that dolomitic lime alone, applied pre-planting on a strongly acid heavy clay soil up to a rate of 64.8 t ha^{-1} had very little effect on leaf calcium content, appreciably increased leaf magnesium content and did not reduce bitterpit.

In Brazil where boron has been diagnosed as deficient 30 kg ha^{-1} of borax is incorporated in the top 40 cm of soil at planting (Wilms and Basso 1988).

It has been reported that thorough incorporation of zinc sulphate to provide 136 kg ha^{-1} of zinc has a protective effect. Zinc chelates are also recommended for soil application.

Problems of Manganese toxicity are most commonly avoided by liming to increase the soil pH, preferably pre-planting.

Necessary nitrogen and potassium dressings are applied at planting and not before because of their solubility and ready leaching. In South Africa in the first year 8.5g N per tree is applied monthly from one month after planting until leaf senescence and on sandy soils 15 g K is applied in alternate months. The fertiliser is broadcast over an area 0.6 m^2 to 1.0 m^2 around the tree. In the second year 17 g N per tree is applied each month from budbreak to leaf senescence. Fifteen g *K* is applied in alternate months if the soil is sandy. The fertiliser is broadcast over an area of 2 m^2. In the third year 25 g N is applied monthly as before and the K dressing is as in previous years. The area of application is increased to 2 to 3 m^2 (Kotzé and du Preez 1988).

4.7.2 *NPK for bearing orchards*

In South Africa fertiliser schedules dependent on soil nutrient levels (Table 6) and orchard production levels have been published (Kotzé and du Preez 1988). The highest level of application, with a 50 t ha^{-1} crop, deficient soils for all elements and poor tree vigour, is:

Table 6. Fertilizer programmes in relation to soil analysis

Phosphorus	*Extractable P (Bray) mg/kg*			*P Fertilizer*
		<30		Full programme if leaf P < 0.13
		>30		Nil
Potassium	*Extractable K (Bray) mg/kg*			*K Fertilizer*
	Sand	*Loam*	*Clay*	
	<30	<50	<70	Full programme
	30-50	50-70	70-100	Only at budbreak
	>50	>70	>100	None
Potassium	*% saturation of CEC*			*K Fertilizer*
		<3		Full programme
		3-4		Only in spring
		>4		None

From Kotzé and du Preez 1988.

Nitrogen: 80 kg ha^{-1} between budbreak and bloom, 80 in mid-summer and 80 post-harvest.
Phosphorus: 25 kg ha^{-1} between budbreak and full bloom, and 25 postharvest
Potassium: 150 kg ha^{-1} between budbreak and full bloom and 150 post-harvest.

In Brazil leaf analysis showed that most orchards in Paraná State and Rio Grande do Sul State had above normal or even excessive levels of nitrogen and quite a large proportion (37 and 24% respectively) had above normal potassium levels. The problems of excess fertiliser application, especially in relation to calcium availability, must be taken very seriously.

4.7.3 *Other elements for bearing orchards*

Deficiencies of other elements commonly cause much more severe problems, particularly but not exclusively to do with post-harvest quality.

Calcium. In a study on overcoming calcium deficiency in mature trees in Brazil, Wilms and Basso (1988) carried out a trial on an existing mature orchard on strongly acidic heavy clay soil in which 0, 1, 2 and 4 t ha^{-1} of gypsum were applied annually in a factorial experiment with 0, 2.5 and 5 t ha^{-1} of dolomitic lime. Gypsum treatments led to a higher enrichment of the effective cation exchangeable capacity with Ca^{2+} than dolomitic lime in equivalent doses and reduced the Mg^{2+} status whereas use of dolomitic lime increased this. The effects on soil nutrient status were accompanied by similar effects on leaf nutrient status with dolomitic lime giving no increase in leaf Ca but increasing Mg whereas gypsum applications increased leaf Ca significantly and slightly reduced Mg and K. As a result of the improved (reduced) fruit K/Ca ratio, bitterpit after 6 months cold storage was reduced by gypsum treatment from 58 to 12%. Fruit rots in fruits without post harvest fungicide treatments were reduced from 26 to 10%.

Terblanche, Gurgen and Hesebeck (1980) concluded that in South Africa the fruit calcium threshold could be reached at a soil pH of 5.5 ($CaCl_2$ scale) where the 1 N NH_4Cl exchangeable calcium, potassium and magnesium represent 70-80, 3-4% and 10-15% of the cation exchangeable capacity. They found that soil-surface applied calcitic lime (15 t ha^{-1}) on a soil with a $CaCl_2$ pH of 4.2 and a calcium saturation level of 50% increased the leaf calcium status (though it was still below the desirable level) and reduced bitter pit incidence from 30.7 to 20.3% in Golden Delicious and from 10.4 to 4.6% in Starking Delicious.

Calcium uptake from the soil and transport to the fruits at the critical early stages of their growth is thought to be influenced by the form of nitrogen applied in early spring. Nitrogen applied as nitrate does not compete with calcium but nitrogen applied as ammonium competes with calcium for uptake by roots and increases amino acid competition for exchange sites in the xylem (Shear 1980, Kotzé 1996). Nitrate nitrogen is therefore preferred for application prior to and shortly after blossoming.

Fruit calcium status can be improved, and bitterpit incidence reduced, although to a limited extent, by use of sprays of calcium nitrate (0.65%) or calcium chloride (0.5 %) at fortnightly intervals from mid December onwards in South Africa (Terblanche, Gurgen and Hesebeck 1980).

Boron. In South Africa boron is applied as a spray of 100 g Solubor/100 litres at full bloom to supplement soil sources. If a deficiency is known to exist the application should be repeated 2 weeks later (Kotzé 1996). Boron can be toxic at relatively low leaf concentrations so continual monitoring of boron status is essential.

Zinc. Once the orchards are established surface dressings of zinc sulphate have been found ineffective and foliar application is the standard approach. Zinc sulphate (20 or 36% zinc) is applied at the dormant to silver tip stage of apple bud development at a rate of 1.6 to 2.25 kg of actual zinc per 454 litres so as to fully wet the buds and shoots. In Israel a single zinc oxide spray has been found to be a suitable alternative to zinc sulphate, remaining longer on the leaf and being absorbed over a longer period (Hoffman and Samish 1976). EDTA zinc chelate can be applied at later stages of tree growth *e.g.* with 2 or more sprays at 10-14 day intervals after petal fall. The zinc status of apple can also be improved by implantation of zinc bentonite or zinc bentonite plus Zn EDTA in spring (Navrot and Banin 1982). On naturally zinc deficient soils prevention of zinc deficiency requires annual treatment, not just one-off attempts to cure symptoms. Anna apple can show serious zinc deficiency symptoms at leaf zinc content levels at which other cultivars are not visibly affected. There is need to define threshold levels for this cultivar in view of its importance in low-latitude apple growing and the prevalence of zinc deficiency in tropical/sub-tropical soils and under low-latitude climatic conditions.

5 Irrigation

Irrigation in general is covered in Chapter 3. The following emphasises work on apple.

The crop factor for apples (kc which is the coefficient by which open-pan evaporation is multiplied to obtain estimated water use) is higher at low latitudes than at high latitudes (Doorenbos and Pruitt 1984) for two reasons. Firstly the actual season over which the trees are leafy may be much longer, so kc is high for longer periods, and secondly if there is a ground cover crop this too may not be killed by winter frost so may be using water for a long season. Assuming that the trees cover about 70% of the orchard surface the kc values in areas with cold winters with light frost (*e.g.* Mediterranean or southern African highveld climates) range from 0.85 in the early season to 1.35 in mid summer under dry conditions with strong winds if they have cover crops and from 0.5 to 1.05 if they do not (Table 7) In Israel excellent apple yields have been obtained using a pan evaporation coefficient of 0.9 to 1.2 during July and August with water applied by drip irrigation and averaging 1000 mm per season (Levin, Assaf and Bravdo 1980). In Queensland, Australia, in a summer rainfall climate, a crop coefficient of 0.7 is used for closely planted trees in full leaf (Crew 1978). It should be noted that under Israeli conditions the wetted soil volume and therefore the root system is about 30-50 percent as large if trickle irrigation is used as is the case where sprinklers are used and most of the roots are within 60 cm depth of the surface.

The soil depth (root zone) for irrigation for deciduous fruits is generally considered to be 1-2m (Vermeiren and Jobling 1984). The Israeli data given above, which is coupled

Table 7. Kc values for mature apple orchards assuming infrequent wetting by irrigation or rain

	With ground cover crop									*Without ground cover crop (clean cultivated, weed free)*								
	Mar	*Apr*	*May*	*Jun*	*Jul*	*Aug*	*Sept*	*Oct*	*Nov*	*Mar*	*Apr*	*May*	*Jun*	*Jul*	*Aug*	*Sep*	*Oct*	*Nov*
COLD WINTER KILLING FROST: GROUND COVER STARTING IN APRIL																		
humid, light to mod. wind	-	.5	.75	1.1	1.1	1.1	1.1	.85	-	-	.45	.55	.75	.85	.85	.8	.6	-
humid, strong wind	-	.5	.75	1.1	1.2	1.2	1.15	.9	-	-	.45	.55	.8	.9	.9	.85	.65	-
dry, light to mod. Wind	-	.45	.85	1.15	1.25	1.25	1.2	.95	-	-	.4	.6	.85	1.0	1.0	.95	.7	-
dry, strong wind	-	.45	.85	1.2	1.35	1.35	1.25	1.0	-	-	.4	.65	.9	1.05	1.05	1.0	.75	-
COLD WINTER WITH LIGHT FROST: NO DORMANCY IN GRASS COVER CROPS																		
humid, light to mod. wind	.8	.9	1.0	1.1	1.1	1.1	1.05	.85	.8	.6	.7	.8	.85	.85	.8	.8	.75	.65
humid, strong wind	.8	.95	1.1	1.15	1.2	1.2	1.15	.9	.8	.6	.75	.85	.9	.9	.85	.8	.8	.7
dry, light to mod. Wind	.85	1.0	1.15	1.25	1.25	1.25	1.2	.95	.85	.5	.75	.95	1.0	1.0	.95	.9	.85	.7
dry, strong wind	.85	1.05	1.2	1.35	1.35	1.35	1.25	1.0	.85	.5	.8	1.0	1.05	1.05	1.0	.95	.9	.75

Data taken from Doorenbos and Pruit (1984) p. 49.

with a comment that some of the orchard soils in Israel are only 60 cm deep, and the reference in the South African and Brazilian work cited earlier in this chapter to soil amelioration to only 40 cm in depth shows that in many low-latitude soils the effective depth may be nearer to 0.5 m. Especially in winter rainfall climates, dependent on irrigation during the growing season, this shallow rooting depth coupled with high evaporative demand makes it essential to have only a short interval between irrigations if the soil to the rooting depth is not to dry out and the trees suffer from stress. This problem is made worse if the soils are sandy with low water holding capacity. An Australian study on sandy granite-belt soils showed that a watering frequency of four times a week gave optimum growth and yield (Chapman and Crew 1978). The need for very frequent irrigation has obvious implications with regard to the type of irrigation systems which are preferred, with a bias towards trickle irrigation or other micro-irrigation systems. The major argument against sole reliance on such systems at low latitudes is in the circumstances where evaporative cooling for prevention of de-chilling in winter or reduction of excessive temperatures and respiration rates in summer involves the need for overhead sprinklers for purposes other than irrigation.

Irrigation is, however, determined not only by evaporative demand but also by the physiological requirements of the tree. Flower buds need to attain high moisture content at bud burst and the flowering and the early period of fruitlet development is very sensitive to water stress (Salter and Goode 1967). In Zimbabwe, where bud-break occurs towards the end of winter but prior to the onset of the summer rains irrigation water is applied ten days before flowering to bring the soil up to field capacity and then again at petal-fall to ensure adequate moisture for fruit set and cell division (Kemp 1971).

6 Hail damage control

Hail is a catastrophic event for apple growers (Corelli and Randi 1986), sometimes being severe enough to shred or even strip leaves but more frequently resulting in major loss simply because of surface damage to fruits, with sunken areas or pits developing on the surface at maturity where there had been hailstones impacting earlier in the life of the fruits.

Cloud seeding with silver iodide from aircraft or by use of rockets had given variable results under European conditions (Mereix 1987) and use of protective nets is of more general interest especially in mountain areas.

Nets are becoming much more widely used under low latitude conditions *e.g.* in South Africa (Avenant 1989, Garner and Fletcher 1990) and Australia. Although they create some shade, Scott (1989) concluded that in Victoria (Australia) any adverse effects on tree growth or fruit quality are markedly outweighed by the protection offered to the crop. In the Granite Belt of Queensland Australia 35% of the area cropped to apples was covered by hail netting in 1995 (Middleton, personal communication). The nets used for protection against hail reduce P.A.R. by 12-16% if white and up to 27% if black but when support structures are in place they could be used to support nets giving more shade if required. Potential uses under the high irradiance conditions characteristic of many low-latitude areas include protection from sunscald and reduction of temperatures to improve the balance of photosynthesis to respiration under extreme conditions in the cropping season, and reduction of daytime temperatures in winter to minimise de-chilling effects and improve subsequent budbreak.

7 Control of waterlogging effects

The problems caused by inadequate soil aeration under water-logging conditions in late winter and early spring in Israel on heavy clay soils were reduced by tri-K-citrate placed in a concentrated solution beneath the emitters of the drip irrigation system. Apple tree growth was enhanced, yields doubled and fruit size improved (Gur, Moyal and Gabai 1993). The citrate is thought to work by suppressing the ethanol production which occurs under anaerobic conditions.

8 Acknowledgements

Thanks are due to Dr J.W. Palmer and Dr S.G. Middleton for providing data.

9 References

Alum, N. and Magherini, R. (1995) Apple production in strictly tropical zones: First observations in Nigeria. *Acta Horticulturae*, 409, 25-38.

Anderson, J.L., Ashcroft, G.L., Richardson, A.E., Alfaro, J.F., Griffin, R.E., Hanson. G.R. and Keller, J. (1975) Effects of evaporative cooling on temperature and development of apple buds. *Journal of the American Society for Horticultural Science*, 100 (3), 229-231.

Avenant, J.H. (1989) Construction of trellising systems with hail protection. *Deciduous Fruit Grower*, 39 (6), 185-190.

Basso, C. and Wilms, F.W. (1988) Nutritional status of apple orchards in Southern Brazil. *Acta Horticulturae*, 232, 187-192.

Beattie, B.B. and Folley, R.R.W. (1978) Production variability in apple crops II. The long term behaviour of the English crop, *Scientia Horticulturae*, 8, 325-332.

Bederski. K.A. (1988) Apple growing in the coastal areas of Peru. *Acta Horticulturae*, 232, 51-55.

Bepete, M. and Jackson, J.E. (1995) Apple cultivar performance and response to a chemical dormancy breaking spray under "marginal" winter-chilling conditions in Zimbabwe. *Acta Horticulturae*, 409, 121-124.

Bernardi, J. (1988) Behaviour of some apple cultivars in the subtropical region of Santa Catarina, Brazil. *Acta Horticulturae*, 232, 46-50.

Bijhouwer, J. (1924) De periodicitiet van de knopontwikkeling bij den appel. *Med. Landb. Hoogesch., Wageningen* 27, 1.

Bramlage, W.J., Drake, M. and Lord, W.J. (1980) The influence of mineral nutrition on the quality and storage performance of pome fruits grown in North America. In *Mineral Nutrition of Fruit Trees*, (eds D. Atkinson, J.E. Jackson, R.O. Sharples and W.M. Waller) Butterworths, London 29-39.

Chandler, W.J., Kimal, M.H., Philip, G.L., Tufts, W.P. and Weldon, G.P. (1937) Chilling requirements for opening of buds on deciduous orchard trees and some other plants in California. *Bul. Calif. Agri. Expt. Sta.*, 601.

Chapman K.R. and Crew, P.S. (1978) Influence of five watering frequencies regulated by trickle irrigation on the growth and cropping of apple trees in Queensland. *Queensland J. Ag. An. Sci.* 35 (2), 105-119.

Childers, N.F. (1983) *Modern Fruit Science*, Horticultural Publications, Gainesville, Florida p. 9.

Corelli, L. and Randi, M. (1986) Remedial treatments in hail damaged fruit plantations. *Rivista di Frutticolutura e di Ortofloricolutura*, 48 (11), 20-26.

Crassweller, R.M., Ferree, D.C. and Stang, E.J.(1981) Effect of Overtree Misting for bloom delay on pollination, fruit set and nutrient element concentration of Golden Delicious Apple Tree. *Journal of the American Society for Horticultural Science*, 106 (1), 53-56.

Crew, P.S. (1978) Water requirements of apples. *Queensland Agricultural Journal*, 104 (1), 2-5.

Cripps J.E. (1978) Apple Orchards take a new look. *West Australian Journal of Agriculture*, 19 (2), 50-53.

Cummins, J.N., Aldwinkle, H.S., and Hauagge, R. (1988) Improved rootstocks for apple trees in subtemperate regions. *Acta Horticulturae*, 232, 26-36.

Delver, P. (1980) The use of leaf analysis in predicting storage quality of apples. In *Mineral Nutrition of Fruit Trees*, (eds D. Atkinson, J.E. Jackson, R.O. Sharples and W.M. Waller) Butterworths, London, 367-368.

Denardi, F. Hough, L.F. and Bonetti, J.I. da S. (1988) Low chilling and disease resistance as main objectives of apple breeding in Santa Catarina Brazil. *Acta Horticulturae*, 232, 15-25.

D.R. and S.S. (1979) Nursery Investigations. *Annual Report Horticulture and Coffee Research Institute, Department of Research & Specialist Services*, Harare 48-51.

Diaz, D.H. and Romo R. (1988) Nutritional status and root system of Anna apple under alkaline soil in a warm dry climate of Mexico. *Acta Horticulturae*, 232, 177-186.

Dourenbos J. and Pruitt, W.O. (1984) Crop water requirements. F.A.O. *irrigation and drainage* paper 24. F.A.O. Rome 144 pp.

Ebert, A. (1988) Hand thinning in the apple cultivars Gala, Golden Delicious and Fuji in the state of Santa Catarina South Brazil. *Acta Horticulturae*, 232, 147-158.

Ebert, A and Kreuz, C.L. (1988) Technical and economic practicability of chemical thinning of 'Fuji' apples in the state of Santa Catarina South Brazil. *Acta Horticulturae*, 232,127-136.

Ebert, A. and Raasch, Z.S. (1988). Recommended practices for training and pruning apple trees in South Brazil. *Acta Horticulturae*, 232,159-166.

Edwards. G.R. (1987) Producing temperate zone fruits at low latitudes: Avoiding Rest and the Chilling Requirement. *HortScience*, 22 (6), 1236-1240.

Erez, A. (1979) The effect of temperature on the activity of oil + dinitrocresol spray to break the rest of apple buds. *HortScience*, 14, 141-142.

Erez, A. (1987) Chemical control of budbreak. *HortScience* 22 (6), 1240-1243.

Erez, A and Lavee, S. (1974) Recent advances in breaking the dormancy of deciduous fruit trees. *Proceedings 19th International Horticultural Congress*, Warsaw 3, 69-78.

Erez, A. and Zur, A. (1981) Breaking the rest of apple buds by narrow-distillation-range oil and dinitro-o-cresol. *Scientia Horticulturae*, 14, 47-54.

Erez, A. and Couvillon, G.A. (1983) Evaporative cooling to improve rest breaking of nectarine buds by counteracting high daytime temperatures. *HortScience*, 18, 480-481.

Erez, A. Gil, N., Lerner, H. and Yablowitz, Z. (1993) Container grown peach trees; evaluation of a commercial endeavour. *Acta Horticulturae* 349, 43-47.

Faust, M. (1989) *Physiology of temperate zone fruit trees*. Wiley, New York 338 pp

Ferree, D.C. and Thompson, A.H. (1970) Internal bark necrosis of the apple as influenced by calcium placement and soil manganese. *Maryland Agr. Exp. Stat. Bull.* A-166

Ferree, D.C. and Carlson, R.F. (1987) Apple rootstocks. In Rootstocks for Fruit Crops, (eds R.C. Rom and R.F. Carlson) Wiley. New York 107-143.

Fishman, S., Erez, A. and Couvillon, G.A. (1987) The temperature dependence of dormancy breaking in plants: Two step model involving a cooperative transition. Journal of Theoretical Biology, 124, 473-483.

Fouad, M.M., Khalil, M.A., Mohamed, S.M and Hussein A.M. (1995) Effect of rootstock on flowering behaviour, fruit set, yield and fruit characters of Anna apple. *Acta Horticulturae*, 409, 251-257.

Fuchigami, L.H. and Nee, C.-C. (1987) Degree Growth Stage Model and Rest-breaking mechanisms in temperate Woody Perennials. *HortScience* 22 (5), 836-845.

Fulford, R.M. (1966) The morphogenesis of apple buds III. The inception of flowers. Annales of Botany, London. 30, 207-19.

Gardner, R.A.W. and Fletcher, C.A. (1990) Hail protection systems for deciduous fruit trees. *Deciduous Fruit Grower*, 40 (6), 206-212.

Greenham, D.W.P. (1980) Nutrient cycling: the estimation of orchard nutrient uptake. In *Mineral Nutrition of Fruit Trees*, (eds D. Atkinson, J.E. Jackson, R.O. Sharples and W.M. Waller) Butterworths, London, 345-352.

Gur, A., Bravdo, B and Mizrahi, Y. (1972) Physiological responses of apple trees to supra-optimal root temperature. *Physiologia Plantarum*, 27,130-8.

Gur, A., Hepner, J. and Shulman, Y. (1979) The influence of root temperature on apple trees IV. The effect on the mineral nutrition of the tree. *Journal of Horticultural Science*, 54 (4), 313-321.

Gur, A., Mizrahi, Y. and Samish R.M. (1976) The influence of root temperature on apple trees II Clonal differences in susceptibility to damage caused by supraoptimal root temperature. *Journal of Horticultural Science*, 51 (2), 195-202.

Gur, A., Moyal, J. and Gabai, A. (1993) Prevention of damage caused to fruit trees by inadequate soil aeration. *Acta Horticulturae*, 349, 215-218.

Hasseeb, G.M. and Elezaby, A.A. (1995) Timing of hydrogen cyanamide application and full bloom, fruit maturity and yield of two apple cultivars. *Acta Horticulturae*, 409, 185-189.

Hoffman, M. and Samish, R.M. (1976) The control of zinc deficiency in apple. *Israel Journal of Agricultural Research*, 16 (3), 105-114.

Howard, B. H. (1968) The influence of 4 (indolyl-3) butyric acid and basal temperature on the rooting of apple rootstock hardwood cuttings. *Journal of Horticultural Science*, 43, 23-31.

Jackson, I.J. (1989) *Climate, water and agriculture in the tropics*. Longman. Harlow. 377 pp.

Jackson, J.E. (1990) Cropping and multiple cropping of apples in medium altitude tropical Zimbabwe. *Abstracts of the 23rd International Horticultural Congress*. Firenze Abstract # 4325.

Jackson, J.E. (1997) Branch induction using hydrogen cyanamide. *Acta Horticulturae*, 451, 679-681.

Jackson, J.E. and Hamer, P.J.C. (1980) The causes of year-to-year variation in the average yield of Cox's Orange Pippin apple in England. *Journal of Horticultural Science*, 55, 149-156.

Jackson, J.E., Hamer, P.J.C. and Wickenden, M.F. (1983) Effects of early spring temperatures on the set of fruits of Cox's Orange Pippin Apple and year-to year- variation in its yields. *Acta Horticulturae*, 139, 75-82.

Jackson, J.E. and Bepete, M. (1995) The effect of hydrogen cyanamide (Dormex) on flowering and cropping of different apple cultivars under tropical conditions of sub-optimal winter chilling. *Scientia Horticulturae*, 60, 293-304.

Jacobs, G. and Strydom, D.K. (1993) Spacing trends of free-standing central leader deciduous fruit trees. Journal of the South African Society for Horticultural Science, 3 (2), 79-80.

Janick, J. (1974) The apple in Java. *HortScience*, 9, 13-15.

Javaraya, H.C. (1943) Biannual cropping of apple in Bangalore. Indian Journal of Horticulture, 1 (1), 31-34.

Joose, M.L. and Besseling, P.A.M. (1995) Kwantitatieve Informatie fruitteelt 94/95. *IKC-Land*, Postbus 482, 6710 BL Ede, Netherlands.

Kemp, P. (1971) *Deciduous Fruit in Rhodesia.* Technical Bulletin No. 13 Rhodesia Ministry of Agriculture Harare.

Kondo, S. and Takahashi, Y. (1987) Effects on high temperature in the nighttime and shading in the daytime on the early drop of apple fruit 'Starking Delicious'. *J. Japan Soc. Hort. Sci.* 56, 142-150.

Kondo, S. and Takahashi, Y. (1989) Relation between early drop of apple fruit and ethylene evolution under high night-temperature conditions. *J. Japan Soc. Hort. Sci.* 58, 1-8.

Kotzé, W.A.G. (1976) Growth and nutrition of apple rootstocks and seedlings in the presence of aluminum. *Agrochemophysica*, 8, 39-44.

Kotzé, W.A.G. (1996) Soil and nutritional factors affecting apple fruit quality. In *Integrated management of postharvest quality* (ed J.C. Combrink) Infruitec, Stellenbosch. 14-24.

Kotzé, W.A.G. and Deist J. (1972) Aluminium in the soil. *Deciduous Fruit Grower*, 22,298-300.

Kotzé, W.A.G. and Joubert, M. (1979) Exchange acidity as criterion for the liming of orchard soils in the Western Cape. *Agrochemophysica*, 12,15-20.

Kotzé, W.A.G. and du Preez, M. (1988) Soil amelioration and fertilisation for apple production in South Africa. *Acta Horticulturae*, 232, 167-176.

Kuden, A. and Kaska, N. (1995) Extension of budding period of apples by using chip budding, T-budding and English grafting. *Acta Horticulturae*, 409, 125-131.

Lakso, A.N. (1994) Apple. In *Handbook of environmental physiology of fruit crops* (eds B. Schaffer and P.C. Andersen) Vol I, Temperate Crops C.R.C. Press Florida. 3-42.

Lakso, A.N. and Johnson, R.S. (1990). A simplified dry matter production model for apple using automatic programming simulation software. *Acta Horticulturae*, 276, 141-148.

Lamb, R.C. (1948) Effect of temperatures above and below freezing on the breaking of rest in the Latham raspberry. *Proceedings of the American. Society for Horticultural Science*, 51, 313-315.

Lang, G.A. (1987) Dormancy: A New Universal Terminology. *HortScience*, 22, 817-820.

Lee, C.H. (1985) Relationship between water logging tolerance and ethanol accumulation in young apple rootstocks. Journal of the Korean *Society for Horticultural Science*, 26, 327-332.

Levin, I., Assaf, R., and Bravdo, B. (1980) Irrigation, water status and nutrient uptake in an apple orchard. In *Mineral Nutrition of Fruit Trees*, (eds D. Atkinson, J.E. Jackson, R.O. Sharples and W.M. Waller) Butterworths, London, 255-264.

Linsley-Noakes, G.C., Allan, P. and Matthee, G. (1994) Modification of rest completion prediction models for improved accuracy in South African stonefruit orchards. *Journal of the South African Society for Horticultural Science*, 4(10), 13-20.

Luckwill, L.C. (1974) A new look at the process of fruit bud formation in apple. *Proceedings 19th International Horticultural Congress*, 3, 237-245.

Luckwill. L.C. (1984) Apple growing around the World, Problems and Prospects. In Apples and pears: report of the Royal Horticultural Society Conference, 1983, (ed Elspeth Napier) The Royal Horticultural Society, London, 1-10.

Mereix, J.F. (1987) La prevention de la grele, realite, echec ou espoir, *Arboriculture Fruitiere*, 34, 31-34.

Middleton S.G. (1986) Apple production in suboptimal environments. *Queensland Department of Primary Industries Bulletin* QB 86001. Brisbane.

Navrot, J. and Banin, A. (1982) Trunk implanted Zn -Bentonite as a source of zinc for apple trees. *Plant and Soil*, 69 (1), 85-95.

Niegel, W. (1988) History, actual advance and future of the equatorian apple growing. *Acta Horticulturae*, 232, 67-73.

North, M.S. (1995) New rest breaking agents for the control of delayed foliation of apples. *Acta Horticulturae*, 409, 151-154.

Nucci, L.S. and Gurnevski, V. (1989) Development of pollen tubes following pollination and fertilization in some apple varieties. *Rasterniev'dni Nauki*, 26 (2), 89-92. Plovdiv. (In Horticultural Abstracts 1991 61:3 Abstract 1736).

Nyamapfene, K. (1991) *Soils of Zimbabwe*. Nehande Publishers. Harare 179 pp.

Panhwar, M.H. (1995). Temperate zone fruits in the subtropical area of Sindh Pakistan. *Acta Horticulturae*, 409, 155-164.

Petri, J.L. and Dittrich, R.C. (1984) The effect of DNBP concentration on breaking the dormancy of apple (*Malus domestica* Borkh) *7th Brazilian Congress on Fruit Culture*, 2, 801-812.

Petri, J.L. and Stuker, H. (1988) Effect of temperature and length of cold storage of nursery plants on the subsequent development of young apple trees. *Acta Horticulturae*, 232, 117-126.

Petri, J.L., Pola, A.C. and Stuker, H. (1990) Effect of mineral oil and hydrogen cyanamide on the dormancy breaking of apples. *23rd International Horticultural Congress*. Firenze. Abstract 4326.

Powell, D.B.B. (1974) Some effects of water stress in late spring on apple trees. *Journal of Horticultural Science*, 49, 257-272.

Ramirez, H., Galvan, F. and Salazar, R. (1990) The use of cyanamide as a chilling compensator in apple under Mexican climatic conditions. *23rd International Horticultural Congress*, Firenze, Abstract 4241.

Richards, S.J., Hagan, R.M. and McCalla, T.M. (1952) Soil temperature and plant growth. In *Soil Physical Conditions and Plant Growth*, (ed B.T. Shaw) Academic Press. New York: 303-480.

Richardson, E.A., Seeley, S.D. and Walker, D.R. (1974) A model for estimating the completion of rest for "Redhaven" and "Elberta" peach trees. *HortScience*, 9, 331-332.

Rom, R.C. and Brown, S.A. (1979) Water tolerance of apples on clonal rootstocks and peaches on seedling rootstocks. *Compact Fruit Tree*, 12, 30-33.

Sadowski, A. Alcalde-Blanco, S. and Enriques-Reyes, S. (1980) Effects of different levels of manganese and boron upon the nutrient status and the incidence of internal bark necrosis in apple tree. In *Mineral Nutrition of Fruit Trees*, (eds D. Atkinson, J.E. Jackson, R.O. Sharples and W.M. Waller) Butterworths, London, 378.

Salter, P.J. and Goode, J.E. (1967) Crop responses to water at different stages of growth. *Research Review* No. 2. Commonwealth Agricultural Bureaux. Farnham Royal.

Samish, R.M. (1945) The use of dinitrocresol mineral oil sprays for the control of prolonged rest in apple orchards. *Journal of Pomology and Horticultural Science*, 21, 164-179.

Sanchez, P.A. (1976) *Properties and management of soils in the tropics*. Wiley. New York.

Scott, B. (1989) The use of netting for hail protection, design and management considerations for pome fruit orchards. *Acta Horticulturae*, 240,147-150.

Shear, C.B. (1974) Interaction of calcium and nitrogen and time of calcium availability in relation to the development of apple disorders. In: *Proceedings of the 7th International Colloquium on Plant Analysis and Fertilizer Problems*. Hanover. German Society for Plant Nutrition, 427-436.

Shear, C.B. (1980) Interaction of nutrition and environment on mineral composition of fruits. In *Mineral Nutrition of Fruit Trees*, (eds D. Atkinson, J.E. Jackson, R.O. Sharples and W.M. Waller) Butterworths, London 41-50.

Soria, N. (1988) Influence of different concentrations of DNBP+Mineral Oil and two methods of application on flowering of Melly Jonathan apple. *Acta Horticulturae*, 232, 116.

Stino, F.G.R. (1995) Aspects related to temperate zone fruit production in Egypt. *Acta Horticulturae*, 409, 203-205.

Subhadrabandhu, S. (1995) Induction of budbreak in apple trees that received insufficient chilling by hydrogen cyanamide. *Acta Horticulturae*, 409, 171-178.

Tagwira, F. (1995) Chemical and Physical Factors Associated with Micronutrient Status and Availability in Zimbabwean Soil. In *Dambo Farming in Zimbabwe*, (eds R. Owen, K. Verbeek, J.E. Jackson and T. Steenhuis) University of Zimbabwe Publications, Harare 141-148.

Tamura T. Fukui, H., Imakawa, S. and Mino, Y. (1981) Effect of temperatures at early stage of fruit development on the fruit and seed in apple. *Journal of the Japanese Society for Horticultural Science*, 50, 287.

Taylor, E.M. and Burns, A.C. (1924) Soil temperatures during the shagari period and their agricultural significance. *Egypt Min. Agr. Tech. Sci. Serv. Bull.*, 31, 1-46.

Terblanche, J.H., Gurgen, K.H. and Hesebeck, I. (1980) An integrated approach to orchard nutrition and bitter pit control. In *Mineral Nutrition of Fruit Trees*, (eds D. Atkinson, J.E. Jackson, R.O. Sharples and W.M. Waller) Butterworths, London, 71-82.

Tromp, J. (1975) Growth and mineral nutrition of apple fruits as affected by temperature and relative air humidity. In *Environmental Effects on Crop Physiology* (eds J.J. Landsberg and C.V. Cutting), Academic Press, London 101-116.

Tromp, J. (1980) Mineral absorption and distribution in young apple trees under various environmental conditions. In *Mineral Nutrition of Fruit Trees*, (eds D. Atkinson, J.E. Jackson, R.O. Sharples and W.M. Waller) Butterworths, London, 173-182.

Tromp, J. and Borsoom, O. (1994) The effect of autumn and spring temperature on fruit set and on the effecttive pollination period in apple and pear. *Scientia Hort.* 60, 23-30.

Tukey, L.D. (1956) Some effects of night temperatures on the growth of McIntosh apples I. *Proceedings of the American Society for Horticultural Science*,. 68, 32-43.

Tukey, L.D. (1960) Some effects of night temperatures on the growth of McIntosh apples II *Proceedings of the American Society for Horticultural Science*, 75, 39-46.

Vermeiren, L. and Jobling, G.A. (1984) Localized irrigation. F.A.O. *Irrigation and drainage* paper 36, F.A.O. Rome 203 pp.

Visser, T., Verhaegh, J.J. and de Vries, D.P. (1976) A comparison of apple and pear seedlings with reference to the juvenile period I. Seedling growth and yield. *Acta Horticulturae*, 56, 205-214.

Wagenmakers, P.S. 1994. *Light relations in orchard systems*. Thesis Wageningen.

Wample, R.L. (1982). Water and the tree. In *Water Management and irrigation of tree fruits*, (ed R.B. Tukey), Washington State University, Pullman 95-108.

Weinberger, J.H. (1950) Chilling requirements of peach varieties *Proceedings of the American Society for Horticultural Science*, 56, 122-128.

Weischet, W. and Caviedes, C.N. (1993) *The persisting ecological constraints of tropical agriculture*. Longman. Harlow 319 pp.

Westwood, M.N. (1993) *Temperate zone pomology, physiology and culture*. Timber Press Inc. Portland Oregon. 523 pp.

Wickenden, M.F. (1982) Meteorological Records 1980-1981. *Report of East Malling Research Station for 1981* Maidstone. 51-55.

Williams, R.R. (1970) Pollen tube growth rate. In *Towards regulated cropping*. (eds R.R. Williams and D. Wilson), Grower Books, London: 61.

Williams, W.T. and Menegazzo, G. (1988) Apple culture in the highlands of Venezuela. *Acta Horticulturae* 232, 57-66.

Wilms, F.W.W. and Basso, C. (1988) Soil amelioration for apple production in southern Brazil. *Acta Horticulturae*, 232, 193- 203.

Zimmerman, R.H. (1971) Flowering in crab apple seedlings. Methods of shortening the juvenile phase. *Journal of the American Society for Horticultural Science*, 96 (4), 404-411.

12 GRAPEVINE (VITIS VINIFERA) GROWTH AND PERFORMANCE IN WARM CLIMATES

SHIMON LAVEE

Institute of horticulture, Volcani center,
Bet Dagan and faculty of agriculture,
Hebrew university of Jerusalem, Rehovot, Israel.

1 Introduction

Although the Vitis vinifera is assumed to have originated in moderate temperate climates, archeological findings indicate the use of grapevine products in south and central Europe during the Bronze age. Grape culture probably originated in Asia minor between the Black and Caspian seas. However, from ancient times it was well established also in warm sub-tropical regions. Grapevines and their products, particularly wine, were a companion to human life, travel and religion (Kirschhaimer 1938, Gur 1974). The original establishment of vinifera vines in warm regions was in the east and south Mediterranean regions characterized by hot and dry summers and moderate winters. The mechanisms of adaptation of grapevines to such condition is diverse and include both morphological and physiological parameters. The ability of the grapevine to develop both deep and wide root systems allowed a relatively easy adaptation to a large array of edaphic conditions. In rocky calcareous soils, grapevine roots were found to penetrate to a depth of 12 meters. This is the case also in very sandy soils and particularly on sand dunes. On the other hand, on heavy wet soils, shallow and wide root systems are usually found. Different grapevine varieties and rootstocks have different adaptability to various soil conditions. The high adaptation potential of the vinifera plants to different soil conditions allows the grapevine to develop well both in the dry mountain region and heavy valley soils typical in the warm subtropical regions. Shoot growth polarity and high vigor are other means for adaptability and survival under extreme environmental conditions. This is well expressed in warm regions were bud opening is limited to apical regions and have a higher potential for continuous development. On the other hand, the high vigor and polarity within buds allows inhibited secondary buds of the grapevine 'eye' to develop, after frost damage to the young growth from the major buds. Thus, the conglomerate of buds in the grapevine 'eye' is controlled by an interaction between environment and the major shoot development.

2 Developmental cycles

The grapevine is a deciduous species and as such, similar to most other deciduous plants originated in temperate zone climates and needs a specific sequence of environmental

Amnon Erez (ed.), Temperate Fruit Crops in Warm Climates, 343-366.
© 2000 *Kluwer Academic Publishers. Printed in the Netherlands.*

conditions for normal completion of its annual cycle (Samish 1954). The beginning of the annual active cycle of grapevine depends on the readiness of its buds to respond to spring conditions. Uniform bud opening will occur only if the vines were subjected to enough winter chilling or treated chemically to compensate for the lack of chilling. The rate of grapevine growth in the spring right after bud opening is slow and compact but becomes most rapid and linear thereafter (Lavee 1980). The rate and duration of the rapid growth is a function of the environmental conditions such as temperature, light, water availability etc. In accordance with the increase of the day temperatures in the summer, reduction of light penetration due to increased canopy shading and build up of water stress, growth rate is reduced. The cessation of elongation growth depends on either endogenous factors or climatic conditions particularly supra optimal temperatures (Fig. 1). Fruit development during the growing season is dependent both on the environmental conditions and on fruit load, particularly the relative fruit load in relation to vegetative growth, vine size and the trellising system. During the second half of the fruit's growth period and ripening, vegetative development slows or even stops.

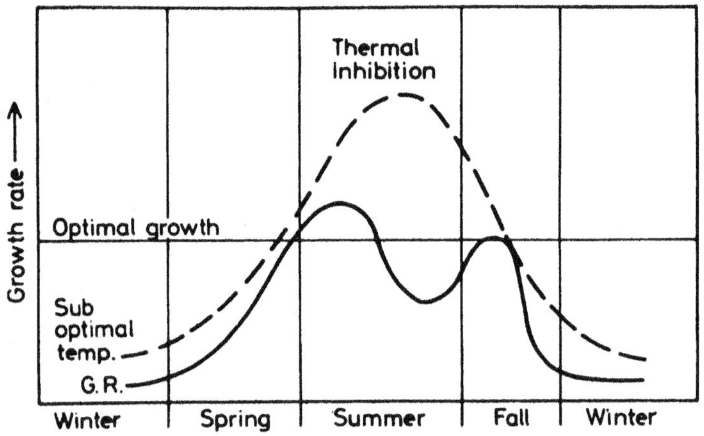

Figure 1. Schematical description of the relation between summer temperatures in hot regions and the growth rate of grapevines.

In the fall, leaf drop and dormancy development are governed by seasonal environmental changes such as day length reduction, temperature decline and change in relative growing conditions. The length of the dormant period, uniformity of its release and rate of bud burst for next seasons growth is controlled by a variety of genetic and climatic (mainly thermal) conditions and the interaction between them (Lavee and May 1997).

Introducing the grapevines into warm areas started thousands of years ago by gradually moving it into the subtropical Mediterranean region (Kirschheimer 1938). During the last 450 years vinifera grapevines were introduced also to other warm regions in the world such as in South Africa, North and South America, South East Asia (Indonesia) and Australia. About 50 years ago a grapevine industry has developed also in the tropics - a warm region with entirely different climatic and growing conditions than those in the "traditional" subtropical warm regions.

2.1 Grapevine development in the tropics

In the traditionally used subtropical warm regions, the climate is characterized by long dry hot summers and relatively short wet winters with moderate cool temperatures. On the other hand in the tropics, the weather is rather uniform year round, thus, less hot in the "summer" and warmer in the "winter". However, the major and probably most significant difference between the two "warm" regions is a marked difference in the prevailing photoperiodic conditions. In the subtropical region the seasonal change in day length, with elongating days in spring and shortening in the fall, are of the same pattern as in moderate temperate climates. In the tropics however, day length is rather uniform the whole year round. Although most deciduous fruit trees are considered day length neutral, shortening of day length was suggested to be one of the major factors to induce dormancy in some vitis species (Fennell and Hoover 1991).

On the other hand, grapevines growing under tropical conditions, with a uniform annual photoperiod do not develop a defined dormant period. Grapevines under such conditions grow as evergreen plants without synchronization of leaf drop, and other developmental stages such as growth flashes, flowering, fruit ripening, etc. Therefore growing grapevines under warm tropical conditions requires a different approach and methodology than when grown in warm subtropical regions (Corzo 1987). The lack of enough chilling hours for 'normal' grapevine development is prevailing in both types of warm climates.

In the tropics due to the uniformity of climatic conditions, the environmental signals for the induction of some of the developmental stages are not expressed. Under such conditions the annual growth cycle and expression of developmental stages of the grapevine changes. The most significant factor which changes the entire development of grapevines under tropical conditions is the lack of signals for dormancy induction. Nevertheless in spite of that missing section of the 'normal' growth cycle, grapevines can grow well and adapt to such conditions. Dormancy avoidance (see Chapter 2) leads to a complete change in the developmental habit and growth cycle of the vines. Grapevines under such conditions stop to be deciduous, become evergreen and lose synchronization in their development. Thus without horticultural manipulation, flowering, different stages of developing fruits as well as vegetative growth from bud sprouting to mature canes are all present at any given time in the vineyard and even on the same vine (Aubert 1973, Corzo 1987). The feasibility of grapevines to grow without dormancy induction by developing evergreen growth cycles devoid of dormancy, enabled the development of a new different, efficient tropical viticulture. Due to the lack of dormancy, 2-3 complete developmental cycles per year can be induced when appropriate methods are applied.

In most cases synchronization is achieved by defoliation, pruning and application of temporary stress conditions where possible. In some cases application of the dormancy breaking agent, Cyanamide is also used not for a release from dormancy but to enhance a more uniform and rapid bud opening (Lin 1984). The moderate-high temperature thereafter allows a rapid uniform fruit development. In most cases growth is too rapid for high quality wine grapes but good quality table grapes can be obtained. Early maturing cultivars with a relative short growing period are more suitable for such condition.

The rapid completion of a full growth cycle enables in many cases to obtain 2-3 yields per year on the same vines. Vigorous cultivars are needed for maximal utilization of these conditions. Still the control of excess vigor during the active growth period is critical to prevent shading which reduces reproductive bud differentiation and enhances abortion of floral primordia. In the wet tropical regions the trellising system must be well adapted to efficient and continuous disease control treatments which are enhanced under such conditions.

2.2 Grapevine growth cycle in subtropical climate

In subtropical climates the seasonal changes are well defined, with shortening days leading into a cool winter. Grapevines as other deciduous species from temperate climatic zones have a defined dormant period which is part of the annual developmental cycle. Therefore, once dormancy is induced, environmental conditions which will allow the vines to complete their dormant development are required. Lack of enough chilling during the fall and winter will cause a delay in leaf drop and more critically a delayed uneven and poor bud opening in the following spring (Pouget 1972). The length and degree of the expression of environmental conditions in each portion of the annual developmental cycle might be different in accordance with the specific conditions during each stage and at different locations. The defined seasonal changes however, will cause the grapevine to develop all stages of its annual life cycle as changes in day length and temperature cycles are clearly defined in the subtropical climate. Still, problems might evolve characterized by either excess or deficiency of conditions for normal development of the vines during specific stages of the growth cycle. As the conditions needed for the fulfillment of the developmental requirements in any of the seasons might not be sufficient or adequate, specific disorders or inability to complete one of the developmental stages of the growth cycle might occur. Such developmental disorders induced by the non adequate environmental conditions in any particular season of the warm subtropical climate have to be considered and overcome by physiological and horticultural means in order to achieve satisfactory vineyard development. The adoption of suitable horticultural techniques is required in order to enable the completion of each stage of the annual growth cycle (Lavee and May 1997). The problems developing in subtropical region due to insufficient chilling are considerable more severe than in the tropics where the entire dormant period is being avoided. This is true also for the early vegetative growth which might be excessive due to the rapid temperature increase in the spring. Furthermore, early growth cessation in the summer and massive reinduction of growth in the fall might jeopardize the normal sequence of the annual developmental stages of the vines.

3 Dormancy and means to overcome it

In grapevines, as in most other deciduous fruit species when dormancy has been induced its fulfillment is required for the completion of its annual growth cycle. Although, grapevines do not have a high chilling requirement, its fulfillment is critical for normal development of the vines. The delay, reduction and lack of uniformity in opening of grapevine buds due to insufficient chilling causes unbalanced growth, reduction in yield and non-uniform maturation. The response of grapevines to dormancy breaking agents is unique. Many of the standard dormancy breaking agents such as mineral oils, DNOC and Thiourea have only a weak or no effect on grapevine bud opening (Nir and Lavee 1993a). On the other hand, hydrogen cyanamide *(*Dormex*)* which was originally developedfor breaking dormancy in grapevines (Shulman *et. al.*, 1983, Lavee *et. al.*, 1984) is most efficient. This dormancy breaking agent is able even to induce growth of lateral buds on old wood (Fig. 2). In warm climates growth induction of latent buds is of particular importance due to the frequent decline of the center arms on the cordon, caused by the high polarity of growth under those conditions. Timing of 'Dormex' application in relation to other growth stimuli such as pruning, the stage of bud development and climatic regional conditions, have a major effect on the responsiveness of the buds to the treatment. Recently it was shown in Southern Spain (Lavee and Klein - unpublished) and in Israel that the growth induction effect by early pruning is very small. Application of 'Dormex' to induce bud opening could be applied independently from an earlier pruning with favorable results. Furthermore early pruning (December) followed a month later by a 'Dormex' treatment produced a considerable higher yield than if sprayed at pruning time.

Figure 2. Latent bud opening on old grape vine cordon due to a winter treatment with hydrogen cyanamide (Dormex).

The reduction in the number of opening buds after warm winter with insufficient chilling and its lack of uniformity is not only due to a complete inability of the buds to break but also to a correlative inhibition induced by the erratic early sprouting buds.

In general, late pruning just before bud opening in the spring will result in more uniform bud opening. A larger number of buds are at that stage responsive to the pruning stimulus. However, in order to benefit from the warm climate and achieve maximal precocity, the earliest possible growth induction in the spring is desired.

In each climatic region the optimal pruning time or the earliest pruning date which will induce a uniform bud opening of a reasonable number of buds could be defined. This date will vary in accordance with chilling accumulation causing the physiological changes needed for dormancy evolution. These physiological changes bring the buds to a responsive state to growth inducing stimuli differently in accordance with the region and each specific year (Fig. 3).

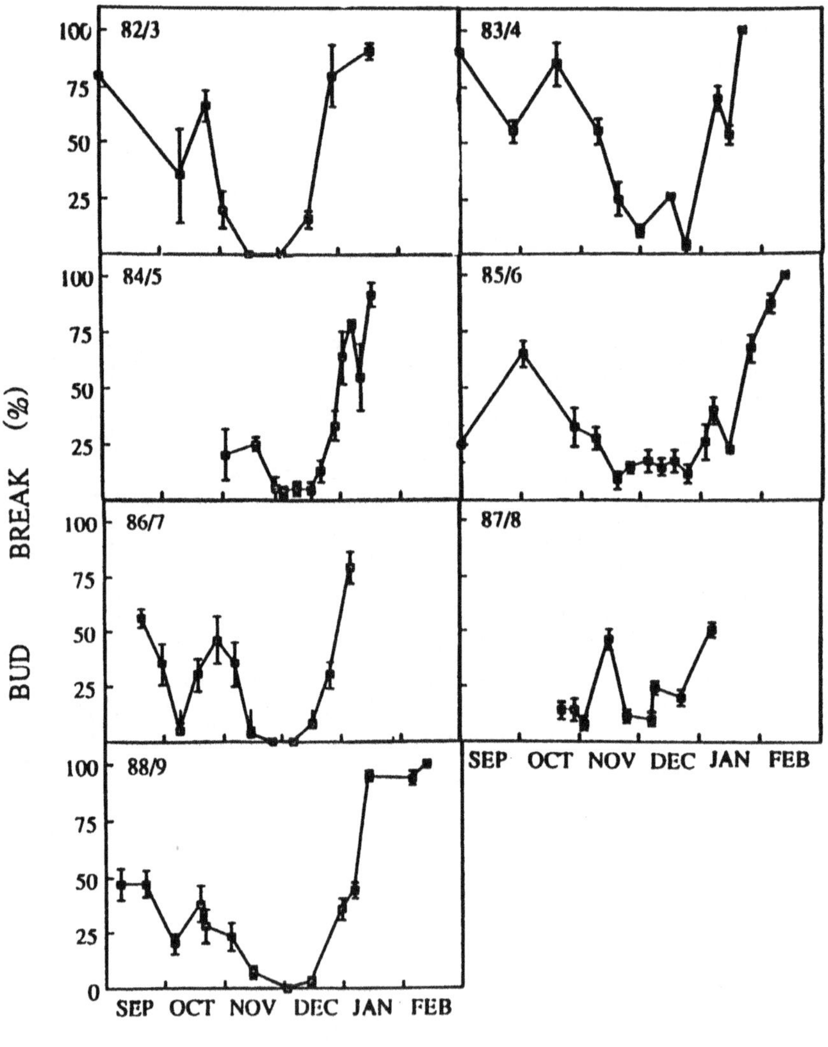

Figure 3. Bud break potential of cv. Perlette during seven different years. using single bud cuttings sampled and prepared along the autumn and winter. Bud opening was measured after 20 forcing days at 23° C.

Prior to responsiveness of buds to favorable growing conditions, a reduction in Catalase activity occurs. This reduction is stimulated on one hand by chilling temperatures and by hydrogen cyanamide on the other (Nir *et al.,* 1986). Pruning in general and particularly spur pruning, can affect bud break in grapvines. Pruning during the deep dormancy stage will usually result in a delayed and non-uniform bud opening (Lavee 1977). Later pruning during the slow endogenous release from dormancy will result in early bud opening. However, the time required from pruning to bud opening will be shorter the later in the spring the pruning is performed (Fig. 4). In warm climates the monthly thermal average as well as minimum and maximum temperatures vary from region to region (Table 1). The accumulation of chilling was calculated and expressed as hours below 10°C during the winters of seven years in the climatically warm Jordan Valley (Fig. 5). The number of hours with temperature above 20°C was also recorded during those months and years. The ratio between the number of hours at high and low temperatures was calculated and used as an index for chilling efficiency. The cumulative ratios for each year gave a reliable indicator for the chilling efficiency of the different years. The monthly indices were correlated with dormancy development, the main dormant period and dormancy release. A low index in October-November resulted in an early induction of dormancy with an earlier and more pronounced peak (Fig. 3). The rate and time of dormancy release was clearly related to the total annual calculated index. In years with a low index (more hours below 10°C and less above 20°C) the peak of the endodormancy and the time of 50% dormancy release occurred earlier.

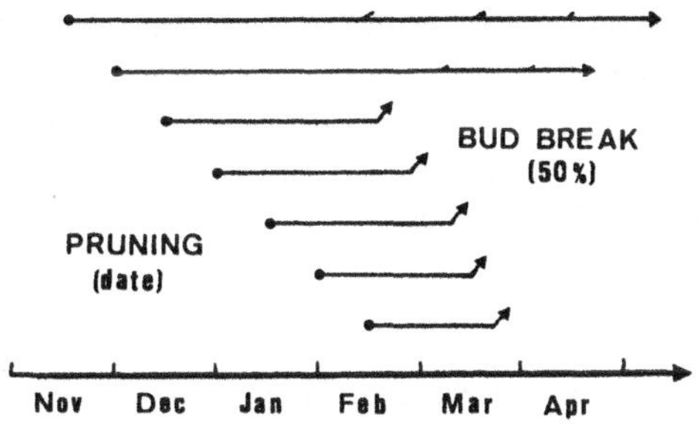

Figure 4. A general description of the effect of pruning date on length of the period to 50% bud opening.

Table 1. Mean monthly minimum and maximum temperatures in various regions of Israel (°C)

Month	*mountains*		*Internal plain North*		*Coastal plain*		*Jordan Valley South*		*Jordan Valley*	
	Min.	Max.	Min.	Max.	Min.	Max.	Min.	Max.	Min.	Max.
Jan.	5.2	12.5	7.5	18.2	8.5	17.9	8.9	18.3	9.8	20.1
Mar.	6.5	15.7	8.2	20.4	9.9	19.2	10.7	22.4	11.4	23.8
Apr.	10.3	20.7	10.1	23.8	11.6	21.4	13.2	26.7	14.8	29.3
Jul.	18.0	29.2	19.4	31.5	20.4	29.0	22.8	36.6	23.6	38.8
Nov.	12.0	20.7	12.5	25.5	13.2	24.2	14.8	25.7	15.9	21.8
Dec.	7.3	14.6	9.1	20.1	9.8	19.6	11.2	20.1	11.6	21.8

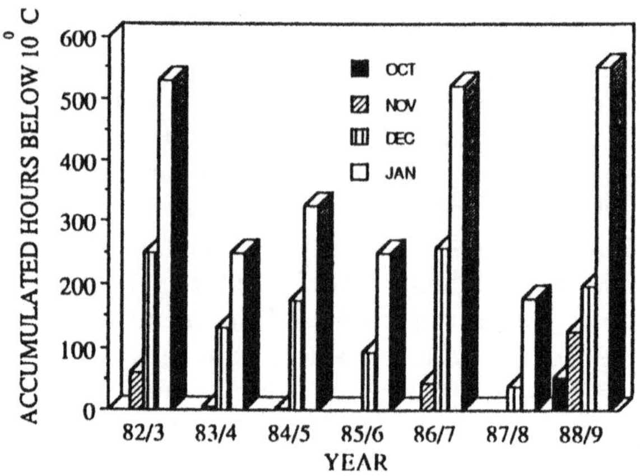

Figure 5. Quantitative accumulation of chilling hours below 10°C - during each of the four months Oct.-Jan. over a period of seven years in the Jordan Valley.

Application of the potent dormancy breaking agent Dormex will also be most effective when applied after the reduction of the endodormancy peak. An earlier application of 'Dormex' will not overcome the lack of uniformity and delayed bud opening (Table 2). Thus, in most cases the optimal time for stimulating the vines to grow by pruning or application 'Dormex' is the same and most efficient when given together (Fig. 6) (Nir and Lavee 1993b). In a number of cases particularly in years or regions with somewhat higher chilling accumulation, pruning time is not critical and can be performed independently prior and experimentally even after to the optimal date for growth stimulation when the 'Dormex' should be applied. This was clearly the case in southern Spain (Almeria) even when the vines were grown under polyethylene (Lavee and Klein, unpublished).

Table 2. The effect of pruning date hydrogen cyanamide application during a relative cool winter on percent bud break of cv. Perlette in the Jordan Valley

Treatment		*Per cent budburst*					
		February 20		*March 1*		*March 10*	
Pruning date	*% H_2CN*	*Mean*	*SE*	*Mean*	*SE*	*Mean*	*SE*
December 26	2.5	1.5	1.0	17.8	5.8	76.6	5.9
	5.0	10.8	2.8	16.7	4.0	69.0	4.2
January 5	2.5	72.2	3.1	77.9	3.1	84.2	3.1
	5.0	56.7	3.1	65.2	2.4	74.2	3.5
January 12	0	12.6	7.1	21.3	9.0	84.8	4.5
	2.5	68.1	2.0	83.4	6.0	86.5	4.5
	5.0	37.8	3.9	68.8	3.6	87.8	4.5
January 21	2.5	15.9	5.1	58.0	3.5	82.4	4.0
	5.0	7.0	1.4	51.0	5.5	76.2	6.6
February 2	0	0	-	0	-	83.0	4.9

SE = Standard error of mean

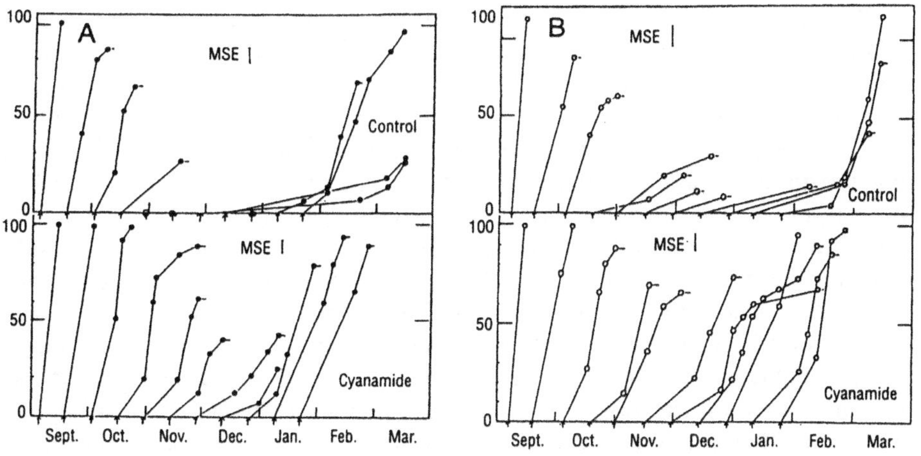

Figure 6. Percent budburst on container-grown Perlette (A) and Barlinka (B) grapevines pruned and concurrently sprayed or not sprayed (Control) with cyanamide between September and January. Cyanamide was applied after pruning, using 2% 'Dormex' (50% active ingredient). The vines were transferred to a semi-heated glass-house after pruning.

After dormancy is induced in the fall it reaches its peak relatively early in the winter (November) in the northern hemisphere. In many cases (Nir *et al.*, 1986) by mid December the grapevine buds are already responsive to growth inducing conditions and thus benefit at that stage also from the application of a dormancy breaking agent. The optimal date for the readiness of response might vary somewhat among cultivars but this variation is relatively small and the period is rather uniform for the Vitis vinifera species. Still each cultivar will respond to the application of dormancy breaking agents differently (Fig. 7).

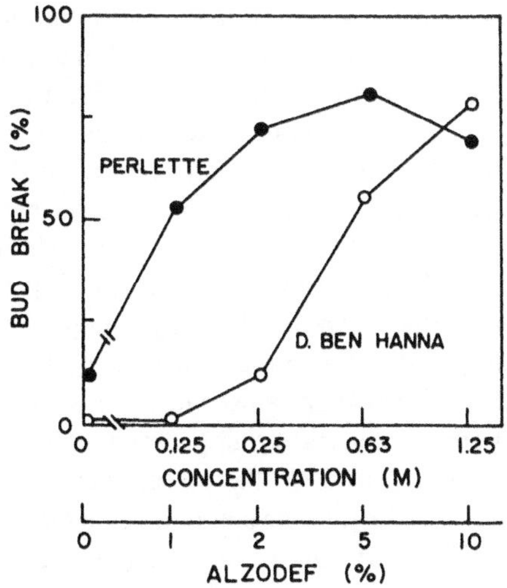

Figure 7. The response of cvs. Perlette and Dan Ben-Hanna to different concentrations of hydrogen cyanamide (Dormex or Alzodef) in the Jordan Valley.

The ability to induce growth already in December even in warm winters allows Israel and other warm countries to utilize the naturally high temperatures in early spring, or even in late winter by using a polyethylene *cover* above the vines, to achieve early growth and subsequently early flowering, fruit development and maturation (Lavee 1988). By 'Dormex' applications at the appropriate time, usually after mid winter, it was possible to induce a larger number of buds to open simultaneously and thus reduce the polarity effect. This could be further induced by applying evaporative cooling to the vines from November to the end of January enhancing the accumulation of chilling by lowering the temperature of the buds (Nir *et al.,* 1988).This lowering of bud temperature by 5-10°C causes on one hand an increase in effective chilling hours and on the other a reduction of the amount of, and period with, unfavorable (reversing) high temperatures of 21°C and higher (Fig. 8). A sectorial application of 'Dormex' even enables to maintain in many cases the viability of lower growing points on vertical vines. (Lavee 1994b). This is based on earlier application of the 'Dormex' to the lower buds inducing them to start growing in the spring before the upper and terminal ones. The earlier outgrowth of the lower buds allows them to develop in full light with no competition of the upper buds. The upper buds which due to the polarity are more vigorous will develop normally in spite of the growth of the lower ones. This methodology has been successfully applied to vines grown under polyethylene to induce precocity (Lavee and Haskal 1982).

4 Vegetative growth and control of excessive vigor

Spring growth of all grapevine cultivars is usually rapid and particularly so in the warm subtropical climate. The growth rate of the vines is highly dependent on soil moisture and the prevailing temperatures. With the exhaustion of the available water and the rapid increase of spring temperatures in warm climates, the growth rate is reduced leading to a complete cessation of growth in the summer.

In dry hot regions as in many Mediterranean zones with long dry summers, irrigation will enhance and prolong the active growing season (Smart and Coombe 1983). In regions with rapidly increasing temperatures in the spring, irrigation is essential for developing a large enough canopy for a high fruiting potential before the thermal inhibition in mid summer takes place. The initial high vigor of grapevines can develop a good canopy within a short active growing period in the spring. Thereafter, during the hot summer this canopy will support fruit development and provide storage metabolites for the vine. The high polarity in grapevine shoots helps during the initial growth period to reduce the outgrowth of secondary shoots and partially control this way canopy size and density. In warm arid non-irrigated regions the rapid initial shoot growth will also take place supported by the soil moisture provided by the winter rains. However, canopy development under such conditions will be short and a limited fruiting potential will develop. This limitation is due both to a direct effect of water stress on the present developing fruit and an indirect one on reducing the level of reproductive differentiation of the buds for the following year's crop (Kliewer and Antcliff 1970). On the other hand intensified culture (fertigation) in warm climates might create an excess vigor due to the prolonged growing period causing a massive increase in shoot length, excess shade and reduced fruiting.

Figure 8. Enhancement of the 'Dormax' effect on bud opening by evaporative cooling in an intensive table grape vineyard in the Jordan valley.Top: Control. Bottom: Evaporative cooling, Nov.-Jan. Photo in mid March.

In grapevines the amount of fruit is mainly governed by the number of shoots and not by their length. Therefore, without proper pruning, irrigated vines in hot regions tend to become very vigorous. Such vigor, if not controlled, might create excess shade leading to non ripening green shoots in the autumn with low lignification and low fruit primordia differentiation which in turn will lead to even more vigor, forming a vicious cycle of high vigor and low production (Fig. 9), (Erez and Lavee 1974, Lavee 1990).

In general, the high vegetative vigor of irrigated grapevines in warm environments requires special care in training and pruning the vines in order to achieve a reasonable ratio between fruit load and the vegetative capacity. Fruit load is determined by the weight of yield per vine while the vegetative capacity is based on the weight of the pruning wood in winter. This ratio when below 4 (excess growth) or above 8 (over cropping) represents an unbalanced situation. (Bravdo and Hepner 1987). The shade produced on the lower portions of the canes of vigorous vines causes not only low reproductive differentiation in the buds but also induces abortion of bunch primordia in those buds which differentiated early in the season before they were overshadowed (Lavee and Rauchberger 1993). The reduction in reproductive differentiation of the lower, early in the season shaded buds, creates the need to apply more complicated and expensive pruning and training methods in order to maintain an economical commercial crop.

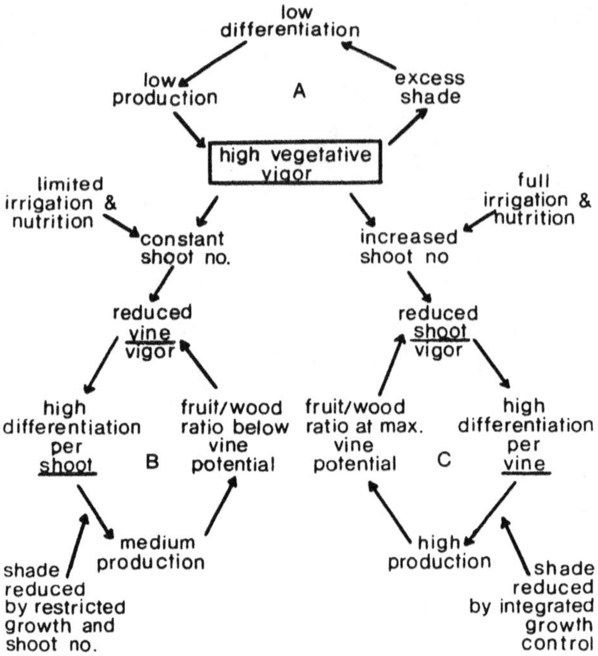

Figure 9. A description of the feedback relations between vigor, shoot number yield and environmental conditions.

As 10-12 leaves above the bunch are enough for normal fruit development (Coombe 1959, Lavee 1982) topping the growing shoots was useful to maintain a more balanced canopy. Because of the rapid vegetative growth in the hot regions and under greenhouse culture or polyethylene tunnels (see 12.7) topping is required already very early in the summer. The non- dormant buds on the shoots at that stage and the favorable temperatures for growth resulted in rapid brunching near the cut as well as a rapid development of the already existing secondary shoots. Under such conditions, the total amount of vegetative growth in topped wines is even larger than in the non-topped ones (Table 3). When this occurs a second and some times third topping is needed to control the canopy size (Lavee and Haskal 1982). A spray treatment based on application of ethephon to the upper buds on the topped shoots can be used to inhibit their opening (Lavee *et al.,* 1977, Lavee 1982) and thus prevent massive regrowth of the canopy and need for continuous topping (Fig. 10). In somewhat cooler regions the excess vigor could be controlled even by one spray directly on the growing point without prior shoot topping. Under even cooler conditions with a slower growth rate, no spray application is needed as the shoots reach the optimal length for topping in mid summer and out growth of the lateral buds does not occur or is only very weak at that time. Under warm and dry growing conditions, the excess vigor can be reduced by controlling the water availability, developing a temporary slight water stress. This method is usually not applied in extremely hot regions because of the high evapotranspiration and thus difficulty to control stress.

Table 3. The effect of topping after fruit set on wood and fruit production of *cv.* Cabarnet souvignon (10 year old single cordon trained vines)

Treatment	*Shoot and laterals length* *cm*	*Wood /vine* *kg*	*Yield /vine* *kg*	*Bunch size* *g*
Control	203	3.1	12.3	189
Topped	213	3.3	10.8	166
MSE ±	8	0.2	0.4	10

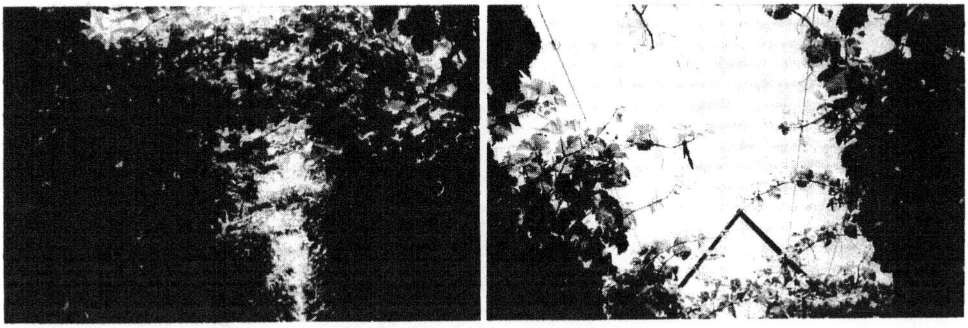

Figure 10. Growth control by application of Ethephon in a vigorous growing factory roof trained Alphonse Lavallée vineyard in the coastal plain. Left, control; Right, ethephon treated canopy.

Shoot topping and subsequent control of lateral bud outgrowth was lately adopted in wine vineyards in warmer regions with excess vegetative growth. This method of controlling shoot length enhanced their lignification and prevented shoot bending. Thus canopy supporting wires in the standard cordon method could be avoided maintaining a non-positioned self supporting free canopy on the horizontal cordons with a better light penetration (Lavee *et al.*, 1993, Lavee 1988). Leaf removal in the bunch region is usually not required or only a very light one is needed when this method is applied. The self supported canopy which developed, had also a considerable advantage for simplifying mechanical pruning as interfering poles and supporting wires are not needed.

Light penetration to the bunches is critical in both cool and hot region viticulture. However, in the hot regions with usually high light intensities the deleafing is important mainly for better penetration of pesticides. Excess deleafing might lead in the hot regions with high radiation to an over exposure causing an uneven coloration and development of sun burns.

Polarity of growth and apical dominance are also more expressed in warm climates. Due to the high vigor during summer the ability of the lateral buds to respond to growth stimuli is reduced and bud opening along the cane in the spring is highly inhibited. The reduced bud opening along canes due to high polarity and *apical dominance* is true not only for annual shoots but is expressed also on the whole vine level causing a decline of the central arms of horizontal cordon trained vines. This decline is expressed both in marked reduction of bud opening in the spring on the central arms and by stunted growth of those shoots from buds which did open.

5 Training and pruning

In subtropical climate with defined seasons, the general performance of grapevines is similar to that in cooler regions thus vine training and pruning are based on similar principles. Only "unfruitful" cultivars requiring long pruning reveal problems of bud opening and demand appropriate pruning techniques. In that climate under irrigation and fertigation, the growth potential of the vines changes considerably and cultivation methods had to be reconsidered in order to develop the appropriate load capacity ratio. As mentioned earlier, vine vigor in warm climates with no water limitation is extremely high with linear growth during the main elongation period (Lavee 1980). Due to the polar growth, excess vigor is primarily expressed by marked elongation of the annual growing shoots. As the amount of yield in grapevines is controlled by the number of growing shoots, an increased number of growing point is needed to utilize the potential vigor more efficiently. The increase in the number of shoots causes a better distribution of the vegetative vigor and by competition reduces the potential length of each individual shoot. Photosynthetic efficiency of grape leaves is drastically reduced under shaded conditions. Thus a better distribution of the canopy is required. Therefore, in the warm regions inducing vigorous growth, larger vines with spread canopy are usually developed. For table grapes harvested manually, trellis systems which enable wide spreading of the canopy such as veranda, Y forms (Fig. 11) and factory-roof-systems are being most efficient (Lavee 1994a). For wine grapes, the classical horizontal cordon system with vertical shoots (Fig. 12) which enables efficient mechanization of the harvest is the common system for the irrigated vineyards (Lavee *et al.*, 1993).

Enlarging the trellis system in comparison with the traditional European stacked or unsupported Goblet training has to consider the planting distances and training systems to be used. Due to the high polarity and apical dominance of the shoots in warm climates, the active portions of the trunks (cordons) have to be kept short. Vigorous growth will develop from the buds near the bending point and at the distal end of the cordon. This will create unbalanced vines and a reduction in fruitfulness (Table 4). The degree of polarity is negatively correlated with the availability of chilling hours in the vineyard. Insufficient chilling not only reduces the number of buds which are induced to grow but also enhances the growth rate of the fewer shoots which do develop (Pouget 1972, Lavee *et al.*, 1993). This problem can be partially overcome by reducing the cordon's length so that the region with buds for annual development is short. Shortening of the cordons is achieved both by developing two sided horizontal cordons from the central trunk and in the wide double canopy factory roof trellis systems developing two lines of double sided cordons. Controlling the length of vegetative growth help to develop the suitable ratio between canopy size and fruit production. This will result in partial "correction" of the wine quality in warm regions by somewhat minimizing the acid reduction and enhancing accumulation of anthocyanins and aromatic compounds.

Figure 11. An intensive table grape vineyard trained in a modified Y system. Left, in winter before pruning; Right, during summer growth.

Figure 12. A modern vineyard with horizontal cordon trained vines for wine cultivars. Left, in winter before pruning; Right, in summer before harvest.

Table 4. The effect of vine vigor on yield and capacity of mature *cv.* Queen of vineyard grapevines near the lake of Galilee

Vineyard vigor	*Wood per vine kg*	*Fruit yield per vine kg*	*Fruit/wood ratio*	*Bunches per shoot No.*	*Berries per bunch No.*
Normal	2.7	18.5	6.7	1.8	81
Vigorous	4.8	12.0	2.5	1.2	70
MSE ±	0.3	2.0	-	0.15	5

Relatively dense planting (1-1.5 m along the raw) resulting in a larger number of smaller vines, helps in shortening the cordons and developing more compact vines (Lavee 1994b). A short cordon might be very similar to head trained vines. Head training is physiologically the most efficient system in warm climates and particularly in sites with considerable insufficient chilling. In the head training system, all the shoots are emerging from a virtually terminal position thus exploiting the better break of terminal buds as well as the high polarity of the growth. Using head training for spreading the canopy on wide trellis systems and prevention of density can be achieved by developing a number of heads per vine distributed on the trellis in accordance with space availability. In cooler regions with more winter chilling, head training is usually associated with cane pruning. In the warm regions however, cane pruning is problematic due to the high polarity allowing only the two distal buds below the pruning cut to open. When canes are bent horizontally, downwards or twisted, one or two additional lower buds might open but those are often not fruitful. Therefore under such conditions relatively large numbers of canes are needed to develop enough fruitful shoots for a commercial yield. Furthermore, under such conditions enough spurs for the development of the vegetation for next year canes is required as the opening and growth of basal shoots on the canes is not reliable enough. Thus cultivars demanding cane pruning are less suitable for warm regions and demand more elaborate and sophisticated growing techniques in order to achieve and maintain a commercial economical yield. In the warm region viticulture, the multi short cordons or heads are used to avoid polarity damage also for fruitful cultivars maintained by the standard spur pruning. The use of vertical training methods of vines or double cordons including the Scott-Henry method (Smart and Robinson, 1991) are difficult to maintain in warm climates because of the strong polarity. In most cases the lower cordon layers will decline within a few growing seasons. Low and high vertical vines could be maintained only when they terminate with a compact head. It should be pointed out however, that with the development of Dormex treatment for breaking grapevine dormancy the severity of the non-uniform and low bud opening along the cane and vines was markedly reduced (Shulman *et al.*, 1986).

6 Irrigation and fertigation

Most of the commercial vineyards in the subtropical regions with a long dry summer are irrigated. In the past, vineyards were irrigated by furrow and later by sprinkler irrigation methods. Today the prevailing irrigation system used is drip fertigation both for table and wine grapes. Only in South Africa microjet irrigation is the major system utilized. The drip system was found most efficient both for controlling vine development and increasing the phytosanitary conditions in the vineyard. The amount of water applied depends on the region, climate and soil type and on the designation of the vineyard. With table grapes the peak of irrigation is reached just before harvest (Hardie and Concidine 1976) while with wine grapes irrigation is stopped about 4 weeks prior to harvest to ensure high wine quality. The amount of water applied in each region is usually based on a factor related to the evapotranspiration of the region (see Chapter 3). Detailed information on grapevine irrigation was described in reviews by Bravdo and Hepner (1986, 1987). To reach the highest quality for wine, of any given irrigated vineyard, a suitable balance between the fruit load and vegetative capacity has to be achieved. Thus, when the vines are well balanced the effectiveness of the induced stress at the final stages of fruit development is seriously questionable. The climatic conditions however, have a major effect on fruit composition even in balanced vines. The thermal conditions and radiation effects on the metabolism of the vine and metabolites accumulation in the fruits.

7 Protected cultivation in warm climates

In accordance with today's economy, table grape viticulture is being developed both in extremely warm climates and in plastic greenhouses developing artificially such conditions. Originally, yields under such condition were low but the high prices compensated for it. The low yields were mainly due to low bud break or reduced reproductive differentiation induced by the high temperature and excessive shading of the massive vegetative growth obtained (Buttrose 1969). Bunches under such conditions are smaller due to unfavorable conditions during the secondary (Berry) differentiation in the hot spring. This also contributes to the low yields. 'Dormex', dense planting and adequate pruning are used to increase the number of developing shoots and thus the potential yield (Fig. 13). On the other hand, the increase in shoot number with potential high vegetative vigor under such growing conditions, creates in many cases too dense a canopy. The density of the canopy has to be reduced in order to prevent early leaf drop and low reproductive differentiation (May 1965, Kliewer and Antcliff 1970).

Figure 13. The effect of 'Dormex' treatment on bud break of vertically trained *cv.* Perlette vines under a polyethylene covered structure. Left: Treated with 4% Dormex Jan.1. Right: Untreated control.

The growing of grape vines in warm regions under temporary cover of polyethylene is rapidly developing for the early season world market. The cultural practice is based on pruning and treating the vines with 'Dormex' after termination of the main endodormancy and covering them with polyethylene sheets on a supporting structure right thereafter (Lavee 1987). Keeping the vines under polyethylene till after veraison, shortens to various degrees all previous developmental stages of the vine (Table 5). After veraison, the whole cover or the side walls of the covering structure are removed to prevent overheating (in excess of 35°C) that will lead to a reduction in photosynthesis and general metabolic activity. Fruit maturation can be advanced in this system by up to two weeks even in warm growing regions. In subtropical regions with somewhat cooler and more moderate summers, as in the coastal plains of Israel, the effect of the polyethylene cover on ripening is usually even stronger. Maturation of covered vines in those regions is enhanced by about 3 weeks with early cultivars *i.e. cvs.* Perlette and Queen of vineyards as well as with later ripening cultivars such as cv. Alphonse Lavallee (Table 6).

The early developmental stages of the vines after pruning and Dormex treatments are enhanced under polyethylene. The earlier stages of vine development are relatively shortened to a higher degree than the later ones such as the time span from veraison to ripening. This might be due to the increase of the ambient temperatures with advancement of the season reducing the advantage of the polyethylene cover. After veraison in many cases the environmental temperatures are high enough for optimal development so that the polyethylene cover has no further enhancing effects. In some cases however, at that stage only the polyethylene walls are removed and the over head cover is left intact for protection from occasional rain.

Recently it has been shown (Nir *et.al.*, unpublished), mainly with early cultivars, that by pruning the vines right after harvest a new canopy usually with no or negligible amount of fruit develops during the second half of the summer. This second canopy, when repruned in the winter results in a better sprouting and more fruitful bud population in the following growing season. In order to achieve the maximal precocity and continuous vine development, supraoptimal temperatures under the polyethylene cover structure have to be avoided (Lavee 1988a). Prevention of high temperatures, particularly during the early stages of annual development, is achieved by controlled, partial, opening of the polyethylene covered structure. Too rapid temperature increase in the covered structure can be buffered by using perforated polyethylene sheet for the covering. During the lower temperatures of the early spring the perforation has only a negligible effect on reducing air heating under the polyethylene cover. When the temperatures in the tunnels reaches 30°C, the perforation has a buffering effect and slows down the further increase of temperature. This allows for more flexibility in timing ventilation by opening the sides or windows of the covered structures.

8 Grape quality and climate

Under warm subtropical conditions sugar accumulation in the fruit is quicker and higher than in the temperate climate. The accumulation starts after veraison and proceeds

Table 5. Phenological behavior of *cv*. Perlette vines under a polyethylene covered structure. (Pruning mid January)

Treatment	*Bud break date*	*Sprouting date*	*Full bloom date*	*Fruitset date*	*Harvest 50%*	*Total days*
Control	Mar. 1	Mar. 7	Apr. 25	May 7	June 16	108
PPE cover	Feb. 18	Feb. 25	Apr. 12	Apr. 25	June 5	107
Difference in days	11	10	13	12	11	1

PPE: Perforated Polyethylene.

Table 6. Effect of polyethylene (PE) tunnels on fruit maturation rate of four grapevine cultivars in the coastal plane of Israel

Treatment	*Fruit removed at each harvest date (percent)*						*yield/vine kg*
Cv. *Queen of Vineyard*							
	May 27	June 3	June 11	June 16	June 23	June 30	
Control	0	0	0	32	41	27	9.0
PE cover	15	66	19	-	-	-	6.1
Cv. *Perlette*							
	May 25	June 2	June 9	June 16	June 23	June 30	
Control	0	0	9	24	60	7	8.7
PE cover	27	37	22	14	-	-	9.5
Cv. *Cardinal*							
	May 31	June 6	June 16	June 21	June 29	July 8	
Control	0	0	0	33	57	10	8.3
PE cover	28	22	27	23	-	-	9.6
Cv. *Alphonse Lavallée*							
	June 24	July 3	July 10	July 17	July 24		
							11.7
Control	0	0	43	41	16	-	11.7
PE cover	27	27	41	5	-	-	8.5

continuously till fruit senescence. Thereafter, the percent of sugar continues to increase due to water loss from the fruit. When the vines are exposed to extremely high temperatures (>40°C) sugar accumulation stops and the fruit might become watery with a low TSS content (Lavee 1988a). Acidity in the fruits drops rapidly in warm climatic regions and stabilizes in the fruit on a considerable lower level than in the cooler temperate zones (Fig. 14). The relative acid composition is different in warm and cool climates. The accumulation of Malic acid in warm climates is reduced considerably more than tartarate (Winkler 1962). The change in ratio between these two acids has a major effect on the nature of fermentation during wine production and thus on wine quality. Until a few years ago, table wines from warm growing regions were of rather low quality. In the past most of the wines originating from those regions contained a considerable amount of residual sugars. Furthermore, the level of phenolic and other aromatic compounds particularly volatile ones were also considerably lower in the juice of ripe grapes originating from hot regions than from cooler ones. The accumulation of anthocyanins is also markedly reduced under high temperature conditions in spite of the more rapid maturation. Thus in many cases the coloration of the fruit is not sufficient for the quality red wine industry. However, utilization of modern temperature control fermentation as well as other advanced enological technology resulted in a marked improvement of wines produced from warm climate vineyards. By effective balancing of the fruit load in relation to vine capacity and utilization of suitable cultivars such as Cabarnet Souvignon and Merlot, adequate anthocyanin accumulation in the grapes for quality red wines was achieved. Still generally the viticulture in regions with warm bright summers as well as most of the tropical viticulture is more suitable and should be aimed towards early table grape production (Lavee 1994b).

The whole ripening process for any single cultivar is significantly shorter in the warm regions. This is due to the quicker accumulation of heat units in the summer of hot regions which is a great advantage for the production of 'early' marketable table grapes. With some table cultivars however, such as Muscat of Hamburg, reduced coloration in the warm climate might cause problems for the quality market. The reduction of anthocyanin biosynthesis is particularly expressed under the combination of high temperature and over cropping common in warm climate viticulture.

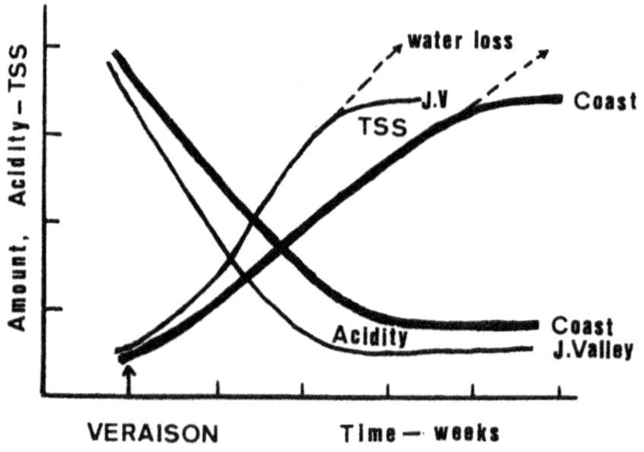

Figure 14. berries during summer in warm (Jordan valley) and cooler climates.

9 Diseases and pests in warm climates

Generally in warm and dry climates pest problems are considerably more severe while diseases less potent than in temperate zones with summer rains. In grapevines as well, pest infestation is usually higher particularly those related to the fruits. On the other hand phyloxera infestation is about the same although the biological cycle in warmer climate is usually not complete. Both Powdery (Oidium) and Downy (Plasmopora) mildews are most potent in subtropical and tropical growing regions. Botrytis was until recently less common in the hot regions but in the last years its incidence increased considerably. In general the pest and disease problems in warm climate viticulture, although somewhat different, are rather similar to those in viticulture of cooler regions. This is particularly the case in modern viticulture cultivated under intensive irrigation conditions. In the closed polyethylene structures, in spite of the high humidity particularly in the mornings, the vines are usually considerably less infected by diseases. Pest infestation is also less severe under the polyethylene cover. It has been shown that this is due to a marked reduction in the spore population on the vines under the protection of a polyethylene cover. The lower pest infestation could also be explained on basis of the physical barrier of the polyethylene. Of special interest was the effect of the polyethylene cover on dead arm (Phomopsis viticola) diseased vines. Diseased *cv.* Perlette vines covered consecutively for 3 years with polyethylene in the spring gradually overcame the disease. No new infections occurred and the yield increased continuously over that period. It has been suggested that this is due to an induced resistance developing during the forced very early spring conditions under polyethylene cover. This early growth might also cause an out-of-phase development between the life cycles of the vine and the fungus. Furthermore, the infection by the fungus occurs in the spring and requires humid conditions, while under the polyethylene during bud opening and the initial growth, the plants are protected from rain and thus the young canopy remains most of the time dry.

10 Rootstocks and cultivars

The need for rootstock and a gallery of different cultivars is similar in warm and cool region viticulture. Under both conditions phyloxera is abundant and the variety of different soil type have to be considered. The behavior of both table and wine cultivars on various rootstock in hot semidesert soils has been studied (Spiegel-Roy *et al.*, 1971a, 1971b). The rootstocks 161/49 and 140R were most efficient under such conditions when grown with full irrigation. In rootstock selection for the vineyards no specific trend based on climatic conditions can be pointed out. The consideration for the use of rootstock, in addition to philoxera resistance dependence on the soil type, lime and pH, quality of water, its availability and the cultivar - rootstock interactions. Only one rootstock, 1202, was reported to have an effect on early ripening due to an earlier bud opening of the cultivars grafted on it.

Generally the overall trend in warm regions is leaning toward the development of an early maturing table grape industry. Though late ripening cultivars are also grown mainly for the local consumption. In extremely warm or hot regions as well as in very wet tropical ones only table grapes are being developed as the quality of wines achievable in such region is low and uneconomical while high quality precocious table grapes can be achieved and receive high prices on the market. Thus cultivars chosen, and new ones which are being developed, for hot regions consist mainly of early maturing ones. In the last 15 years the use of seedless cultivars such as Perlette, Superior, Flame seedless, Thompson seedless and an array of new experimental cultivars was widely emphasized. The collection of good early cultivars and particularly seedless ones is rather limited. Thus considerable efforts had and are being devoted to the breeding of new precocious cultivars. In most centers working in or for warm regions the emphasis is put on early maturing seedless varieties. To increase the "yield" of seedlessness in the F1 progeny at least one and often two seedless parents are being utilized using a tissue culture embryo rescue technique (Spiegel-Roy *et al.*, 1985). In this way five new early, one mid, and one late season seedless varieties have been released from a long term breeding program (Perl *et al.*, 1998). It should be noted that four of the new early seedless cultivars, Spring blush, Vered, Mystery and Hod, the first two red and the other two white, produce large berries (6-8 g) even without gibberellic acid application. Still, in many locations, precocious seeded cultivars are also widely used. Presently genetic engineering methods leading to transformed improved varieties, are being developed (Perl and Eshdat, 1998) Stable transgenic grapevine plants were obtained using agrobacterium mediated genetic transformation of embryogenic cell lines derived from *cvs.* Superior seedless and Redglobe (Perl *et al.*, 1996). Currently, the major use of transformation techniques in viticulture is aimed to introduce and express resistance toward fungal diseases. Various antifungal genes were introduced into embryogenic grape cells. The use of vectors containing multiple genes is being presently tested and plants from such transformed cells are being evaluated (Perl *et al.*, 1998). Another important approach of the same group is an attempt to introduce seedlessness into favorable seeded cultivars by means of genetic engineered transformation. Embryo rescue technique is presently the major method for breeding of new seedless grapevine varieties. The presently developing biotechnological approaches might be critical in reducing the time needed for the development of new grapevine varieties.

11 References

Aubert, B. (1972). Viticulture en région tropicale pour la production de raisin de table. Aspects et possibilités. *Fruits* 27, 513-517.

Bravdo, B.A. and Hepner, Y. (1986). Water management and effect on fruit quality in grapevines. *Proceeding of the Sixth Aust. Wine Ind. Tech. Conf.* p. 150-158.

Bravdo, B.A. and Hepner, Y. (1987). Irrigation management and fertigation to optimize grape composition and vine performance. *Acta Hortic.* 206, 49-67.

Buttrose, M.S. (1969). Fruitfullness of grape vines: Effects of light intensity and temperature. *Bot.Gaz.* 130, 166-173, 173-179.

Coombe, B.G. (1959). Fruit set and development in seeded grape varieties as effected by defoliation, topping, girdling and other treatments, *Amer. J. Enol. Vitic.* 10, 85-100.

Corzo, P. (1987). Tropical Viticulture in Venezuela. *Acta Hortic.* 199, 27-29.

Erez, A. and Lavee, S. (1974). Recent advances in breaking the dormancy of deciduous fruit trees. In: *Proc. 19th Inter. Hort. Cong. Warzawa.* Vol. 3: 69-78.

Fennell A. and Hoover, E. (1991). Photoperiod influences growth, bud dormancy and cold acclimatization in vitis labruscana and V. riparia. J. *Am. Soc. Hort. Sci.* 116, 270-273.

Gur A. (1974). The grapevine. In: *Fruits of the land of Israel.* Hakibutz Hameuhad press, Tel Aviv 25-56 (in Hebrew).

Hardie, W.J. and Considine, J.A. (1976). Response of grapes to water-deficit stress in particular stages of development, *Amer. J. Enol. Vitic.* 27, 55-61.

Kirschheimer, F. (1938). Aus der Geshichte der Rebengewächse. *Wein und. Rebe* 20, 188-192.

Kliewer, W.M. and Antcliff, A. J. (1970). Influences of defoliation, leaf darkening and cluster shading on growth and composition of Sultana grapes, *Amer. J. Enol. Vitic.* 21, 26-36.

Lavee, S. (1977). The response of vine growth and bunch development to elevated winter spring and summer day temperatures, in *Int. Symp. Quality of Vintage, O.I.V.*, Cape Town, South Africa 209-226.

Lavee, S. (1980). IAA reversible growth inhibition of grapevine shoots (vitis venifera) by Maleic hydrazide compared to gibberellic acid induced growth. *Vitis*, 19, 207-215.

Lavee, S. (1982). Control of grapevine growth and development, using ethylene releasing substances and some aspects of their action. *Proc. UCD Grape and Wine Centennial Symposium* 1980. p. 126-131. Davis, Calif.

Lavee, S. (1987). Usefulness of growth regulators for controlling vine growth and improving grape quality in intensive vineyards. *Acta Hortic.* 206, 89-108.

Lavee, S. (1988). Training, growth control fruit load and pruning an integrated approach for future mechanized industrial vineyard. Proc. 2nd Inter. Seminar on Mechanical pruning of vineyards. *Revista di Ingegneria Agraria* paper 9 p. 95-109.

Lavee, S. (1988a). Quality of grapevine fruit in protected cultivation, parameters and problems. *Appl. Agr. Res.* 3, 288-292.

Lavee, S. (1990). Chemical growth regulation as a tool for controlling vineyard development and production. *Proc. 7th Australian Industry. Wine Tech Conf.* 142-149.

Lavee S. (1994a). Training in trellis systems and canopy management of table grapes in Israel. *Amer. J. Enol. Vitic.*, Special vol. Ed. J.M. Ranz 179-188.

Lavee, S. (1994b). Table grape production in Israel - past, present and trends for the future, *Amer. J. Enol. Vitic.*, Special volume Ed. J.M. Ranz 8-17.

Lavee, S. and Haskal, A. (1982). An integrated high density intensification system for table grapes. *Proc. UCD Grape and Wine Centennial Symposium* 1980. p. 390-398. Davis, Calif.

Lavee, S. and May D. (1997). Dormancy of grape vine buds - facts and speculation. *Aust. J. Grape and Wine Res.* 3, 31-46.

Lavee, S. and Rauchberger, S. (1993). The effect of cv. Perlette (Vitis vinifera) vine form on light interception, fruit temperature and its maturation. *Acta Hortic.* 349, 49-53.

Lavee, S., Erez, A. and Shulman, Y. (1977). Control of vegetative growth of grapevine (Vitis vinifera) with chloroethylphosphonic acid (Ethephon) and other growth inhibitors. *Vitis* 16, 89-96.

Lavee, S., Shulman, Y. and Nir, G. (1984). The effect of cyanamide on bud break of grapevines *Vitis vinifera. Proc. Inter. Seminar on Dormancy of grapevines. UC.* Davis P. 17-29.

Lavee, S., Ziv, M., Melamud H. and Berenstein, Z. (1993). The involvement of gibberellins in controlling bud development of grapevines (Vitis vinifera L.) *Acta Hortic.* 329, 177-182.

Lin, C.O. (1984). Viticulture in Taiwan and experiments on dormancy in: *Proc. Inter. Seminary on Dormancy of grapevines*. (Eds R.J. Weaver, J.O. Johnson and A.S. Wicks). Davis, Calif. p. 30-35.

May, P. (1965). Reducing inflorescence formation by shading individual Sultana buds. *Aust. J. Biol. Sci.* 18, 463-473.

Nir G. and Lavee, S. (1993a). The common and different responses of dormant grapevines buds to chemicals of various groups. *HortScience* 28, 571

Nir, G. and Lavee, S. (1993b). Metabolic changes during cyanamide induced dormancy release in grapevines. *Acta Hortic.* 329, 271-274.

Nir, G., Shulman, Y., Fanberstein, L. and Lavee, S. (1986). Changes in the activity of Catalase (EC 1.11.1.6) in relation to the Dormancy of Grapevine (Vitis vinifera) buds. *Plant Physiol.* 81, 1140-1142.

Nir, G., Klein, I., Lavee, S., Spieler, G., Barak, U. (1988). Improving grapevine bud break and yield by evaporative cooling. *J. Amer. Soc. Hort. Sci.* 113, 512-517.

Perl, A. and Eshdat, Y. (1998) DNA transfer and gene transfer in transgenic grapes. In: *Biotechnology and Genetic Engineering Reviews*. (ed. Tombes M.). Vol 15, 365-386.

Perl, A., Lotan, O., Abu-Abeid, M. and Holland, D. (1996). Establishment of an Agrobacterium mediated genetic transformation system for grape (Vitis vinifera L): The role of Antioxidants during grape-Agrobacterium interaction. *Nature Biotechnology* 14: 624-628.

Perl, A., Sahar, N., Spiegel-Roy, P., Gavish, S., Elyasi, R., Or, E. and Bazak, H. (1998). Conventional and Biotechnological approaches in breeding of seedless table grapes. *VIIth International Symposium on Grapewine Genetics and Breeding*, Montpellier, France.

Pouget, R. (1972). Considérations générales sur le rythme végétatif et la dormance des bourgeons de la vigne. *Vitis* 11,198-217.

Samish, R.M. (1954). Dormancy in woody plants. *Ann. Rev. Pl. Physiol.* 51, 183-203.

Shulman, Y., Nir, G. and Lavee S. (1986). Oxidative processes in bud dormancy and use of hydrogen cyanamide in breaking dormancy. *Acta Hortic.* 179, 141-148.

Shulman, Y., Nir, G., Fanberstein, L and Lavee, S. (1983). The effect of cyanamide on the release from dormancy of grapevine buds. *Scientia hortic.* 19, 97-104.

Smart, R.E. and Coombe, B.G. (1983). Water relations of grapevines, in *Water Deficits and Plant Growth*, Vol 7. (ed Kozlowski, T.T.). Academic Press, New York, 138-196.

Smart, R.E. and Robinson, M. (1991). *Sunshine into wine*. A handbook for winegrape canopy management. Winetiteles, Adelaide, pp. 88.

Spiegel-Roy, P., Kochba, J. and Lavee, S. (1971a). Performance of table grape cultivars on different rootstocks in an arid climate. *Vitis* 10, 191-200.

Spiegel-Roy, P., Kochba, J. and Lavee, S. (1971b). Quality and yield of Alicante Grenache and Semillon on various rootstocks in an arid climate. *Vitis* 11, 99-107.

Spiegel-Roy, P., Sahar, N., Baron, J. and Lavi, U. (1985). *In vitro* culture and plant formation from grape cultivars with abortive ovules and seeds. *J. Amer. Soc. Hort. Sci.* 110, 109-112.

Winkler, A. J. (1962). *General Viticulture*. University of California Press, Berkeley and Los Angeles.

13 KIWIFRUIT

GARTH S. SMITH *and* ERIC F. WALTON

The Horticulture and Food Research Institute of New Zealand
Ruakura Research Centre, Private Bag 3123, Hamilton, New Zealand

1 Introduction

The kiwifruit of commerce are large-fruited selections of Actinidia deliciosa which have been subjected to little genetic selection and are essentially indistinguishable from the plants which grow wild in their natural habitat in China (Ferguson, 1990a). There are more than 50 species and over 100 taxa in the genus Actinidia. Actinidia species can be found from cold-temperate forests to the tropics *i.e.* from latitude 50° N to the equator (Ferguson, 1990a). However, only a few taxa occur in the north, with most confined to a narrow band between latitudes 25 and 30° N. Ferguson (1990a) concluded that the complexity of varieties and forms in this zone indicates that it is the centre of current evolution. All Actinidia species are perennial climbing or straggling plants.

Kiwifruit has been known by a variety of botanical names. For a long time Actinidia chinensis Planch was accepted, but it was reclassified as Actinidia deliciosa (A. Chev.) C.F. Liang et A.R. Ferguson (Ferguson, 1985). Kiwifruit cultivars such as 'Hayward', 'Bruno', 'Monty' and 'Abbot' belong to this species. A large fruited closely related species retained the name of Actinidia chinensis, different to A. deliciosa in that it has soft epidermal hairs and smooth-skinned fruit.

Although A. chinensis and A. deliciosa are widely distributed throughout the mainland of China, they are separated geographically (Ferguson, 1990b). A. chinensis is found in the eastern, warmer regions of China, while A. deliciosa is found inland in cooler regions. Where the two species occur together, A. deliciosa is usually found at higher altitudes.

The majority of commercial plantings throughout the world are of the cultivar 'Hayward'. This cultivar was selected in New Zealand just over 50 years ago from an extraordinarily small group of about 40 seedlings, which themselves can be traced back to a small quantity of seed introduced from China in 1904 (Ferguson and Bollard 1990). The dramatic increase in the quantity of kiwifruit being produced worldwide has seen the total annual trade in fruit rise from approximately 195,000 tonnes in 1985 to the current level of production of 1.14 million tonnes (World Kiwi Fruit Review1999). New Zealand Italy and China are now the main kiwifruit producing countries.

The rapid increase in plantings of kiwifruit in many parts of the world has resulted in vines being grown under a diverse range of soils and climatic conditions. In their natural habitat in Southern and Central China, kiwifruit vines are confined to the sides of forest-covered hills and mountains where they grow in well-illuminated environments at forest margins (Ferguson, 1984). Annual rainfall in these areas is evenly distributed and

Amnon Erez (ed.), Temperate Fruit Crops in Warm Climates, 367-379.
© 2000 *Kluwer Academic Publishers. Printed in the Netherlands.*

ranges from 1045 to 1950 mm, maintaining a relative humidity of 75-85% (Li *et al.*, 1985). The climate is continental with winter temperatures often falling well below 0°C. There are long frost-free periods during the growing season of 210-280 days, while summer temperatures occasionally reach 40°C. By contrast, conditions under which kiwifruit vines are now grown commercially range from the cool humid environment of New Zealand where rainfall during the growing season often exceeds 800mm and the average temperature rarely exceeds 17°C (Smith and Buwalda, 1994), to the hot dry environment of the Central Valley of California where the average rainfall is often less than 300 mm and the average temperature usually exceeds 23°C (Walton and DeJong, 1990). Associated with the increased diversity of these growing areas has been a range of physiological disorders, of which many can be linked to environmental stresses (Smith and Buwalda, 1994).

2 Phenology and growth

2.1 Vine Development

Growth and development of the kiwifruit vine is described briefly to provide a basis for examining the physiological responses of the vine to environmental stresses associated with a warm climate. A more detailed description of vine development is given by Smith and Buwalda (1994).

In commercial orchards, a mature vine typically has two cordons trained in opposite directions from the trunk along a support structure. A number of canes are trained perpendicularly from the cordon. The current commercial winter-pruning system for kiwifruit vines in New Zealand is one of cane replacement. In this system the previous growing season's canes (principally those with shoots that bore fruit) are removed and the 'new' canes (former shoots) are retained to carry the next growing seasons crop. Approximately 25-40 axillary buds (first-order) m^{-2} of vine area are retained after winter pruning (Smith and Buwalda, 1994). Under New Zealand conditions, fewer than 50% of these axillary buds develop into shoots (Brundell, 1975), although, the proportion of buds that break can be greater in other countries such as Italy (Costa *et al.*, 1990). Only the first flush of shoots carry flowers; any later-breaking buds produce only vegetative shoots (Grant and Ryugo, 1982).

The seasonal development of the first-order buds that develop into floral shoots has been described in detail by Snowball (1995) and Walton *et al.*, (1996). Evocation (the inductive process whereby a meristem becomes committed to reproductive development through a change in developmental state, McDaniel 1994), of the developing first-order axillary buds is generally considered to occur during late summer through autumn of the season prior to inflorescence differentiation (Davison, 1990). However, more recent work suggests that evocation may start as early as the beginning of summer (Snelgar and Manson, 1992). This earlier timing is also supported by the observation of summer flowering in the closely related Actinidia species, A. eriantha Benth. (Walton, 1995). Flower differentiation in kiwifruit commences just prior to budbreak (Brundell, 1975b; Polito and Grant, 1984; Watanabe and Takahashi, 1984), up to 10 months after floral evocation.

The proportion of buds that break producing floral shoots can be modified through the use of the dormancy breaking chemical hydrogen cyanamide, which not only increases the uniformity of shoot development and compresses the bud break period (Walton, 1985; Henzell and Briscoe, 1986; Linsley-Noakes, 1989) but also increases the number of flowers per shoot (Walton and Fowke, 1993). A large increase in the concentration of proline has been measured in the buds of vines treated with hydrogen cyanamide (Walton, Clark and Boldingh, 1991). It was suggested by Walton *et al.* (1991) that the observed increase in proline placed the buds in a more favourable bio-energetic state which resulted in an increase in budbreak and flower numbers.

2.2 Effect of temperature

Kiwifruit, like many deciduous fruit species, has a winter chilling requirement (Ferguson, 1984). Estimations of the chilling requirements of kiwifruit vines to break dormancy have been based on both laboratory and field observations. A significant delay in bud break was found for vines that received less than 950-1000 hours of chilling at 4°C (Brundell, 1976; Lionakis and Schwabb, 1984). These authors also showed an increase in flower numbers per shoot with an increase in the duration of chilling during the early to mid stages but not in the later stages of dormancy. Constant temperature conditions were less effective in breaking dormancy than where there was a change in the temperature between day and night (Lionakis and Schwabb, 1984; Lawes, 1984).

Field observation initially indicated that kiwifruit vines required approximately 700-800 Richardson Chill Units (RCUs) to break dormancy (Lötter, 1984). In New Zealand, RCUs calculated for the winter period typically range from 1033 in the warmest kiwifruit growing district to 1892 units in the coolest district (Davison, 1990). Bud break was reported to be approximately 10% greater in the cooler regions than in the warmer regions (Davison, 1990). However, recent research has shown that RCUs cannot be used to reliably predict bud break and flowering in kiwifruit vines as the response of the vines to winter chilling occurs over a wide range of accumulated RCUs (McPherson, Warrington and Stanley, 1989). The main effects of increased winter chilling are an increase in the proportion of dormant buds that break in spring and a reduction in the number of developing flowers that abort.

The rate of leaf appearance (leaves per shoot per day) in spring increases approximately lineary with temperatures above 10°C to a maximum rate at 20°C, above which the rate declines (Morgan, Warrington and Halligan, 1985). The time from bud break to flowering also appears to be temperature dependent. Full bloom was shown to occur after approximately 360 growing degree days (Morley-Bunker and Salinger, 1987). Controlled environment studies using container-grown vines have shown that bud break was 16 days later when the temperature was maintained at 10°C then it was at 19°C (McPherson *et al.*,1988). The duration from bud break to flowering increased approximately 10 days for every 1°C reduction in temperature. In the same experiment, flowering was found to be spread over 34 days with a mean temperature of 12°C, but only over 13 days where the temperature was 19°C.

Temperature also influences fruit maturity. Harvest maturity is currently determined in kiwifruit by the concentration of soluble solids (°Brix) of expressed juice (Beever and Hopkirk, 1990). In New Zealand, the required minimum °Brix for harvest is 6.2 (Harman, 1981). Cool temperatures during autumn enhance the rate of increase in the °Brix level of the fruit in the field (Walton and DeJong, 1990) in controlled environments (Seager, 1991). At a mean temperature of 11°C, starch in the fruit is hydrolysed very quickly with a coincidental increase in total sugar. In contrast, at a mean temperature of 17°C there is little degradation of starch and a proportionally smaller increase in total sugars (Seager, 1991).

The seasonal pattern of new root growth has been shown to be broadly related to the annual changes in soil temperature (Buwalda and Hutton, 1988). Experiments where the temperature of the root and shoot environments were controlled independently have shown that vine growth ceased when the temperature of the root zone was maintained at 10°C, whereas maximum root growth occurred at 20°C (Smith *et al.*,1989). Root growth of Actinidia deliciosa therefore appears to have a temperature optimum lower than that for other deciduous fruiting plants. Root growth of apples, for example, has an optimum root temperature of 25°C (Voorhees, Allmoral and Johnson, 1981). Soil temperature could have a major influence on the commencement of root growth in spring, or soil temperatures in New Zealand, for example, do not exceed 10°C until late spring or early summer (Smith *et al.*, 1989). By this stage of the season the vine has undergone several major growth phases, including bud break, leaf growth, anthesis, and early fruit growth (Smith and Buwalda, 1994).

2.3 Effect of light

Measurements made throughout the season have shown that kiwifruit leaves exhibit considerable temporal variation in the irradiance response of photosynthesis (Buwalda, Meekings and Smith, 1991). The time within the growing season was found to be less important than the age of the leaves at the time of measurement. Photosynthesis increased from about 10 μmol CO_2 m^{-2} s^{-1} at 30 days after leaf emergence to a peak about 17 μmol CO_2 m^{-2} s^{-1} at 120-180 days after leaf emergence, and declined with leaf senescence at the end of the growing season. The radiation level required to saturate photosynthesis appeared to peak at about 800-900 μmol m^{-2} s^{-1} at about 90 days after leaf emergence.

The slow development of photosynthesis after leaf emergence limits whole-plant photosynthesis during canopy expansion for many species. Maximum photosynthetic capacity for any leaf is attained at or prior to full leaf expansion; *e.g.* within 15-17 days for strawberry (Jurik, Chabot and Chabot, 1979), and 11 days for bean (Catsky and Tricha, 1980). While photosynthesis generally develops more slowly for leaves of perennial plants compared to leaves of annual plants (Harper, 1988), the development of photosynthesis for leaves of kiwifruit vines is still lower than that reported for other perennial species such as cherry (Sams and Flore, 1983). Stepwise multiple regression analysis indicated that the slow development of photosynthesis for kiwifruit vines was related more closely to leaf nitrogen (per unit of leaf area) than to other lead attributes such as leaf chlorophyll and stomatal conductance (Buwalda, Meekings and Smith, 1991). In fact, the limited photosynthetic capacity in young kiwifruit leaves is physiologically similar to that usually associated with nitrogen deficiency (Field and Mooney, 1986). It is possible to affect directly the radiation saturation rate of photosynthesis (DeJong, Tombesi and Ryugo, 1984).

A canopy photosynthesis model enabling estimation of the incident radiation at any point on surfaces with complex shapes, and the distribution of photosynthetic rates within canopies with non-homogenous distribution of leaf area, has been developed (Buwalda, Curtis and Smith, 1993). For vines on T-bar trellises, net CO_2 assimilation near the cordon was greater than that near the ends of the laterals. On a clear day (*i.e.* photosynthetically active radiation <1800 μmol m^{-2} s^{-1} at midday), this difference was about 60% for vines with a high leaf area index (2.74 m^2 m^{-2}) and 85% for vines with a lower leaf area index (1.65 m^2 m^{-2}). On cloudy days photosynthesis across the canopy was more uniform, but still 10-20% greater near the centre of the vine than near the ends of the laterals. This difference in photosynthetic capacity within the canopy appears to impact directly on the developing fruit, with the greatest proportion of fruit with superior attributes being located in the denser parts of the canopy close to the cordon (Smith *et al.*, 1994).

Direct measurements of canopy photosynthesis also revealed midday declines in photosynthesis, typically after 09.00-10.00 hours on days with high radiation levels (Buwalda *et al.*, 1992). Such declines led to canopy photosynthesis rates in the afternoon of only 70-80% of those recorded at similar radiation levels in the morning. Increased respiration with increasing temperature was insufficient to account for the decline in canopy photosynthesis. Feedback inhibition of photosynthesis has been suggested as a possible contributor to this phenomenon (Buwalda *et al.*, 1992).

2.4 Effect of water

The typical diurnal pattern of stomatal conductance to water vapour for leaves of unstressed vines shows a rapid rise in early morning with little change during the day until radiation levels decline in the evening (Smith and Buwalda, 1994). However, partial stomatal closure about midday has recently been reported, which resulted in a reduction in stomata conductance by 20-30% and was associated with a reduction in leaf and canopy photosynthesis (Succi, 1992). Relatively high values for stomatal conductance have been recorded at night, especially for vines growing in advective conditions (Judd, McAneney and Trought, 1986), and are consistent with nocturnal water use measurements for whole canopies (Buwalda *et al.*,1992). As a consequence, up to 20% of the water use by vines may occur at night (Green and Clothier, 1988).

3 Environmental stress

3.1 Effects of high temperatures

Temperatures greater than 35°C have been shown to cause significant reductions in photosynthesis and excessive rates of transpiration (Tombesi, Ryugo and DeJong, 1988). For vines growing in the field, increasing the average air temperature during late winter from 12.2 to 17.1°C increased the rate of growth of the apical shoot, which in turn reduced the number of subtending shoots that produced flowers (Snelgar, Bayley and Manson, 1988). Elevated temperatures during spring, however, advanced the date of flowering. At harvest fruit from the vines grown at the higher temperature were significantly larger and had higher soluble solids concentrations than fruit from vines grown at the ambient temperature.

Root temperatures in excess of 25°C reduced both shoot and root growth (Smith *et al.*, 1989).

3.2 Effects of low temperatures

The minimum temperature tolerated by kiwifruit vines depends on the stage of growth. In spring, young growth is seriously damaged by temperatures of -1.5°C or less for durations as short as 30 minutes (Hewett and Young, 1981). In autumn, following leaf fall, temperatures as low as -5.0°C have been found to cause no obvious shoot damage or reduction in the number of buds that developed following spring, but temperatures of -7.0°C were damaging (Pyke *et al.*, 1985). By mid-winter, vines can tolerate temperatures of -7.0°C but exposure to -10.0°C for 1 hour has been shown to damage dormant canes (Hewitt and Young, 1981).

Leaves of kiwifruit vines suffer severe damage when exposed to moderate light (650 (μmol m^{-2} s^{-1}) at 10°C, including photobleaching, yellowing of the leaf margins, reduced expansion rates of the leaf, and death of the stem apices (Morgan, Warrington and Halligan, 1985). However, these symptoms were not apparent when vines were grown at lower light 280 (μmol m^{-2} s^{-1}) at 10°C suggesting that the vines were affected more from high-light stress at low temperatures than directly from low temperature stress per se. Photosynthesis has been cited as the cause of this leaf damage (Greer, 1988; Greer and Kipins, 1988). Photoinhibition has been shown to occur at all temperatures examined (5-35°C), but was greatest at low temperatures. Recovery from the effects of photoinhibition was also found to be temperature dependent with little or no recovery occurring below 20°C, but rapid recovery at 30-35°C (Greer, 1988; Greer and Liang, 1989). Visual leaf symptoms typical of high-light low-temperature stress have been observed during early spring in New Zealand (Smith and Buwalda, 1994).

3.3 Effects of reduced irradiance

For vines growing in a controlled environment where radiation was reduced from 600 to 220 (μmol m^{-2} s^{-1}), there were a marked reduction in shoot growth and leaf weight, while bud break and flowering in the following year were also reduced (Morgan, Stanley and Warrington, 1985). Overhead shading of vines to 45% of ambient radiation for three consecutive seasons reduced the average fruit weight by 14 g and the return bloom by 56% (Snelgar and Hopkirk, 1988). During cool storage, fruit from shaded vines had lower soluble solids concentrations and were only slightly less firm, than the fruit from unshaded vines. Shading the canopy during the most rapid phase of fruit growth immediately after anthesis, had a greater effect on first fruit size than shading later in the growing season (Snelgar and Hopkirk, 1988).

The optimum leaf area reduces with decreasing radiation (*e.g.* cloudy weather or shading from shelter trees or shelter cloth). Net carbon acquisition by the canopy is reduced slightly at excessive leaf area, but is reduced strongly by low radiation (Smith and Buwalda, 1994). For kiwifruit in New Zealand conditions, the optimum leaf area index appears to be about 3.0-3.5 (Buwalda, Meekings and Smith, 1991). This leaf area index is typically attained about 100 days after bud break. During canopy expansion, canopy photosynthesis is limited more by low leaf area (*i.e.* low radiation interception) than by irradiance.

3.4 Effects of drought

Measurements of leaf water potential at dawn have been found to be useful for predicting drought stress of kiwifruit vines. The degree of wilting during peak evaporative demand at midday has been related to dawn measurements of leaf water potential (Van Oostrum, 1985). Vines with leaf water potentials of less than -0.12 MPa at dawn readily wilt during peak periods of evaporative demand during sunny days, but were shown to regain tugor at night. Wilting however, has been shown to be irreversible where the leaf water potential at dawn was less than -0.65 MPa.

Drought stress reduces dry matter accumulation in all parts of the vine (Wilson, 1988). Fruit are particularly sensitive to drought stress, with an immediate reduction in growth. Relieving the vine of drought stress restores the growth rate of the fruit to that of fruit on non-stressed vines, but growth lost during the period of stress is never recovered (Prendergast *et al.*, 1987; Judd and McAneney, 1987). Although fruit size can be restricted by drought stress, beneficial effect on post-harvest storage characteristics of the fruit can result. Fruit from non-irrigated vines were found to be firmer and have greater concentration of soluble solids at harvest and during storage than fruit from irrigated vines (Brown and Brash, 1991). Storage life of fruit has also been found to be enhanced when vines were drought stressed during early summer, during the cell division phase of fruit growth. The response may be related to the unusually high accumulation of the sugar alcohol myo-inositol in kiwifruit and other Actinidia species (Klages, Smith and Bieleski, 1996). Inositol is directly involved in the production of UDP-glucuronic acid used in the synthesis of cell wall material, particularly polyglucuronic and polygalacturonic acids (Loewus, 1990).

3.5 Effects of root anoxia

Kiwifruit are not well adapted to tolerate low concentrations of oxygen in their root zone. There are a number of physiological and morphological features of kiwifruit vines which make them especially vulnerable to root anoxia. The rate of consumption of oxygen (9.1×10^{-10} mol m^{-1} s^{-1} at 20°C) by the roots is relatively high, and it has been estimated that the time taken to deplete the oxygen supply in a well aerated soil to an anaerobic condition would be approximately 5 hours assuming no replenishment of oxygen (Smith *et al.*, 1989). Generally, plant species with root porosities of less than 5% are highly sensitive to anaerobic conditions in the soil (Justin and Armstrong, 1987). The average air space between the cells of kiwifruit roots accounts for 2% or less of the total root volume (Smith *et al.*, 1989). Unlike plant species more tolerant of root anoxia kiwifruit roots do not develop aerenchyma tissue in response to decreasing oxygen concentrations in the root zone (Smith and Miller, 1991a; Smith *et al.*, 1989). Aerenchyma formation has been shown to be triggered by enhanced internal concentrations of ethylene (Drew and Jackson, 1979). Anoxia is known to cause a stimulation of the ethylene precursor, 1-amino-cyclopropene-1-carboxylic acid (ACC) in the roots of many plants (Bradford and Yang, 1981). A keysubstrate for ACC production is methionine (Yang and Hoffman, 1984) and it is interesting to speculate that the absence of free methionine in the tissue of kiwifruit vines (Clark and Smith, 1991) may be responsible in part for the lack of formation of aecrenchyma in the roots. The recent failure to find a clone for the key enzyme, ACC synthase, for ethylene production (Whittaker, Smith and Gardiner, 1995) adds further weight to this supposition.

Field studies have shown that the damaging effects of root anoxia on the leaves of kiwifruit vines were greater in early summer when the ambient temperature was higher and the evaporative demand was greater than in late summer or early autumn (Smith and Miller, 1991a). Naturally low concentrations of oxygen in the soil during mid-summer following increased consumption of oxygen by soil micro-organisms as a result of the higher temperatures in the soil (Glinski and Stepnicwski, 1985), are also likely to have contributed to the susceptibility of the vine to root anoxia during this period. By contrast with the effects on the leaves, a greater percentage of fruit was found to be soft at harvest on those vines that had been stressed during late summer or early autumn than on non-stressed vines or those that had been stressed for a similar length of time in early summer (Smith and Miller, 1991). The nitrogen status of the vines was also shown to influence the extent to which root anoxia damaged the vine and the speed with which the vine recovered from the stress. Vines that were nitrogen deficient were much more susceptible to root anoxia than vines that had received adequate quantities of nitrogen fertilizer (Smith and Miller, 1991).

The reported death of over 30,000 vines in New Zealand as a result of excessive rainfall (six times the seasonal average) during early summer, graphically demonstrates the sensitivity of kiwifruit vines to root anoxia (McAneney *et al.*, 1989). Large numbers of vines in Italy have also died due to high water tables (0-0.5 m) in winter (Magliulo *et al.*, 1991)

3.6 Effects of excess salts in the root zone

Kiwifruit vines are generally tolerant of relatively high concentrations of soluble salts in the root zone, with the exception of sodium ions (Smith, Asher and Clark, 1987). Severe damage to kiwifruit vines has resulted from the use of irrigation water containing sodium concentrations in excess of 230 mg l^{-1} (Smith, Asher and Clark, 1987). Suggestions that kiwifruit vines are sensitive to chlorine toxicity (Limmer, 1988) have not been supported by direct experimentation. An unusual feature of kiwifruit vines is their relatively high requirement for chlorine. To maintain healthy growth they require at least 2-6 g kg^{-1} dry weight of chlorine in their leaves (Smith, Clark and Holland, 1987; Smith *et al.*, 1988). These levels are at least 10 times higher than those required by other non-halophytic plants (Flowers, 1988). Where in excess of 1.2 tonnes per hectare of chlorine has been applied in one application in spring to kiwifruit vines in the field, no adverse effects were observed on the vines despite the concentration of chloride in the soil being up to 10-15 times those usually found in commercial kiwifruit orchards in New Zealand (Smith and Miller, 1991b). A recent comparison of potassium fertilizer applied with chloride or sulphate as the accompanying anion showed fruit yields to be approximately 28% greater for vines receiving potassium chloride rather than potassium sulphate (Buwalda and Smith, 1991).

When exceptionally large quantities of soluble fertilizer (up to 10 tonnes per hectare of a 12:10:10 N:P:K mixture) were used to increase the soluble salt levels in the root zone to high levels (the electrical conductivity of the soil increases to over 2 dSm at 25°C), there was a marked increase in the number of buds that developed in spring along with flower numbers, and root growth (Smith and Miller, 1991b). Measurements at harvest showed fruit yields were doubled for the high fertilizer treatment compared to the control vines which received typical annual input of fertilizer. The results of this study suggest that osmotic effects may play an important part in the bud-breaking process in spring.

4 Conclusion

Growing kiwifruit in warm climates can pose major constraints to productivity. While the gross effects of stresses associated with warm climates are relatively easy to identify and describe, it is the subtle effects of these stresses that are less well understood. Frequently these effects are overlooked and often confused with natural variation in the vine itself. As the history of research into the physiology of the kiwifruit vine is very recent compared to that for most other fruit crops, much has still to be documented in relation to the impacts of these environmental stresses on growth and development.

From existing knowledge the following practices are options for maximising productivity of kiwifruit vines in warm climates:

Ensuring a uniform bud break in spring through the selection of cultivars with low chill requirements and using bud breaking chemicals such as hydrogen cyanamide.

Reducing the dependence of the vine on irrigation by encouraging the growth of a large root system. The greater volume of soil explored by a large root system means that the transpiration requirements of the vine can be met at a greater soil moisture tension because of the overall increase in the quantity of water available to the vine (Buwalda and Smith, 1990). Because the root system of kiwifruit vines is a poor competitor for resources of growth, indirect effects of limited supplies of these resources increase the sensitivity of the vine to below ground stresses such as water and mineral nutrient deficiencies. Maintaining a leaf area index of 3.0-3.5 m^2 m^{-2} will ensure that a large root system is maintained (Smith and Buwalda, 1994).

Avoiding the risk of root anoxia and the adverse effects which excess water has on the storage properties of the fruit by refraining from flood irrigation on the application of large quantities of water after infrequent intervals to slowly draining soil types.

Ensuring that irrigation water contains low concentrations of sodium and boron (Smith, Buwalda and Clark, 1988).

Selecting rootstock that are better adapted to local environments. Kiwifruit production has traditionally involved "Hayward" grafted on to seedling rootstocks or "Hayward" grown on its own roots (Sale, 1990). Neither root system confers any particular advantage to the 'Hayward' scion. However, the recent selection of the clonal rootstock of Actinidia hemsleyanna ("Kaimai") for "Hayward" (Lowe *et al.*, 1992) has shown what can be achieved in terms of improvements in productivity and fruit quality.

5 References

Beever, D.J. and Hopkirk, G. (1990). Fruit development and fruit physiology, in *Kiwifruit Science and Management*, (eds I.J. Warrington and G.C. Weston), Richards Publisher, Auckland, pp. 97-126.

Bradford, K.J. and Yang, S.F. (1981). Physiological responses of plants to waterlogging. *HortScience*, 16, 25-30.

Brown, N.S. and Brash, D.W. (1991). Effect of various irrigation regimes on kiwifruit storage quality, The Second International *Symposium on Kiwifruit*, Massey University, Palmerston North, New Zealand, 127. (abstract)

Brundell, D.J. (1975). Flower development of the Chinese gooseberry (Actinidia chinensis Planch.). I. The development of the flowering shoot. *New Zealand Journal of Botany*, 13, 473-83.

Brundell, D.J. (1976). The effect of chilling on the termination of rest and flower bud development of the Chinese gooseberry. *Scientia Horticulturae*, 4, 175-82.

Buwalda, J.G., Green, T.G.A. Meekings, J.S. (1992). Measurement of canopy gas exchange of kiwifruit vines using a suite of whole-canopy cuvettes. *Environmental and Experimental Botany*, 32, 425-38.

Buwalda, J.G. and Hutton, R.C. (1988). Seasonal changes in root growth of kiwifruit. *Scientia Horticulturae*, 36, 251-60.

Buwalda, J.G., Meekings, J.S. and Smith, G.S. (1991). Seasonal changes in photosynthetic capacity of leaves of kiwifruit (Actinidia deliciosa) vines. *Physiologia Plantarum*, 83, 93-8.

Buwalda, J.G. and Smith, G.S. (1990). Acquisition and utilisation of carbon, mineral nutrients, and water by the kiwifruit vine. *Horticultural Reviews*, 12, 307-47.

Buwalda, J.G. and Smith, G.S. (1991). Influences of anions on the potassium status and productivity of kiwifruit (Actinidia deliciosa) vines. *Plant and Soil*, 133, 209-18.

Clark, C.J. and Smith, G.S. (1991). Seasonal variation of nitrogenous compounds in components of the kiwifruit vine. *Annals of Botany*, 68, 441-50.

Costa, G., Succi, F. and Biasi, R. (1990). Vegetative and cropping performance of kiwifruit (*cv.* Hayward) as related to pruning length and bud number. *Acta Horticulturae*, 282, 113-8.

Davison, R.M. (1990). The physiology of the kiwifruit vine, in Kiwifruit Science and Management (eds I.J. Warrington and G.C. Weston), *Richards Publisher*, Auckland, pp. 127-54.

DeJong, T.M., Tombesi, A. and Ryugo, K. (1984). Photosynthetic efficiency of kiwi (Actinidia chinensis Planch.) in response to nitrogen deficiency, *Photosynthetica* 18, 139-45.

Drew, M.C., Jackson, M.B. and Giffard, S. (1979). Ethylene-promoted adventitious rooting and development of cortical air spaces (aerenchyma) in roots may be adaptive responses to flooding, in Zea mays L, *Planta*, 147, 83-8.

Ferguson, A.R. (1984). Kiwifruit: a botanical review. *Horticulturae Reviews*, 6, 1-64.

Ferguson, A.R. (1990a). The genus Actinidia. In *Kiwifruit Science and Management*, (eds I.J. Warrington and G.C. Weston), Richards Publisher, Auckland, 15-35.

Ferguson, A.R. (1990b). Botanical nomenclature: Actinidia chinensis, Actinidia deliciosa and Actinidia setosa, in *Kiwifruit Science and Management.* (eds I.J. Warrington and G.C. Weston) Richards Publisher, Auckland, pp. 36-57.

Ferguson, A.R. and Bollard, E.G. (1990). Domestication of the kiwifruit, in *Kiwifruit Science and Management.* (eds I.J. Warrington and G.C. Weston), Richards Publisher, Auckland, pp. 165-246.

Field, C.B. and Mooney, H.A. (1986). The *photosynthesis*-nitrogen relationship in wild plants, in *On the Economy of Plant Form and Function*, (ed. T.J. Givnish), Cambridge University Press, Cambridge, pp. 25-55.

Flowers, T.J. (1988). Chloride as a nutrient and osmoticum. *Advances in Plant Nutrition*, 3, 55-65.

Glinski, J. and Stepniewski, W. (1985). *Soil Aeration and Its Role for Plants*, CRC Press, Florida.

Green, S.R. and Clothier, B.E. (1998). Water use of kiwifruit vines and apple trees by the heat-pulse technique. *Journal of Experimental Botany*, 39, 115-23.

Greer, D.H. (1988). Effect of temperature of photoinhibition and recovery in Actinidia deliciosa. *Australian Journal of Plant Physiology*, 15, 195-205.

Greer, D.H. and Laing, W.A. (1989). Photoinhibition of photosynthesis in intact kiwifruit (Actinidia deliciosa) leaves: Effects of growth temperature on photoinhibition and recovery. *Planta*, 180, 32-9.

Greer, D.H., Laing, W.A. and Kipnis, T. (1988). Photoinhibition of photosynthesis in intact kiwifruit (Actinidia deliciosa) leaves: Effect of temperature. *Planta*, 174, 152-8.

Harman, J.E. (1981). Kiwifruit Maturity. *The Orchardist of New Zealand*, 54, 126-130.

Henzell, R. F. and Briscoe, M.R. (1986). Hydrogen cyanamide: a tool for consistently high kiwifruit production. *New Zealand Kiwifruit Special Publication*, No. 1, pp. 8-11.

Hewett, E.W. and Young, K. (1981). Critical freeze damage temperatures of flower buds of kiwifruit (Actinidia chinensis Planch.). *New Zealand Journal of Agriculture Research*, 24, 73-5.

Judd, M.J. and McAneney, K.J. (1987). Economic analysis of kiwifruit irrigation in a humid climate. *Advances in Irrigation*, 4, 307-30.

Judd, M.J., McAneney, K.J. and Trought, M.C.T. (1986). Water use by sheltered kiwifruit vines under advective conditions. *New Zealand Journal of Agricultural Research*, 29, 83-92.

Jurik, T.W., Chabot, J.F. and Chabot, B.F. (1979). Ontogeny of photosynthetic performance in Fragaria virginiana under changing light regimes. *Plant Physiology*, 63, 542-7.

Justin, S.H.F.W. and Armstrong, W. (1987). The anatomical characteristics of roots and plant response to soil flooding. *The New Phytologist*, 106, 465-95.

Klages, K., Smith, G. and Beileski, R. (1996). myo-Inositol is a major carbohydrate in species of Actinidia. *Acta Horticulture*, 444, 361-7.

Lawes, G.S. (1984). Winter temperatures and kiwifruit bud development. *The Orchardist of New Zealand*, 57, 100-10.

Li, R., Huang, C., Liang, M. and Huang, Z. (1985). Investigation of germplasm resources of Actinidia in Guangxi. *Guihaia* 5, 253-67.

Linsley-Noakes, G.C. (1989). Improving flowering of kiwifruit in climatically marginal areas using hydrogen cyanamide. *Scientia Horticulture*, 38, 247-59.

Lionakis, S.M. and Schwabes, W.W. (1984). Some effects of daylength, temperature and exogenous growth regulator application on the growth of Actinidia chinensis Planch. *Annals of Botany*, 54, 485-501.

Lötter, J. de V. (1984). An evaluation of the climatic suitability of various areas in Southern Africa for commercial production of Hayward kiwifruit (Actinidia chinensis). *Deciduous Fruit Grower*, 34, 122-300.

Lowe, R.G., Wang, Z.Y. and Patterson, K.J. (1992). Rootstocks - enhancing the vine's potential, *New Zealand*

Kiwifruit Special Publication, No.4, pp. 33-6.

McAneney, K.J., Clough, A. and Green, A. (1989). Waterlogging and vine death at Kerikeri. *New Zealand Kiwifruit*, No.56, 15.

McPherson, H., Stanley, J. and Warrington, I. (1988). Dynamics of budbreak and flowering. *New Zealand Kiwifruit Special Publication*, No.2., pp. 9-11.

McPherson, H., Warrington, I.J. and Stanley, J. (1989). Winter and spring temperature effects on budbreak and flowering, Kiwifruit Cropping with Hi-Cane. *Nufarm and MAFTech North*, pp. 5-9.

Maglinlo, V., d'Andria, R. and Gioriom P. (1991). Cumulative effects of winter shallow water-table conditions on stem growth and yield of kiwifruit. *New Zealand Journal of Crop and Horticultural Science*, 19, 375-80.

Manson, P.J., Snelgar, W.P. and Seelye, R.J. (1991). Relationship between timing of cane tipping and yield of kiwifruit (Actinidia deliciosa). *Journal of Horticultural Science*, 66, 415-22.

Morgan, D.C., Stanley, C.J. and Warrington, I.J. (1985). The effects of simulated daylight and shade-light on vegetative and reproductive growth in kiwifruit and grapevine. *Journal of Horticultural Science*, 60, 473-84.

Morgan, D.C., Warrington, I.J. and Halligan, E.A. (1985). Effect of temperature and photosynthetic photon flux density on vegetative growth of kiwifruit (Actinidia chinensis). *New Zealand Journal of Agricultural Research*, 28, 109-16.

Morley-Bunker, M.J. and Salinger, M.J. (1987). Kiwifruit development, the effect of temperature on bud burst and flowering. *Weather and Climate*, 7, 26-30.

Polito, V.S. and Grant, J.A. (1984). Initiation and development of pistillate flowers in Actinidia chinesis. *Scienta Horticulturae*, 22, 365-71.

Prendergast, P., McAneney, K.J. and Astill, M.S. (1987). Water extraction and fruit expansion by kiwifruit. *New Zealand Journal of Experimental Agriculture*, 15, 345-50.

Sale, P.R. (1990). *Kiwifruit Growing*, GP Books, Wellington. 84.

Sams, C.E. and Flore, J.A. (1983). Net photosynthetic rate of sour cherry (Prunus cerasus L. "Montmorency') during the growing season, with particular reference to fruiting. *Photosynthesis Research*, 4, 307-16

Smith, G.S., Asher, C.J. and Clark, C.J. (1987). *Kiwifruit Nutrition: Diagnosis of Nutritional Disorders.* Ag Press Communications, Wellington.

Smith, G.S., Buwalda, J.G. and Clark, C.J. (1988). Nutrient dynamics of a kiwifruit ecosystem. *Scientia Horticulture*, 37, 87-109.

Smith, G.S., Clark, C.J. and Holland, P.T. (1987). Chlorine requirement of kiwifruit (Actinidia deliciosa). *New Phytologist*, 106, 71-80.

Smith, G.S., Clark, C.J. and Buwalda, J.G. (1988). Influence of light and form of nitrogen on chlorine requirement of kiwifruit. *The New Phytologist*, 100, 5-12.

Smith, G.S., Judd, M.J. and Miller, S.A. (1990). Recovery of kiwifruit vines from transient waterlogging of the root system. *New Phytologist*, 115, 325-33.

Smith, G.S. and Miller, S.A. (1991a). Effects of root anoxia on the physiology of kiwifruit vines. *Acta Horticulturae*, 297, 401-7.

Smith, G.S. and Miller, S.A. (1991b). Osmotic effects on vine performance and fruit quality, *Acta Horticulturae*, 297, 331-6.

Snelgar, W.P, Bayley, G.S. and Manson, P.J. (1988). Temperature studies on kiwifruit vines using relocatable greenhouses. *New Zealand Journal of Experimental Agriculture*, 16, 329-39.

Snelgar, W.P. and Hopkirk, G. (1988). Effect of overhead shading on yield and fruit quality of kiwifruit (Actinidia deliciosa). *Journal of Horticultural Science*, 63, 731-42.

Succi, F. (1992). Studio di fotosintesi e transpirazione in Actinidia deliciosa nel corso della stagione vegetativa, *PhD Thesis*, Università degli Studi di Bologna, Bologna.

Tombesi, A., Ryugo, K. and De Jong, T.M. (1988). L'influenza della intensita di illuminazione, della temperature e della concentrazione di CO_2 sulla fotosinteri de Actinidia deliciosa (Liang e Ferguson). *Rivista di Frutticoltura* - N, 11, 63-8.

Voorhees, W.B., Allmaras, R.R. and Johnson, C.E. (1981). Alleviating temperature stress, in *Modifying the Root Environment to Reduce Crop Stress*, (eds Arkin, G.G. and Taylor, H.M.), American Society of Agriculture Engineers, Michigan, pp. 217-66.

Walton, E.F. (1985). Winter-chilling requirements of kiwifruit, in 1983/84 *Agriculture Research Division Annual Report*, Ministry of Agriculture and Fisheries, Wellington, New Zealand, p. 30.

Walton, E.F. and DeJong, T.M. (1990). Growth and compositional changes in kiwifruit berries from three Californian locations. *Annals of Botany*, 66, 285-98.

Walton, E.F., Clark, C.J. and Boldingh, H.L. (1991). Effect of hydrogen cyanamide on the amino acid profiles of kiwifruit buds during dormancy. *Plant Physiology*, 97, 1256-9.

Walton, E.F. and Fowke, P.J. (1993). Effect of hydrogen cyanamide on kiwifruit shoot flower number and position. *Journal of Horticultural Science*, 68, 529-34.

Walton, E.F. (1995). Occurrence of secondary floral shoots on Actinidia eriantha Benth. during summer. *New Zealand Journal of Crop and Horticultural Science*, 23, 341-343.

Walton, E.F., Fowke, P.J. and Weis, K. (1997). Shoot axillary bud morphogenesis in kiwifruit (Actinidia deliciosa). *Annals of Botany*, 80,13-21.

Wilson, K.S. (1988). Water and energy relationships of kiwifruit, *MSc Thesis*, University of Waikato, Hamilton.

World Kiwi Fruit Reviews (1999) Belrose Inc. Pullman, Wa, USA.

Yang, S.F. and Hoffman, M.E. (1984). Ethylene biosynthesis and its regulation in higher plants. *Annual Review of Plant Physiology*, 35, 155-89.

14 PECAN IN WARM CLIMATE

DARRELL SPARKS

Department of Horticulture, University of Georgia
Athens, GA 30602-7273

1 Introduction

Pecan [*Carya illinoensis* (Wangenh. C. Koch)] production can be divided into two categories, production in dry and production in humid climates. Production in dry climates can be subdivided into production in low elevation climates and production in high elevation climates. The major pecan growing areas with a dry, low (DL) elevation climate are the Crystal City-Eagle Pass area of Texas, the Red Rock-Picacho area of Arizona, the Fresno-Visalia area of California, and the Hermosillo, Sonora and eastern Coahuila areas of Mexico. These areas are <460 m in elevation. The major pecan areas with a dry, high (DH) elevation climate are the El Paso-Las Cruces area of Texas-New Mexico, the Tucson area of Arizona, and the Chihuahua area of Chihuahua, Mexico. Elevation varies from ≈730 to 1220 m. Georgia is the major producer of pecans in humid areas with ≈40% of the United States production (Sparks, 2000). Native pecan groves (pecan forests from which other tree species have been removed) make up most of the total pecan production in some states, for example, Texas and Oklahoma. Because of the high water requirement of pecan (Miyamoto *et al.*, 1995), native pecans are primarily in humid areas although groves occur in semi-arid areas subject to overflow from floods.

2 Climate

The truly warm areas of pecan production are Tucson and Red Rock, Arizona and Crystal City, Texas areas (Table 1) with growing degree-days that are at least one third higher than other pecan regions. Similarly warm areas are Eagle Pass, Texas (west of Crystal City), and eastern Coahuila and Hermosillo, Sonora, Mexico. These areas are warm because their growing season is longer and day and night temperatures are substantial higher than that of other pecan areas. Tucson and Red Rock are arid and Crystal City is semi-arid. Las Cruces and Visalia are arid but, as for growing degree-days, these areas are only about one half as warm as Tucson, Red Rock, and Crystal City. Regardless, these areas have problems in common with the warmer Tucson, Red Rock, and Crystal City regions and, therefore, are considered in the same category for this paper.

Growing degree days are higher in the humid Albany, Georgia area than in desert areas of Las Cruces, New Mexico and Visalia, California (Table 1). However, because

Amnon Erez (ed.), Temperate Fruit Crops in Warm Climates, 381-403.
© 2000 *Kluwer Academic Publishers. Printed in the Netherlands.*

Georgia and adjacent pecan states are humid, the region generally does not have the same production problems as warm, dry climates. Thus, in this discussion, pecans in humid climates are excluded. Regions with native pecan groves, for example, Chetopa, Kansas, are also excluded because they are primarily in humid areas. Consequently, the discriminating climatic characteristic used for classifying pecan climates is humid versus arid. This climatic characteristic is either directly or indirectly related to very different soil and water management problems, nut quality problems, diseases, and, to some extent, insects.

Tucson, Red Rock, and Crystal City areas have higher December-February temperatures (Table 1) than other areas which result in low chilling. Staggered budbreak (Sparks, 1993b), characteristic of these mild winter areas, can create pollination problems. Additionally, staggered budbreak results in staggered nut maturity which requires either delayed or multiple harvests. Low chilling is also a problem in Visalia, California but it is a minor problem in the cool winters of Georgia and chilling is maximized in the cold winter areas, such as Chetopa, Kansas and Las Cruces, New Mexico. Pecan budbreak is under the interactive control of heating and chilling (Sparks, 1993a). Heat required for budbreak varies inversely with chill accumulation. Variability in budbreak increases abruptly when there are fewer than ¿ 100 chilling degree-days (base 3.9°C).

Standard pecan cultivars need about 190 to 200 frost-free days for successful production (Sparks, 1992). Thus, Las Cruces and similar high elevation areas are marginal for commercial pecan production because of the minimal growing season (Table 1), the length of which varies greatly among years. Late spring and early fall freezes are perennial threats to production (Sparks, 2000).

Pecan production in DH elevation climates in the United States is well established. Consequently, many of the production problems have been resolved. In contrast, production in DL elevation climates is comparatively young with a few orchards about 30-years old but most are much younger. Many production problems have been encountered (Sparks, 1993b) which have had to be resolved. During the interim, some orchards have not been profitable or only marginally profitable.

2.1 Implication of climate to marketing niche

The early spring associated with DL elevation climates is a major advantage because pecans grown in these areas mature their fruit early. For example, nut maturity in Crystal City, Texas is 2 to 4 weeks earlier than that of the same cultivar grown in Albany, Georgia. The nut matures earlier in a warm climate because spring occurs earlier than in cooler climates (March-February temperatures, Table 1) and higher springtime temperatures decrease the time until nut maturity (Sparks, 1989b). Nuts mature earlier in the Crystal City-Eagle Pass region than in any other region of the United States. Nut maturity is similarly early in DL elevation areas of Mexico and slightly later in Red Rock, Arizona. Early nut maturity of pecans grown in DL elevations is a major marketing advantage because the nuts sell on the early market. This market is characteristically the highest priced in the US due to the demand from the gift, Thanksgiving, and Christmas trade (Sparks, 1992). In contrast, nut maturity in DH elevation climates, for example, Las Cruces, New Mexico is too late for the profitable

Table 1. Climatic characteristics of geographical areas of pecan production

Geographical area	*Elevation (m)*	*Growing season (frost-free days)*[a]	*Temperature C (max./min.)* Dec.-Feb.	Mar.-Apr.	May-Aug.	Sept.- Oct.	*Growing degree days*[b]	*Annual rainfall (cm)*[a]
Chetopa, Kan.	274	195	7.2/-3.9	17.8/ 4.4	30.0/17.8	24.4/11.1	904	108
Cruces, N. M.	1183	215	15.0/-1.7	22.8/ 4.4	32.8/16.1	27.8/11.1	969	26
Visalia, Calif.	101	312	13.9/ 3.9	21.7/ 8.3	32.2/16.1	28.9/13.3	987	27
Albany, Ga.	63	228	16.7/ 1.7	28.9/ 7.7	32.2/18.9	28.9/15.0	1251	133
Tucson, Ariz.	753	334	19.4/ 6.1	26.1/11.1	36.1/22.2	32.2/18.9	1918	31
Red Rock, Ariz	431	262	20.6/ 3.3	27.8/ 7.2	40.0/19.4	35.0/15.5	1938	26
Crystal City, Tex.	177	293	19.4/ 6.7	27.2/14.4	34.4/22.7	31.1/19.4	1908	54

[a] Data for 1975-1993.

[b] Data for 1980- 1993. Growing degree-days were calculated from the daily mean temperature determined by averaging the maximum and minimum for the day minus base temperature 18.3°C. Calculation of growing degree-days was the sum of the degrees from 1 Apr. to 31 September.

early market. Whereas, nut maturity is typically mid-September in Crystal City, it is normally mid-November or later in Las Cruces (Herrera, 1995a). Nut maturity in Visalia, California and in the DH elevation area of Tucson, Arizona is intermediate to maturity in Crystal City and Las Cruces.

2.2 Effect on vegetative growth

Tree growth is typically very vigorous in DL elevation climates of Visalia, California and Crystal City, Texas in contrast to decidedly less vigorous growth in the DH elevation area of Las Cruces. Saline/sodic soils, which are a problem in Las Cruces and some areas of Mexico but are minor in most other pecan areas, greatly suppress tree growth (Miyamoto and Gobran, 1983). Also, difference in night temperatures (Table 1) and length of growing seasons may contribute to differential growth. Rapid tree growth is advantageous as it results in the tree coming into commercial production in minimal time (Ware and Johnson, 1957). In DH elevation areas with slower tree growth, delay in commercial production is partially offset by high density planting (9.1 x 9.1 m). Orchards planted in soils especially high in Salinity may be planted 4.6 x 9.1 m. In contrast, the typical planting in the DL elevation area of Crystal City area is 15.2 x 15.2 m or 12.2 x 18.3 m, both on the diamond. Although tree spacing is greater in Crystal City, tree thinning is needed earlier, 10th-12th year, in the life of the orchard than in Las Cruces, 14th-20th year, (Herrera. 1995b; Miyamoto and Storey, 1995).

3 Cultivars

'Western Schley' and 'Wichita', are the main cultivars planted in the warm, dry climates of the United States and Mexico. 'Western Schley' is the predominant cultivar in DH elevation areas. Dean Stahmann Sr., who pioneered the pecan industry in Las Cruces, New Mexico, made a large planting of 'Western Schley' beginning in 1934 (Kilby and Herrera, 1997). Subsequent plantings by other growers were 'Western Schley' simply because it was successful on Stahmann Farms. 'Wichita' is not as well suited to DH elevation climates because of its susceptibility to winter injury (Sparks, 1992). In Arizona, where elevation is less and winters are not as severe as Las Cruces (Table 1), 'Wichita' is more widely planted (Herrera, 1995a). 'Wichita' is extensively planted in DL elevation climates. 'Western Schley' is the second leading cultivar. 'Desirable', 'Choctaw', 'Cheyenne', and 'Pawnee' are planted to a limited extent. As indicated later, neither 'Western Schley' nor 'Wichita' is well suited to DL elevation areas.

3.1 'Western Schley'

Budbreak is relatively late (Sparks, 1992) which reduces the chance of late spring freeze damage. The tree has good resistance to freeze injury (Goff and Tyson, 1991; Sparks, 1992). Thus, this cultivar is particularly suited to DH elevation climates, such as Las Cruces, in which late and early fall freezes are perennial problems (Sparks, 1997) associated with a short growing season (Table 1).

'Western Schley' is precocious and prolific and the tree can be in commercial production within five years from planting. The fruit is borne on shoots from short branches along the length of long, slender branches; therefore, 'Western Schley' is said to be a lateral bearer. Fruiting shoots are maintained along the branches because the canopy is open. 'Western Schley's open tree canopy and fruiting pattern make the tree particularly suited to hedging (Fig. 1). Unless fruit load is controlled by hedging or fruit thinning, the tree is a severe alternate bearer. Well managed 'Western Schley' orchards average ≈ 1700 to 2200 kilograms of nuts per hectare.

Figure 1. 'Western Schley' trees hedged to maintain tree size and nut quality. Typical regrowth is on an adjacent row.

Nut size is about 141 nuts per kilogram and is too small to sell well in the early market and, therefore, is not desirable for DL elevation climates. Percentage kernel is high with a maximum of about 60. On excessive fruiting trees, kernel percentage can decrease to the level where the kernel is unmarketable (Sparks, 1992). Kernel quality is further decreased by shuck decline (premature death of the shuck) (Sparks *et al.*, 1995), a major problem on extremely heavy fruiting trees. Premature germination (vivipary) can also be a massive problem in DL elevation climates.

The nut is not a good cracker. The packing tissue tends to stick in the deep kernel grooves and the kernel often breaks during cracking. Kernel color is about average (Sparks, 1992).

'Western Schley' is susceptible to many insects and diseases (Sparks, 1992). Those that are problems in DL and DH elevation areas are discussed in section 4.
If fruiting is extremely excessive, the tree dies back during the following growing season and, in extreme cases, may die (Sparks, 1993b). Dieback has been especially severe in the Crystal City - Eagle Pass area. 'Western Schley' is more tolerant to low leaf Zn than some other cultivars (Gray, 1974b). 'Western Schley' is much more salt tolerant than 'Wichita' and, consequently, does better than 'Wichita' in the saline/sodic soils of El Paso - Las Cruces.

Overall, 'Western Schley' is not an ideal cultivar. However, 'Western Schley's high yielding ability, suitability to hedging, late budbreaking habit, resistance to winter injury, and tolerance to saline/*sodic soils* makes it a very suitable cultivar to DH elevation climates.

3.2 'Wichita'

'Wichita' budbreak is earlier than 'Western Schley' (Sparks, 1992) and the difference is accentuated (Fig. 2) following exceptionally mild winters. Early budbreak increases the chance for damage from late spring freezes in DH elevation climates with a marginal growing season. Additionally, the tree is susceptible to freeze damage as such (Sparks and Payne, 1978: Smith and Couch, 1984). Thus, 'Wichita' is not well suited to the marginal growing season of DH elevations (Sullivan and Bradford, 1978). At DL elevations, early budbreak is not a risk and instead is an advantage because of the emphasis on early nut maturity.

Figure 2. Delayed budbreak in 'Western Schley' (right) compared to 'Wichita' (left). The trees were growing in a region (Crystal City, Tex.) with extremely mild winters. In this region, the delayed budbreak of 'Western Schley' (due to insufficient chilling) makes 'Western Schley' a poor pollinizer for 'Wichita'.

'Wichita' is precocious and prolific and the tree can be in commercial production within five years. 'Wichita', in contrast to 'Western Schley', has a moderate dense canopy and is a terminal bearer, consequently, the tree is not suited to mechanical hedging. When hedged, regrowth is dense and the interior of the tree shades out, thus reducing potential production (Sparks, 1992). Unless fruit is mechanically thinned, the tree is a severe alternate bearer. Typical nut production is ≈1700 to 2200 kilograms per hectare in well-managed orchards.

Nut maturity is early and, in the Crystal City-Eagle Pass region, is 15 to 20 days before the standard 'Stuart' (Sparks, 1992). Early nut maturity is especially advantageous because the attractive kernel sells very well in the early market. lnherently, nut maturity is staggered in 'Wichita' which is accentuated by the low chilling winters of DL elevation climates (Sparks, 1993a).

Nut size is about 125 per kilogram and a good kernel percentage is 60. When fruit set is extreme, kernel development can be very poor. Poor kernel quality is greatly accentuated by shuck decline which is massive on excessive fruiting trees (Sparks *et al.*, 1995). Premature germination (Fig. 4) is a major problem in DL elevation climates. The kernel is tightly packed in the shell. As a result, the percentage of intact halves is usually less than 90 following mechanical cracking (Sparks, 1992). Kernel color is above average (Sparks, 1992) but with increasing time within the shuck, the color progresses to an unusual bronze. Consequently, nuts must be harvested early for good kernel color and to obtain premium prices.

'Wichita' is susceptible to many insects and diseases (Sparks, 1992). Those that are problems in DL and DH elevation climates are discussed in section 4.

Following a year of excessive fruiting, the tree dies back (Fig. 5). In severe cases, the tree dies. Tree dieback is much worse in 'Wichita' than 'Western Schley' and, in the Crystal City - Eagle Pass area, dieback is a major problem (Sparks, 1993b). The tree needs careful training to reduce breakage. Even then, breakage can be massive during a heavy crop year. Compared with 'Western Schley', 'Wichita' is extremely susceptible to Zn deficiency. 'Wichita' is more sensitive to excess Cl and Na in the soil than 'Western Schley'.

'Wichita' is not well suited to warm, dry climates. At DH elevations, winter injury is a threat. In DL elevation regions, premature germination can be and is often a disastrous problem. Universally, over production and associated poor kernel quality, shuck decline, and tree dieback are problems. The tree is not suited to conventional hedging which limits control over tree size and excessive production.

3.3 Other Cultivars

No other cultivar has proven to be more suited to DH elevation climates than 'Western Schley'. 'Pawnee' and 'Desirable' (Sparks, 1992) are well suited to DL elevation climates. Premature germination is not a problem with either cultivar. The shuck of both cultivars open better than and opening is not staggered as in 'Wichita'. Uniform shuck opening and lack of premature germination allow once-over harvesting. Both cultivars produce large and excellent quality nuts as demanded by the early market. 'Desirable' produces large crops but rarely over produces because of massive fruit abortion that almost invariably occurs during the second drop (Sparks and Madden, 1985). If over production becomes a problem in 'Pawnee', the fruit can be mechanically thinned (Reid and Smith, 1992). In the Crystal City - Eagle Pass area, 'Pawnee' fruit maturity occurs 2 weeks before 'Wichita', a major plus for the early market. Fruit maturity in 'Desirable' is 10-14 days after 'Wichita'. Later nut maturity is not a major disadvantage because of the high and universal demand for the premium nut of 'Desirable'.

3.4 Pollinizers

Pecan is a monoecious, dichagamous, wind pollinated tree. In native groves and in Georgia, where multiple cultivars and seedlings often exist in the same orchard, pollen is usually sufficient for fruit set. In such situations, pollinizers as such are not necessarily critical. However, in isolated areas where only one or two cultivars are in the orchard (typical of most DL and DH elevation regions), suitable pollinizers are essential (Marquard, 1988; Sparks and Madden, 1985).

'Ideal' is a good pollinizer for 'Western Schley' in Las Cruces. 'Western Schley', in turn, pollinizes 'Ideal' (Nakayama, 1967). 'Ideal's late budbreak diminishes damage from late spring freezes (Brison, 1946: Gray, 1974a). Early nut maturity reduces the chance of damage from fall freezes. Thus, 'Ideal' is suited to the same marginal pecan climates as 'Western Schley'. 'Wichita' is the pollinizer in some 'Western Schley' orchards (Mielke and Kuykendall, 1977).

Extensive overlapping of pollen shedding and pistillate receptivity often occurs in 'Western Schley' (Sparks, 1992). Consequently, selfing is common so that kernel quality is suppressed (Marquard, 1988). Regardless, some 'Western Schley' orchards either lack a pollinizer or have insufficient pollinizers. Mass supplemental pollination is practiced in some orchards (Bennett *et al.*, 1986). Success is dependent upon proper pollen storage (Yates *et al.*, 1991).

'Western Schley' was often planted as the pollinizer for 'Wichita' in DL elevation climates. 'Western Schley' is a good pollinizer in climates where cold winters ensure delayed budbreak is not a problem (Madden and Brown, 1975). However, when planted in warm winter areas, 'Western Schley' proved to have a higher chilling requirement than 'Wichita'. Because of the high chilling requirement and the mild winters of Crystal City, 'Western Schley' sometimes breaks buds too late (Fig. 2) to effectively pollinize 'Wichita'. Subsequently, 'Cape Fear' proved to be a better pollinizer for 'Wichita' in DL elevation climates (Sparks, 1993b). 'Cheyenne' is commonly used as a pollinizer for 'Wichita' in California.

4 Pecan pests

Presently, the number of pecan insect pests is relative small in DL and DH elevations regions mainly due to geographical isolation. Diseases are few due to lack of substantial rainfall.

4.1 Insect control

The major insect pests associated with pecan production in DL and DH elevation climates are yellow pecan (*Monelliopsis pecanis* Bissell) and blackmargined [*Monellia caryella* (Fitch)] aphids, black pecan aphids [*Melcanocallis caryaefoliae* (Davis)], hickory shuckworm [*Laspeyresia caryana* (Fitch)], and pecan nut casebearer (*Acrobasis nuxvorella* Neunzig).

Aphids are a universal problem. Once, insecticides were recommended for control. Since this practice was questioned for yellow pecan and blackmargined aphids (Sparks, 1991), most growers no longer apply insecticides and instead rely on natural occurring beneficial insects for control. Some growers release beneficial insects (Stahmann, 1991) but the value is dubious in orchards with a sod. Release may be valuable in orchards with clean cultivation. Beneficials do not control black aphids and insecticides are applied against this insect. Otherwise, severe premature defoliation suppresses nut production the following year (Moznette, 1934).

Presently, pecan nut casebearer and hickory shuckworm are not problems in California, Arizona, New Mexico, and west Texas. However, both have recently been found in west Texas and Las Cruces, New Mexico (Glogoza *et al.*, 1989; Glogoza and Davis, 1991; Davis and Lucero, 1993) and are probably destined to become problems. Hickory shuckworm and pecan nut casebearer are destructive pests in the Crystal City-Eagle Pass area. Control is by insecticides. Pecan nut casebearer produces three - four generations per year. The most effective insecticide application is targeted at first-generation larvae before they burrow into the fruit. Only two or three days may lapse between larvae hatching from the egg and entering into the fruit (Bilsing, 1927). Sprays are carefully timed because foliar insecticides are ineffective once larvae enter the fruit. Two models (Ring and Harris, 1983; Sparks, 1995a) predict the first entry of the larvae into the fruit and can be used to time the insecticide spray against the first generation.

4.2 Disease control

Downy spot (*Mycospharerella caryigena* Demaree and Cole) and cotton or Texas root rot [*Phymatotrichum omnivorum* (Shear) Duggar] are the major pecan diseases in arid and semi-arid areas. Rainfall is too low for downy spot in arid climates but in semi-arid climates it can be a major disease in some years, especially with 'Western Schley.' Unless downy spot is controlled, 100% defoliation can occur. The usual practice is to scout for downy spot. Following initial infection, the disease is controlled by one application of Orbit (propiconazole) (Sparks, 1993b).

Texas root rot is by far the most severe disease. If the disease is not treated, the tree dies. The fungus survives in the soil for years (Streets, 1937; Streets and Bloss, 1973) and orchards are often unsuspecting planted in infected soils. High pH and high soil temperature are required for survival and success of the fungus (Streets and Bloss, 1973). In the high temperature region of Crystal City and in similarly warm parts of Coahuila and Sonora, Mexico, the fungus is a major problem but it is minor in the cooler El Paso-Las Cruces and Arizona areas.

Texas root rot is extremely difficult to control because the two conditions necessary for its survival, high soil pH and high soil temperature, are not easily managed. Reducing soil pH by the addition of acidulents, such as sulfuric acid, is short-term due to the high buffering capacity of inherently high pH soils. Soil temperature can be reduced a few degrees by growing a sod, but this practice alone will not control Texas root rot.

The control program for Texas root rot (Sparks, 1993b) is based on the work of Streets (1937) and Streets and Bloss (1973). A major thrust is to replenish soil organic matter as organic matter promotes microorganisms which suppress the Texas root rot fungus. During the first 3 to 4 years following orchard establishment, cover crops are grown and turned under. Before commercial nut production begins, the orchard floor is converted to common Bermuda grass. Bermuda grass is a good sod for minimizing Texas root rot for two reasons. One, the massive root system annually replenishes soil organic matter and, two, soil temperature is reduced over much of the growing season as the grass grows until frost, usually November or December.

The organic program is preventive maintenance for the orchard as a whole, but trees infected with Texas root rot require individual treatment. Treatment includes severely cutting back the tree, massive application of ammonium sulfate, and soil fumigation (Fig. 7). Early detection is essential for control. Consequently, in better managed orchards, trees are inspected monthly during the growing season for initial symptoms of Texas root rot. The control program outlined has maintained an annual tree loss in Crystal City from Texas root rot at less than 0.2%. In contrast, tree losses in untreated orchards have been more than 60% (Sparks, 1993b).

5 Orchard management

Management in DL and DH elevation areas is largely directly or indirectly dictated by the dry, hot climate. Climate intensifies tree related problems, requires irrigation, and makes time of harvest more critical. The calcareous, saline/sodic soils require special nutritional, water, and orchard floor management.

5.1 Tree related problems

The cultivars grown, primarily 'Western Schley' and 'Wichita', are prolific which require crop control through either hedging, mechanical fruit thinning, or both. Unless fruit load is carefully controlled, alternate bearing and nut quality are major problems. Additionally, in DL elevation areas, premature germination is a perennial problem which must be managed.

a. Premature germination

Premature germination (Fig. 3) is nut germination while on the tree. Kernel flavor is objectionable and the embryo and surrounding area are black, a condition called "embryo rot". The disorder is associated with delayed shuck opening and high temperature during

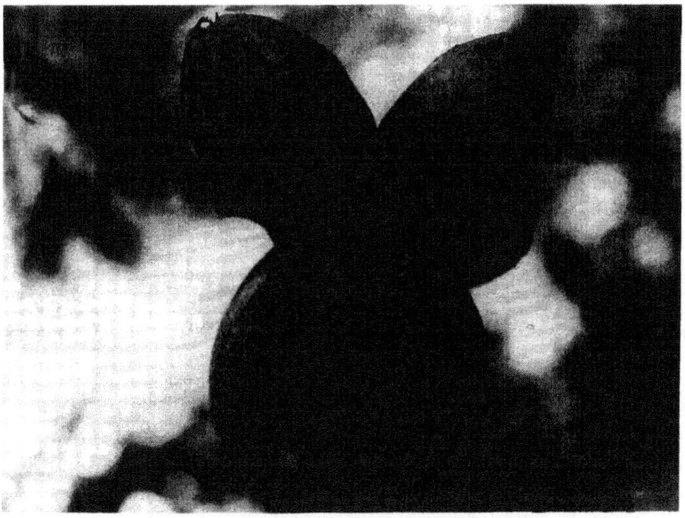

Figure 3. Premature germination in 'Wichita'. The nuts were removed from shucks that had not dehisced.

nut ripening (Sparks, 1993b). Delayed shuck opening prevents the nut, which is about 30% moisture, from drying. High temperatures, besides promoting germination, apparently retard shuck opening. Premature germination is a major problem in areas (Crystal City-Eagle Pass, Red Rock-Picacho and Tucson) with high temperatures during the nut ripening period (September-October temperatures, Table 1). Losses may be 50% or more. Premature germination is not a problem in the El Paso-Las Cruces area where cool temperatures prevail during the ripening period. 'Western Schley' and 'Wichita' are susceptible. 'Wichita' is particularly susceptible because the shuck barely cracks open at maturity (Fig. 4) and often does not open.

Figure 4. Shuck opening in 'Wichita'. The shuck has barely cracked open. Shuck opening in 'Wichita' characteristically remains at this stage for a prolonged period. Slow opening delays nut drying and increases the chance for premature germination.

The preventive program (Sparks, 1993b) consists of reducing kernel moisture as quickly as possible following nut maturity. The program is based on work of Finch (1937) and Finch and Van Horn (1936). Removing the shuck tip near the time of nut maturity (Finch and Van Horn, 1936) reduced nuts germinating prematurely (5%) compared with the control (38%), implicating delayed shuck opening/high nut moisture as a factor(s) in premature germination. The dramatic increase in germination with time (Table 2) suggests early nut harvest would reduce premature germination as practice proved to be the case (Sparks, 1993b).

Practically, the program consists of accelerating abscission of the shuck from the shell followed by rapid nut harvest. Shuck-shell abscission is a perquisite to mechanical harvesting. Abscission is delayed by soil moisture deficit (Stein *et al.*, 1989). Consequently, soil moisture is optimized during the nut ripening period (Sparks, 1993b). Although the shuck separates from the shell, the shuck quarters do not necessarily dehisce or else they barely crack open (Fig. 4). Both nondehiscence and marginal shuck opening slows nut drying and the nut remains susceptible to germination, thus, the need for rapid harvest. The mechanical forces of tree shaking, windrowing, harvesting, and dehulling "knock off" the shuck allowing the nut to dry. Often, premature germination is not a problem during the 10-14 days following initial nut maturity, but can increases rapidly afterwards. Therefore, harvest must begin at initial nut maturity and be completed as quickly as possible. Toward this end, a model for predicting nut maturity is used to time the last irrigation before harvest and estimate the earliest time that harvest can begin (Sparks, 1993b). Optimum soil moisture and early harvesting keeps premature germination at <1% provided fruiting is not excessive. Excessive fruiting complicates premature germination as discussed in 5.1b.

Changes in specific gravity of the nut (Table 2), a measure of nut development (Smith *et al.*, 1948), show early harvest results in a 5% loss in kernel weight which reconfirms other data (Smith and Loustalot, 1944). The 5% loss in kernel weight is the price paid for early harvest.

Table 2. Percent premature germination with time, 'Halbert' pecan, Yuma, Arizona, 1936

Date	*Preharvest germination (%)*	*Specific gravity of nuts*
Oct. 9	3.4	0.75
Oct. 15	15.9	0.78
Oct. 21	29.6	0.79
Nov. 2	35.7	0.79

Source: Finch (1937).

b. *Excessive fruit production*

Excessive fruit production results in poor kernel quality by directly suppressing kernel development and indirectly by inducing shuck decline and accelerating premature germination (Sparks *et al.*, 1995). The following year pistillate bloom is reduced and the crop is off (Malstrom *et al.*, 1984). At inceptive levels of excessive fruiting, the kernel is directly suppressed without shuck decline and premature germination. As fruiting increases, shuck decline is induced, premature germination is accelerated, and the tree may dieback (Sparks, 1993b; Fig. 5).

Figure 5. Regrowth of 'Wichita' trees in the spring following excessive fruiting (A) and mechanical thinning the fruit to ≈20% of a full crop (B).

Shuck decline (Sparks *et al.*, 1995) indirectly affects kernel quality because the shuck dies before kernel development is completed. Shuck decline begins as a fine dark, necrotic line on the inner shuck surface. Deterioration quickly spreads toward the outer shuck surface. The inner shuck turns green and slimy. At this stage, the shuck surface appears water soaked and has a glassy-green sheen. Later the inner shuck and shuck surface turn black. Fruits fall prematurely, die and remain in the cluster, or the shuck opens prematurely.

Premature germination is prevented by adequate soil moisture and early nut harvest only if the tree is not fruiting excessively. Once the tree fruits excessively, time of premature germination is accelerated and begins before the nut is mature enough for harvest (Sparks, 1993b). Thus, management of premature germination becomes more complex.

The adverse effect of fruit load on premature germination, shuck decline, kernel quality, and tree dieback can be resolved by mechanical fruit thinning (Table 3). The fruit is thinned as early as mechanically possible (ovular wall is one half expanded) to maximize quality and return bloom (Reid *et al.*, 1993). Commercially, maintaining adequate soil moisture, harvesting early, and fruit thinning result in high quality nuts with little, if any, premature germination or shuck decline (Sparks, 1995b). Furthermore, tree dieback (Fig. 5) is prevented and an adequate return bloom is ensured. In DH elevation areas, nut quality is maintained and shuck decline is prevented by fruit control from hedging. Presently, quality and shuck decline are mainly problems in nonhedged orchards and especially if they are crowded (personal communication, M. W. Kilby).

Table 3. Effect of mechanical fruit thinning 'Wichita' on shuck decline, premature germination,nut quality, and tree decline

		Fruits thinned %				
		0	*25*	*41*	*56*	*77*
Shuck decline	%	79.2	41.4	24.4	10.9	3.1
Germination	%	39.4	32.4	22.3	15.3	4.5
Kernel	%	40.7	50.7	57.9	60.0	63.7
Edible kernel	kg/tree	.09	2.68	8.12	8.84	8.49
Tree dieback	%[a]	64.7	20.5	5.8	3.0	0.0

[a] Tree dieback in the year following fruit thinning.
Source: Adapted from Sparks et al. (1995).

c. *Harvesting*

In the United States, all phases of harvesting, tree shaking, harvesting, dehulling, cleaning, and shelling are mechanical. In Mexico, mechanical harvesting is not universal.

Harvesting in DH elevation areas is often delayed until the first hard freeze (Herrera, 1995a), allowing a once-over harvest. In contrast, in DL elevation areas, harvesting is done as soon and as quickly as possible to capitalize on the profitable early market and by necessity to prevent or minimize losses from premature germination (Sparks, 1993b).

Early harvest is successfully used to minimize premature germination and capitalize on the early market. However, success, from the standpoint of nut quality, depends on precise definition of nut maturity. For early harvest, the nut is mature after the shuck-shell abscission and the shell color changes from white to brown with full development of the "natural markings," except a small area on the base of the nut. At this stage, the shuck is green and may or may not be open. The nut (shell plus kernel) contains ≈30% moisture, and the kernel has not developed full flavor (the flavor develops during drying) and is at ≈95% of its maximum dry weight (Smith and Loustalot, 1944; Thor and Smith, 1935). Following moderately heavy shaking, ≈95% of the nuts that shake off the tree should have a brown ground color and full development of the natural markings. If not, the tree was shaken too hard or else the nuts are not sufficiently matured for early harvest. Judging maturity accurately is critical, otherwise the result will be immature nuts with more than the normal 5% loss in kernel weight and increase fuzzing (adherence of packing tissue) on the kernel.

Early harvest requires a major capital expense for dehulling and nut drying facilities plus expertise in nut drying. In addition, two harvests are needed versus a once-over harvest if the nuts are allowed to mature naturally. Additional harvests are needed because early harvest is limited only to the first nuts to mature on the tree.

Besides the increased expense associated with early harvest and the 5% loss in kernel weight, early harvesting has another disadvantage, shell cracking of the moist nut (Fig. 6). Nuts of cultivars with thin shells and that have most of the shell cavity filled with kernel, such as 'Wichita', crack within 15 minutes if shaken from the tree on sunny days. The shell cracks because it shrinks at a faster rate than the kernel. Cracking is diminished by shaking nuts from the tree during late afternoon and early evening. The nuts partially dry during the night and are harvested the following morning. Sometimes nuts are harvested at night but this is difficult (Sparks, 1993b).

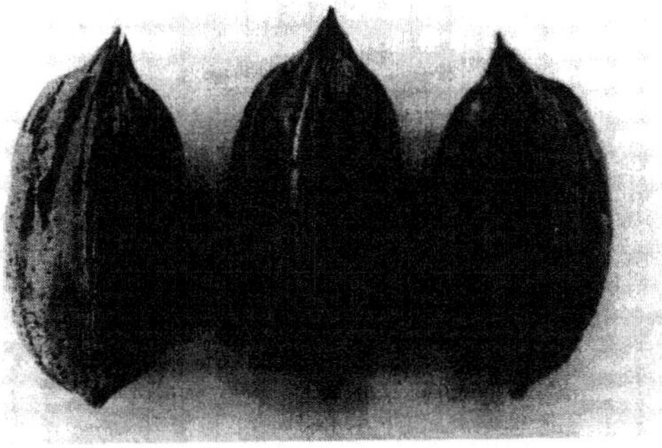

Figure 6. Shell cracking of 'Wichita' nuts following early harvesting on a sunny day.

5.2 Nutrition

Fertilization based on leaf and soil analysis (Sparks, 1977b) is widely used in humid areas but, in many orchards in dry warm areas, only N and Zn are applied. N is applied annually and varies from 112 to 224 kilograms per hectare. Application is usually split two or three times and sometimes more if the fertilizer is distributed in the irrigation water. K and P are applied by some growers and especially those in the Crystal City, Texas and Visalia, California areas. K and P are applied to minimize poor kernel quality and tree dieback (Hunter, 1948; Sparks, 1993b) from fruit-induced deficiencies of N, P, and K (Sparks, 1977a). Four to five foliar sprays of Zn are applied annually during leaf expansion to control deficiency (Herrera, 1995a). In addition, foliar sprays of Mn and Mg are routinely applied in some orchards (Sparks, 1993b). Foliar sprays are required because uptake of Zn, Mn, and Mg is insufficient from alkaline soils. Some growers apply foliar sprays containing macro nutrients although there is no evidence that these sprays have any appreciable benefit (Sparks, 1986).

Iron deficiency is a major problem in the Eagle Pass-Crystal City area and especially during the first 1-5 years following planting. Deficiency symptoms appear during the leaf expansion stage. Iron deficiency develops on trees if irrigated during leaf expansion but not if irrigated before and after this developmental stage. Iron deficiency is avoided by irrigation scheduling. The soil is irrigated to field capacity before budbreak (mid-March) to a soil depth of 2.5 to 3 m or more. The soil is not irrigated again until after the first cycle of shoot elongation is completed, about 1 May. Irrigation managed in this manner does not suppress leaf efficiency because of the high water holding capacity of the deep soil. Once leaf expansion is completed, irrigation is applied as needed without inducing iron deficiency. Foliar iron sprays do not prevent or correct deficiency (Sparks, 1993b).

5.3 Soil related problem

Soils of DL and DH elevations climates are characteristically calcareous, saline, or sodic. The need to leach excess salts require different irrigation strategies than for humid soils. High water tables can be a problem which further intensifies soil and water management.

a. Calcareous, saline and sodic soils

Saline and especially sodic soils are frequent occurring problems in El Paso-Las Cruces and to a lesser extent in the pecan growing regions of California, Arizona, and some areas of Mexico. Calcareous soils are widespread.

Calcareous soils contain free lime. The major concern with these soils, unless they are also saline, is high pH and associated unavailability of micro nutrients, especially Zn. Localized application of acidulents can increase micro nutrient availability, but broadcast application of acidulents is not cost effective (Sibbett, 1995). The usual procedure is to supply micro nutrients via foliar sprays.

Saline soils have an electrical conductivity (EC_e) of $\geq 0.4\ S \cdot m^{-1}$ with an exchangeable Na of $<50\ g \cdot kg^{-1}$ and a pH usually <8.5. Principal ions are Ca^{++}, Mg^{++}, Na^{+}, Cl^{-}, and SO_4^{--}. Excessive Na^{+} and Cl^{-} are detrimental to tree growth and production. If soil EC_e exceeds $0.2\ S \cdot m^{-1}$, pecan growth is impaired (Miyamoto and Gobran, 1983). Due to the predominance of Ca^{++} and Mg^{++}, saline soils are generally well flocculated which permit good water penetration (Sibbett, 1995).

In some orchards, Salinity is an inherent condition of the soil; but in most, Salinity was induced by saline water applied as infrequent, light irrigations, or by heavy irrigations without provisions for subsurface drainage. Excessive salts are removed from a saline soil by applying enough nonsaline irrigation to leach them out of the root zone. Correction is dependent upon adequate subsurface drainage. Salinity is prevented from becoming a problem by frequent irrigation with all irrigations containing slightly more water than is needed by the tree. Further leaching of salt from the soil is accomplished by applying an excessive irrigation one or twice during the winter (Sibbett, 1995).

Sodic soils have exchangeable Na of $>150.kg^{-1}$ and an electrical conductivity of $<0.4\ S \cdot m^{-1}$ with a pH usually >8.5. Excessive Na degrades soil structure causing it to become impervious to air and water (Sibbett, 1995). Soil structure degradation becomes a major production problem due to pecan's sensitivity to water and salt stress, waterlogging, poor aeration, and soil hardness (Alben, 1958; Loustalot, 1945; Miyamoto and Gobran, 1983; Miyamoto *et al.*, 1985, 1986a, 1986b; Rieger and Daniell, 1988; Smith and Ager, 1988; Smith and Bourne, 1989; Sparks, 1989). Before soluble salts in sodic soils can be leached

Figure 7. Pecan tree dead from Texas root rot (A). Tree that has been cut back and treated for the fungus (B). Regrowth from a cut back and treated tree (C).

out, the sodic conditions that impair water penetration must be corrected. In California, correction is most commonly by incorporating Ca into the soil, usually gypsum. Ca replaces the Na and the soluble sodium sulfate formed is leached from the soil profile. If the soil contains large amounts of Ca, gypsum can be former indirectly by applying an acidulent (sulfur, lime-sulfur, of sulfuric acid) which reacts with free lime to form gypsum (Sibbett, 1995). Correcting permeability of sodic soils and/or leaching excess salt from the soil is relatively easy if the irrigation water is nonsaline. If the irrigation water is high in soluble salts, correction depends on soil type. Leaching of sandy soils works well. However, salt clay and silty clay loam (the problem soils in El Paso-Las Cruces) require chiseling or other types of soil-profile modification before leaching (Miyamoto, 1989). Generally in the Las Cruces area, and in contrast to California (Wildman *et al.*, 1988), chemical amendments are secondary to soil profile modifications.

b. High water tables

Three types of high water tables are commonly encountered in arid pecan growing regions. One, is a clay subsoil under sandy soils which causes a perched water table. The problem is usually solved by deep trenching to break through the clay layer. Two, are old depressions, in which drainage was originally poor, that were filled in during the development of irrigation projects. This problem is solved by installation of subsurface tile drains or open ditches. The third type occurs from canal seepage or periodical excessive flow of rivers. Canal lining or interception drains are used to manage this problem (Miyamoto, 1989). After a water table recedes, leaching is required if high levels of salts are left in the soil.

During establishment of pecan orchards in the Crystal City area, a perched water table (from a deep lying caliche layer) was a major problem (Sparks, 1993b). The perched water table developed when the land was used for intensive vegetable production. Excessive irrigation for years created a water table that was often within 0.5 to 1 m of the soil surface. Following planting to pecans, tree growth varied inversely with water logging. Chloride toxicity and fungal leaf scorch (Sparks, 1992) were massive. Installation of tile drainage at a depth of 2.5 m and subsequent leaching eliminated the problem (Sparks, 1993b).

5.4 Irrigation

Evapotranspirational loss of water (consumptive use) from a pecan orchard has been well studied (Miyamoto, 1982, 1983, 1985; Worthington *et al.*, 1987). Peak summer consumptive use of a mature orchard is about equal to pan evaporation rates. Flood is the predominant irrigation system in DL and DH elevation climates. Sprinkler, drip, and microjet systems are used in special situations (Miyamoto *et al.*, 1995).

Water quality is important for salt leaching. Water containing 1500 ppm salts is successfully used in well drained sandy soils. However, on low permeable clay soils, water with 1000 ppm salts can cause salt damage (Miyamoto, 1991). Irrigation water with a sodium adsorption ratio (SAR) > 5 or with B > 1 ppm are difficult to manage if the soil is not highly permeable (Miyamoto, 1991; Picchioni *et al.*, 1991).

Because of problems unique to different production areas, timing of irrigation varies. In areas subject to iron deficiency, irrigation is not applied during shoot elongation as discussed earlier. In regions with premature germination, soil moisture is carefully monitored during nut ripening. Irrigation is timed so that soil moisture is optimum during ripening but leaves the surface dry enough to support harvest equipment. In this regard, a sod is essential for quick return following irrigation. In many areas, Salinity is not a major problem in the soil nor in the irrigation water. Soils in these areas are often sandy. The sandy soil and the lack of Salinity reduce the need to irrigate for control of excessive soil salts. However, areas with major salt problems, for example El Paso-Las Cruces, demand irrigations for leaching of salts as such. Large volumes of water are applied in the spring when the trees are dormant and when the soils are most permeable. When needed, several consecutive irrigations are applied after appropriate soil structural improvements have been made (Miyamoto, 1991; Miyamoto and Storey, 1995).

Many methods are available to decide irrigation scheduling (Miyamoto *et al.*, 1995). In practice, two methods are dominant, a fixed schedule and the appearance and feel of the soil. The "feel" method is better than the fixed schedule method as the latter often results in under or over irrigation and especially if variable soil types exist in the orchard. Variable soil type epitomizes alluvial soils on which pecans are often planted.

5.5 Orchard floor management

Historically, most pecan orchards in DL and DH areas have been clean cultivated. However, within recent years orchards in DL elevation areas have been converted to sod (primarily common Bermuda grass). A sod is especially beneficial in these areas. The sod helps replenish organic matter which aids in cotton root rot control and sod allow quick return of equipment following rain or irrigation. Quick return and thus minimizing harvest delays is especially important in preventing premature germination as discussed previously. A sod is less expensive to maintain than clean cultivation, provides a natural habitat for beneficial insects, improves and maintains soil structure and thus permeability, and facilitates harvest. Clean cultivation is the dominant orchard floor system in El Paso-Las Cruces (Miyamoto and Storey, 1995). Compaction from trafficking is a common problem and is alleviated by shallow chiseling in young orchards and by spiking in mature orchards. Sod culture is increasing. In theory, a sod should improve the characteristically poor permeability of soils in this area. The hesitancy to use a sod in the El Paso-Las Cruces area (Miyamoto and Storey, 1995) is apparently due to a negative report by an area grower (Fletcher, 1974). The report was not based on a controlled study.

6 Outlook

Thousands of hectares of pecans are planted in DL elevation areas of Texas, Arizona, and Mexico. Climate of these areas promotes early nut maturity, allowing nuts to be sold on the high priced early market. Consequently, the profit potential is outstanding

(Sparks, 1993b). Profit potential is critically dependent on controlling premature germination and, in some areas, Texas root rot. The dependence of the industry on controlling these two disorders is evidenced by the history of the pecan industry in the Crystal City region. Pecan orchards were planted in the 1920s and 30's but were abandoned because of Texas root rot and premature germination. Vestiges of these orchards remain. Texas root rot and premature germination can be controlled, but intense, carefully executed management is required. Without doubt, success and/or survival of orchards will depend on management quality. Even now, trees in poorly managed orchards in the Eagle Pass-Crystal City area and in Mexico are rapidly deteriorating. Information for controlling over production and associated premature germination, shuck decline, and poor quality (Sparks, 1993b; Sparks *et al.*, 1995) is being used in California and to limited extent in Texas, but lack of harvesting and cleaning equipment limit information adaptation in Mexico. Very few new plantings are being made in any DL elevation area. Outlook for increased plantings is poor. Any new plantings should consist of cultivars ('Desirable' and 'Pawnee') suitable to DL elevation areas.

The predominance of pecan plantings in the warm climates of the United States and Mexico is in DH elevation areas (Herrera, 1995a). Except increased hectares from thinning mature orchards and transplanting the mature tree removed, new planting in DH elevation areas is limited primarily to the state of Chihuahua, Mexico. Production costs in all DH elevation areas will increase with the advance of pecan nut casebearer and hickory shuckworm. In nonhedged (Kilby, 1996; Sparks, 1993b) and hedged orchards (E. A. Herrera; N. Sachs, communications), shuck decline is becoming a problem as production increases with increasing tree maturity. Controlling fruit load (Sparks *et al.*, 1995) will be required if these orchards are to be fully profitable.

Herrera (1995a) predicted that pecan production in warm climates will increase ≈65% by 2005. Many established orchards are now producing at maximum capacity. Accuracy of his estimate will depend on the number of new plantings made.

7 References

Alben, A.O. (1958). Water logging of subsoil associated with scorching and defoliation of Stuart pecan trees. *Proceedings of the American Society for Horticultural Science*, 72, 219-223.

Bennett, J., Kofanovich, T. and Stahmann, III, W. (1986). Pecan growers' experiences with artificial pollination. *Western Pecan Conference Proceedings*, 20, 8-10.

Bilsing, S.W. (1927). Studies on the biology of the pecan nut casebearer (*Acrobasis caryae*). *Texas Agriculture Experiment Station Bulletin*, 347.

Brison, F.R. (1946). Pecan varieties for central Texas. *Proceedings Texas Pecan Growers Association*, 25, 30-37.

Davis, J.H. and Lucero, V.S. (1993). Update on the status of pecan nut casebearer and hickory shuckworm in El Paso County, Texas and Dona Ana County, New Mexico. *Western Pecan Conference Proceedings*, 27, 6-16.

Finch, A.H. (1937). The use of ethylene to improve pecan harvesting. *Proceedings of the American Society for Horticultural Science*, 34, 74-77.

Finch, A.H. and Van Horn, C.W. (1936). The physiology and control of pecan nut filling and maturity.

Arizona Agriculture Experiment Station Technical Bulletin, 62.

Fletcher, L.F. (1974). Influence of sod-culture on compaction in irrigated pecan orchards. *Western Pecan Conference Proceedings*, 8, 36-37.

Glogoza, P. and Davis, J.H. (1991). The discovery of hickory shuckworm, Cydia caryana, in El Paso County. *Western Pecan Conference Proceedings*, 25, 2-6.

Glogoza, P., White, J. and Helmers, S. (1989). Pecan nut casebearer. *Western Pecan Conference Proceedings*, 23, 97-104.

Goff, W.D. and Tyson, T.W. (1991). Fall freeze damage to 30 genotypes of young pecan trees. *Fruit Variety Journal*, 45, 176-179.

Gray, O.S. (1974a). The O.S. Gray varieties. The Pecan Quarterly, 8 (4), 30.

Gray, O.S. (1974b). Looking at the outside and inside of some pecan varieties. *Western Pecan Conference Proceedings*, 8, 73-81.

Herrera, E.A. (1995a). Pecan growing in the western United States. *HortTechnology*, 5, 200-201.

Herrera, E.A. (1995b). Economic comparison of removing pecan trees and planting young trees and transplanting established, mature trees. *HortTechology*, 5, 212-214.

Hunter, J.H. (1948). Nutritional deficiencies associated with poor filling of pecan nuts. *Proceedings of the Southeastern Pecan Growers Association*, 41, 16-34.

Kilby, M.W. (1996). 1995 pecan situations required all management skills. *The Pecan Grower*, 7(3), 17.

Kilby, M.W. and Herrera, E.A. (1997). Pecan industry in the western U. S. *Proceedings of the Southeastern Pecan Growers Association*, 90, 126-128.

Loustalot, A.J. (1945). Influence of soil moisture conditions on apparent photosynthesis and transpiration of pecan leaves. *Journal of Agricultural Research*, 71, 519-532.

Madden, G.D. and Brown, E.J. (1975). Budbreak, blossom dates, nut maturity and length of growing season of the major varieties grown in the west. *Pecan South*, 2(3), 96,97,112,113.

Malstrom, H.L, Fenn, L.B. and Riley, T.R. (1984). Methods of zinc fertilization. *Western Pecan Conference Proceedings*, 18, 18-25

Marquard, R.D. (1988). Outcrossing rates in pecan and the potential for increased yields. *Journal of the American Society for Horticultural Science*, 113, 84-88.

Mielke, E.A. and Kuykendall, J.R. (1977). Pecan pollination in Arizona. *Western Pecan Conference Proceedings*, 11, 25-32.

Miyamoto, S. (1982). Consumptive water use and irrigation scheduling of pecans in far west Texas. *Proceedings Texas Pecan Growers Association*, 62, 44-55.

Miyamoto, S. (1983). Consumptive water use of irrigated pecans. *Journal of the American Society for Horticultural Science, 108*, 676-681.

Miyamoto, S. (1985). Water consumption. Growers need to know this to plan irrigation. *Pecan South*, 19(3), 8, 10, 12, 13.

Miyamoto, S. (1989). Reclamation of salt-affected pecan orchards. *Proceedings Texas Pecan Growers Association*, 68, 1-14.

Miyamoto, S. (1991). Salinity management in irrigated pecan orchards. *Western Pecan Conference Proceedings*, 25, 63-69.

Miyamoto, S. and Gobran, G. (1983). Assessment and potential remedies of salinity problems in pecan orchards of the middle Rio Grande Basin. *Western Pecan Conference Proceedings*, 17, 1-12.

Miyamoto, S. and Storey, J.B. (1995). Soil management in irrigated pecan orchards in the southwestern United States. *HortTechology*, 5, 219-222.

Miyamoto, S., Gobran, G. and Piela, K. (1985). Salt effects on growth and ion uptake of three pecan rootstocks cultivars. *Agronomy Journal*, 77, 383-388.

Miyamoto, S., Henggeler, J. and Storey, J.B. (1995). Water management in irrigated pecan orchards in the southwestern United States. *HortTechnology 5*, 214-218.

Miyamoto, S., Riley, T., Gobran, G. and Petticrew, J. (1986a). Effects of saline water irrigation on soil salinity, pecan tree growth and nut production. *Irrigation Science*, 7, 83-95.

Miyamoto, S., Picchioni, G. and Storey, B. (1986b). Salinity: a major factor of poor tree performance in irrigated pecans. *Pecan South*, 20(4), 14-18.

Moznette, G.F. (1934). Experiments in control of pecan black aphids under orchard conditions. *Proceedings of the Southeastern Pecan Growers Association*, 28, 55-61.

Nakayama, R.M. (1967). Pecan variety characteristics. *New Mexico Agriculture Experiment Station Bulletin* 520.

Picchioni, G., Miyamoto, S. and Storey, B. (1991). Growth, boron uptake of five pecan cultivar seedlings. *HortScience*, 26, 386-388.

Rieger, M. and Daniell, J.M. (1988). Leaf water relations, soil-to-leaf resistance, and drought stress in pecan seedlings. *Journal of the American Society for Horticultural Science*, 113, 789-793.

Reid, W. and Smith, M.W. (1992). Thinning pecans reduces alternate bearing. *Annual Report of the Northern Nut Growers Association*, 83, 91-94.

Reid, W., Huslig, S.M. and Smith, M.W. (1993). Fruit-removal time influences return bloom in pecan. *HortScience*, 28, 800-802.

Ring, D.R and Harris, M.K. (1983). Predicting pecan nut casebearer (Lepidoptera: Pyralidae) activity at College Station, Texas. *Environmental Entomology*, 12, 482-486.

Sibbett, G. S. (1995). Managing high pH, calcareous, saline, and sodic soils of the western pecan-growing region. *HortTechology*, 5, 222-225.

Smith, M.W. and Ager, P.L. (1988). Effects of soil flooding on leaf gas exchange of seedling pecan trees. *HortScience*, 23, 370-372.

Smith, M.W. and Bourne, R.D. (1989). Seasonal effects of flooding on greenhouse-grown seedling pecan trees. *HortScience*, 24, 81-83.

Smith, M.W. and Couch, G. (1984). Early fall or late spring freezes pose threat of injury to pecan trees. *Pecan South*, 18(5), 11-14.

Smith, C.L. and Loustalot, A.J. (1944). Effects of harvest date and curing on the composition and palatability of pecan nuts. *Journal of Agricultural Research*, 68, 395-403.

Smith, C.L, Hamilton, J. and Romberg, L.D. (1948). Specific gravity and percentage of kernel as criteria of filling of pecan nuts. *Proceedings of the American Society for Horticultural Science*, 51, 157-170.

Sparks, D. (1977a). Effects of fruiting on scorch, premature defoliation, and nutrient status of 'Chickasaw' pecan leaves. *Journal of the American Society for Horticultural Science*, 102, 669-673.

Sparks, D. (1977b). Methods of predicting the nutrient needs of nut crops. *Annual Report of the Northern Nut Growers Association*, 68, 25-30.

Sparks, D. (1986). Heavy nut production can suppress nutrients in pecan trees. *Pecan South*, 20(4), 22-24.

Sparks, D. (1989a). Drought stress induces fruit abortion in pecan. *HortScience*, 24, 78-79.

Sparks, D. (1989b). Predicting nut maturity of the pecan from heat units. *HortScience*, 24, 454-455.

Sparks, D. (1991). Cultural practices, in *Pecan Husbandry: Challenges and Opportunities* (eds B.W. Wood and J.A. Payne), *USDA, ARS*-96, pp. 22-33.

Sparks, D. (1992). *Pecan cultivars: The Orchard's Foundation*. Pecan Production Innovations. Watkinsville, Ga.

Sparks, D. (1993a). Chilling and heating model for pecan budbreak. *Journal of the American Society for Horticultural Science*, 118, 29-35.

Sparks, D. (1993b). Growing pecans at low elevations in a hot climate. *Annual Report of the Northern Nut Growers Association*, 84, 127-141.

Sparks, D. (1995a). A budbreak-based chilling and heating model for predicting first entry of pecan nut casebearer. *HortScience*, 30, 366-368.

Sparks, D. (1995b). Mechanical fruit thinning — a pecan success story. *Annual Report of the Northern Nut Growers Association*, 86, 108-112.

Sparks, D. (1997). A model for predicting pecan production under arid conditions at high elevations. *Journal of the American Society for Horticultural Science*, 122, 648-652.

Sparks, D. (2000). Pecan, in *Compendium of Temperature Nut Crop Diseases*. (ed. J.J. Michalides) APS press.

Sparks, D. and Madden, G.D. (1985). Pistillate flower and fruit abortion in pecan as a function of cultivar, time, and pollination. *Journal of the American Society for Horticultural Science*, 110, 219-223.

Sparks, D. and Payne, J.A. (1978). Winter injury in pecans — a review. *Pecan South*, 5(2), 56-60, 82-88.

Sparks, D., Reid, W. and Yates, I.E. (1995). Fruiting stress induces shuck decline and premature germination in pecan. *Journal of the American Society for Horticultural Science*, 120, 43-53.

Stahmann, R. (1991). Grower's experience in using beneficials for aphid control. *Western Pecan Conference Proceedings*, 25, 10-41.

Stein, L.A., McEachern, G.R. and Storey, J.B. (1989). Summer and fall moisture stress and irrigation scheduling influence pecan growth and production. *HortScience*, 24, 607-611.

Streets, R.B. (1937). Phymatotrichum (cotton or Texas) root rot in Arizona. *Arizona Agriculture Experiment Station Technical Bulletin 71*.

Streets, R.B. and Bloss, H.E. (1973). Phymatotrichum root rot. *American Phytopathology Society Monograph* 8, St. Paul, MN.

Sullivan, D.T. and Bradford, A.R (1978). Winter injury to 'Western Schley' and 'Wichita' pecan trees. *Western Pecan Conference Proceedings*, 12, 62-63.

Thor, C.J.B. and Smith, C.L. (1935). A physiological study of seasonal changes in the composition of the pecan during fruit development. *Journal of Agricultural Research*, 50, 97- 121.

Ware, L.M. and Johnson, W.A. (1957). The effect of cultural and fertilizer practices on the nitrate and moisture levels of the soil and on growth of young pecan trees. *Proceedings of the Southeastern Pecan Growers Association*, 50, 39-50.

Wildman, W.E., Pecock, W.L. and Wildman, A.M. (1988). Soluble calcium compounds aids low-volume water application. *California Agriculture* (Nov.-Dec.), 7-8.

Worthington, J.W., Lasswell, J. and McFarland, M.J. (1987). Irrigation-the trees' prospective. *Pecan South*, 21(1), 4-8.

Yates, I.E., Sparks, D. and Connor, K. (1991). Reducing pollen moisture simplifies long-term storage of pollen. *Journal of the American Society for Horticultural Science*, 116, 430-434.

15 WALNUTS (*JUGLANS REGIA* L.) IN MEDITERRANEAN WARM CLIMATES

DIEGO FRUTOS TOMÁS

Centro Regional de Investigación y Desarrollo Agroalimentario
Consejeria de Medio Ambiente, Agricultura y Agua
La Alberca, Murcia, Spain

1 Introduction

World walnut production is estimated at nearly a million metric tons (MT), most of which is in the Northern Hemisphere. The most important producing countries in 1997 were, in thousands of MT, China (240), the United States (209), Turkey (115), Iran (89), Ukraine (77), Yugoslavia (25), India (26), France (23), Greece (20), Bulgaria (14), Italy (12), Chile, Austria and Germany (13), and Spain, Belarus and Argentina (9). Total production has been estimated in Europe, Asia and North America to reach 324, 501 and 227 thousand metric tons respectively (FAO, 1997). Total World export reached 197.4 thousand MT, 102 from the United States, 54 from China, 15.3 from France, 15 from India, 1.1 from Turkey and 1 from Italy. The E.U. countries import 75% of the U.S. in-shell walnut production. The United States exports walnut to the E.U., Japan and other countries in roughly equal amounts (Hutin, 1996). These data emphasize the importance of the world walnut industry.

2 Plant material

2.1 Origin of walnut

During the last glaciation (Würm glaciation), walnuts disappeared in Southern Europe and Southern Turkey, but survived in the warmest areas of the Black Sea and of the Caspian Sea. According to available pollen data walnuts were reintroduced into the former areas by Neolithic men during the second millennium BC (Zohary and Hopf, 1988). In northern Spain, walnut pollen has been found in sediments from the Münsterian period, of the Low Paleolithic, when temperatures fell (Sánchez Goñi, 1988). Walnut pollen has also been found in southern Spain, in sediments of the Carihuela's Cave 28.000 years BC, when the Würm glaciation occurred (Carrión and Sánchez-Gómez, 1992). Pollen from 7,000-11,000 years BC has also been found in Central Italy. Walnut pollen in Southern Alps and Balkans have been dated 5,000 years BC. The area in which walnuts occurred was expanded by Neolithic man (Huntley and Birks, 1983), and in the Serra da Estrela (Portugal), where the oldest walnut pollen was dated to 4,300 years BC (Van den Brink and Janssen, 1985), which is older than the fossil pollen

Amnon Erez (ed.), Temperate Fruit Crops in Warm Climates, 405-427.
© 2000 *Kluwer Academic Publishers. Printed in the Netherlands.*

found in Turkey and in Greece (Bottema, 1980). Other pollen deposits indicate that walnuts were present during the Low and Middle Holocene in North Africa (Pons and Quezel, 1958; Balouche, 1986).

Some questions have arisen concerning Vavilov's theory about the centers of origin of the cultivated species (Zohary, 1970), given the existence of walnuts in both Central Asia and in the Iberian Peninsula during the Würm glaciation.

Most authors agree on the importance of man's activities in expanding walnuts in Europe (Carrión and Sánchez Gómez, 1992). Iberian walnuts may have been first introduced into Europe during the Azilien - Tardenois times of the Mesolithic period, from the 12th to the 10th millennium B.C., during the migration of the Iberian people to northern France, Belgium and Germany (Grimberg and Svanström, 1963). A second reintro-duction of Iberian walnuts into Europe could have occurred between the 3rd and the 2nd millennium BC, when migrations of Neolithic people of southern Iberia occurred.

Walnuts in the northern Himalayas would have been killed by low temperatures during the last glacial period. However, those in the southern Himalayas would have spread south to reach other warmer areas. In eastern Asia, mountain orientation allowed passage of walnuts to milder areas, permitting survival during the last glacial period.

2.2 Native walnut populations in Spain in relation to environment

Native walnuts are found in very diverse environments in Spain (Table 1), leading to selection of new cultivars between 1977 to 1985 from the mountain highlands to warm areas of the South. For these trees, a wide soil adaptation range was observed (Table 2). Soil texture ranged from sandy to heavy clay. The accompanying flora was also diverse. Halophytic flora, however, has never been observed in conjunction with walnuts (Frutos, 1990b).

The Spanish walnut populations includes about 500,000 old seedlings. Western and northwestern forests of native walnuts have a very high percentage of trees with erect canopies, whereas in the eastern mountains the native walnuts usually have a round canopy. Trees with erect shape are usually accompanied by trees such as chestnuts and oaks. In these forests sunlight seems to be the limiting factor. In the Mediterranean eastern forests, on the other hand, the main limiting factor is usually rainwater and the round-shape canopy is better adapted to the sparse forest. Such trees are found besides other round-shaped forest trees, as evergreen oaks (*Quercus ilex* L.). Eastern and western trees grow similarly when planted together.

2.3 Introduction of Spanish walnuts into the New World

As occurred with other fruit species, walnuts were introduced to America by the Spanish missionaries, and were successfully established in the subtropical climate of Chile and California.

Table 1. Climatic data for locations in Spain where old productive and healthy walnut seedlings have been observed (Elias and Gimenez, 1965)

Location	*County*	*Average temperatures (°C)* Jan	Dec.+Jan.+Feb. / 3	Jul.+Aug. / 2	*Annual hydric balance (mm) by Thornwhite*	*Altitude (m)*
San Javier (1)	Murcia	9.8	10.5	24.6	-573.6	16
Caravaca	Murcia	7.6	8.1	24.7	-479.7	675
Nerpio	Albecete	5.0	5.6	23.0	-409.4	1,082
Santiago de la Espada	Jaén	3.3	4.1	23.0	-388.7	1,328
Pontones	Jaén	2.4	3.3	20.9	-347.9	1,350
Olivenza	Badajoz	8.3	8.9	27.3	-598.7	260
El Tiemblo	Avila	4.8	5.5	22.6	-401.9	691
Barco	Avila	3.7	4.4	20.4	-334.4	1,007
Sancti-Spiritu (2)	Salamanca	3.8	4.8	20.0	-358.9	756
Zamora	Zamora	3.9	4.7	21.5	-436.9	649
Castrocontrigo	León	3.3	4.1	18.7	-314.0	920
La Bañeza	León	3.1	4.1	19.6	-365.1	800
Ponferrada (3)	León	4.8	5.6	21.6	-381.4	544
Lugo(3)	Lugo	5.0	5.6	23.0	-202.6	454

(1) Nearest stations toTorre Pacheco and Cartagena (Murcia)

(2) Nearest stations to La Alberca (Salamanca)

(3) Nearest stations to Becerrea, Doncos and Los Nogales (Lugo)

Table 2. Soil analysis of samples from 0, 12 or 20 cm beneath old productive and healthy walnut seedlings

Location	*Organic matter (%)*	*Total carbonates (%)*	*Active limestone (%)*	*pH (H_2O)*	*Conductivity (in Est. 1:5) mhos/cm*	*P_2O_5 ppm*	*K_2O ppm*
Caravaca (MU)	0.74	69.6	23.1	8.1	262	12.8	Trace
La Olla (MU)	1.24	70.4	18.5	7.6	232	12.1	Trace
Campo de San Juan (MU)	2.77	28.2	9.9	8.1	381	13.6	Trace
Nerpio (AB).	2.36	17.3	7.08	8.1	252	13.6	Trace
El Tiemblo (AV)	2.41	2.4	0.62	7.5	189	12.8	Trace
Burgohongo (AV)	3.01	6.2	1.72	7.0	250	24.0	Trace
La Eliseda(AV)	4.10	0.2	-	6.5	250	12.4	Trace
Los Llanos - El Barco(AV)	4.60	4.7	1.4	7.1	151	13.0	Trace
Villafalé (LE)	6.90	2.1	0.7	7.8	358	12.4	Trace
Los Nogales (LU)	3.60	17.3	2.3	7.9	211	14.8	Trace

Chile. During the summer of 1986, Frutos and Astorga (unpublished) evaluated the Chilean walnut population from the Ovalle area (latitude 33°S), to Angol, 800 km southward (latitude 38°S). The northern Chilean walnuts required very little chilling and succeeded well where papayas are grown. Astorga (personal communication) selected a few dwarf walnut forms that were very productive. They were introduced to California during the early 1980's and to Murcia, Spain, in 1986. Seedlings with columnar canopy and both gray and smooth bark, in Chile, like those found in the western and northern mountains of Spain, were found (Frutos, 1990b), although seedlings with round canopies were also found. The dry climate conditions of northern Chile keep walnuts free of bacteriosis caused by *Xanthomonas campestris p.v. juglandis* (Pierce) Dowson, even though this disease is very well known in Chile as "peste negra". On the other hand, walnut trees in which the stigmas remained green after the fruit had completed its growth were also common. Walnut orchards were planted in the fertile areas between the Coastal mountains and the high Andes, or in the interior valleys of the mountains. A few trees 140 or more years-old, yielded between 200 and 250 kg of nuts per year (Astorga, 1969).

Grafted materials from California were introduced in the Quinta Normal, near Santiago. The seedlings coming from this introduction are given the name of the cultivar of origin, followed by the word 'Seedling' (*e.g.*'Payne', 'Eureka', 'Franquette', 'Mayette', 'Placentia', and 'Ehrhardt Seedlings'), as they produce nuts that resemble the parental cultivars, and are known as 'Californian' type. Kernel percentage vary widely, and can reach up to 59 %. Flowering is normally late, the 'Payne' type is usually the first to leaf out, and the 'Franquette' type the last. The most variable character of the Chilean walnuts seems to be their flowering date (Gil and Saavedra, 1979). The 'Mayette' and 'Franquette' types behave better in the higher places in the mountains and in the cool area of Angol, as late blooming allows them to escape injury from spring frosts. In frost-free sites, other types yield more.

California. The walnut industry in California began when Joseph Sexton introduced seeds from Chile in 1887, although some believe that it originated in Wolfskill, near Los Angeles, from walnuts introduced from Spain. Some of the seedlings produced nuts known as 'papershell' because of their thin shells. They were open-pollinated with thick shelled Spanish walnuts introduced by the missionaries, and thus developed the 'Santa Barbara' or 'softshell' type (Beede, 1985). The oldest Californian cultivars, such as 'Placentia', 'Pride of Ventura', 'Ehrhardt', 'Wasson' and 'Chasse', developed from the 'Santa Barbara' type (Serr, 1969). The papershell character may have been introduced either from Chile or Wolfskill, but its origin could have been a Spanish walnut related to a cultivar with no shell found in Argentina, near Mendoza (Welkerling de Tacchini, 1960).

2.4 Introductions of other European walnuts into the United States

Another introduction of walnuts from Europe to Pennsylvania in the 18th century (Grimo, 1979) led to the development of the old varieties 'Hansen', 'Breslau', 'Jacob', 'Desming' and others (Brook and Olmo, 1972). On the other hand, Paul Crath later introduced walnuts from the Carpathian mountains to the Northern United States (Grimo, 1979). At present, all walnut cultivars adapted to the Northern United States are known as 'Carpathian walnuts' (O'Rourke, 1979).

2.5 Walnut cultivars of California

The appearance of the cv. 'Payne' represented an important step because it replaced all the 'Santa Barbara' cultivars (Serr, 1969). On the other hand, in a few Northern California's counties and in Oregon 'Franquette' was successfully introduced during the middle of the 19th century (Beede, 1985). It has good nut quality and resistance to spring frosts, although its production was low (Serr, 1969). Another seedling introduction from Iran to California during the last decade of the 19th century produced the cultivar 'Kaghazi'; one of its descendents, 'Eureka', was planted in many orchards. 'Eureka' was the origin of 'Waterloo' and 'Marchetti'; 'Marchetti' is a hybrid between 'Eureka' and 'Payne'. Another important variety, 'Hartley', was selected in 1915 in Napa, and is probably a seedling of 'Mayette'. Because of its good quality, 'Hartley' has remained the most important Californian variety (Beede,1985), in spite of its sensitivity to 'deep bark canker' (Moller and Ramos, 1982).

The cultivars released by the University of California were based upon these prior materials plus some few more. They mainly incorporate the characters of lateral bearing, later flowering than 'Payne', good production and yield, and clear kernel color. Three hybrids between 'Franquette' and 'Payne' were introduced as 'Midland', 'Vina' and 'Pioneer'; and 'Waterloo' crossed with 'Payne' gave the cultivars 'Lompoc', 'Gustine' and 'Thehama'. Likewise, a cross between one introduction from Afghanistan, 'P.I. 159568', with 'Payne' produced the cultivar 'Serr'. The 'P.I. 159568' mother plant grew well at 2,400 m in Afghanistan. A 'Payne' x 'Conway Mayette' cross produced the cultivar 'Pedro'. The cultivars 'Chico', 'Amigo' and the selection 'UC 56-224' were obtained from the cross between 'Sarkey' and 'Marchetti' (Brooks and Olmo, 1972). 'UC 56-224' was crossed with 'Pedro' and yield 'Chandler' and 'Howard'. They can be pollinated by 'Cisco', a late leafing cultivar from the cross between 'Meylan' x 'Pedro' (McGranahan *et al.*, 1990). Another important cultivar, 'Sunland', originated from 'Lompoc' and 'P.I. 159568', (Hendricks *et al.,* 1985), and the last introduction, 'Tulare', is a hybrid between 'Tehama' and 'Pedro' (McGranahan *et al.*, 1992).

Most of the cultivars from California have been introduced in Mediterranean countries and some information about their behavior is available. So, in Caserta (Italy), twelve Californian cultivars came into production in the second year, earlier than 'Sorrento' (Manzo and Tamponi, 1973). Period of harvest in 'Serr' lasted for twenty four days, and in 'Franquette' for only ten. When the fallen seeds were left on the ground for twenty five days, the kernels of 'Pedro' and 'Serr' didn't darken, whereas those of 'Eureka', 'Trinta', and 'Vina' did (Radicatti *et al.*, 1993). Production of 'Serr', 'Chico' and other Californian cultivars has been excellent in Reus, Spain (Tasias, 1982). In Puerto Lumbreras, Murcia, with a dry and hot summer and mild winter conditions, an isolated tree of 'Serr' bears well without any pollinizer. Likewise, in the Alpujarras's mountains of Almeria, in Southern Spain, some Californian cultivars are planted from 400 to 800 m above the sea. In this area, the trees differentiate flowers even in the year of planting.

Rapid entry into production and high yield of 'Amigo', 'Pedro', 'Payne' and 'Serr' were noted in Alcobaça, Portugal (Gomes, 1988). 'Tehama' and 'Pedro' perform well in Southern Bulgaria (Anadoliev and Anadoliev, 1988).

In Mediterranean countries two main factors restrict introduction of Californian walnut cultivars: bacteriosis and spring frosts. Bacteriosis is the most important walnut disease in France, and may reduce the crop by up to 80% in years with heavy spring rains, its control requires four to seven copper spray treatments (Gardan *et al.*, 1993). On the other hand, in Reus, Spain, with its high air humidity due to its proximity to the sea, and its mild temperatures, ten or more treatments are usually necessary to control bacteriosis (Tasias, personal communication). Honey dew of aphids helps to spread bacteriosis, even during hot and dry summers, by maintaining high humidity due to the hygroscopic characteristics of the molasse. For this reason, Californian varieties fail in the Central Coast of Portugal (Germain, 1987).

Spring frosts also limit the spread of Californian cultivars in the mountain regions of Italy (Minotta, 1990), although Southern and Central Italy seem satisfactory (Barone *et al.*, 1990; Barone and Zappia, 1993; Belisario, 1992). Local microclimate defines the upper limit of walnut culture especially with regard to frosts risk.

2.6 Walnuts in other warm climates

Diverse international cooperation programs have advanced walnut selection in all the Southern European countries. Two Portuguese native selections, 'Arco' and 'Rego', have been introduced (Gomes, 1988). Fourteen clones from Northwestern Spain are at present under study in Tarragona, Spain. This plant material, which seems to be moderately sensitive to bacteriosis, leafs out before the 10th of April. Four of these clones are lateral bearers. Other Spanish selections, mainly from Catalonia, are also included in this collection (Aletá and Ninot, 1991). Selections of the Centro de Investigación y Desarrollo Agroalimentario (CIDA) are reported below.

Selections from Feltre and Cassiomaggiore, Italy, named 'Feltrine', 'Comune' and 'Grossa de Vigni', produced from 30 to 35 kg of nuts per adult tree (Rizzotto, 1960). The characteristics of 'Noce di Cerreto' and 'Gigante di Volpara' have been also reported (Fregoni and Zioni, 1964). The name 'Sorrento nut' is considered synonymous with the general characteristics of the walnut fruits of Campania (Limongelli, 1986). The characteristics of 'Tonda di San Martino' (Sannino and Pilone, 1986), and 'Malizia' (Sannino and Pilone, 1986; Limongelli, 1991) have also been pointed out. On the other hand, in Chico (California), 'Sorrentina', an introduction from Italy, was evaluated as a productive tree that seemed compatible with *Pterocarya stenoptera* C.DC. The fruits of 'Sorrentina' are similar to those of 'Eureka' (Smith and Allinger, 1969).

Walnut selection in Fokida, Evritania and Ithiotis, Greece, introduced 'Big of Morilos', a new variety that produces in clusters, and several others that bear laterally. These selections usually have early leafing out, good productivity, fast entry into production and a good tolerance to bacteriosis (Rouskas and Katranis, 1988). On the other hand, 'Valbona', 'Cipëholla', 'Gashit 1' and 'Gashit 2' varieties were selected in the Tropoj's region, Albania (Berisha, 1987).

Walnut selection work in Turkey led to preselection of 116 old seedlings, but at present the attention is focused on nine of them: 'Yalova 1', 'Yalova 2', 'Yalova 4', 'Kaplan 86', 'Sebin', 'Bilecik', '24KE 25', '04 KH 2' and '04 KR 1'. Some of the last named ones are considered to be resistant to *Xanthomonas* and *Carpocapsa* (Feratoglu, 1991). Also, in the Darende region of Turkey, 95 trees have been preselected from a population of about 9,000 old seedling trees; 69 were protandrous, 13 protogynous and 13 homogamous. Only two had light shell and yellow kernel color (Sen, 1991).

Native Iranian walnut populations are mainly found in the northern forests, from Astara to Golidaghi, from the Caspian Sea to 1,500 m elevation, and in the western forests, from Marivan to Sanandaj. Ten superior clones have been selected out of 250 preselected ones (Atefi, 1990). 'K21' and 'K 312' come into production quickly; 'G 3' has 100% of lateral bearing; 'Z 63', that has a second flowering period and their flowers are in female catkins, and 'Z 51', with two flowering period also. Other Iranian selections are 'Z 60', 'Z 53', 'B 21' and 'B 1/3'. Grafted walnuts are at least three times more productive than seedlings ones (Atefi, 1998). Finding two years-old seedlings in the nursery bearing nuts is not unusual, as was seen in Nazarabadh. The quality of some walnut selections is very good, with large size and light color kernel (Frutos, 1995).

Walnut selections from India differ greatly in their percentage of kernel. 'Govinda' reached a kernel percentage of 52.8 (Chauan and Sharma, 1979). Other Indian varieties, including 'Hannual Kagzi', 'Siri Kagzi' and 'Kuma Kagzi', were recommended by Lal and Sing (1978). Botha (1998) pointed out that three groups of cultivars have been established in Himachal Pradesh: a. recommended, ('Govinda', 'R.K. Selection', 'Kahsmir Budded' and 'Partap'), b. introduced, ('Lake English', 'Tutle 16', 'Kx Giant', 'Payne' and 'Hartley'), and c. under selection ('K.S.Bagh', 'Partap' and 'Normal Selection'). Likewise, a program of propagation and selection of walnuts is yielding good results in Jammu and Kahsmir State, in which no bacteriosis symptoms in native trees have been found (Frutos, 1997 and 1998). The best quality walnuts from India come from the hilly regions of Jammu and the Kashmir valley. Nuts produced over 1500 m. elevation are considered superior (F.A.O., 1995). Indian introductions leaf out early to mid spring at Chico, California, and their nuts are of good size, but the kernels are difficult to separate from the shell (Smith and Allinger, 1969).

3 Cultivation practices

3.1 Rootstocks and grafted walnuts: the blackline problem

The main walnut rootstocks belong to species *Juglans regia* L., *J.nigra* L., *J.hindsii* (Jesp) Rehder, 'Paradox hybrid' (*J. hindsii x J. regia*) and *Pterocarya stenoptera* (C. D.C.). This latter species is widely spread across Northern China and is used mostly on poorly drained soils, its compatibility with most walnut cultivars is poor (Germain *et al.*, 1975). 'Paradox' is the hybrid rootstock most widely used. However, other hybrids, such as *J. regia* x *J. Mandshurica* Maxim. (Balashov, 1978), and *P. stenoptera* x *J. regia* (Mc Granahan *et al.*, 1986) have been reported.

Several investigations have examined rootstock adaptability to soils (Aletá, 1987; Besarov, 1973; Catlin, 1978; Catlin and Olson, 1986; Catlin *et al.*, 1977; Charlot,1987; Chernobay, 1971; Girona *et al.*, 1993; 1975; Kairov, 1975; Kovalenko *et al.*, 1976; Lagerstedt and Robert, 1972; Maurer, 1960; Meliksetyan, 1987; Rusu, 1974; Serr, 1960; Solignat, 1974; Tkhagushev and Klopenkova, Thomson and McComb, 1962; Zielinsky, 1952), resistance or tolerance to soil diseases (Green and Pratt, 1970; Matheron and Mircetich, 1985; Mircetich *et al.*, 1978), nematodes (Lounsberry *et al.*, 1974; McGranahan *et al.*,1986; Scotto, 1976, 1988), and effect on growth of grafted walnut varieties (Bauckman, 1974; Browne *et al.*, 1977; Lagerstedt and Robert, 1972; Maurer, 1974; Nedev, 1972 and 1973; Nedev and Mileva, 1973; Vishanka *et al.*, 1980).

Early works reported an incompatibility between rootstock and scion in walnut, called blackline (Glenn, 1966; Germain *et al.*, 1975; Hamilton and Rizzi; 1966; Miller *et al.*, 1952; Serr and Forde, 1959; Zielinski, 1952, 1953, 1955). A study carried out in Oregon, USA, on a population of 10,000 walnut trees showed that 45% of the trees were lost due to blackline. This disorder occurred on *J. hindsii* rootstocks but never on *J. regia* rootstocks (Miller *et al.*, 1952 and 1953; Miller *et al.*, 1975). This led to the recommendation to always use *J regia* c.v. 'Manregian' seedlings in Oregon, the high vigor of this stock permitted grafting in the nursery. In California, blackline was first detected in 1929, near San Francisco (Mircetich, 1979), and later was found in orchards in the Sacramento and San Joaquin valleys (Hamilton and Rizzi, 1966). All the affected trees were grafted on *J. nigra*, *J. hindsii* or *J. californica* species, or on 'Royal' (*J. nigra* x *J. regia*) or 'Paradox' hybrids (Serr and Forde, 1959; Martin and Forde, 1975). 'Eureka' and 'Payne' were the most susceptible cultivars (Mircetich *et al.*, 1980). This disease was also observed on *J. nigra* in England (Glenn, 1964), in Germany (Maurer, 1967) and in France (Gagnaire and Valier 1968).

Blackline occurs as a reaction to Cherry leafroll virus (CLRV) when it infects hypersensitive cells of rootstocks grafted with tolerant Persian walnut (*J. regia*) cultivars, (Mircetich *et al.*, 1980). This explains why blackline never has been found on J. regia rootstocks. The CLRV occurs in Italy (Savino *et al.*, 1977), France (Dosba *et al.*, 1986), Hungary (Nemeth *et al.*, 1990), Bulgaria, (Lazarova, 1990) and Spain (Frutos and Hita, 1993). This led to the use of *J.regia* in Spain as the only walnut rootstock, as in Oregon (Zielinski, 1958; Lagerstedt and Robert, 1972). This resulted in selection of *J. regia* rootstocks, and in development of means to improve seed germination (Frutos, 1983a).

Selection of *J. regia* rootstocks for heavy soils was successful (Rusu, 1976; Aletà, 1987; Frutos, 1990), but selection for resistance or tolerance to salinity seems to be more difficult (Browne *et al.*, 1977; Bessarov, 1973; Girona *et al.*,1993). However, J. regia is the most drought-resistant of the Juglans species (Meliksetyan, 1987).

Tolerance to *Armillaria mellea* (Vahl) Kummer, seems better in *J. hindsii* than in *J. regia*, (Mircetich *et al.*, 1978). However, three *J. regia* seedlings from a sowing of an old seedling were observed as tolerant to *A. mellea* in Tarragona, Spain, were tested also tolerant in Murcia. Another fungus that kills walnuts in orchard, *Armillaria tabescens* (Scop. ex Fr.) Sing., has been observed in Ronda, Spain. Phytopththora has never been reported infecting walnuts in Spain (Frutos, 1990 b). However, this fungus has killed *J. regia* trees in Chile.

Crown gall (*Agrobacterium tumefaciens* Smith and Towsend) is able to infect and to produce tumors in *J. hindsii*, 'Paradox' and *J. regia* (Anonymous). This tumors have been also observed even in the union of *J. regia* grafted trees, when wounds were infected by grafting tools.

Nematodes *Pratylenchus vulnus* Allen and Jenson, are able to infect all Juglans species, whereas 'Paradox' seems to present diverse degrees of susceptibility, and *Pterocarya. stenoptera* is immune (Lownsberry et. al., 1974). In France, *P. vulnus*, *Meloidogine* sp., *Cacopaurus pestis* (Thorne) and *Criconemella xenoplax* (Raski) have been found in walnuts, but none of them has produced significant damage (Scotto, 1988). In Greece, the presence of *Tylenchus davainei* (Bastian) has been found in walnut (Vlachopoulos, 1991).

3.2 Fruits bearing pattern

The juvenile period of *J. regia* seedlings is normally 10 years or more. However, reports from the former Soviet Union have described walnut seedlings with Short Juvenile period (SJP) (Petrosian and Antonenko, 1972). Similar reports have come from California (Smith *et al.*, 1960), from Spain, in seedlings of 'Chico' and 'Serr' sent from California (Frutos, 1983 b), and from Ukraine, in seedlings from Central Asia (Kanivets and Tyzh, 1990). Seedlings with the SJP character produce nuts during the 1st, 2nd and the 3rd years after sowing. In Karaj, Iran, seedlings were selected in the nursery one or two years after sowing with the purpose of planting orchards with ungrafted, but productive, walnut trees (Atefi, personal communication; Frutos, 1995).

The SJP character was also observed in Torreblanca, Spain, in 1981 and 1982 after planting seeds of 'Chico', 'Serr', 'Manregian' and 'Spurgeon' obtained from the United States in the open field in 1980. Likewise, nuts from 'Franquette' and 'Lozerone' were obtained from France, and seeds of 'Garganta' from Spain. One- and 2-year-old seedlings of 'Chico' and 'Serr' had pistillate flowers. The SJP character was transferred by 'Chico' to 50% of its progeny, and by 'Serr' to only 7.8 %. Seedlings of all other cultivars failed to produce any SJP seedling.

The source of this characteristic was thought to be 'Payne', a common parent of 'Chico' and 'Serr' (Frutos, 1983 b), but a later study did not verify this. However, the other parent of 'Chico', 'Kaghazi', is an introduction from Iran. 'Serr' is a descendant of 'P.I. 159568', a seedling from a tree from Afghanistan. Therefor, both parents, 'Kaghazi' and 'P.I. 159568', came from the Central Asian walnut population, in agreement with Kanivets and Tyzh (1990) and with the observations made in Iran during the summer of 1995 (Frutos, 1995).

A high percentage of lateral bearing favors early production. However, lateral bearing is not so important in adult orchards, because adult trees have more bearing buds than fruits they can support. When thinning young walnut trees, cultivars that bear many lateral flowers require limbs thinning and also heavy heading-back cuts to improve vigor. In cultivars with few lateral flowers, only thinning is recommended (Freeman and Aldrich, 1985). These pruning procedures are recommended when trees are 5- to 10- years-old. At this age, light is still not a limiting factor, and competition for nutrients seems to be the more important factor limiting growth in fruit-bearing shoots. However, when the orchard becomes older, light incidence inside and in the lower part of the tree must be improved to avoid serious loss of production. Cultivars with many lateral flowers perform well in hedgerow systems (Ramos *et al.*, 1985)

3.3 Pollination

J. regia is monoecious and dichogamous species, but total or partial homogamy (coincidence of female receptivity and pollen shedding) can occur. Self-pollination results in normal fruits, but dichogamy favors outcrossing.

During the first stages of receptivity pistillate inflorescences can abscise. This phenomenon, known as Pistillate Flower Abscission (PFA), was first observed in the *cv.* 'Serr' in California in 1978. One year later, PFA of 'Serr' was evaluated in 11 locations in California, and ranged from 8 to 87 %. There was no correlation between PFA and location, variability was high even among limbs of the same tree. No association could be established between PFA and nutrition, diseases, low winter chilling, cold injury, wind, rain, copper sprays, mites, rootstock, organ competition, pruning or shading (Catlin and Ramos, 1985; Catlin *et al.*, 1987). Later, pollen source was show to be a factor directly involved in PFA (Por and Por, 1990; McGranahan *et al.*, 1994).

In 1995, PFA was recorded in two walnut collections of the CIDA located in Abarán and in Torreblanca, Murcia, Spain, at 38° N latitude. Climatic data for these locations are described in section 3.4. In both places walnuts were free of CLRV according to ELISA test. Nevertheless, PFA as described by Catlin and Ramos (1985) was observed in both plots (Table 3). In Abarán, 22 out of 30 cultivars exhibited PFA, while in Torreblanca, only 7 of 27 did. 'Alcalde 1', 'Alcalde 2' and 'Alcalde 3' selections and their mother plant, 'Taibilla Madre', exhibited PFA in Abarán, but only two of them, 'Alcalde 1' and 'Alcalde 3', did so in Torreblanca. The PFA percentage was always higher in Abarán, except for 'Franquette' and 'Algaida' (Table 3).

Table 3. Pistillate flower abscission (% of total), in 21 walnut cultivars in Abarán and in Torreblanca, Murcia, during 1995

	Location			*Location*	
Cultivar	*Abarán*	*Torreblanca*	*Cultivar*	*Abarán*	*Torreblanca*
Arriba	52.6	0	Serr	23.5	0
Badajoz	50.0	22	Nugget	16.0	0
Chandler	42.5	0	Hartley	15.0	0
Amigo	40.0	0	VZ-6	12.7	8.8
Taibilla Madre	40.0	0	Alcalde 2	12.9	0
Payne	40.0	0	Howard	12.5	0
Chico	36.0	0	Pedro	0	0
Alcalde 3	34.1	8.3	Carril	0	0
Alcalde 1	31.1	9.5	Franquette	0	8.8
Sundland	29.0	0	Algaida	0	48.9
Tobilla	27.3	0			

Comparative data (Table 3) suggest that PFA is a complex phenomenon involving factors other than pollen density on the stigma, as suggested by Por and Por (1990) and McGranahan *et al.* (1994), because some cultivars had much lower PFA after warm winter conditions. Nevertheless, PFA increased in a few cultivars, including 'Algaida' and 'Franquette' in Torreblanca, in warmer conditions. In both collections the amount of pollen at the time of pistillate bloom presumably was abundant. But the latest cultivar in flowering, 'Franquette', a protandrous cultivar with short period of homogamy in Torreblanca, only had available for fecundation its own pollen, in short amount, and in spite of it, PFA was recorded in this cultivar. On the other hand, 'Serr', with a high PFA in California (Catlin *et al.,* 1987), had PFA values of 23.5% and 0% in Abarán and Torreblanca, respectively (Table 3). This suggests that the behavior of walnut cultivars should be re-evaluated in warm climates even though they have high PFA levels in cooler locations.

Another type of pistillate flower drop occurs in walnut. In this case, the base of the ovary on some flowers abscises whereas those of other flowers on the same peduncle do not (Catlin and Ramos, 1985).

3.4 Chilling and heat requirement

To evaluate a wide range of walnut cultivars, two locations were chosen in Murcia at latitude 38° N. The first in Torreblanca, near the Mediterranean Sea, has a mild winter with about 120 hours below 7°C. In the exceptionally cold winter of 1991/92 a total of 150 hours below 7°C were recorded. Temperature averages of January, of the three coldest months (December, January and February), and of the two warmest ones (July and August) were 11.1, 11.8 and 25.5°C, respectively. At the second location in Abaran, the corresponding three temperature averages were 10.2, 10.7 and 25.4°C, respectively. The average number of hours at temperatures below 7°C in Abarán is 260, but in the cold 1990/91 winter 1300 hours were recorded. The two locations are 70 km apart. In Abarán frosts are frequent in winter, but winter frosts seldom occur in Torreblanca, with its subtropical Mediterranean climate. Summers in Torreblanca are milder and longer than in Abarán because of its maritime climate. Both locations are free of spring frosts. The trees were grafted on J regia seedlings. The older walnuts were planted in 1981 and 1984 in Torreblanca and Abarán respectively. Spanish and Californian selections were included. Spanish selections were collected from 1600 to 1300 m elevation ('Carril', 'Arriba', 'Gran Jefe', 'Enano', 'Cabra 1' and 'Cabra 2'); between 1300 and 1000 m ('Taibilla Madre' and its seedlings 'Alcalde 1', 'Alcalde 2', 'Alcalde 3', 'Alcalde 4'; and the selections 'Mollar de Germán' and 'Tobilla'), all of them from Nerpio, Albacete; and the series 'VZ', which originated in Ventas de Zafarraya, Granada. Below 300 m elevation ('Callao', selected in Agramón, Albacete, and 'Algaida' and 'Carcagente', selected by Luna in Murcia and Valencia, respectively, and 'Badajoz', selected by Silva near Olivenza) (Table 4). Short Juvenile period (SJP) walnuts were added to the collection: 'Mar Menor'a seedlings of Chico, that flowered in the second leaf after sowing; '15 A-1', '33 A-1', '14 A-1', '15 A-3', '16 B-6', '15 A-2', '14 B-1' and '15 B-2', also seedlings of 'Chico'; 'and '9 A-1', '9 C-1' '23 B 1' and '23 B-2', seedlings of 'Serr' that bore flowers in the third leaf too. All these seedlings were nominated as Short Juvenile period (SJP) On the other hand, 'Payne', 'Hartley' and 'Franquette' were included as standard varieties. The rest of the varieties in the collection were the Californian 'Sunland', 'Howard', 'Amigo', 'Chico', 'Pedro', 'Serr', and 'Theama', and the Portuguese 'Arco' and 'Rego'. Later, additional Spanish selections were introduced.

a. Bud break and flowering

In Torreblanca, most selections began leafing out during the last decade of March and finished in the beginning of the first decade of April. 'Franquette' leafed out during the 2nd decade period of May. Some Spanish selections as 'Taibilla Madre' and 'Carril', coming from the Nerpio's mountains, leafed out in the collection during the 3rd decade of March, before 'Hartley' and after 'Payne', the first cultivar to break dormancy. This indicates a relatively low chilling requirement in 'Taibilla Madre' and 'Carril', which avoid spring frosts in their native habitat probably by requiring a high sum of heat accumulation before bud break. This usually happens late enough to reduce damage from last spring frosts. Similar behaviors have been reported in Chico, in California, where 'P.I.159568' broke dormancy early, although its mother tree was located at 2440 m above sea level in Afghanistan. Other genotypes introduced from Manchuria, Iran and India also leafed out early, as did 'P.I.159568' (Smith *et al.*, 1960; Smith and Allinger, 1969). This indicates a lack of direct connection between cold location and a high chilling requirements. 'Franquette', on the other hand, has a high winter chilling requirement, and is slow in leafing out if placed in a greenhouse from September to December (Mauget, 1983; Mauget and Germain, 1988). 'Franquette' exhibits symptoms of insufficient chilling, like dry branch ends (Rizzi, 1977; Forde, 1979), in Torreblanca, Murcia.

Another mean of fighting spring frosts is by having a very long flowering period. That occurs in 'Cabra 1', a selection from 1600 m elevation that was the first to leaf out in Abarán, even before 'Payne', although it is the last to finish its flowering period, as is true of 'Franquette' (Fig. 1).

In many mountain environments, like the Alpujarra (Almeria, Spain), the presence of catkins on one-year-old scaffold branches have been recorded in young 'Howard' trees, and in other cultivars. This early catkins production is considered as a sign of good adaptation.

b. Leaf drop

'Carril' and 'Franquette' were the first to show leaf drop in Torreblanca. This occurred around December 5 (Fig. 2). Likewise, 'Taibilla Madre' and its seedlings 'Alcalde 1' and 'Alcalde 2' finished their leaf drop on December 10, as did 'Payne', 'Hartley' and 'Howard'. The early leaf drop could be related to a higher physiological temperature threshold or a longer photoperiod threshold in short cycle cultivars. This may favor adaptation to environments where early frosts occur. However, other varieties that keep their leaves longer, like 'Sunland', 'Badajoz', '15A-1', '23B-1', 'Chico', 'Serr' and 'Pedro', also have leafed out earlier in Torreblanca. A longer productive cycle

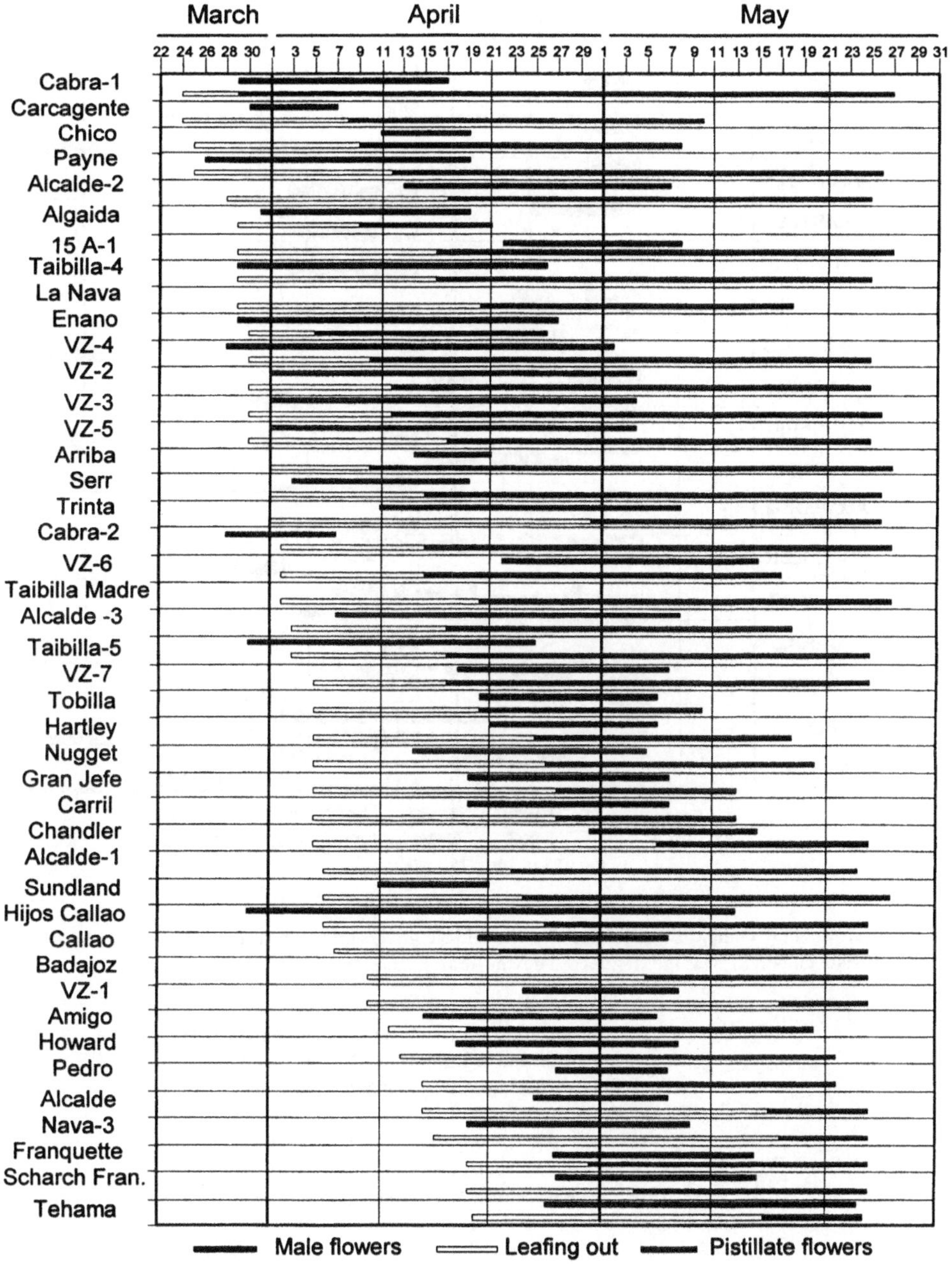

Figure 1. Leafing out, male and female pistillate flowering periode of walnut cultivars in Abarán (Murcia) in 1995.

should be advantageous in warm climates, as this improves competition for light and nutrients.

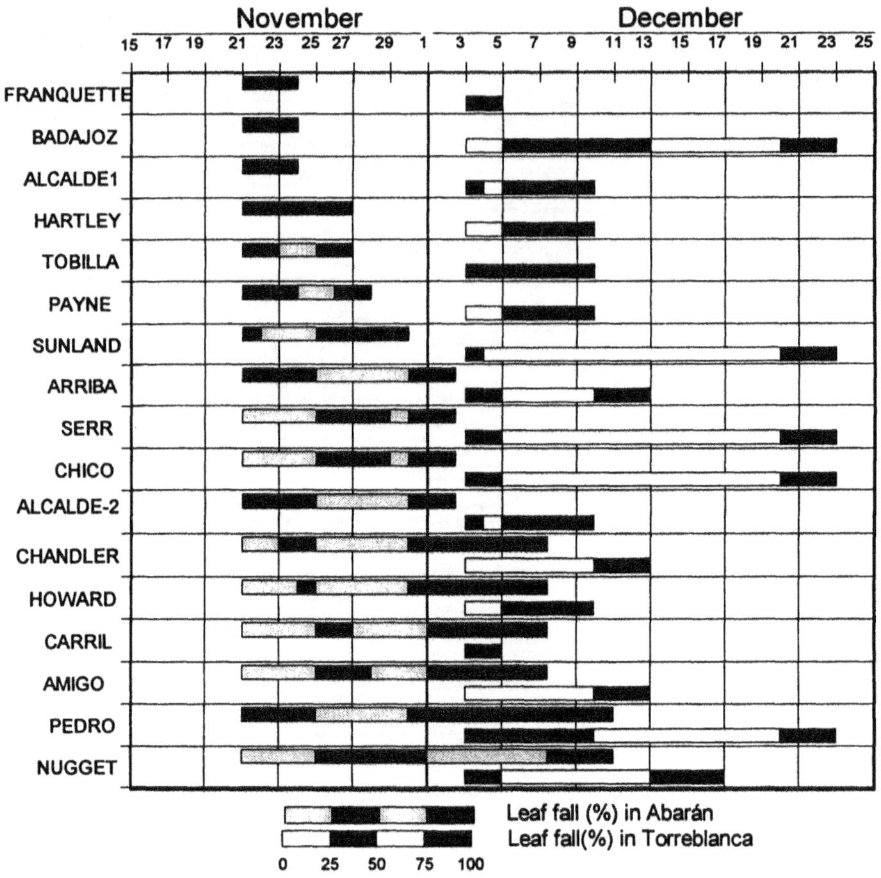

Figure 2. Leaf fall comparison, in 1991, of some walnut cultivars grown in Torreblanca and in Abáran, Murcia (Spain). November, 21, was the first record day in Abáran, and December, 3, in Torreblanca.

Table 4. Average production (kg/tree) of seven walnut cultivars in Abarán and in Torreblanca (Murcia) in 6th, 7th and 8th leaves, separated by Duncan's Multiple Range test at the 5% level

Cultivar	*Abarán*	*Torreblanca*
Chico	68.0	12.9
15 A-1	43.3	33.3
Serr	49.2	28.3
23 B-1	46.1	28.8
Pedro	59.6	3.0
Amigo	40.1	1.9
Hartley	36.5	1.2
Total	344.8 a	109.4 b

c. Production

The cooler Abarán location markedly enhanced production (Table 4). 'Serr' and one of its seedlings, '23B-1', performed well in Torreblanca, although the largest production was obtained from '15A-1', an open-pollinated seedling of 'Chico'.

Table 5. Age of first harvest, cumulative yield, trunk cross-section area and productivity per tree, in 17 cultivars of walnut at Abarán, Murcia, Spain

Variety	*Year of*		*First harvest*	*Accumul. harvests*	*Trunk cross section area*	*Productivity*	
	Planting	*First harvest*	*Kg /tree*	*Untill 1991 (Kg / tree)*	*(cm²)*	*g /cm²*	*Kg/ year*
Chico	1984	88	0.6	68.9	184.3	372.3	17.2
Pedro	1985	88	0.2	59.8	392.0	152.5	14.9
Nugget	1986	88	0.2	59.1	233.0	253.3	14.8
Serr	1984	88	0.8	50.0	297.7	167.8	12.5
Amigo	1984	-	-	40.1	-	-	-
Hartley	1984	-	-	36.5	-	-	-
Sunland	1986	89	2.5	25.0	115.5	119.5	8.3
Payne	1986	89	1.2	24.8	233.3	106.3	8.2
Arriba	1986	89	0.6	18.9	151.3	125.2	4.7
Carcagente	1986	88	0.6	18.9	151.3	125.2	4.7
Howard	1986	89	1.1	15.1	112.5	134.2	5.0
Gran Jefe	1984	89	2.0	14.8	414.7	35.7	4.9
Chandler	1986	90	4.6	14.5	128.0	44.1	7.2
Algaida	1984	89	0.9	13.8	246.4	56.0	4.6
Nava 3	1986	90	3.0	12.9	147.9	87.0	6.4
Trinta	1986	89	0.3	10.2	151.1	67.2	3.4
Franquette	1984	90	0.1	2.6	121.7	0.9	1.3

Production of 'Chico' itself was poor in this last location. All varieties performed better in Abarán, with 'Chico' and 'Pedro' the most productive cultivars. 'Serr', 'Nugget', 'Amigo', 'Hartley' and 'Sunland', also were good producers, whereas 'Franquette', 'Chandler' and 'Howard' cropped poorly. The Spanish selections 'Arriba' and 'Carcagente' were promising in Abarán (Table 5).

3.5 Pests and diseases

a. Insects

Moths. The most important pest of this group is the codling moth (*Laspeyresia pomonella* L.). Chemical control affects the natural fauna and may favor other pests, such as mites or aphids, that appear later. In the future, resistant cultivars of *J. regia* may be produced by introducing *Bacillus turingiensis* genes able to produce toxins to protect this species against codling moth, navel orange worm (*Amyelois transitella* Walker) and Indian meal (*Plodia interpunctella* Hübner) (Wood,1991; Dandekar *et al.*, 1992). In 1977, a new pest of walnut fruits in warm climates, *Apomielois ceratoniae* Zeller, was detected in Murcia (Lacasa, personal communication).

Aphids. At the end of summer the yellow walnut aphid, *Chromaphis juglandicola* Kaltenbach, usually appears. A parasitic insect, *Tryopsis pallidus* Halliday, resistant to methyl-azinophos and able to overcome the winter in walnut orchards in California, may control this pest (Hoy *et al.*, 1991; Hoy and Cave, 1991).

Walnut Husk Fly. This fly, *Rhagoletis completa* Cresson, well known on walnuts in the United States, appeared in Europe (Merz, 1991). In some walnut orchards in the Veneto area of Italy, damaged by this fly may reach 50% of the crop (Duso, 1991).

b. Diseases and resistance

The most injurious disease in South and Central Italy, one of the most important areas where walnuts are grown, is bacteriosis, caused by *Xanthomonas campestris* pv. Juglandis (Pierce) Dowson. Other diseases, such as anthracnose (*Gnomonia leptostyla* (Fr.) Ces. Et De Not.), *Microstoma juglandis* (Bér.) Sacc, *Ascochyta juglandis* Boltsh, *Gloeosporium epicarpii* Thüm., *Melanconium juglandinum* Kze., *Cytospora*, *Phoma* and *Phomosis* spp., are much less important (Belisario, 1992).

Bacteriosis can be controlled with copper applications that also controls other diseases as well. Continuous application of copper can produce some necrosis and drop of fruits in sandy and shallow soils (Radix and Seigle-Murandi, 1992). Unfortunately, copper-resistant strains of this bacteria have been reported in orchards (Gardan *et al.*, 1993).

Resistance to bacteriosis in Spain was found in some few old seedlings. One of these selections from Nerpio's mountains was introduced in the collection as CA-1.

The warm Mediterranean climate in Spain is characterized by a long and dry summer with a low incidence of bacteriosis, even for a very sensitive cultivar like 'Amigo'. Still the orchard must be treated with copper in spring and early summer because of the incidence of rains in the Spanish Mediterranean areas. The deep bark canker caused by the bacterium *Erwinia rubifaciens* Wilson *et al.*, seems not to be a problem in dry summer condition, even for the most sensitive cultivar 'Hartley'.

4 Summary

The successful culture of *J. Regia* L. in warm Mediterranean climates is possible using the diverse selections available that adapt well to such conditions.

The Iberian Peninsula and Central Asia sheltered *J. regia* during the last glaciation. It spread to Europe from the Iberian Peninsula during the Mesolithic and the Neolithic periods, and to the Mediterranean climates of the New World through Spanish missionaries. The Californian walnut industry began in the middle of the 19th century. Since then, new walnut materials and technologies have been developed in California that can contribute to walnut development in other warm regions

The performance of a wide selection of walnut cultivars of Spanish and foreign origin has been studied in Murcia, Spain, in two different Mediterranean climates. Important differences in behavior among cultivars have been reviewed in this study.

Cultivars that avoid damage by spring frosts have developed different strategies, such as chilling and heat accumulation, and a long period of flowering.

Pistillate flower abscission (PFA) is lower in most of the walnut cultivars tested in warmer Mediterranean conditions than in colder climates. PFA remained high only in 'Algaida' and 'Franquette' in warmer areas. This points to the influence of environment on incidence of PFA.

Special environmental conditions in some places in the Almeria mountains, in southern Spain, have allowed us to observe the differentiation of catkins in grafted trees during their first year of growth.

The sole use of *J. regia* seedlings as rootstock was recommended to avoid the potential threat of blackline, a scion-rootstock incompatibility caused by Cherry Leafroll Virus (CLRV). Selection work within this species revealed resistance or tolerance to Armillaria, and to waterlogged soils, but not to saline soils or water.

Phytophthora is not known to be harmful in the native walnuts in Spain, but an Armillaria like fungus able to kill walnut trees has been observed.

Among other pests and diseases, codling moth, aphids and *Xanthomonas campestris pv. juglandis* (Pierce) Dowson are important. A new moth, *Apomielois ceratoniae* Zeller has been found lately in walnut orchards in Spanish Mediterranean climates.

5 References

Aletá, N. (1987). El nogal, un cultivo tradicional con futuro. *Fruticultura Profesional*, 11, 55 -59.

Aletá, N. and Ninot, A. (1991). Exploration and evaluation of some Spanish native walnut (*J. regia* L.) populations from Catalonia and Galicia. *Acta Horticulturae*, 311, 17-23.

Aletá, N., Vargas, F., Romero, M.and Ninot, J. (1986). Nogal y pecanero: situación, avances e investigaciones en curso en el Centro Agropecuario Mas Bové. *I Congreso Español de Frutos Secos, Reus,* 1, 299-310.

Anadoliev, G.and Anadoliev, Z. (1988). Position and role of nut crops in fruit production (In Russian). *Selskostapanska Nauka*, 26 (6), 84-93.

Anonymous.- Walnut rootstocks. Univ. California Agr. Ext. Service, *Pub. AXT.* 120.

Astorga, E. (1969). El nogal. *El campesino*, n/ Octubre,54.

Atefi, J. (1990). Evaluation of walnut genotypes in Iran. *Acta Horticulturae*, 311, 24-27.

Atefi, J. (1998). Situation of walnut industry in Iran: past and current programmes of the walnut industry. *International workshop on walnut production in Jammu and Kashmir.* Srinagar, 15-17 June, oral communication.

Balashov, P. K. (1978). Biological characteristics of hybrid walnuts in lower Volga conditions (In Russian). *Bulletin'VNII Agrolesmelior.* 1(26), 41-42.

Ballouche, A. (1986). Paleoenvironnements de l'homme fossile holocéne au Maroc. Apport de la palynologie.

134 pp, *Thesis, Univ. de Bordeaux.*

Barone, E., Bounous, G., Inglese, P., Zappia, R., and Sepe, L. (1990). English walnut cultivars testing in southern Italy. *Acta Horticulturae* 284, 105-110.

Barone,E, Zappia, R. (1993). Some phenological observations on vegetative bud phases in English walnut cultivars. *Acta Horticulturae,* 311, 172-181.

Bauckman, N.M. (1974). Die lange mannlicher blüten (kätchen) auf veredelten. Mitteilungen Rebe und Wein, *Obstbau und Früchtewervertung* 24(6), 463-466.

Beede, R.H. (1985). California history. In *Walnut Orchard Management,* (ed. Ramos D.E.). Pub. 21410, Univ. of California, 3-4.

Belisario, A. (1992). Problematiche fitopatologiche del noce comune. *Informatore Agrario* 48 (28), 51-53.

Bessarov, S. F. (1973). The resistance of walnut to salinity. (In Russian). *Lesovedenic* (2), 25-33.

Berisha, R. (1987). Assessing the best forms of walnut in the district of Topoj'.(in Russian). *Buletini y Shkencave Bujq'sore* 26(4), 35-40.

Bottema,S. (1980). On the history of the walnuts (*Juglans regia* L.) in southeastern Europe. *Acta Bot. Neerl.* 29, 343-349.

Brooks, H. and Olmo, H.P. (1972). Register of new fruit and nut varieties. 2nd. ed., *Univ. of Calif. Press,* 613-640.

Browne, L.T., Brown, L.C. and Ramos, D.E. (1977). Walnut rootstocks compared. *California Agriculturae* 31 (7), 14 -15.

Carrión, J.S. and Sanchéz Gómez, P. (1992). Palinological data in support of the survival of walnut (*Juglans regia* L.) in the western Mediterranean area during last glacial times. *Journal Biogeography,* 19, 623-630.

Catlin, P.B. (1978). Root-soil interrelationships. *Walnut Short Course,* Univ of California, Davis, pp 53 57.

Catlin, P.B. and Ramos, D.E. (1985). Pistillate flower abscission. *Walnut Orchard Management.* Univ. of California. Div. Agric and Nat. Resources, 87 -90.

Catlin, P.B., Ramos, D.E., Sibbett, G.S., Olson, W.H.and Olson, E.A. (1987). Pistillate flower abscission of the Persian walnut. *HortScience,* 22 (20), 201 -252.

Catlin, P.B. and Olsson, E.A. (1986). Response of Eastern black walnut and Californian black walnut seedlings to waterlogging. *HortScience,* 21, 1379-1380.

Catlin, P.B., Martin, C.G., and Olsson, E.A. (1977). Differential sensitivity of *Juglans hindsii, J.regia Paradox* hybrid and *Pterocarya stenoptera* to waterlogging. *Journal Amer. Soc. Hort. Sci.* 102 (2), 101-104.

Charlot, G. (1987). Variétés de noyer. Que planter actuellement dans le Sud-Est. *Infos. CTIFL,* (35), 23-24.

Chauan, J.S. and Sharma, S.D. (1979). Phenotypic variability in walnut of Kinnaur district of Himachal Pradesh. *Indian Journal of Agricultural Science,* 49 (6),420-422.

Chernobai, G.M. (1971). The root system of walnuts under unirrigated condition in the Crimean steppe (In Russian). *Trudy Gosudarstiennogo Nikitskogo Botanicheskogo Sada* 52,119-123.

Ciampolini, M.and Trematerra, P. (1992). Diffusa presenza di mosca delle noci (*Rhagoletis completa* Cresson) nel nord d'Italia. *Informatore Agrario,* 48, 52-56.

Dandekar, A.M., McGranahan, G.H., Uratsu, S.C., Leslie, C.,Vail, P.V., Tebbets, J.S., Hoffmann, D., Drive, J., Viss, P. and James, D.J. (1992). Engineering for apple and walnut resistance to codling moth. *Proceedings* of *Brighton Crop Protection Conference, Pest and Diseases.*

Dosba, F., Lansac, M., Mazy, K. and Rovira, M. (1986). Le virus du Cherry Leaf Roll (CLRV): relations avec le dépérisement du noyer et comportement de differentes especes ou hybrides interspecifiques de Juglans. *1er Colloque Noyer-Noisetier AGRIMED,* Roma, 25 -27 of June 1986.

Duso, C. (1991). Sulla comparasa in Italia de un Tetrifide Neartico del noce: *Rhagoletis completa* Cresson.

Bolletino di Zoologia Agraria e di Bachicoltura, 23,2, 203-209.

Elias, E. and Gimenez, R. (1965). Evapotranspiraciónes potenciales y balances de agua en España. *Mapa Agonómico Nacional Ministerio de Agricultura.*

FAO. (1997). Production Yearbook Vol 51. Rome, Italy. *Food and Agriculture Organization of the United Nations.*

FAO. (1995). Assitance for transfer of technology on vegetative propagation of walnuts (*Juglans Regia* L.). *Project TCP/IND*/4554. 13 pp.

Feratoglu, Y. (1991). The characteristics of walnut cultivars obtained though selection. *Acta Horticulturae*, 311, 34-36.

Freeman, M.W. and Aldrich, T.M. (1985). In *Walnut orchard management*, Coop. Ext. Serv., University of California, 57-61.

Fregoni, M. and Zioni, E. (1964). Contributo allo studio di due cultivar di noce della Val Borbera (Alessandria). *Ann. Fac. Agr. Univ-catol.* S. Cuore, Milano, 4, 286-94.

Forde, H.I. (1979). Persian walnuts in the Western United States. In *Nut tree culture in Noth America*, (ed. Jaynes R.A.). North Nut Growers. Ass., 84 -87.

Frutos, D. (1983a). Siembra directa en campo de semillas de *Juglans regia* L. tratadas con ácido giberélico (GA_3) y comportamiento en vivero de sus descendencias. *I Congreso Nacional, Soc. Española de Ciencias Hortícolas* 561-572.

Frutos, D. (1983b). Corto periodo juvenil en plantas de semilla de nogal (*Juglans regia* L.) *I Congreso Nacional, Soc. Española de Ciencias Hoticolas*, 573 -578.

Frutos, D. (1990). Selection of Spanish walnuts (*Juglans regia* L.), *Acta Horticulturae*, 284, 111-119.

Frutos, D. (1995). Walnut propagation research: Hypocotyl grafting in Juglans regia L. Consultation task report in Karaj, Iran. *Project FAO IRA / 89 /* 029, 27 pp.

Frutos, D. (1997). Report about a walnut consultation visit to Jammu and Kashmir State, India. *Project FAO: TCP/IND/4554.* 28 pp.

Frutos, D. (1998). Consulta sobre nogal en Jammu and Kashmir, India. *Informe FAO, proyecto TCP/ IND/8821*: 14 pp.

Frutos, D. and Hita, I. (1993). Nota sobre detección del virus del enrollado de la hoja del cerezo (CLRV) en selecciones españolas de *Juglans regia* L.. *Investigación Agraria, Producción y Protección Vegetales*, 8 (2), 275-277.

Gagnair, J. and Vallier, C. (1968). Variation des taux de potasium dans les feuilles du noyer Juglans regia/ J.nigra cultivés dans les conditions naturels. *Compte rendu Academie Agric. Francaise.*, 54, 81-86.

Gardan, L., Brault,T., and Germain, E. (1993). Copper resistance of *Xanthomonas campestris* p.v. *juglandis* in French walnut orchards and its association with conjugative plasmids. *Acta Horticulturae,* 311, 259-265.

Germain, E., Leglise, P. and Bayol, M. (1975). Les porte-greffes du noyer. In *Le noyer*, INVUFLEC, V, 53-72.

Germain. E. (1987). Compte rendœ de missión au Portugal. *AGRIMED.*

Gil, S.G. and Saavedra, E. (1979). Variabilidad de algœn caracter agronómico del nogal en huertos chilenos. *Ciencia e Investigación Agraria* 6 (3), 201-212.

Girona, J., Cohen, M., Rodriguez, I. and Mata, M. (1993). Walnut (*J. regia*) seedling responses to different levels of Cl and Na in irrigation water. *Acta Horticulturae* 311, 191-200.

Glenn, E.M. (1966). Incompatibility in the walnut. *Annual Report East Malling Research Station*, 49,102.

Gomes, J. (1988). Comportement de quelques varietés de noyer au Portugal. *2ème* Colloque *AGRIMED* noyer - nisettier. Bordeaux.

Green, R. Jr. and Pratt, R.G. (1970). Root rot of black walnut seedlings caused by *Phytophthora citricola. Plant Disease Reports* 54, 583-585.

Grimo, E. (1979). Carpathian (Persian) walnuts. *In: Nut tree culture in North America.* (ed. Jaynes R.A.). North Nut Growers Ass., 74-83.

Grimberg C.and Svaström, R. (1983). *El alba de la civilización.* Ediciones Diamond, pp 58 and 76.

Hamilton,W.D.and Rizzi, A.D. (1966). *Blackline in walnut trees.* One sheet-answer. Univ. of California, Extension Service.

Hendricks, L.C., McGranahan, G.H., Ramos, D.E., Iwakiri, B., and Forde, H.I. (1985). Selection of varieties. In *Walnut orchard management*, Coop. Ext. Service, University of California, 46-51.

Hoy, M.A. and Cave, F.E. (1991). Genetic improvement of a parasitoid: response of *Trioxys pallidus* to laboratory selection with azin-phosmethyl. *Biocontrol Science and Technology* 1 (1), 31-41.

Hoy, M.A., Cave, F.E. and Caprio, M.A. (1991). Guthion resistant parasite ready for implementation in walnuts. *California Agriculture* 4515, 29 -31.

Huntley, B.and Birks, H.J.B. (1983). *An atlas of past and present pollen maps for Europe: 0-13.000 years ago.* 667 pp, Cambridge University Press, Cambridge.

Hutin, C. (1996). Il mercato europeo delle noci. *Frutticoltura,* 12, 29 -32.

Kanivets, V.I. and Tyzh, R.M. (1990). Precocious type of walnut. *Sadovostvo y Vinogradarstvo,* 12, 30-34.

Kelly, M.G. and Huntley, B. (1991). An 11000-year record of vegetation and environment from Lago di Martino, Latium, Italy. *J. Quat. Sci.* 6, 209-224.

Lagerstedt, H.B., and Robert, W.W. (1972). Walnut grafting in Oregon. Problems and solutions. In *63th Ann. Rep. North Nut Grwers Ass.*

Lal, H. and Sing, R.D. (1978). Some promising walnut strains in Chacrata Hills of Uttar Pradesh. *Progressive Horticulture,* 10 (1),61-65.

Lazarova, N. (1990). Detecting Cherry Leaf Roll virus in common nut (*Juglans regia* L.). (In Russian). *Rasteniev 'dni Nauki,* 27 (6), 78 -83.

Limongelli, F. (1986). Selezione clonale de la noce di Sorrento. *I colloque noyer- noisetier, AGRIMED.* Rome.

Limongelli, F. (1991). A new walnut cultivar: 'Malazia'. *Acta Horticulturae* 311,46-48.

Lounsberry, B.F., Martin, G.C., Forde, H.I. and Moody, E.H. (1974). Comparative tolerance of walnut species, walnut hybrids and wingnut to the root lesion nematode *Pratylenchus vulnus. Plant Disease Reporter* 58 (7), 630 -633.

Manzo, P. and Tamponi, G. (1973). Prime osservazioni su alcune cultivar di noce introdotte dalla California. *Ann. del Ist. Sperim. per la Frutticoltura de Roma,* IV, 89-107.

Maringer, J. (1971). *Los dioses de la Prehistoria.* Ed. Destino, Barcelona. 280 pp

Matheron, M.E.and Mircetich, S.M. (1985). Relative resistance of different rootstocks of English walnuts to six Phytophthora spp. that causes root and crown rot in orchard trees. *Plant Diseases,* 69 (12), 1039-1041.

Mauget, J.C. (1983). Etude de la levée de dormance et du debourrement des bourgeons de noyer (*Juglans regia* L. cv. Franquette*)* soumis à des temperatures superieures à 15°C au cours de leur periode de repos apparent. *Agronomie,* 3 (8),745-749.

Mauget, J.C. and Germain, E. (1988). Analyse de la dormance des bourgeons chez les hybrides de noyer à debourrement trés tardif. *2ème Colloque Noyer-Noisetier.* Bordeaux.

Maurer, K.J. (1960). Studies on *Juglans nigra* seedlings for rootstock breeding. (In german). *Mitt. Klosterneuburg. Ser. B,* 10B, 121-124.

Maurer, K.J. (1967). Affinitäts und Verwachsungsprobleme bei der Walnussveredlung. *Mitt. Klosterneuburg,* 17, 481-491.

Maurer, K.J. (1974). Graft combination in walnut (In German). *Baumchulpraxis*, 4 (6), 164-170.

McGranahan, G.H., Demetrios, G., Catlin, P.B. and, Polito, V.S. (1994). High pollen loads can cause pistillate flower abscission in walnut. *J. Amer. Soc. Hort. Sci. 119*, 505-509.

McGranahan, G.H. and Forde, H.I. (1985). Genetic improvement. In *Walnut Orchard Management*, (H. Seney ed.) Univ. California. Davis, pp.8-12.

McGranahan, G.H., Forde, H.I., Snyder, R.G., Sibbett, G.S., Reil, W., Hasey, T. and Ramos, D.E. (1992). 'Tulare' Persian walnut. *HortScience*, 27 (2), 186 -187.

McGranahan, G. and Leslie, Ch. (1990). Walnuts *(Juglans). Acta Horticulturae,* 290, 907-951.

McGranahan, G.H., Ramos, D.E., Forde, H.I. and Snyder, R.G. (1990). 'Cisco' Persian walnut. *HortScience* 25 (3), 371-372.

McGranahan, G.H., Tulecke, W., Arulsekar, S. and Hansen, J.J. (1986). Intergeneric hybridization in the Juglandaceae: *Pterocarya stenoptera* x *Juglans regia. J. Amer. Soc. Hort. Sci.* 111 (4), 627-630.

Martin, G.C. and Forde, H.I. (1975). Incidence of Blackline in *Juglans regia* L. propagated on various rootstock species. *Jour. Amer. Soc. Hort. Sci.* 100 (3), 246-249.

Meliksetyan, N.A. (1987). Threshold resistance to dehydration as an index of adaptation of North American nuts in the Tbilisi Botanic Garden. (In Russian) *Soobcheniya Akademii Nauk Gruzzinskoi SSR.* (3), 629-632.

Merz, B. (1991). *Rhagoletis completa* Cresson und *Rhagoletis indifferens* Curran, zwei wirtschaftlich Bedentende neu for Europa. *Mitteilungen der Schwizerischen Entomologischen Gesselshaft.* 64, 1-2, 55-57.

Miller, P.W., McSwan, I.C. and Bluhm, W. (1975). Blackline of walnut. *Oregon State University Extension Service*, F.S. 216.

Miller, P.W., Rawling, C.O. and Painter. J.H. (1952). Causes of the decline of walnuts in Oregon: Second report. *Proc.Ore. State Hort. Soc.* 44th *Ann. Rep.*, 183-185.

Miller, P.W., Rawling, C.O. and Painter, J.H. (1953). A survey of the causes of the decline in vigour and death of some walnut trees in Oregon; a four year report of progress. *Proc. Oregon State Hort. Soc.* 45th *Ann. Rep.*, 4, 202-207.

Minotta, G. (1990). La coltura del noce da frutto de a duplice attitudine produttiva in collina e in montagna. *Monti e Boschi*, 41, 1, 27-33.

Mircetich, S.M. (1979). Blackline research. *Diamond and Sunsweet News*, Feb.1979. pp.8,9 and 24.

Mircetich, S.M., Moller, W.R. and Ramos, D.E. (1978). Root and crown rot disease of walnut trees. In: Walnut Short Course, *University of California*, Davis.

Mircetich, S.M., Sanborn, R.R. and Ramos, D.E. (1980). Natural spread, graft transmission and possible etiology of walnut Blackline disease. *Phytopathology*, 70, 962-968.

Moller, W.J. and Ramos, D.E. (1972). A guide to walnut diseases. *Diamond Walnut News*, Dec. 1972.

Nedev, N.V. (1972). The effect of common, black and Manchurian walnut rootstocks on the scion during the juvenile period and start of bearing. I. The growth behaviour of the trees.(In Bulgarian) Grad. i *Lozarska Nauka*, 9 (8), 15-25.

Nedev, N.V. (1973). The effect of common, black and Manchurian walnut rootstock on the scion during the juvenile period. II. The reproductive behaviour of the trees. (In Bulgarian). *Grad.i lozarska Nauka*, 10, 1-8.

Nedev, N.V. and Mileva, D. (1973). Studies on some physiological processes in walnut cultivars grafted onto different rootstocks.(In Bulgarian) *Grad. i Lozarska Nauka*, 10, 15-22.

Nemeth, M., Kolber, M. and Szentivanyi, P. (1990). Cherry Leaf Roll virus in Hungarian walnut orchards. *Acta Horticulturae*, 284, 299.

O'Rourke, F.L. (1979). The Carpathian (Persian) walnuts. In*: Handbook of the North American Nut Trees.*

North Nut growers Ass., (ed. Jaynes R.A.) 233 pp.

Petrosyan, A.A. and Antonenko, G.A. (1972). Results of investigating walnut plantation and selections of best forms in the Krasnodar region.(In Russian) *Nauchnye Trudy, Svero-Kavkazskii Zonal'nyi Naucho Issledovatel'skii* Institut Sadovostvo y Vinogradarstvo, 12, 30 -34.

Pons, A. and Quézel, P. (1958). Premières remarques sur l'étude palynologique d'un guano fossile du Hagger. *Compte rendú. Acad. Sci., Paris*, 244, 2290 -2292.

Por, A. and Por, J. (1990). The effect of the excess of pollen on the fruitset of walnuts in Balatonboglar. *Acta Horticulturae*, 284, 253-256.

Ramos, D., Ryugo, K., Olson, W., Osgood, J.and Sibbet, G. (1985). Pruning and tree thinning. In: *Walnut Orchard Management*. Coop. Extension Service University of California, (Ramos, D. ed.) 96-102.

Radicati, L., Vergano, G.and Martin. Y. (1993). Evaluation of fruit characters in different walnut cultivars. *Acta Horticulturae* 311, 49-55.

Radix, P.and Seigle-Murandi, F. (1992). Hazards of copper accumulation in soils of walnut orchards. *Fresenius Environmental Bulletin* 2, 10, 565 -569.

Rizzi, A.D. (1977). *Walnut Orchard Management Short Course.* (ed. Ramos D.E.) University of California, Davis.

Rizotto, N. (1960). Noce Feltrina: pianta da salvare!. *Agric. Venezie*, 14, 142-149.

Rouskas, D. and Katranis, N. (1988). Prospections realisées dans les populations locales de noyers de Grece Centrale. *2éme Colloque noyer - noisetier. AGRIMED*, Bordeaux.

Rusu, E.,(1976). An ecotype of *Juglans regia* suitable for growing on river sides. *Proceeding 19 Congress, International Society Horticultural Science*, 1 (A), 375.

Sánchez Goñi, M.F. (1988). A propós de la presence du pollen de *Castanea* et de *Juglans* dans les sédiments archéologiques würmiens anciens du Pays Basque espagnol. *Int. fr. Pondichery trav.sec.tech.* 25, 73-82.

Sánchez, A. (1984). Razas bovinas españolas. *Publicacones de Extensión Agraria*, M. de Agricultura, p 136.

Sannino, G. and Pilone, N. (1979). Il noche: une culture da salvare. *Informatore Agrario*, 35 (2), 4189-4193.

Savino, V., Quacquarelli, A., Gallitelli, D., Piazolla, P. and Martinelli, G.P. (1977). Il virus dell'accortoc-ciamento fogliare del ciliegio dolce. I. Identificazione e caratterizzazione. *Phytopathologia Mediterranea*, 16 (2/3), 96-102.

Scotto, C. (1976). Repercusion de la modernisation de l'explotation de noyer en France sur les problèmes nematologiques de la culture. *Rev. Zool. Agric. Pathol. Veg.* 73 (3), 103-116.

Scotto, C. (1988). Les problémes nematologiques du noyer et du noissetier. *2ème Colloque noyer-noiseties AGRIMED*. Bordeaux.

Sen, S.M. (1991). A study on the selection of Darande walnuts. *Acta Horticulturae*, 311, 57-60.

Serr, E.F. (1960). Walnut orchards on volcanic soils deficient in phosphorus. *Cal. Agric.* 14 (6), 6-7.

Serr, E.F. (1969). Persian walnuts in the Western States. In: *Handbook of North American Nut Trees.* (ed. Jaynes R.A.) p. 257.

Serr, E.F. and Forde, H.I. (1959). Blackline, a delayed failure at the union of *Juglans regia* trees propagated on other *Juglans* species. *Proc. Amer. Soc. Hort. Sci.* 74, 220-231.

Siebert, J. (1993). Economic and policy considerations in marketing walnuts. *Acta Horticulturae* 311, 249-255.

Smith, R.L. and Allinger, H. (1969). Evaluation of Persian walnuts introduced from Russia, Hungary and Korea. *North Nut Growers Ass., 60th Ann. Report*, 61 -65.

Smith, R.L., Joley, L.E. and Whitehouse, W.E. (1960). Evaluation of Persian walnut introduction. *U.S. Plant Introduction Station, Chico, California. USDA, CR* 44 -60.

Solignat, G. (1974). Note sur le comportement à l'asphyxie radiculair de porte-greffes du noyer *Juglans regia*.

La Pomologie Francaise, XVI, 4, 75 -77.

Tamponi, G. (1988). Fruit sensitivity to *Xanthomonas juglandis* of walnut cultivars. *2nd colloque AGRIMED*, Bordeaux.

Tasias, J. (1982). Trabajos sobre nogal, pistacho y pecanero en curso de realización en el Centro Agropecuario Mas Bové. *Conferencia I Feria Agricola y Nacional Frutera de San Miguel*, Lerida.

Teviotdale, B.L., Schroth, M.N. and Mulrean, E.N. (1985). Bark, fruit, and foliage diseases. p. 153-157. In: *Walnut Orchard Management.* (ed. Ramos D.E.) Pub. 21410, University of California.

Thomson, G.W. and McComb, A.L. (1962). Growth of plantation black walnut in relation to pH and certain chemical factors of the soil. *For. Sci.* 8, 322-333

Tkhagushev, V.A. and Khlopenkova, Z.F. (1975). The root system of walnut trees growing in the Kubanskaya region of the Krasnodarskii Krai (In Russian) *Trudy kubanskaya S. Kh. Instituta.* 111(139), 56-65.

Van den Brink, L.M. and Janssen, C.R. (1985). The effect of human activities during cultural phases on the development of montane vegetation in the Serra da Estrela, *Portugal Rev. Paleobot. Palynol.* 44, 193-215.

Vlachopoulos E.G. (1991). Nematode species in nurseries in Greece. *Ann. Institut Phytopathol. Benaki*, 16, 2, 115-122.

Vishanskaya, Yu., Nedev, N.V. and Petrova,V. (1990). Influence of rootstock on fatty acid composition of walnut oil.(In Bulgarian) *Gradinarska i Lozarska Nauka*, 17 (3/4), 3-7.

Welkerling de Tacchini, E.M.L. (1960). A variety of walnut with no shell (In Spanish). Rev. *Fac. Cien. Agrar., Mendoza*, 8, 27 -30.

Wood, M. (1991). Mothproofing walnut trees of the future. *USDA Agric. Res. Wash.* 39, 6, 10-11.

Zielinski, Q.B. (1952). A three-year comparison of walnut rootstocks. *Proc. Ann. Nut Growers Society, Wash.* 38, 215-216.

Zielinski, Q.B. (1953). A report on walnut rootstock test at Oregon State College. *Proc. Oregon State Hort. Soc.* 45th Ann. Rep. 212-215.

Zielinski, Q.B. (1955). New walnut rootstocks best answer to blackline problems. *Oregon Agricultural Progress.* 2 (1), 5-6.

Zielinski Q.B. (1958). Comparison of the growth status of walnut rootstock varieties in Oregon. *Proc. Nut Growers Soc. Oregon and Washington*, 12, 214-216.

Zohary, D. (1970). Centers of diversity and centers of origin. In *Genetic Resources in plants* (eds. Frankel and Bennett) 33-42.

Zohary, D., Hopf, M. (1988). *Domestication of plants in the Old World.* Clarendon Press, Oxford, 249 pp.

16 BLUEBERRIES

REBECCA L. DARNELL

Horticultural Sciences Department
University of Florida, Gainesville, FL 32605

1 Introduction

Blueberries are in the family Ericaceae and the genus *Vaccinium*, section *Cyanococcus*. *Vaccinium* includes about 400 species, the majority of which are found on open mountain slopes in the tropics (Camp, 1945). Approximately 40% of the species are native to southeast Asia, 25% to North America, and 10% to Central and South America, with the rest scattered world-wide (Luby *et al.*, 1991). Four species (*V. angustifolium* Ait., *V. ashei* Reade, *V. corymbosum* L., and *V. darrowi* Camp) have contributed most of the germplasm to current blueberry cultivars (Ballington, 1990). Highbush (*V. corymbosum*), lowbush (*V. angustifolium*), rabbiteye (*V. ashei*), and southern highbush (*V. corymbosum* interspecific hybrid) blueberries are grown commercially. More recently, half-high blueberry genotypes (*V. corymbosum* x *V. angustifolium*) have been developed that are similar in growth and fruiting characteristics to highbush blueberry, yet exhibit greater cold hardiness (Luby *et al.*, 1986; Wildung and Sargent, 1989).

There is increasing interest in blueberry production in warm temperate, sub-tropical, and even tropical areas of the world (Luby *et al.*, 1991; Phatak and Austin, 1990). Current low chilling cultivars (*e.g.* 'Sharpblue', 'Gulf Coast', 'Wannabe', 'Misty', 'Beckyblue', 'Bonita', and 'Climax'), grown in traditional production systems, are not generally amenable to production in areas with extremely low or no chilling temperatures during the dormant season. However, the advent of cultivars with very low chilling requirements, the possibility of chemical regulation of dormancy, as well as the movement towards non-traditional, non-dormant production system may soon increase the feasibility of growing commercial blueberries in areas such as these. Even if the chilling/dormancy problem is circumvented; however, there are additional physiological and cultural factors that must be taken into account before the feasibility of growing blueberries in these climates can be properly assessed. These include the physiological/morphological responses of existing cultivars and germplasm to high temperatures, varying soil types, and increased photoperiods.

2 Physiological/Morphological Responses

2.1 High temperature tolerance

In general, leaf photosynthetic rates in cultivated blueberry increase as temperatures increase from 10 to 25°C (Davies and Flore, 1986; Forsyth and Hall, 1965; Moon *et al.*, 1987a).

Amnon Erez (ed.), Temperate Fruit Crops in Warm Climates, 429-444.
© 2000 *Kluwer Academic Publishers. Printed in the Netherlands.*

Temperature optima for both highbush and rabbiteye are similar, averaging 20 to 25°C (Davies and Flore, 1986; Moon *et al.*, 1987a). As temperatures increased to 30°C, leaf photosynthetic rates declined in both species. This result is somewhat surprising considering that rabbiteye blueberries are native to the southeastern United States and therefore might be expected to be better adapted to higher air temperatures than highbush blueberry.

V. darrowi, a wild, evergreen blueberry species also native to the southeastern United States, is well adapted to high temperatures and dry areas (Camp, 1945). Moon *et al.* (1987a) reported that leaf photosynthetic rates increased in *V. darrowi* as leaf temperatures increased from 20 to 30°C, although the maximum photosynthetic rate for *V. darrowi* was lower than it was for the highbush cultivar, 'Bluecrop' (8.8 *vs.* 10.7 μmol m^{-2} s^{-1}, respectively) [Table 1]. Transpiration rates were also lower for *V. darrowi*, averaging 1.4 mmol m^{-2} s^{-1} compared to 2.7 mmol m^{-2} s^{-1} in 'Bluecrop' at leaf temperatures of 20°C, and 2.4 *vs.* 3.0 mmol m^{-2} s^{-1}, respectively, at leaf temperatures of 30°C. Leaf conductance to water vapor was lower for *V. darrowi* at all temperatures compared to highbush. Thus, *V. darrowi* may possess a greater tolerance to high temperatures and drought through decreased stomatal opening and subsequent restriction of water loss. The decrease in water loss may compensate for the lower leaf photosynthetic rates under conditions of high temperature and/or dry conditions. In a companion study, Moon *et al.* (1987b) found that this high temperature tolerance of *V. darrowi* appeared to be heritable. Leaf gas exchange characteristics were compared among highbush ('Bluecrop'), *V. darrowi* (Fla-4B), and US75 (an F_1 hybrid). US75 had significantly higher leaf photosynthetic rates than either parent at all temperatures examined (15 to 35°C) [Fig. 1]. The temperature response curve of US75 paralleled that of Fla-4B, with an optimum at 30°C.

Table 1. Effect of leaf temperature on the gas exchange characteristics[z] of 'Bluecrop' and V. darrowii blueberry

	Gas exchange parameter									
	A (*μmol·s⁻¹·m⁻²*)		*g_l* (*mmol·s⁻¹·m⁻²*)		*E* (*mmol·s⁻¹·m⁻²*)		*g_r* (*mmol·s⁻¹·m⁻²*)		*WUE* (*μmol CO_2/ mmol H_2O*)	
Genotype	20°C	30°C	20°C	30°C	20°C	30°C	20°C	30°C	20°C	30°C
Bluecrop	10.7a[y]	7.6	237.3a	155.8a	2.69a	3.04	42.5	32.3	4.0	2.5b
V. darrowii	8.1b	8.8	117.5b	93.1b	1.45b	2.39	29.6	49.9	5.6	3.7a

[z] CO_2 assimilation (A), leaf conductance to water vapor (g_l), transpiration (E), residual conductance to CO_2 (g_r) and water use efficiency (WUE) were measured at a saturating PPF of 1012 μmol•s^{-1}•m^{-2} (±22.3), ambient CO_2 concentrations of 332-343 μl•$liter^{-1}$ and at vapor pressure deficits <1 kPa.

[y] Mean separation within columns by Tukey's w-test, 5% level.

Reproduced with permission from Moon et al., 1987a.

Hancock *et al.* (1992) examined the photosynthetic heat tolerance of several progeny generated by various crosses between highbush blueberry and Fla-4B. They found that all highbush cultivars exhibited a significant decrease in leaf photosynthetic rates when

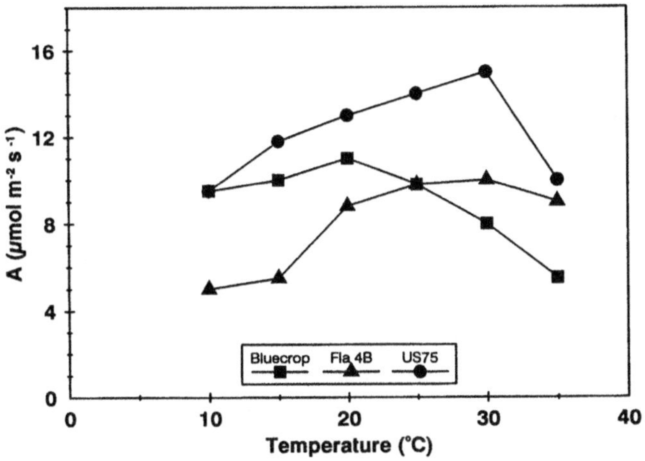

Figure 1. Effect of leaf temperature on net CO_2 assimilation for 'Bluecrop', *V. darrowi* (Florida 4B) and US75. Measurements were made at saturating PPFD (1000 µmol s^{-1} m^{-2}), at ambient CO_2 levles (322-343 µl $liter^{-1}$) and at leaf-to-air vapor pressure deficits <1 kPa. Each symbol represents the mean of 20 determinations and each curve is representative of a typical response for its respective genotype. Reproduced with permission from Moon *et al.*, 1987a.

air temperatures increased from 20 to 30°C, with reductions ranging from 22 to 47%. All hybrid populations contained some individuals that were more heat tolerant than highbush. Selfed populations of US75 yielded one genotype that exhibited a 10% increase in leaf photosynthetic rates as temperatures increased from 20 to 30°C. The coefficients of variation in the populations ranged from 46 to 101%, indicating substantial genetic variability in the populations. This suggests that the photosynthetic heat tolerance of *V. darrowi* can be transferred to commercial blueberry cultivars, which may be an important factor in maintaining and/or increasing yield under high temperature conditions. However, all of the above mentioned studies were done with containerized plants that were not previously acclimated to high temperatures; thus, the relevance of these findings to field-grown plants is unknown. Additonally, correlations between leaf photosynthetic rates and yield in most other crops are often poor, and yield increases are better correlated with increasing carbohydrate partitioning to the harvestable organ. Field testing of blueberry progeny with increased photosynthetic heat tolerance will be required before it can be determined if this will have any meaningful effect on yields in warm climates.

In commercial blueberry cultivars, high air temperatures limit not only photosynthetic rate, but reproductive development as well. Williamson *et al.* (1995) found that warm night temperatures (21°C) decreased fruit set and final size in rabbiteye blueberry compared to cool night temperatures (10°C) [Table 2]. These responses may be related to high temperature effects on decreasing ovule longevity (Darnell *et al.*, 1992). Warm day temperatures (29°C) did not affect fruit set, but decreased final fruit size compared to day temperatures of 26°C (Williamson *et al.*, 1995). Carlson and Hancock (1991) suggested that fruit development of highbush blueberry may be inhibited to some extent by temperatures between 21 and 30°C, depending on cultivar. These effects on fruit development may reflect the adverse effects of high temperatures on leaf photosynthetic rates. Fruit quality is also adversely affected by high temperatures (Kender and Brightwell, 1966), presumably due to high temperature effects on decreasing photosynthetic rates and increasing respiration rates, leading to a decrease in carbohydrates available for fruit, and therefore a reduction in soluble sugar and/or organic acid content. Additionally, high temperatures during harvest would decrease berry firmness. Currently, there are limited cultural techniques that could alleviate the problems associated with excessively high air temperature on blueberry growth and development. Several wild species may be useful in transferring heat tolerance traits to commercial blueberry cultivars, as indicated previously. Besides the increased photosynthetic heat stability of *V. darrowi*, this species may also be important in conferring superior fruit quality and firmness under warm conditions (Ballington, 1990). Additional sources of genes for heat tolerance include *V. tenellum* Ait., *V. myrsinites* Lamar, *V. elliottii* Chap., *V. ashei*, and *V. amoenum* Ait., all of which are native to low elevations in the southeastern United States (Camp, 1945; Luby *et al.*, 1991).

Table 2. Temperature and pollination effects on fruit set and development of 'Beckyblue' rabbiteye blueberry

Day/night temp. (°C)	*Fruit set (%)*[z]	*Average fruit wt. (g)*	*FDP*[y] *(days)*
26/21	63.9b[x]	1.5b	85b
26/10	83.2a	1.7a	88ab
29/10	71.4ab	1.4b	90a

[z] Data were arc sin transformed prior to analysis.
[y] FDP = fruit development period.
[x] Mean separation by Duncan's Multiple Range test, 5% level.
Reproduced with permission from Williamson *et al.*, 1995.

Not only air temperatures, but root temperatures as well may have a profound effect on blueberry growth and development. The fine, shallow root system of blueberry is especially sensitive to high soil temperatures. Abbott and Gough (1987) found that root growth of highbush blueberry declined at soil temperatures above 18°C. Spiers (1995) reported that root, shoot, and total plant dry weight of containerized southern highbush and rabbiteye cultivars decreased linearly when roots were exposed to increasing soil temperatures from 16 to 38°C [Table 3]. Both species responded similarly to increasing soil temperatures. High soil temperatures have been reported to decrease leaf P, K, Fe, Mn, and Zn content in rabbiteye blueberry, without affecting root nutrient content (Basiouny *et al.*, 1985). This suggests that high soil temperatures may decrease translocation from roots to shoots.

Soil temperatures (at a 10 cm depth) in the warm temperate areas of the southeastern United States average 28 to 31°C during the summer (Darnell *et al.*, 1992), and soil temperatures in subtropical and tropical areas are likely to average at least as great or greater. Thus, it is likely that supraoptimal soil temperatures in warm areas may limit root and shoot growth in blueberry, unless cultural practices that ameliorate high soil temperatures are implemented. Although mulching may encourage confinement of what is already a shallow root system, it is also effective in decreasing soil temperatures. Organic mulches, such as uncomposted pine sawdust and wood chips, decreased soil temperatures by 6-10°C (Magee and Spiers, 1995; Pliszka *et al.*, 1997) and increased plant growth and yield of both rabbiteye and southern highbush blueberry plants (Clark and Moore,1991). Thus, the beneficial effects of mulches on blueberry growth and yield may be due, in part, to amelioration of high soil temperatures.

Table 3. Influence of substrate temperature on growth of southern highbush (SHB) and rabbiteye (RE) blueberry

Variable Temp.	*New growth (dry wt. in g)*			*Root length*	*Vigor rating*[z]	
(°C)	*Shoot*	*Root*	*Total plant*	*(cm)*	*Shoot*	*Root*
16	4.8	15.2	20.0	22.8	3.5	4.0
27	3.2	9.7	12.9	20.4	2.5	2.5
38	1.5	7.3	8.8	14.7	1.1	1.4
Linear Type	***	***	***	***	***	***
SHB	3.8a[y]	14.6a	18.4a	19.8a	2.4a	2.7a
RE	2.6a	6.9b	9.5b	18.9a	2.3a	2.6a

[z] Vigor rating: 0 = dead, 1 = worst, 5 = best vigor.
y Mean separation within columns by LSD, 0.05.
*** Significant at $P < 0.001$.
Reproduced with permission from Spiers, 1995.

2.2 Soil adaptation

Blueberries, probably more than any other fruit crop, have strict soil requirements for optimal growth. In general, blueberry production is limited to acidic, well-drained soils, high in organic matter (Erb *et al.*, 1990; Kender and Brightwell, 1966). Optimum soil pH ranges from 4.5 to 5.5, while optimum organic matter averages greater than 3%. Growth and yield on other types of soils are markedly reduced, and production can be extended to these types of soils only with extensive soil modifications. Korcak *et al.* (1982) grew blueberry seedlings from several species over a range of soil conditions. For all species, dry matter production increased on low pH, high organic matter, sandy soil compared to high pH, high organic matter, sandy-clay soil. Plants grown at high soil pH were not only smaller, but exhibited leaf tip burn and chlorosis. Even when soil pH is within the optimum range for blueberry production, low organic matter content and/or unsuitable soil texture may limit growth. Erb *et al.* (1993) examined growth of various blueberry genotypes on a typical blueberry soil (Berryland, pH 4.0, 3% organic matter, 95% sand) compared to a mineral soil (Galestown, pH 4.3, 1.3% organic matter, 59% sand/24% clay). The genotypes included *V. corymbosum*, *V. ashei*, *V. darrowi*, and several interspecific hybrids. They found that shoot growth decreased significantly for all genotypes grown in the mineral soil compared to the blueberry soil (Table 4). Root dry weight also decreased for clones grown in the mineral soil; however, the root dry weight ratio (*i.e.* the proportion of dry weight allocated to roots) was significantly higher in this soil, indicating that plants grown in mineral soils partitioned a greater percentage of their biomass into root growth. The authors suggested that growth reduction on mineral soil was due to the combination of decreased leaf area and increased root biomass leading to a decreased ratio of photosynthetic tissue to respiratory tissue.

Table 4. Biomass characteristics of blueberry clones grown for 18 weeks in one of two soils

Variable *Soil*	*No. of samples*	*TLA* (cm^2)	*LW* (*g*)	*SW* (*g*)	*RW* (*g*)	*S:R* (*g/g*)	*RWR* (*g/g*)	*SWR* (*g/g*)	*LWR* (*g/g*)
Berryland	62	2750a[z]	23.9a	15.7a	9.6a	7.1a	0.166b	0.332a	0.502a
Galestown	88	1616b	13.5b	9.8b	6.3b	4.6b	0.202a	0.333a	0.466b

[z] Mean separation within columns by F-test, P = 0.05.
TLA, total leaf area; LW, leaf dry weight; SW, stem dry weight; RW, root dry weight; S:R, shoot: root ratio; RWR, root dry weight ratio; SWR, stem dry weight ratio; LWR, leaf dry weight ratio.
Reproduced with permission from Erb *et al.*, 1993.

Soil texture alone appears to have a profound effect on growth, and dry weight gain in blueberry has been positively correlated with the percent sand and negatively correlated with the percent clay in soils (Korcak *et al.*, 1982). Although incorporation of peatmoss has been found to ameliorate the detrimental effects of low organic matter on blueberry growth, this appears to be true only for predominantly sandy soils. Korcak (1986a, b) examined the effects of different soils, with or without the incorporation of peatmoss, on growth of several different blueberry genotypes. The soils included a typical blueberry soil (96% sand, 1% silt, 4% clay), and three other soils in which the pH was within the optimum range for blueberry growth. These other soils; however, contained a lower percentage sand (44 to 59%) and a higher percentage clay (24 to 30%) than did the blueberry soil. Plant volume was greatest on the typical blueberry soil compared to the other soil types. Furthermore, addition of peatmoss, which increased soil organic matter content, did not consistently improve growth on any of these soils. The reductions in growth on soils other than the sandy blueberry soil may have been due to induced nutrient imbalances involving primarily Ca, Fe, and Mn. Korcak suggested that these imbalances may be related to the decreased organic matter reactivity he found in the mineral soils, which resulted in decreased cation exchange capacity in these soils (ca. 14 meq/100g) compared with the typical blueberry soil (ca. 21 meq/100g).

The fact that incorporation of peatmoss had very little beneficial effect on blueberry growth in these mineral soils suggests that it may be economically more efficient to develop blueberry cultivars that are vigorous and productive on unamended mineral soils, rather than attempt to amend the soils themselves. There are several potential sources available for adaptation to mineral soils, including *V. darrowi*, *V. myrsinites*, *V. atrococcum* Heller, *V. elliottii* Chap., *V. simulatum* Small, *V. amoenum* Ait., *V. angustifolium*, *V. myrtilloides* Michaux, *V. pallidum*, *V. constablaei* Gray, and *V. ashei* (Ballington, 1990; Chandler *et al.*, 1985, Hancock *et al.*, 1995; Luby *et al.*, 1991).

Several studies have shown that plant vigor on mineral soils is heritable. Chandler *et al.* (1985) found that general and specific combining ability effects for canopy volume were significant in a diallel set of progenies involving one highbush clone and three interspecific hybrids grown on an unamended mineral soil. Seedling progenies of NJUS 11 (*V. ashei* x *V. atrococcum)* and US75 (*V. darrowi* x *V. corymbosum*) grew vigorously on the mineral soil. In a later study using the same clones, Erb *et al.* (1990) concluded that the highly significant general combining ability for plant canopy volume, total fruit weight:canopy volume, and fruiting traits indicated that genetic diversity for mineral soil adaptation was present and heritable in the *Vaccinium* clones studied. Additionally, they concluded that, of the progenies studied, US75 contained the best combination of genes for transferring mineral soil adaptation. None of these studies, however, compared the relative growth of plants on blueberry soils vs. mineral soils. That is, clones that grew most vigorously on blueberry soils also grew most vigorously on mineral soils. This does not, in itself, indicate selection for adaptation to mineral soils, rather, it indicates selection for overall plant vigor.

There may also be sources available in *Vaccinium* for adaptation to higher soil pH. Genotypes from high pH sites have been identified in *V. angustifolium* and *V. myrtilloides* (Luby *et al.*, 1991). Finn *et al.* (1993) identified selected populations derived from *V. angustifolium* that were tolerant of high soil pH. *Vaccinium arboreum* Marsh. also tolerates soil pH above 6.0 and is adapted to a wide range of soil types (Luby *et al.*, 1991; Lyrene, 1993). Adaptation to *V. arboreum* may be related, in part, to relatively higher rates of iron uptake compared with cultivated *Vaccinium* species (Darnell, unpublished data). Although *V. arboreum* is in section *Batodendron*, fertile tetraploid breeding lines that combine highbush and *V. arboreum* genes have been developed and backcrossed to the highbush cultivars (Lyrene, 1991). Thus, *V. arboreum* may be a good source of adaptation to both high pH and mineral soils.

In predominantly sandy, well-aerated soils, deficiencies in organic matter content and/or soil pH problems may be more successfully changed with soil amendments than is the case with mineral soils. The major role of organic matter in blueberry growth may be its effect on increasing the water holding capacity of the soil. In rabbiteye blueberry, root dry weight was enhanced by irrigation, incorporation of peatmoss, and/or mulching (Spiers, 1986) [Table 5]. Mulching resulted in the greatest increase in root dry weight but the shallowest root system compared to the other two treatments. Spiers suggested that the growth increase was due to improved soil water relations, and that in sandy soil, moisture appears to be the limiting factor to blueberry growth. Although no data on soil moisture, water use, or temperature were presented in Spier's work, the fact that blueberries can be successfully grown in 100% silica sand containing no organic matter, as long as irrigation and fertilization are adequate (Merhaut and Darnell, 1995), supports Spier's conclusion. More recently, Pliszka *et al.* (1997) reported that mulch significantly increased both soil moisture and yield in highbush blueberry grown in sandy soils. These studies indicate that blueberries can be successfully grown on sandy, well-aerated soils under conditions of adequate soil moisture and fertilization. Methods that improve water availability include incorporation of peatmoss, mulching, and, of course, irrigation. Additionally, as indicated previously, mulching can ameliorate high soil temperatures, further benefitting blueberry growth.

Table 5. Influence of irrigation, mulching and incorporated peatmoss on dry weight of roots (g) found at various depths in 'Tifblue' rabbiteye blueberry

Treatment	*Soil depth (cm)* 0-15	15-30	30-45	45-60
Irrigation alone	16.4b[z]	46.8a	22.8b	10.8b
Peatmoss alone	22.4b	34.4ab	51.2a	20.8b
Mulch alone	143.6a	51.6b	30.4bc	12.0c
Irrigation + peatmoss	77.2a	87.6a	90.8a	42.8b
Irrigation + mulch	313.2a	41.2b	26.4b	18.0b
Peatmoss + mulch	213.6a	102.4b	64.8c	75.6c
Irrigation, peatmoss and mulch	426.8a	201.6b	137.2c	62.4d
Overall	151.6a	70.8b	53.2bc	30.4c

[z] Mean separation within rows by LSD, 5% level.
Reproduced with permission from Spiers, 1986.

2.3 Photoperiod

Unlike most fruit crops (strawberry being the notable exception), blueberries are photoperiodically sensitive with respect to flower bud initiation. In lowbush blueberries, flower bud initiation generally required photoperiods less than 12 h (Aalders and Hall, 1964; Hall and Ludwig,1961). Although some clones were able to produce flower buds at 14 and 16 h, the number of flower buds produced was usually not as extensive (Hall and Ludwig, 1961). Additionally, a few clones appeared to be photoperiodically insensitive. For the photoperiodically sensitive clones, at least six weeks of the required photoperiod was needed for normal flower bud formation. Concomitant with an increase in flower bud production was a decrease in vegetative growth.

Highbush blueberries exhibit a similar photoperiodic response. Flower bud initiation occurred after eight weeks of 8, 10, or 12 h photoperiods (Hall *et al.*, 1963). Fewer flower buds were formed under 14 h and no flower buds were formed under 16 h photoperiods. Vegetative growth was greatest under 16 h photoperiods and least under the 8 h photoperiod. Flower bud initiation was greater under 10 h photoperiods than under 8 h. This was attributed to increased shoot growth under 10 h photoperiods, which provided more shoots on which to produce flower buds. All three cultivars examined in this study responded similarly to photoperiod.

Photoperiodic effects in rabbiteye blueberry have also been documented. Phatak and Austin (1990) found that six weeks of 10 or 12 h photoperiods increased flower bud initiation compared to 14 or 16 h photoperiods. Photoperiodic durations of four weeks at 10 or 12 h induced some flower bud initiation, but inflorescences were abnormal. No flower bud initiation occurred at durations less than four weeks, regardless of photoperiod. Under conditions conducive to flower bud initiation, flower and fruit development occurred normally on plants that were not allowed to enter dormancy, supporting previous work indicating dormancy and chilling are not required for normal vegetative and reproductive development in blueberry (Perlmutter and Darrow, 1942;

Sharpe and Sherman, 1971). The authors suggested that these data indicated the suitability of rabbiteye blueberry cultivars for production in tropical and subtropical areas of the world; however, this conclusion may be premature, even in areas with appropriate soils, since the developmental responses to high temperatures are unknown.

There is genetic variability for sensitivity to photoperiod within rabbiteye cultivars. 'Beckyblue' rabbiteye blueberry, exposed to gradually decreasing daylengths from 10 to 8 h over a five-week period in the fall initiated twice as many flower buds as those plants exposed to natural daylengths (i.e., gradually decreasing daylengths from 12 to 11 h) during that same time period (Darnell, 1991) [Table 6]. Additionally, 'Beckyblue' plants exposed to shortened photoperiods the previous fall had a greater percentage of floral budbreak and a shorter, more concentrated bloom period than did the control plants. This suggests that short photoperiods increase not only the number of flower buds initiated, but also flower bud development. 'Climax' rabbiteye blueberry did not respond to the photoperiod treatments in any way, indicating that the critical daylength (or nightlength) for flower bud initiation in rabbiteye blueberry is cultivar dependent. Variation in response to photoperiod in blueberry suggests that careful consideration must be given to cultivar selection to ensure that daylength does not limit flower bud initiation in areas of low latitude.

Temperatures have a profound effect on photoperiodic regulation of flower bud initiation in most crops; however, there is no information on the interaction of photoperiod with temperature on blueberry flower bud initiation. Clearly, cultivar evaluation for production in warm climates must include information on temperature x photoperiod interactions on flower bud initiation.

Table 6. Daylength effect on flower bud initiation, bloom period, floral budbreak and total flower number of 'Beckyblue' rabbiteye blueberry

	Flower buds per plant		*Bloom period (days)*		*Floral budbreak (% budbreak/plant)*	*Florets/plant (no.)*	
Daylength[z]	1989	1990	1989	1990	1989[y]	1989	1990
Natural	21.9	183.7	12.7	37.5	43.8	56	1016
SD	48.0	214.9	9.0	32.9	61.7	177	1194
Significance	**	*	**	**	**	**	**

[z] Daylength during the 5 weeks just before defoliation and chilling. Natural = decreasing daylength from 12 to 11 hrs; SD = decreasing daylength from 10 to 8 hrs.
[y] No data for 1990.
* $P < 0.10$, **$P < 0.05$.
Reproduced with permission from Darnell, 1991.

3 Production Systems

There are two major production systems that should be considered in determining the feasibility of growing blueberries in warm climates. The first would be the traditional, deciduous system where the plants undergo dormancy and chilling. This system should be considered in warm temperate, as well as some sub-tropical areas, where the use of low chill cultivars with or without dormancy breaking chemicals would be warranted. The second production system would be more applicable in tropical areas, although it has been used in sub-tropical areas as well. This system involves maintaining the plants in an active growth stage throughout the year, avoiding dormancy, and consequently the need for chilling.

3.1 Deciduous production system

A deciduous production system is defined, for the purposes of this review, as a system where plants enter an endodormancy phase, and therefore have a requirement for chilling to break dormancy. The success of such a system in warm climates depends on the availability of low chilling cultivars. Attempts to develop blueberry genotypes with low chilling requirement began in the late 1940's (Sharpe, 1953). Since that time, a major emphasis of the blueberry breeding program in Florida has been the development of low chilling cultivars. The earliest releases from this program were two southern highbush cultivars, 'Flordablue' and 'Sharpblue' (Sharpe and Sherman, 1976). Both were derived from a series of complex crosses involving primarily *V. darrowi* and *V. corymbosum* (the northern highbush). Despite various production problems, 'Sharpblue' is the most widely planted southern highbush cultivar in Florida and the cultivar that is currently most amenable to deciduous production systems in warm climates. Other cultivars of southern highbush have either higher chilling requirements than 'Sharpblue' or exhibit production problems that would seriously limit their use commercially. Although rabbiteye blueberries are more widely adapted to various climatic and edaphic factors than southern highbush, there are currently no cultivars amenable to a deciduous production system in warm climates.

In some areas and/or years, there may be insufficient winter chilling to satisfy the chilling requirement even in low chilling cultivars of blueberry. Chemical means of satisfying the chilling requirement have been used in many fruit crops, including apple, peach, grape, and others. Use of chemicals to satisfy the chilling requirement in blueberry has not been well studied, and results are preliminary, at best. Use of hydrogen cyanamide as a dormancy breaking chemical has met with limited success in blueberry. Some cultivars of southern highbush blueberry (*e.g.* 'Sharpblue') appear to be extremely sensitive to this chemical, and even low concentrations (1% a.i.) resulted in bud and shoot injury, and in some cases, plant death (Darnell, unpublished). Flower buds were much more sensitive to injury than were vegetative buds; more than 95% of the flower buds were killed on dormant plants sprayed with hydrogen cyanamide. In other southern highbush blueberry cultivars; however, hydrogen cyanamide appears to be more effective. The extent and earliness of vegetative budbreak and canopy establishment in 'Misty' southern highbush blueberry increased linearly with increasing concentrations of 0-2% hydrogen cyanamide (Williamson *et al.*, 1998). Furthermore, fruit size increased and fruit development period decreased in hydrogen cyanamide-treated plants compared with non-treated plants. Application of 1% hydrogen cyanamide to 'Climax' rabbiteye blueberry during late dormancy significantly increased the amount of vegetative budbreak relative to the control (Krewer *et al.*, 1995). Total amount of vegetative budbreak was not increased in a second cultivar, 'Tifblue', but the rate of budbreak was accelerated. Although effects on flower buds were not discussed, the fruit density data suggest that there was little, if any, detrimental effect of hydrogen cyanamide on flower buds. Differences in response to hydrogen cyanamide may reflect differences in species and cultivar sensitivity, and demonstrate the need for careful evaluation of this chemical for use in dormancy release. Nonetheless, use of chemicals to increase or accelerate vegetative budbreak in blueberry would be extremely useful in warm climates, where bloom occurs before vegetative budbreak. Under conditions of insufficient chilling in these areas, the temporal difference in floral and vegetative budbreak is further exacerbated, resulting in the production of small, low quality fruit, and in some severe cases, shoot and/or plant death (Mainland, 1985). Additional studies on the use of chemicals to regulate dormancy and budbreak in blueberry are clearly needed.

3.2 Non-dormant production system

The success of a non-dormant production depends on the ability to maintain healthy leaves and avoid low temperatures that result in defoliation. Thus, this system is limited to tropical areas or subtropical areas with mild winter temperatures that do not fall below freezing. In some cases, this system may be successful in areas that experience one or two frosts a year, but which have adequate overhead irrigation systems to protect the plants (Reeder *et al.*, 1994).

A common feature of temperate fruit production in tropical climates has been the use of defoliation after fruit harvest and flower bud initiation to prevent the onset of dormancy and promote budbreak in the absence of chilling (Edwards, 1987a, b; 1990). Timing of defoliation is critical; if done too early, flower bud initiation will be decreased. If done too late, budbreak will be reduced. Defoliation has been used successfully to grow apple, peach, and grape in tropical areas. The non-dormant production system differs from this in that neither defoliation nor any other cultural technique is used to synchronize specific growth cycles. Instead, plants are grown under conditions where vegetative growth continues throughout the year, albeit at slower rates during the cooler season. The first reported success of this system with blueberries is in Australia (Wright, 1993), although this system is being tried both experimentally and commercially in Florida (Reeder *et al.*, 1994).

In some areas of Australia, the southern highbush cultivar 'Sharpblue' is grown as an "evergreen" (Wright, 1993). The climate where this occurs is an overlap between subtropical and temperate, without extreme temperature conditions of either. Winter temperatures seldom fall below freezing, and annual temperatures are mild enough to allow continued vegetative growth throughout the year. Because growth cycles are not synchronized, production is spread over the year. Peak production occurs between October and December, with between 60 to 70% of the total annual yield harvested during that three month period.

Interest in the non-dormant blueberry production system in Florida began in the late 1980's, when it was found that fall fertilized rabbiteye blueberries, which were not subjected to freezing temperatures, continued to grow throughout the winter, avoiding dormancy (Reeder and Darnell, unpublished). Both vegetative shoot growth and flower bud density increased significantly in plants receiving late fall fertilization compared to plants which were not fertilized in late fall and allowed to go dormant (Table 7). Fall fertilized (non-dormant) plants flowered profusely and set fruit the following spring. There was little difference noted in bloom time or duration between the non-dormant vs. the dormant plants.

Table 7. Shoot length and flower bud density of fall fertilized ('non-dormant') vs non-fall fertilized ('dormant') 'Aliceblue' rabbiteye blueberries

	Shoot length (cm)		*Flower bud density (no./cm shoot length)*	
Production system	*New growth*[z]	*Old growth*[z]	*New growth*	*Old growth*
Evergreen	12.9a[y]	33.6a	0.62a	0.25a
Dormant	4.6b	31.0a	0.41b	0.02b

[z] "New growth" = shoot growth occurring subsequent to fall fertilizer treatments.
"Old growth" = shoot growth occurring prior to fall fertilizer treatments.
[y] Mean separation within columns by t-test, 0.05.

Observations of the few commercial non-dormant blueberry plantings in Florida indicate that plants grow vegetatively and reproductively throughout the year. The flower buds on these plants do not enter dormancy, but instead begin blooming in November and continue through March. Consequently, ripe berries can be harvested from January through June (P.M. Lyrene, personal communication).

Based on both the preliminary and observational data, a non-dormant production system was established in a subtropical area of southwestern Florida in order to determine the feasibility of using such a system to extend the production area for blueberry (Reeder *et al.*, 1994). Three southern highbush cultivars are being evaluated in this system and the non-dormant condition is maintained by constant nitrogen applications throughout the year at three different concentrations (85, 170, and 250 kg N/ha/year during the first year, and 170, 250 and 335 kg N/ha/year during subsequent years). During the first two years, plant growth (as measured by canopy volume), leaf retention, and rate of new vegetative budbreak increased as N concentrations increased (Reeder *et al.*, 1998). Flowering occurred from November through January, and the majority of ripe fruit was harvested by the end of March (Obreza *et al.*, 1997). There has been no evidence that vegetative and/or floral buds enter endodormancy in this system. Supporting this is the fact that chilling temperatures in that area have been inadequate to fulfill the chilling requirement of these cultivars, estimated to range from 300 to 600 hours below 7°C. Furthermore, new vegetative budbreak in all cultivars has been uniform and copious every year. This planting will continue to be evaluated over the next several years for flowering, fruiting, and yield characteristics. Additional research examining long-term yield performance, as well as appropriate cultural/management techniques, is also required before this system can be fully integrated into commercial blueberry production.

4 Conclusions

The choice of production system (deciduous *vs.* non-dormant) will determine the importance of cultivar selection for low chilling requirement and the need for development of chemical or cultural control of dormancy release. Regardless of the production system, however, adaptation to other factors needs to be considered before the feasibility of producing blueberries in warm climates can be assessed. These factors include high temperature tolerance, adaptation to varying soil types, and photoperiodic effects on flower bud initiation. Some facets of these factors may be overcome with cultural manipulations, such as the use of soil mulches to alleviate excess soil temperature and the use of organic matter addition and/or chemical control of soil pH to ameliorate unsuitable soils. However, successful production under conditions of high air temperatures and/or photoperiods too long for optimum flower bud initiation present greater difficulties. The development of new cultivars that are adapted to these, as well as the previously mentioned conditions, will be required for successful, long-term production of blueberries in warm climates. Sources of germplasm that might be useful in developing cultivars with specific adaptation to low chilling, high temperatures, associated droughty conditions, and/or varying soil types are summarized in Table 8. *V. darrowi* appears to be the richest source of germplasm for adaptation to these conditions, although several other species also carry traits of interest. A combination of new cultivars specifically adapted to warm climates, as well as the use of cultural techniques focused towards alleviation of stresses associated with warm climates, should ensure successful production of blueberries in these areas.

Table 8. Sources of *Vaccinium* germplasm with specific adaptations for production in warm climates[z]

V. amoenum	heat and drought tolerance, mineral soil adaptation
V. angustifolium	mineral soil adaptation, high soil pH tolerance
V. arboreum	mineral soil adaptation, high soil pH tolerance, heat and drought tolerance
V. ashei	low chilling requirement, heat tolerance, mineral soil adaptation
V. atrococcum	mineral soil adaptation
V. constablaei	mineral soil adaptation
V. darrowi	low chilling requirement, heat and drought tolerance, mineral soil adaptation
V. elliottii	low chilling requirement, heat and drought tolerance, mineral soil adaptation
V. myrsinites	heat and drought tolerance, low chilling requirement, mineral soil adaptation
V. myrtilloides	mineral soil adaptation, high soil pH tolerance, drought tolerance
V. pallidum	heat and drought tolerance, mineral soil adaptation
V. simulatum	mineral soil adaptation
V. tenellum	heat and drought tolerance, mineral soil adaptation

[z] Adapted from Ballington, 1990; Chandler *et al.*, 1985; Hancock *et al.*, 1995; Luby *et al.*, 1991; Lyrene, 1993.

5 References

Aalders, L.E. and Hall, I.V. (1964). A comparison of flower bud development in the lowbush blueberry, *V. angustofolium* Ait. under greenhouse and field conditions. *Proc. Amer. Soc. Hort. Sci.* 85, 281-284.

Abbott, J.D. and Gough, R.E. (1987). Seasonal development of highbush blueberry under sawdust mulch. *J. Amer. Soc. Hort. Sci.* 112, 60-62.

Ballington, J.R. (1990). Germplasm resources available to meet future needs for blueberry cultivar improvement. *Fruit Var. J.* 44, 54-62.

Basiouny, F.M. and Powell, A.A. (1985). Effects of high temperature, fertilization and irrigation on growth and leaf elemental contents of newly established rabbiteye blueberries. *Proc. Fla. State Hort. Soc.* 98, 155-158.

Camp, W.H. (1945). The North American blueberries with notes on other groups of *Vacciniaceae. Brittonia.* 5, 203-275.

Carlson, J.D. and Hancock, Jr., J.F. (1991). A methodology for determining suitable heat-unit requirements for harvest of highbush blueberry. *J. Amer. Soc. Hort. Sci.* 116, 774-779.

Chandler, C.K., Draper, A.D., Galletta, G.J. and Bouwkamp, J.C. (1985). Combining ability of blueberry interspecific hybrids for growth on upland soil. *HortScience* 20, 257-258.

Clark, J.R. and Moore, J.N. (1991). Southern highbush blueberry response to mulch. *HorTechnology* 1, 52-54.

Darnell, R.L. (1991). Photoperiod, carbon partitioning and reproductive development in rabbiteye blueberry. *J. Amer. Soc. Hort. Sci.* 116, 856-860.

Darnell, R.L., Stutte, G.W., Martin, G.C., Lang, G.A and Early, J.D. (1992). Developmental physiology of rabbiteye blueberry. *Hort. Rev.* 13, 339-406

Davies, F.S. and Flore, J.A. (1986). Gas exchange and flooding stress of highbush and rabbiteye blueberries. *J. Amer. Soc. Hort. Sci.* 111, 565-571.

Edwards, G.R. (1987a). Producing temperate-zone fruit at low latitudes: Avoiding rest and the chilling requirement. *HortScience.* 22, 1236-1240.

Edwards, G.R. (1987b). Conditions of growth, dormancy and rest to produce temperate zone fruits under tropical conditions. *Acta Hort.* 199, 128.

Edwards, G.R. (1990). Ten year's experience with temperate fruits in the tropics. *Acta Hort.* 279, 47-51.

Erb, W.A., Draper, A.D., Galletta, G.J. and Swartz, H.J. (1990). Combining ability for plant and fruit traits of interspecific blueberry progenies on mineral soil. *J. Amer. Soc. Hort. Sci.* 115, 1025-1028.

Erb, W.A., Draper, A.D. and Swartz, H.J. (1993). Relation between moisture stress and mineral soil tolerance in blueberries. *J. Amer. Soc. Hort. Sci.* 118, 130-134.

Finn, C.E., Luby, J.J., Rosen, C.J. and Ascher, P.D. (1993). Blueberry germplasm screening at several soil pH regimes. I. Plant survival and growth. *J. Amer. Soc. Hort. Sci.* 118, 377-382.

Forsyth, F.R. and Hall, I.V. (1965). Effect of leaf maturity, temperature, carbon dioxide concentrations and light intensity on rate of photosynthesis in clonal lines of the lowbush blueberry, *Vaccinium angustifolium* Ait. under laboratory conditions. *Can J. Bot.* 43, 893-900.

Hall, I.V., Craig, D.L. and Aalders, L.E. (1963). The effect of photoperiod on the growth and flowering of highbush blueberry. *Proc. Amer. Soc. Hort. Sci* 82, 260-263.

Hall, I.V. and Ludwig, R.A. (1961). Effects of photoperiod, temperature and light intensity on the growth of the lowbush blueberry. *Can. J. Bot.* 39, 1733-1739.

Hancock, J.F., Erb, W.A., Goulart, B.L. and Scheerens, J. (1995). Utilization of wild blueberry germplasm: The legacy of Arlen Draper. *J. Small Fruit Vitic.* (in press).

Hancock, J.F., Haghighi, K., Krebs, S.L., Flore, J.A. and Draper, A.D. (1992). Photosynthetic heat stability in highbush blueberries and the possibility of genetic improvement. *HortScience* 27, 1111-1112.

Kender, W.J. and Brightwell, W. T. (1966). Environmental relationships, in: *Blueberry Culture* (eds P. Eck and N. Childers), Rutgers Univ. Press, New Brunswick, pp. 75-93.

Korcak, R.F. (1986a). Adaptability of blueberry species to various soil types: I. Growth and initial fruiting. *J. Amer. Soc. Hort. Sci.* 111, 816-821.

Korcak, R.F. (1986b). Adaptability of blueberry species to various soil types: II. Leaf and soil analysis. *J. Amer. Soc. Hort. Sci.* 111, 822-828.

Korcak, R.F., Galletta, G.J. and Draper, A. (1982). Response of blueberry seedlings to a range of soil types. *J. Amer. Soc. Hort. Sci.* 107, 1153-1160.

Krewer, G., NeSmith, S., Mullinix, B., Clark, J., Bruorton, M. and Stanaland, D. (1995). Experiments to accelerate leaf development in blueberries. *Univ. Ga. Hort. Dept. Misc. Public.* H95-023.

Luby, J.J., Ballington, J.R., Draper, A.D., Pliszka, K. and. Austin, M.E. (1991). Blueberries and cranberries (*Vaccinium*), in *Genetic Resources of Temperate Fruit and Nut Crops* (eds J.N. Moore and J.R. Ballington, Jr.), Int. Soc. Hort. Sci., Wageningen, pp. 391-456.

Luby, J.J., Wildung, D.K., Stushnoff, C., Munson, S.T., Read, P.E. and Hoover, E.E. (1986). 'Northblue', 'Northsky' and 'Northcountry' blueberries. *HortScience* 21, 1240-1242.

Lyrene, P.M. (1993). Some problems and opportunities in blueberry breeding. *Acta Hort.* 346, 63-71.

Lyrene, P.M. (1991). Fertile derivatives from sparkleberry x blueberry crosses. *J. Amer. Soc. Hort. Sci.* 116, 899-902.

Magee, J.B. and Spiers, J.M. (1995). Influence of mulching systems on yield and quality of southern highbush blueberries. *J. Small Fruit Vitic.*3, 133-141.

Mainland, C.M. (1985). Some problems with blueberry leafing, flowering and fruiting in a warm climate. *Acta Hort.* 165, 29-34.

Merhaut, D. and Darnell, R.L. (1995). Ammonium and nitrate accumulation in containerized southern highbush blueberry plants. *HortScience* 30, 1378-1381.

Moon, J.W., Jr., Flore J.A. and Hancock, Jr, J.F. (1987a). A comparison of carbon and water vapor gas exchange characteristics between a diploid and highbush blueberry. *J. Amer. Soc. Hort. Sci.* 112, 134-138.

Moon, J.W., Jr., Hancock, Jr., J.F., Draper, A.D. and Flore, J.A. (1987b). Genotypic differences in the effect of temperature on CO_2 assimilation and water use efficiency in blueberry. *J. Amer. Soc. Hort. Sci.* 112, 170-173.

Obreza, T.A., Williamson, J.G., Darnell, R.L. and Lyrene, P.M. (1997). Performance of a young southwest Florida non-dormant blueberry planting. *Proc. Fla. State Hort. Soc.* 110, 175-177.

Perlmutter, F. and Darrow, G.M. (1942). Effect of soil media, photoperiod and nitrogenous fertilizer on the growth of blueberry seedlings. *Proc. Amer. Soc. Hort. Sci.* 40, 341-346.

Phatak, S.C. and Austin, M.E. (1990). The effect of photoperiod on the growth and flowering of two rabbiteye blueberry cultivars. *Appl. Agric. Res.* 5, 350-352.

Pliszka, K., Scibisz, K., Rojek, H. and Zakowicz, S. (1997). Effect of soil management and water stress upon growth and cropping of the highbush blueberry. *Acta Hort.* 446, 487-496.

Reeder, R.K., Darnell, R.L. and Obreza, T.A. (1994). Establishment of an evergreen high density blueberry planting in southwest Florida. *Proc. Fla. State Hort. Soc.* 107, 326-328.

Reeder, R.K., Obreza, T.A. and Darnell, R.L. (1998). Establishment of a non-dormant blueberry (*Vaccinium corymbosum* hybrid) production system in a warm winter climate. *J. Hort. Sci. Biotech.* 73, 655-663.

Sharpe, R.H. (1953). Horticultural development of Florida blueberries. *Proc. Fla. State Hort. Soc.* 66,

188-190.

Sharpe, R.H. and Sherman, W.B. (1971). Breeding blueberries for low chilling requirement. *HortScience.* 6, 145-147.

Sharpe, R.H. and Sherman, W.B. (1976). 'Flordablue' and 'Sharpblue': Two new blueberries for central Florida. *Univ. Fla. Agric. Expt. Sta. Circ.* S-240.

Shutak, V.G. and Marucci, P.E. (1966). Plant and fruit development, in: *Blueberry Culture* (eds P. Eck and N. Childers), Rutgers Univ. Press, New Brunswick, pp. 179-198.

Spiers, J.M. (1995). Substrate temperatures influence root and shoot growth of southern highbush and rabbiteye blueberries. *HortScience.* 30, 1029-1030.

Spiers, J.M. (1986). Root distribution of 'Tifblue' rabbiteye blueberry as influenced by irrigation, incorporated peatmoss and mulch. *J. Amer. Soc. Hort. Sci.* 111, 877-880.

Wildung, D.K. and Sargent, K. (1989). The effect of snow depth on winter survival and productivity of Minnesota blueberries. *Acta Hort.* 241, 232-237.

Williamson, J.G., Maust, B.E. and Darnell, R.L. (1998). Hydrogen cyanamide increase leafing and reduces fruit development period of southern highbush blueberry. *HortScience.* 33,528 (abstr.).

Williamson, J.G., Darnell, R.L., Krewer, G., Vanerwegen, J. and NeSmith, S. (1995). Gibberellic acid: A management tool for increasing yield of rabbiteye blueberry. *J. Small Fruit Vitic.* 3, 203-218.

Wright, G. (1993). Performance of southern highbush and rabbiteye blueberries on the Corindi Plateau N.S.W. Australia. *Acta Hort.* 346,141-148.

17 STRAWBERRIES

JAMES F. HANCOCK

Department of Horticulture
Michigan State University, East Lansing, MI 48824

1 Introduction

Commercial strawberries (Fragaria x. ananassa Duch.) are successfully grown in a broad range of climates including temperate, grassland, mediterranean, taiga, and subtropical. However, most of the current production is limited to the temperate and Mediterranean climates located between latitudes 28 and 60. In these areas, average mid-summer temperatures in July range from 15 to 30°C, with average summer highs of 20 to 40°C. Average mid-winter temperatures in January range from 15 to -20°C, with mean lows of 10 to -40°C (Galletta and Bringhurst, 1990; Hancock and Luby, 1995).

When growing strawberries in hot environments, attention must be paid to temperature and photoperiod patterns across the whole season, not just the summer. The strawberry is composed of several different meristems that are regulated by the interaction between photoperiod and temperature (Darrow, 1966; Guttridge, 1985; Durner and Poling, 1988; Larson, 1994). The strawberry grows upright through specialized structures called crowns, but it also has axillary buds at the base of each leaf that can form inflorescences, runners or remain dormant depending on the environmental cues it receives. Leaf development and root growth are also strongly regulated by temperature. Few cultivars exist that are specifically adapted to temperatures much above 30°C, but summer heat can be avoided by using cultivars adapted to fall and winter planting systems.

2 Climatic factors regulating development

2.1 Flowering

There are two primary types of strawberries now grown commercially, day-neutral and short day plants. Long day ("everbearing") plants are also available, but they are rarely grown outside of home gardens. The short-day types are actually facultative short-day plants and initiate flower buds either under short day conditions (less than 14 hrs of day length) or when temperatures are less than 15°C (Darrow, 1936; Guttridge, 1985; Larson, 1994). Above 15°C, the critical photoperiod for floral induction is between 8-14 hours, depending on the cultivar.

In mild tropical climates, the short-day cultivars developed for temperate climates can be highly productive (Subramanium and Iyer, 1974); although flower bud formation is restricted if temperatures get too hot (Strik, 1985). Durner *et al.* (1984) observed no

Amnon Erez (ed.), Temperate Fruit Crops in Warm Climates, 445-455.
© 2000 *Kluwer Academic Publishers. Printed in the Netherlands.*

floral induction under short days in plants of F. x ananassa held at 26/22 and 30/26°C day/night temperature regimes. Chabot (1978) found that no flowering occurred in the related species, F. vesca, above a 20/10°C temperature regime. In tropical highlands with short days and cool climates, short-day cultivars of F. x. ananassa perform similar to day-neutrals in temperate climates, with cyclical patterns of flowering and fruiting.

The minimum number of photoinductive cycles necessary to induce flowering in short day plants has been variously reported to be 7-24, depending in part on temperature (Hartmann, 1947 a and b; Went, 1957; Guttridge, 1985; Larson, 1994). At higher temperatures, longer photoinduction periods appear to be needed. Ito and Saito (1962) found that under 8-hr photoperiods only 10 cycles were necessary for floral induction at 24°C, while at 30°C more than 20 were required. With 16-hr photoperiods, 10 cycles were necessary at 9°C, but 16 were required at 17°C.

While short-day cultivars adapted to cooler climates can be grown in tropical and sub-tropical regions, they have a tendency to require a chilling period for full productivity. Those cultivars developed for warm regions may not need one. The ability to grow well during the short days of October, November and December in North American greenhouses has classically been used as an indicator of a cultivar's regional adaptation. Those adapted to cooler regions generally grow poorly during this period and enter a rest period, while those adapted to warmer climates continue to grow (Darrow and Waldo, 1934).

A new type of short-day strawberry has recently been developed in Israel for tropical and sub-tropical environments (Izsak and Izhar, 1984; Izhar and Izsak, 1995). These "infra" short-day types do not require chilling for high yields and initiate flower bud primordia in response to longer light regimes (13.5 to 14 hrs.) and higher temperatures (10-26°C) than traditional short-day plants. These genotypes fill an important economic window as they are productive in the fall and early winter when short-day varieties are incapable of bearing fruit. Day-neutral plants have been developed for temperate climates that produce crowns and flower buds approximately 3 months after planting, regardless of the day length (Bringhurst and Voth, 1975; Galletta and Bringhurst, 1990). They have the potential to initiate flower buds cyclically throughout the growing season, although high temperatures appear to inhibit bud formation. Durner *et al.* (1984) found that day/night temperatures of 30/26°C almost completely inhibited flower bud initiation in the day-neutrals 'Hecker' and 'Tristar', while these cultivars were highly productive under 18/14, and 22/18°C regimes.

While most cultivars are now categorized as day-neutrals or short-day plants, some genotypes are hard to rank precisely due to complex interactions between genotype, temperature and photoperiod (Nicoll and Galletta, 1987; Yanagi and Oda, 1989). Darrow (1966) suggested that strawberries actually range from obligate short day to facultative short-day to complete day-neutrals. The infra short-day types developed by Izsak and Izhar (1984) fall in the middle of this range. Darrow also suggested that a range of everbearers types exist from weak, intermediate to strong. Galletta and Bringhurst (1990) have suggested that flowering in modern strawberry cultivars is regulated more by sensitivity to high temperature than photoperiod, with traditional short-day plants being more sensitive to high temperatures than day-neutrals.

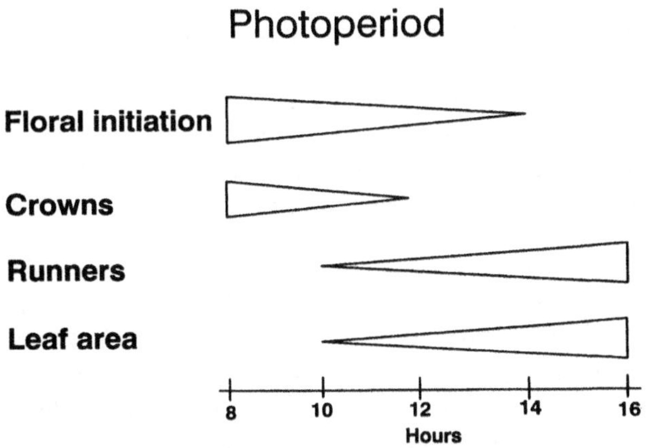

Figure 1. Developmental responses of strawberries to differential photoperiods (gift of Rebecca Darnell).

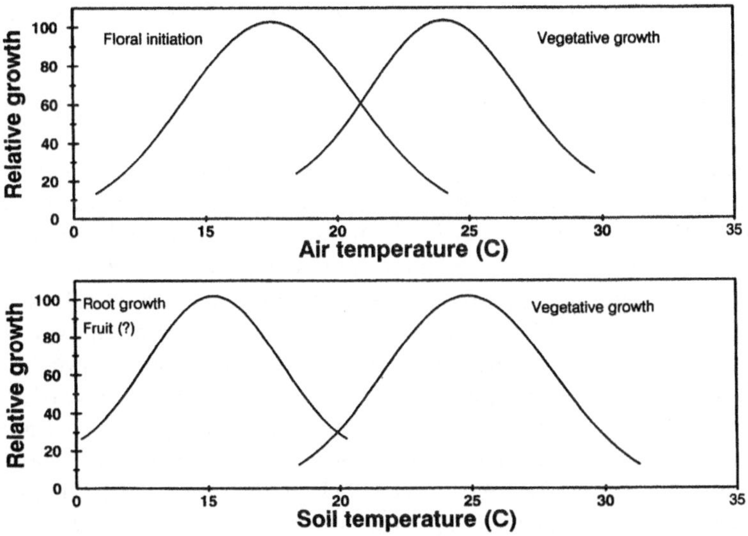

Figure 2. Developmental responses of strawberry to differential air and soil temperatures (gift of Rebecca Darnell).

2.2 Fruit

Elevated air temperatures during fruiting have been shown to have a negative effect on fruit size. Went (1957) showed that daytime temperatures above 15-17°C reduced fruit size and fruit aroma, and optimal fruit growth occurred at 12°C night temperatures. Fruit size in several day-neutral types in Maryland was about 50% smaller in the hot summer than cool spring (Draper *et al.*, 1981). Fruit temperatures can exceed air temperatures by as much as 8°C on sunny days of 26.5°C (Austin *et al.*, 1960), causing tissue damage, softness and breakdown near the berry surface. Temperatures above 25°C can also reduce fruit set (Kronenberg, Braak and Zeilinga, 1959; Abdelrahman, 1984), decrease fruit soluble solids content (Abdelrahman, 1984; Hellman and Travis, 1988) and increase the rate of fruit development (Darrow, 1966; Dana, 1980). Galletta *et al.* (1981) suggested that in day-neutral types, a decrease in soil temperature of 10°C can increase fruit size by 0.9 to 1.6 g/berry.

In warm climates, high air temperatures probably play an important role in restricting growth and fruit development by reducing photosynthetic activity and increasing respiration rate (Larson, 1994; Darnell and Hancock, 1996). The optimal temperatures for photosynthesis in strawberries ranges from 15 to 23°C, depending on cultivar and species (Chabot, 1978; Sruamsiri and Lenz, 1985; Hancock, Flore and Galletta, 1989; Hancock 1991). Temperatures above 30°C often restrict CO_2 assimilation rates by more than 50% (Caldwell *et al.*, 1990; Hancock *et al.*, 1996). High root temperatures have also been shown to have a negative impact on gas exchange rates. Net photosynthesis, stomatal conductance to water vapor and transpiration rates dropped significantly on a per leaf basis as soil temperatures were raised from 15 to 25°C, although whole plant photosynthesis increased due to greater leaf areas (Udagawa *et al.*, 1989).

2.3 Vegetative development

In short-day plants, stolons are produced after flowering as new leaves form. They are formed most readily during long days (>10-hrs) when temperatures fall between 21 and 30°C (Darrow, 1936; Heide, 1977; Durner *et al.*, 1984; Hellman and Travis, 1988). Darrow (1936) found that three times more stolons were produced under 16-hr than 14-hr photoperiods, and at both photoperiods, three times more stolons were formed at 21°C than at 12.8 or 15.6°C. Runner formation in day-neutral cultivars is also highest under long days with moderate temperatures, but it is much more sporadic than in short-day types (Downs and Piringer, 1955; Smeets, 1979; Durner *et al.*, 1984).

Leaf production in both day-neutral and short day plants continues throughout the whole season, although leaf production and growth occurs most readily under long days (Arney, 1953a, 1953b, and 1954; Nishizawa, 1990). Temperature optima range between 15-26°C, depending on cultivar (Darrow, 1936; Arney, 1953a, 1953b and 1954; Abdelrahman, 1984). Arney (1953a) found leaf initiation in 'Royal Sovereign' to be greatly reduced above 35°C. It has also been found that when plants are heat treated for virus elimination before propagation, they stop growing at 35-38°C, but if temperatures are raised a few degrees a day, leaves appear undamaged for at least 6 weeks (Converse, 1987).

Active root growth occurs from 7 to 32°C, but is greatest at the lower end of this range (Roberts and Kenworthy, 1956; Proebsting, 1957; Abdelrahman, 1984). High root temperatures can greatly effect top growth, as Proebsting (1957) found that strawberry shoot growth was highest at root temperatures of 24°C and decreased sharply at higher temperatures. Brouwer (1963) and Cooper (1973) reported that shoot growth in strawberry was unaffected by root temperatures between 18 and 30°C during the growing season, but was markedly reduced at temperatures above 30°C. Nishizawa and

Hori (1993) found that if the roots of plants were kept at 26°C while the rest of the plant was held under chilling conditions for 35 days, subsequent petiole and leaf growth was greatly restricted.

3 Agrotechnical considerations.

There are two major open production systems in the world - annual hills and matted rows. The annual hill or plasticulture system utilizes crowns as the primary yield component (Galletta and Bringhurst, 1990). It features summer or fall/winter planting, raised beds, plastic mulch, trickle irrigation and 1-2 year production seasons (Hochmuth, 1988; Voth and Bringhurst, 1990; Poling, 1991). The annual system is used with short day plants in areas having warm winters and either hot or moderate summers. It is particularly useful in areas where the climate is extreme during part of the year. The matted row system utilizes runners as the primary yield component (Hancock *et al.*, 1984) and is used with short day cultivars in climates with short summers and cold winters. It features spring planting, flat beds, straw mulch, overhead irrigation and 3-5 production seasons (Cooley and Schloemann, 1995). Day-neutrals produce too few runners to form good matted rows and as a result are grown in annual hills in both warm and cold climates.

Annual hills of short-day cultivars are set in the summer or late fall/early winter, depending on summer and winter temperatures (Galletta and Bringhurst, 1990; Larson, 1994). A late fall/ early winter planting date can be used to avoid prohibitively hot summers, where winter temperatures are high enough to sustain crown growth and proliferation. Summer planting dates of short-day types are most appropriate where summers are long and moderate, but the winter is too cold for continued crown growth and proliferation. Day-neutral cultivars are generally planted in the spring in areas with cold winters and cool summers or in the fall/early winter in climates where summers are hot and winter temperatures are mild enough for continued growth.

In the annual production system, both chilled and fresh by dug plants are used, depending on climate and the desired production season (Voth and Bringhurst, 1970; Durner *et al.*, 1986; Voth 1989). Short-day plants that have been partially chilled (300-500 hours below 7.2°C) typically produce flowers a few months after planting, but make little vegetative growth and produce few runners (Galletta and Bringhurst, 1990). Freshly dug short-day plants produce fruit immediately from nursery-initiated flowers and then later produce another round of flowers. Green plants of infra short-day types normally begin flowering 6-8 weeks after floral bud initiation and continue to flower until the end of the summer (Izhar and Izsak, 1995).

In summer planted annual systems and matted rows, plants are usually dug when their chilling requirements have been satisfied and they are cold stored until planting ("frigo plants"), except in the Mid-Atlantic states of the U.S.A., where freshly rooted "plug plants" are utilized in annual plantings (Poling, 1991). Very precise recommendations on digging and planting dates have been developed for many cultivars depending on location (Durner at al., 1986; Bringhurst and Voth, 1990). Spring plantings are used with matted rows to encourage summer production during the long days of summer.

Clear, black, brown, and occasionally white polyethylene mulches are used in the annual hill systems to keep fruit clean and modify soil temperatures for maximized growth and development (Renquist *et al.*, 1982; Fear and Nonnecke, 1989; Galletta and Bringhurst, 1990; Voth and Bringhurst, 1990). Black, brown, and white polyethylene is used in areas where summer temperatures are prohibitive to fruit development and winter temperatures are high enough to allow for good root growth. Transparent plastic is utilized in climates where winter growth temperatures are not sufficient to maximize root growth and summer temperatures are not prohibitive. Straw is used in conjunction with matted row systems in areas where summer temperatures are high and winter temperatures are cold.

The various types of mulch have large effects on soil temperatures. Voth and Bringhurst (1990) observed an increased temperature of 4°C under transparent plastic in January in southern California, while temperatures under black plastic were about 1°C cooler than areas without mulch. Temperatures under clear polyethylene mulch averaged almost 8°C higher than non-mulched plots in March (during fruit production), while temperatures under black plastic were again a degree cooler. Temperatures under white and white/black combinations were 3-6°C cooler than black in both the winter and summer. In mid-summer in Iowa, late afternoon temperatures were highest under clear, followed by white, white-on-black and straw mulches (Fear and Nonnecke, 1989).

Evaporative cooling from mist or sprinkler irrigation can be used to cool plants and fruit during periods of high temperature, although they have not been used commercially. Sprinkling plants with overhead irrigation during the hottest parts of the day in Lousianna (>42°C) reduced leaf temperatures by more than 12°C (Chesness and Braud 1970), while in Pennsylvania, minisprinkler treated canopy and berries were consistently 3-4°C cooler when temperatures exceeded 28°C (Goulart *et al.*, 1996). In Iowa, microsprinklers were found to increase runner production, dry matter accumulation and truss numbers in day-neutral strawberries, although leaf temperatures were not reported (Nonnecke and Fear, 1987).

4 Germplasm resources

Strawberries are most commonly grown in hot climates by utilizing annual production systems that avoid the hot season. Since the cultivars represent a very narrow sample of the wild germplasm available (Sjulin and Dale, 1987; Luby *et al.*, 1991), it seems likely that new sources of heat tolerance can be found that could expand the heat tolerance of both short-day and day-neutral types.

Fragaria virginiana has long been considered to have the highest tolerance to high temperature of all the native species (Darrow, 1966), and may be a rich source of genes for heat adaptation. Few controlled experiments have been performed, but Fragaria virginiana ssp virginiana is found on some of the hottest, driest sites in the central US. F. chiloensis may also have substantial heat tolerance, as it is found on hot, dry dune and scrub sites in California and Chile (Hancock *et al.*, 1990; Hancock and Luby, 1993; Cameron *et al.*, 1993), and recent studies have shown that it has high tolerance to drought through osmotic regulation (Zhang and Archbold 1993a and b). Fragaria orientalis and F. vesca are also believed to have high heat tolerance (Darrow, 1966; Hancock and Bringhurst, 1978; Goncharova and Dobrenkova, 1981), but clones of F. virginiana and F. chiloensis should prove to be better parents as they cross much more readily with the cultivated types and still offer high levels of genetic diversity (Luby *et al.*, 1991; Hancock and Luby, 1993).

In preliminary work, Caldwell, Flore and Hancock (1990) reported that F. virginiana ssp. virginiana appeared to photosynthetically acclimate more readily to high temperature than F. chiloensis and F. x ananassa; but this conclusion was based on a very limited number of clones. More recently, we have found representatives of all these taxa that appear to acclimate to high temperature (Hancock, Sakin and Flore unpubl.) The mean photosynthetic temperature optima of these clones were significantly higher after maintenance at high temperature, and they showed a significantly lower reduction in CO_2 assimilation rate as temperatures were raised from 20 to 30°C. Infact, some clones of F. chiloensis from California and commercial F. x ananassa appeared to be better adapted to heat than F. virginiana from the midwestern and eastern USA, as the maximum CO_2 assimilation rate was significantly higher in F. chiloensis and F. x ananassa at the higher acclimation temperature, while that of F. virginiana decreased.

Finding new sources of heat tolerance in native day-neutral types may prove particularly useful in breeding for hot climates, as the existing germplasm pool is very limited. All of the day-neutral cultivars currently grown utilize the same source of day neutrality derived from a single Utah plant of F. virginiana ssp. glauca (Bringhurst and Voth, 1978) whose progeny do not perform well in summer heat. Sakin, Hancock and Luby (1996) have found that virtually all populations of Rocky Mountain F. virginiana ssp glauca contain high proportions of multiple croppers, and some of these have high photosynthetic heat tolerance (Hancock *et al.*, 1996). Luby and Stahler (1993) and Dale (1993) have also identified multiple croppers in F. virginiana ssp virginiana, but only a few of these come from hot climates. Repeat flowering has not been described in F. chiloensis, except for an occasional late season bloom (Ahokas, 1993; Luffman and MacDonald, 1993).

5 Conclusions

The currently available cultivars of F. x ananassa are in general not well adapted to high temperatures. Most floral and vegetative responses are significantly inhibited by temperatures in excess of 25-30°C. However, fall and winter planting systems have been developed for the warm climates of Florida and California that produce fruit during the cooler winter and early spring seasons and avoid summer heat. In addition "infra short" day types have been developed that produce fruit earlier than typical short-day types because they lack a chilling requirement, are induced by longer days (13.5-14 hrs.) and can tolerate a wider temperature range (10-26°C). Native germplasm is probably available that can be used by breeders to increase the heat tolerance of the commercial strawberry, but this material is only in the early stages of evaluation.

6 Acknowledgements

Thanks to Rebecca Darnell for a thorough review of an early draft of the manuscript, the gift of the figures, and the push generated by her timely submission of her manuscript on blueberries. I also thank Amnon Erez for making me aware of the important work of E. Izsak and S. Izhar.

7 References

Abdelrahman, M.H. (1984). *Growth and productivity of strawberry cultivars at high temperatures.* Ph.D. Diss., Kansas State University, Manhattan (Diss. Abst. Inter. B 46, 17-B).

Ahokas, H. (1993). Two decades of naturalization in Finland of the octoploid strawberries, *Fragaria chiloensis* and *F. virginiana* from the boreal North America. *Memoranda Soc. Fauna Fennica* 69, 33-45.

Arney, S.E. (1953a). Studies of growth and development in the genus *Fragaria.* I. Factors effecting the rate of leaf production in Royal Sovereign strawberry. *J. Hort. Sci.* 28, 73-84.

Arney, S.E. (1953b). Studies of the growth and development of the genus *Fragaria* II. The initiation, growth and emergence of leaf primordia in *Fragaria. Ann. Bot.* 17, 476-492.

Arney, S.E. (1954). Studies of growth and development in the genus *Fragaria* III. The growth of leaves and shoots. *Ann. Bot.* 18, 349-365.

Austin, M.E., Shutak, V.G. and Cristopher, E.P. (1960). Color changes in harvested strawberry fruits. *Proc. Amer. Soc. Hort. Sci.* 75, 382-386.

Bringhurst, R.S. and Voth, V. (1975). Breeding and exploitation of day-neutral strawberries in subtropical environments. *HortScience* 10, 329 (abstract).

Bringhurst, R.S. and Voth, V. (1978). Origin and evolutionary potentiality of the day-neutral trait in octoploid *Fragaria. Genetics* 90, 510 (abstract).

Brouwer, R. (1963). Some physiological aspects of the influence of growth factors in the root medium on growth and dry matter production. *Medelel. Inst. Biol. Scheik. Onderz. Landbouwgewassen* 11-30.

Caldwell, J.D., Hancock, J.F. and Flore, J.A. (1990). Strawberry leaf photosynthetic acclimation to temperature. *HortScience,* 25, 1166 (abstract).

Cameron, J.S., Sjulin, T.M., Ballington, J.R., Shanks, C.H., Munoz, C.E. and Lavin, A. (1993). Exploitation, collection and evaluation of Chilean *Fragaria:* Summary of 1990 and 1992 expeditions. *Acta Hort.* 348, 65-74.

Chabot, B. (1978). Environmental influences on photosynthesis and growth in *Fragaria vesca. New Phytol.* 80, 87-98.

Chesness, J.L. and Braud, H.J. (1970). Sprinkling to reduce heat stressing of strawberry plants. *Agr. Eng.* 51, 140-142.

Converse, R.H. (1987). Detection and elimination of virus and viruslike diseases in strawberry, In *Virus diseases of small fruits* (ed. Converse, R.H.). USDA/ARS. Washington, D.C. p. 2-10.

Cooley, D.R. and Schloemann, S.G. (1995). Integrated pest management for strawberries in the northeastern United States. *University of Massachusetts Cooperative Extension System.* Bulletin C211.

Cooper, A.J. (1973). *Root temperature and plant growth.* Commonwealth Agricultural Bureaux, Franham Royal, England.

Dale, A., Daubeny, H.A., Luffman, M. and Sullivan, J.A. (1993). Development of *Fragaria* germplasm in Canada. *Acta Hort.* 348, 75-80.

Dana, M.N. (1980). The strawberry plant and its environment. In *The strawberry: Cultivars to marketing.* (ed. Childers N. F.) Horticultural Publications, Gainesville, Fla. pp. 32-44.

Darnell, R.L. and Hancock, J.F. (1996). Balancing vegetative growth with productivity in strawberry. *Proc. IV North American Strawberry Conf.* Orlando, Florida. pp. 144-150.

Darrow, G.M. (1936). Interrelation of temperature and photoperiodism in the production of fruit buds and runners in the strawberry. *Proc. Amer. Soc. Hort. Sci.* 34, 360-363.

Darrow, G.M. (1966). *The strawberry: history, breeding and physiology.* Holt, Rinehart and Winston, New York.

Darrow, G.M. and Waldo, G.F. (1934). Responses of strawberry varieties and species to the duration of the daily light period. *U.S. Dep. Agric. Tech. Bull.* 453.

Draper, A.D., Galletta, G.J. and Swartz, H.J. (1981). 'Tribute' and 'Tristar' everbearing strawberries. *HortScience* 16, 794-795.

Downs, R.J. and Piringer, A.A. (1955). Differences in photoperiod responses of everbearing and June-bearing strawberries. *Proc. Amer. Soc. Hort. Sci.* 66, 234-236.

Durner, E.F., and Poling, E.B. (1988). Strawberry developmental responses to photoperiod and temperature. *Adv. Strawberry Prod.* 7, 6-15.

Durner, E.F., Poling E.B. and Albregts, E.A. (1986). Early season yield responses of selected strawberry cultivars to photoperiod and chilling in a Florida winter production system. *J. Amer. Soc. Hort. Sci.* 112, 53-56.

Durner, E.F., Barden, J.A., Himelrick, D.G. and Poling, E.B. (1984). Photoperiod and temperature effects on flower and runner development in day-neutral, Junebearing and everbearing strawberries. *J. Am. Soc. Hortic. Sci.* 109, 396-400.

Fear, C.D. and Nonnecke, G.R. (1989). Soil mulches influence reproductive and vegetative growth of 'Fern' and 'Tristar' day-neutral strawberries. *HortScience* 6, 912-913.

Galletta, G.J. and Bringhurst, R.S. (1990). Strawberry management. In *Small fruit crop management.* (eds. Galletta G.J. and Himelrick D.) Prentice Hall, Englewood Cliffs, New Jersey. pp 83-156.

Galletta, G.J., Draper, A.D. and Swartz, H.J. (1981). New everbearing strawberries. *HortScience* 16, 726.

Goncharova, E.A. and Dobrenkova, L.D. (1981). Growth processes and yield in strawberry under conditions of drought and salinity. *Trudy Priklandoi Bot. Gen. Selektsii* 70, 97-102.

Goulart, B.L., Heinemann, P.H., Strombaugh, T., Demchak, K. and Morrow, C.T. (1996). An automated irrigation for strawberry production. *Adv. Strawberry Research* 15, 34-41.

Guttridge, C.G. (1985). Fragaria x ananassa. In *CRC Handbook of flowering.* (ed. Halvey A.H.) Vol. III. CRC Press, Boca Raton, Fla. p. 16-33

Hancock, J.F. (1991). Photosynthesis in strawberries and the potential for improvement. In *The strawberry into the 21st century.* (eds. Dale A. and Luby J.J.). Timber Press, Portland, Oregon.

Hancock, J.F. and Bringhurst, R.S. (1978). Inter-populational differentiation and adaptation in the perennial, diploid species *Fragaria vesca* L. *Am. J. Bot.* 65, 795-803.

Hancock, J.F. and Luby, J.J. (1993). Genetic resources at our doorstep: The wild strawberries. *BioScience* 43, 141-147.

Hancock, J. F. and Luby, J.J. (1995). Adaptive zones and ancestory of the most important North American strawberry cultivars. *Frt. Var. J.* 49, 85-89.

Hancock, J.F., Flore, J.A. and Galletta, G.J. (1989). Variation in leaf photosynthetic rates and yields in strawberry. *J. Hort. Sci.* 64, 449-454.

Hancock, J.F., Pritts, M.P. and Siefker, J.H. (1984). Yield components of strawberries maintained in ribbons and matted rows. *Crop. Res.* 24, 37-43.

Hancock, J.F., Salin, M., Luby, J.J., Dale, A. and Darnell, R.L. (1996). Germplasm resources in octoploid strawberries: Potential sources of genes to increase yields in northern climates. *Proc. IV. North American Strawberry Conference.* Orlando, Florida. pp. 87-94.

Hancock, J.F., Maas, J.L., Shanks, C.H., Breen P.J. and Luby, J.J. (1990). Strawberries (Fragaria ssp). In *Genetic resources of temperate fruit and nut crops* (eds Moore J.N. and Ballington J.R.) Int. Soc. Hort. Sci., Wageningen. pp. 489-546.

Hartmann, H.T. (1947a). Some effects of temperature and photoperiod on flower formation and runner production in the strawberry. *Plant Physiol.* 22, 401-420.

Hartmann, H. T. (1947b). The influence of temperature on the photoperiodic response of several strawberry varieties grown under controlled environmental conditions. *Proc. Amer. Soc. Hort. Sci.* 50, 243-245.

Heide, O.M. (1977). Photoperiod and temperature interactions in growth and flowering of strawberry. Physiol. Plant. 40, 21-26.

Hellman, E.W. and Travis, J.D. (1988). Growth inhibition of strawberry at high temperature. *Adv. Strawb. Prod.* 7, 36-39.

Hochmuth, G.J. (1988). Strawberry production guide for Florida. *Florida Cooperative Extension Service. Circular* 142C.

Ito, H. And Sato, T. (1962). Studies on the flower formation in the strawberry plant. I. Effects of temperature and photoperiod on the flower formation. *Tohoku J. Agr. Res.* 13, 191-203.

Izhar, S. And Izsak, E. (1995). Infra short-day strawberry plants. *United States Patent #5,444, 179.*

Izsak, E. and Izhar, S. (1984). Breeding and testing of early strawberry varieties in the central and northern Negev regions. *Hassadeh* 64, 1774-1777 (Hebrew original and English translation).

Kronenberg, H.G., Braak, J.P. and Zeilinga, A.E. (1959). Poor fruit in strawberries. II. Malformed fruit in Jucunda. *Euphytica* 8, 245-249.

Larson, K.D. (1994) Strawberry. In *Handbook of Environmental Physiology of Fruit Crops: Volume I. Temperate Crops* (eds. Schaffer B. and Anderson P.C.). CRC Press, Boca Raton, Florida.

Luby, J.J. and Stahler, M. (1993). Collection and evaluation of *Fragaria virginiana* in North America. *Acta Hortic.* 345, 49-53.

Luby, J.J., Hancock, J.F. and Cameron, J.C. (1991). Expansion of the strawberry germplasm base in North America. In *The Strawberry into the 21st century.* (eds. Dale A. and Luby J.J.) Timber Press, Portland, OR. pp. 66-75.

Luffman, M. and MacDonald, P. J. (1993). *Fragaria* at the Canadian Clonal Genebank. *Acta Hortic.* 348, 102-108.

Nicoll, M.F. and Galletta, G.J. (1987). Variation in growth and flowering habits of Junebearing and everbearing strawberries. *J. Am. Soc. Hortic. Sci.* 112, 872-880.

Nishizawa, T. (1990). Effects of daylength on cell length and cell number in strawberry petioles. *J. Jpn. Soc. Hort. Sci.* 59, 533-539 (Japanese, English summary).

Nishizawa, T. and Hori, Y. (1993). Effects of defoliation and root heating during rest on leaf growth in strawberry plants. *Tohoku J. Agric. Res.* 43, 3-4.

Nonnecke, G.R. and Fear, C.D. (1987). Influence of mist irrigation on growth and development of day neutral strawberry. *HortScience* 22, 1044 (abstract).

Poling, E.B. (1991). The annual hill planting system for southeastern North Carolina. In *The strawberry into the 21st century* (eds. Dale A. and Luby J.J.). Timber Press, Portland, Oregon. pp. 258-263.

Proebsting, E.L. Sr. (1957). The effect of soil temperature on mineral nutrition of the strawberry. *Proc. Am. Soc. Hortic. Sci.* 69, 278-281.

Renquist, A.R., Breen, P.J. and Martin, L.W. (1982). Effects of black polyethylene mulch on strawberry leaf elongation and diurnal leaf water potential. *J. Amer. Soc. Hort. Sci.* 107, 640-643.

Roberts, A.N. and Kenworthy, A.L. (1956). Growth and composition of the strawberry plant in relation to root temperature and intensity of nutrition. *Proc. Am. Soc. Hortic. Sci.* 68,157-168.

Sakin, M., Hancock, J.F. and Luby, J.J. (1997). Identifying new sources of genes which determine cyclic flowering in Rocky Mountain populations of Fragaria virginiana. *J. Amer. Soc. Hort. Sci.* 122, 205-210.

Sjulin, T.M. and Dale, A. (1987). Genetic diversity of North American strawberry cultivars. *J. Amer. Soc. Hort. Sci.* 112, 375-385.

Smeets, L. (1979). Effect of temperature and day-length on flower initiation and runner formation in two everbearing strawberry cultivars. *Sci. Hort.* 12, 19-26.

Sruamsiri, P. and Lenz, F. (1985). Photosynthesis and stomatal behavior of strawberries. I. Effect of light. *Gartenbauwissenschaft* 50, 78-84.

Strik, B. (1985). Flower bud initiation in strawberry cultivars. *Fruit Var.* J. 39, 5-9.

Subramanium, M.D. and Iyer, C.P.A. (1974). Performance of temperate strawberry cultivars in the tropics. *Proceedings XIX International Hort. Congress.* p.527.

Udagawa, Y., Ito, T. and Gomi, K. (1989). Effects of root temperature on some physiological and ecological characteristics of strawberry plants 'Reiko' grown in nutrient solution. *J. Japan Soc. Hort. Sci.* 58, 627-632. (Japanese, English summary).

Voth, V. (1989). The effect of nursery location latitude on California winter planted strawberries. *Acta Hortic.* 265, 283-284.

Voth, V. and Bringhurst, R.S. (1970). Influence of nursery harvest date, cold storage and planting date on performance of winter planted California strawberries. *J. Amer. Soc. Hort. Sci.* 75, 373

Voth, V. And Bringhurst, R.S. (1990). Culture and physiological manipulation of California strawberries. *HortScience* 25, 889-892.

Went, F. (1957). The experimental control of plant growth. Vol 17, *Chronica Botanica*, Waltham, MA, Chapter 9.

Yanagi, T. and Oda, Y. (1989). Effects and interactions of pre-chilling history and day-length on successive floral formation in everbearing and non-everbearing strawberry cultivars. *J. Japan. Soc. Hort. Sci.* 58, 635-640.

Zhang, S. and Archbold, D.D. (1993a). Water relations in a *Fragaria chiloensis* and a *F. virginiana* selection during and after water deficit stress. *J. Amer. Soc. Hort. Sci.* 118, 274-279.

Zhang, S. and Archbold, D.D. (1993b). Solute accumulation in leaves of a *Fragaria chiloensis* and *F. virginiana* selection responds to water deficient stress. *J. Amer. Soc. Hort. Sci.* 118, 280-285.

Subject Index

GPSR Compliance

The European Union's (EU) General Product Safety Regulation (GPSR) is a set of rules that requires consumer products to be safe and our obligations to ensure this.

If you have any concerns about our products, you can contact us on ProductSafety@springernature.com

In case Publisher is established outside the EU, the EU authorized representative is:

Springer Nature Customer Service Center GmbH
Europaplatz 3
69115 Heidelberg, Germany

Batch number: 10809439

Printed by Printforce, the Netherlands